Markov Processes for Stochastic Modeling

Markov Processes for Stochastic Modeling

Second Edition

Oliver C. Ibe
University of Massachusetts, Lowell, MA, USA

AMSTERDAM • BOSTON • HEIDELBERG • LONDON • NEW YORK • OXFORD
PARIS • SAN DIEGO • SAN FRANCISCO • SINGAPORE • SYDNEY • TOKYO

Elsevier
32 Jamestown Road, London NW1 7BY
225 Wyman Street, Waltham, MA 02451, USA

First Edition 2009
Second Edition 2013

Notices

Knowledge and best practice in this field are constantly changing. As new research and experience broaden our understanding, changes in research methods, professional practices, or medical treatment may become necessary.

Practitioners and researchers must always rely on their own experience and knowledge in evaluating and using any information, methods, compounds, or experiments described herein.
In using such information or methods they should be mindful of their own safety and the safety of others, including parties for whom they have a professional responsibility.

To the fullest extent of the law, neither the Publisher nor the authors, contributors, or editors, assume any liability for any injury and/or damage to persons or property as a matter of products liability, negligence or otherwise, or from any use or operation of any methods, products, instructions, or ideas contained in the material herein.

British Library Cataloguing-in-Publication Data
A catalogue record for this book is available from the British Library

Library of Congress Cataloging-in-Publication Data
A catalog record for this book is available from the Library of Congress

ISBN: 978-0-323-28295-6

For information on all Elsevier publications
visit our website at store.elsevier.com

This book has been manufactured using Print On Demand technology. Each copy is produced to order and is limited to black ink. The online version of this book will show color figures where appropriate.

Working together
to grow libraries in
developing countries

www.elsevier.com • www.bookaid.org

Contents

Acknowledgments

There is an African adage that it takes a village to raise a child. This second edition of *Markov Processes for Stochastic Modeling* is a typical example of the application of that adage. It was Dr Erin Hill-Parks of Elsevier Insights who initiated the revision at a time I thought I was done with book writing. She persistently encouraged me throughout the time it took to get the project completed. I am sincerely grateful to her, especially for her perseverance in getting me motivated enough to embark on the project. I also wish to thank Ms Sarah Lay, the Editorial Project Manager at Elsevier Insights, for ensuring that the production schedule is met. I am proud to be involved in one of her first projects in editorial production at Elsevier Insights.

Finally, I would like to thank my wife, Christina, and our children, Chidinma, Ogechi, Amanze, and Ugonna, for the joy they have brought to my life. I could not have completed this project without their encouragement.

Acknowledgments

Preface to the Second Edition

Markov processes are the most popular modeling tools for stochastic systems. They are used in communication systems, transportation networks, image segmentation and analysis, biological systems and DNA sequence analysis, random atomic motion and diffusion in physics, social mobility, social networks, population studies, epidemiology, animal and insect migration, queueing systems, resource management, dams, financial engineering, actuarial science, and decision systems. In this second edition, we maintain the same goal as in the first edition, which is to bring into one volume many of the Markovian models that are used in different disciplines.

Many chapters have been thoroughly revised, and new chapters and material have been added in this edition. The following is a summary of the changes that have been made in the second edition:

1. Chapter 1 (Basic Concepts) has been split into two chapters. The new Chapter 1 deals with basic concepts in probability, while Chapter 2 deals with basic concepts in stochastic processes. This allows us to devote more coverage in Chapter 2 on Poisson processes, which are the foundation of Markov processes.

2. Chapter 5 (Markovian Queueing Systems) has been expanded to include a section on networks of queues and another section on priority queueing.

3. Chapter 6 (Markov Renewal Processes) has been expanded to include a discussion on Markov regenerative processes.

4. Chapter 8 (Random Walk) has been expanded to include a discussion on self-avoiding random walk and nonreversing random walk.

5. Chapter 9 (Brownian Motion and Diffusion Processes) has been split into three separate chapters: Brownian Motion (Chapter 9), Diffusion Processes (Chapter 10), and Levy Processes (Chapter 11). This enables us to discuss anomalous diffusion that is widely used in physics, chemistry, polymer science, and financial instruments. Anomalous diffusion is also called fractional diffusion, which requires knowledge of fractional calculus. Thus, an introduction to fractional calculus is provided. There are other fractional processes that require knowledge of fractional calculus, including fraction random walk, fractional Brownian motion, and fractional Gaussian noise. All these processes are discussed in Chapter 11.

6. Chapter 12 (Markov Random Fields) has been merged with Chapter 13 (Markov Point Processes) since the two topics are very closely related. In fact, Markov point process is an extension of the Markov random field with the capability to express inhibition between neighboring points on a finite graph.

7. Chapter 14 (Markov Chain Monte Carlo (MCMC)) has been deleted because newer simulation models that use MatLab are now more prevalent in science and engineering than the MCMC.

This second edition of the book is organized as follows. Chapter 1 deals with basic concepts in probability, and Chapter 2 discusses basic concepts in stochastic processes. Chapter 3 discusses general properties of Markov processes. Chapter 4 discusses discrete-time Markov chains, while Chapter 5 discusses continuous-time Markov chains. Chapter 6 discusses Markov renewal processes including semi-Markov processes and Markov regenerative processes. Chapter 7 discusses Markovian queueing systems, which are essentially those queueing systems where either the interarrival times or service times or both are exponentially distributed. Chapter 8 deals with the random walk. Chapter 9 discusses Brownian motion while Chapter 10 discusses diffusion processes. Chapter 11 discusses Levy process, and Chapter 12 discusses Markovian arrival processes. Chapter 13 discusses controlled Markov processes, which include Markov decision processes, semi-Markov decision processes, and partially observable Markov decision processes. Chapter 14 discusses hidden Markov models, and Chapter 15 discusses Markov point processes, including Markov random fields.

The rationale for this rearrangement of the chapters is as follows. The topics covered in Chapter 1 through Chapter 7, which constitute Part 1 of the book, deal with general principles of Markov processes. Chapters 8, 9, 10, and 11, which constitute Part 2 of the book, deal with models that are popularly used in physical and biological sciences as well as in the analysis of financial instruments. Chapter 12 deals with a subject that is popularly used in teletraffic engineering and control systems. It can be regarded as Part 3 of the book. Finally, Chapters 13, 14, and 15, which constitute Part 4 of the book, deal with models that are popularly used in image processing. Thus, a combination of the topics in Part 1 and any one of the other parts can be used for a one-semester graduate course in stochastic modeling in the relevant area.

Preface to the First Edition

Markov processes are used to model systems with limited memory. They are used in many areas including communication systems, transportation networks, image segmentation and analysis, biological systems and DNA sequence analysis, random atomic motion and diffusion in physics, social mobility, population studies, epidemiology, animal and insect migration, queueing systems, resource management, dams, financial engineering, actuarial science, and decision systems.

Different books have been written specifically for different types of Markov processes. For example, books on bioinformatics discuss hidden Markov models, books on financial markets and economics discuss random walks and Brownian motion, and books on image analysis discuss Markov random fields and Markov point processes.

The purpose of this book is to bring into one volume the different Markovian models that are individually scattered across many books. The book is written for graduate students and researchers in the different fields where Markov processes are used. It is particularly designed for students in traffic engineering, image analysis, bioinformatics, biostatistics, financial engineering, and computational biology. It is a combination of theory and applications of Markov models and presents the essential details of these models. Therefore, any reader who is interested in more information than what is presented in any particular topic discussed in the book is advised to consult any of the specialized books listed in the reference.

The book is organized as follows. Chapter 1 deals with basic concepts in probability and stochastic processes. Chapter 2 discusses general properties of Markov processes. Chapter 3 discusses discrete-time Markov chains, while Chapter 4 discusses continuous-time Markov chains. Chapter 5 discusses Markovian queueing systems, which are essentially those queueing systems where either the interarrival times or service times or both are exponentially distributed. Chapter 6 deals with Markov renewal processes including semi-Markov processes. Chapter 7 discusses Markovian arrival processes, while Chapter 8 deals with the random walk. Chapter 9 discusses Brownian motion and diffusion processes. Chapter 10 discusses controlled Markov processes, which include Markov decision processes, semi-Markov decision processes, and partially observable Markov decision processes. Chapter 11 discusses hidden Markov models, Chapter 12 discusses Markov random fields, and Chapter 13 discusses Markov point processes. Finally, Chapter 14 deals with Markov chain Monte Carlo.

The first nine chapters of the book are appropriate for a one-semester graduate course on applied stochastic processes. Such a course would usually include Markov chain Monte Carlo. The remainder of the book can be used according to the interests of the students.

1 Basic Concepts in Probability

1.1 Introduction

The concepts of *experiments* and *events* are very important in the study of probability. In probability, an experiment is any process of trial and observation. An experiment whose outcome is uncertain before it is performed is called a *random experiment*. When we perform a random experiment, the collection of possible elementary outcomes is called the *sample space* of the experiment, which is usually denoted by Ω. We define these outcomes as elementary outcomes because exactly one of the outcomes occurs when the experiment is performed. The elementary outcomes of an experiment are called the *sample points* of the sample space and are denoted by w_i, $i = 1, 2, \ldots$. If there are n possible outcomes of an experiment, then the sample space is $\Omega = \{w_1, w_2, \ldots, w_n\}$.

An *event* is the occurrence of either a prescribed outcome or any one of a number of possible outcomes of an experiment. Thus, an event is a subset of the sample space. For example, if we toss a die, any number from 1 to 6 can appear. Therefore, in this experiment, the sample space is defined by

$$\Omega = \{1, 2, 3, 4, 5, 6\}$$

The event "the outcome of the toss of a die is an even number" is a subset of Ω and is defined by

$$E = \{2, 4, 6\}$$

For a second example, consider a coin-tossing experiment in which each toss can result in either a head (H) or a tail (T). If we toss a coin three times and let the triplet xyz denote the outcome "x on the first toss, y on the second toss, and z on the third toss," then the sample space of the experiment is

$$\Omega = \{HHH, HHT, HTH, HTT, THH, THT, TTH, TTT\}$$

The event "one head and two tails" is a subset of Ω and is defined by

$$E = \{HTT, THT, TTH\}$$

Markov Processes for Stochastic Modeling. DOI: http://dx.doi.org/10.1016/B978-0-12-407795-9.00001-3

If a sample point x is contained in event A, we write $x \in A$. For any two events A and B defined on a sample space Ω, we can define the following new events:

- $A \cup B$ is the event that consists of all sample points that are either in A or in B or in both A and B. The event $A \cup B$ is called the *union* of events A and B.
- $A \cap B$ is the event that consists of all sample points that are in both A and B. The event $A \cap B$ is called the *intersection* of events A and B. Two events are defined to be *mutually exclusive* if their intersection does not contain a sample point; that is, they have no outcomes in common. Events $A_1, A_2, A_3, \ldots$ are defined to be mutually exclusive if no two of them have any outcomes in common and the events collectively have no outcomes in common.
- $A - B$ is the event that consists of all sample points that are in A but not in B. That is, $A - B = \{x \in A | x \notin B\}$. The event $A - B$ is called the *difference* of events A and B. Note that $A - B$ is different from $B - A$. The difference is sometimes denoted by $A \backslash B$, and $A - B = A \cap \overline{B}$, where $\overline{B}$ is defined as the *complement* of event B, which is the part of Ω that is not contained in B. That is, $\overline{B} = \Omega - B$.
- $B \subset A$ is the event where all the sample points in event B are contained in event A, and we say that B is a *subset* of A.

Consider an abstract space Ω; that is, Ω is a space without any special structure. Let F be a family of subsets of Ω with the following properties:

1. $\emptyset \in F$ and $\Omega \in F$, where $\emptyset$ is the empty space called the null event.
2. If $A \in F$, then $\overline{A} \in F$.
3. F is closed under countable unions and intersections; that is, if $A_1, A_2, \ldots$ are events in F, then $\cup_{k=1}^{\infty} A_k$ and $\cap_{k=1}^{\infty} A_k$ are both in F.

Under these conditions, F is defined to be a σ-algebra (or σ-field). A probability measure defined on a σ-algebra F of Ω is a function P that maps points in F onto the closed interval $[0,1]$. Thus, for an event A in F, the function $P[A]$ is called the *probability* of event A. The probability measure P satisfies the following Kolmogorov axioms:

1. As stated earlier, for any event $A \in F$, $0 \leq P[A] \leq 1$.
2. $P[\Omega] = 1$, which means that with probability 1, the outcome will be a sample point in the sample space.
3. For any set of n disjoint events $A_1, A_2, \ldots, A_n$ in F,

$$P[A_1 \cup A_2 \cup \cdots \cup A_n] = P[A_1] + P[A_2] + \cdots + P[A_n]$$

That is, for any set of mutually exclusive events defined on the same space, the probability of at least one of these events occurring is the sum of their respective probabilities.

The triple (Ω, F, P) is called a *probability space*. The following results are additional properties of a probability measure:

1. $P[\overline{A}] = 1 - P[A]$, which states that the probability of the complement of A is one minus the probability of A.
2. $P[\emptyset] = 0$, which states that the impossible (or null) event has probability zero.
3. If $A \subset B$, then $P[A] \leq P[B]$. That is, if A is a subset of B, the probability of A is at most the probability of B (or the probability of A cannot exceed the probability of B).

4. If $A = A_1 \cup A_2 \cup \cdots \cup A_n$, where $A_1, A_2, \ldots, A_n$ are mutually exclusive events, then

$$P[A] = P[A_1] + P[A_2] + \cdots + P[A_n]$$

5. For any two events A and B, $P[A] = P[A \cap B] + P[A \cap \overline{B}]$, which follows from the set identity: $A = (A \cap B) \cup (A \cap \overline{B})$. Since $A \cap B$ and $A \cap \overline{B}$ are mutually exclusive events, the result follows.

6. For any two events A and B, $P[A \cup B] = P[A] + P[B] - P[A \cap B]$.

7. We can extend Property 6 to the case of three events. If A_1, A_2, A_3 are three events in F, then

$$P[A_1 \cup A_2 \cup A_3] = P[A_1] + P[A_2] + P[A_3] - P[A_1 \cap A_2] - P[A_1 \cap A_3] - P[A_2 \cap A_3]$$
$$+ P[A_1 \cap A_2 \cap A_3]$$

This can be further generalized to the case of n arbitrary events in F as follows:

$$P[A_1 \cup A_2 \cup \cdots \cup A_n] = \sum_{i=1}^{n} P[A_i] - \sum_{1 \le i < j \le n} P[A_i \cap A_j] + \sum_{1 \le i < j < k \le n} P[A_i \cap A_j \cap A_k] - \cdots$$

That is, to find the probability that at least one of the n events occurs, first add the probability of each event, then subtract the probabilities of all possible two-way intersections, then add the probabilities of all possible three-way intersections, and so on.

1.1.1 Conditional Probability

Let A and B denote two events. The conditional probability of event A given event B, denoted by $P[A|B]$, is defined by

$$P[A|B] = \frac{P[A \cap B]}{P[B]} \quad P[B] > 0$$

For example, if A denotes the event that the sum of the outcomes of tossing two dice is 7 and B denotes the event that the outcome of the first die is 4, then the conditional probability of event A given event B is defined by

$$P[A|B] = \frac{P[A \cap B]}{P[B]}$$

$$= \frac{P[\{4, 3\}]}{P[\{4, 1\}] + P[\{4, 2\}] + P[\{4, 3\}] + P[\{4, 4\}] + P[\{4, 5\}] + P[\{4, 6\}]}$$

$$= \frac{(1/36)}{(6/36)} = \frac{1}{6}$$

1.1.2 Independence

Two events A and B are defined to be independent if the knowledge that one has occurred does not change or affect the probability that the other will occur. In particular, if events A and B are independent, the conditional probability of event A, given event B, $P[A|B]$, is equal to the probability of event A. That is, events A and B are independent if

$$P[A|B] = P[A]$$

Because by definition $P[A \cap B] = P[A|B]P[B]$, an alternative definition of independence of events is that events A and B are independent if

$$P[A \cap B] = P[A]P[B]$$

The definition of independence can be extended to multiple events. The n events $A_1, A_2, \ldots, A_n$ are said to be independent if the following conditions are true:

$$P[A_i \cap A_j] = P[A_i]P[A_j]$$
$$P[A_i \cap A_j \cap A_k] = P[A_i]P[A_j]P[A_k]$$
$$\vdots$$
$$P[A_1 \cap A_2 \cap \cdots \cap A_n] = P[A_1]P[A_2] \cdots P[A_n]$$

This is true for all $1 \le i < j < k < \cdots \le n$. That is, these events are pairwise independent, independent in triplets, and so on.

1.1.3 Total Probability and the Bayes' Theorem

A partition of a set A is a set $\{A_1, A_2, \ldots, A_n\}$ with the following properties:

a. $A_i \subseteq A$, $i = 1, 2, \ldots, n$, which means that A is a set of subsets.
b. $A_i \cap A_k = \emptyset$, $i = 1, 2, \ldots, n$; $k = 1, 2, \ldots, n$; $i \neq k$, which means that the subsets are mutually (or pairwise) disjoint; that is, no two subsets have any element in common.
c. $A_1 \cup A_2 \cup \cdots \cup A_n = A$, which means that the subsets are collectively exhaustive. That is, the subsets together include all possible values of the set A.

Let $\{A_1, A_2, \ldots, A_n\}$ be a partition of the sample space Ω, and suppose each one of the events $A_1, A_2, \ldots, A_n$ has nonzero probability of occurrence. Let B be any event. Then

$$P[B] = P[B|A_1]P[A_1] + P[B|A_2]P[A_2] + \cdots + P[B|A_n]P[A_n]$$
$$= \sum_{i=1}^{n} P[B|A_i]P[A_i]$$

This result is defined as the *total probability* of event B.

Suppose event B has occurred, but we do not know which of the mutually exclusive and collectively exhaustive events A_1, A_2, ..., A_n holds true. The conditional probability that event A_k occurred, given that B occurred, is given by

$$P[A_k|B] = \frac{P[A_k \cap B]}{P[B]} = \frac{P[A_k \cap B]}{\sum_{i=1}^{n} P[B|A_i]P[A_i]}$$

where the second equality follows from the total probability of event B. Because $P[A_k \cap B] = P[B|A_k]P[A_k]$, the preceding equation can be rewritten as follows:

$$P[A_k|B] = \frac{P[A_k \cap B]}{P[B]} = \frac{P[B|A_k]P[A_k]}{\sum_{i=1}^{n} P[B|A_i]P[A_i]}$$

This result is called the *Bayes' formula* (or *Bayes' rule*).

1.2   Random Variables

Consider a random experiment with sample space Ω. Let w be a sample point in Ω. We are interested in assigning a real number to each $w \in \Omega$. A random variable, $X(w)$, is a single-valued real function that assigns a real number, called the value of $X(w)$, to each sample point $w \in \Omega$. That is, it is a mapping of the sample space onto the real line.

Generally a random variable is represented by a single letter X instead of the function $X(w)$. Therefore, in the remainder of the book, we use X to denote a random variable. The sample space Ω is called the *domain* of the random variable X. Also, the collection of all numbers that are values of X is called the *range* of the random variable X.

Let X be a random variable and x a fixed real value. Let the event A_x define the subset of Ω that consists of all real sample points to which the random variable X assigns the number x. That is,

$$A_x = \{w|X(w) = x\} = [X = x]$$

Because A_x is an event, it will have a probability, which we define as follows:

$$p = P[A_x]$$

1.2.1   Distribution Functions

Let X be a random variable and x be a number. As stated earlier, we can define the event $[X \leq x] = \{w|X(w) \leq x\}$. The distribution function (or the cumulative distribution function (CDF)) of X is defined by

$$F_X(x) = P[X \leq x] \quad -\infty < x < \infty$$

That is, $F_X(x)$ denotes the probability that the random variable X takes on a value that is less than or equal to x.

1.2.2 Discrete Random Variables

A discrete random variable is a random variable that can take on at most a countable number of possible values. For a discrete random variable X, the *probability mass function* (PMF), $p_X(x)$, is defined as follows:

$$p_X(x) = P[X = x]$$

The PMF is nonzero for at most a countable or countably infinite number of values of x. In particular, if we consider that X can only assume one of the values $x_1, x_2, \ldots, x_n$, then

$$p_X(x_i) \geq 0 \quad i = 1, 2, \ldots, n$$
$$p_X(x_i) = 0 \quad \text{otherwise}$$

The CDF of X can be expressed in terms of $p_X(x)$ as follows:

$$F_X(x) = \sum_{k \leq x} p_X(k)$$

The CDF of a discrete random variable is a staircase-shaped function. That is, if X takes on values $x_1, x_2, x_3, \ldots$, where $x_1 < x_2 < x_3 < \cdots$, then the value of $F_X(x)$ is constant in the interval between x_{i-1} and x_i, and then takes a jump of size $p_X(x_i)$ at x_i, $i = 2, 3, \ldots$. Thus, in this case, $F_X(x)$ represents the sum of all the probability masses we have encountered as we move from $-\infty$ to slightly to the right of x.

1.2.3 Continuous Random Variables

Discrete random variables have a set of possible values that are either finite or countably infinite. However, there exists another group of random variables that can assume an uncountable set of possible values. Such random variables are called continuous random variables. Thus, we define a random variable X to be a continuous random variable if there exists a nonnegative function $f_X(x)$, defined for all real $x \in (-\infty, \infty)$, having the property that for any set A of real numbers,

$$P[X \in A] = \int_A f_X(x)\mathrm{d}x$$

The function $f_X(x)$ is called the *probability density function* (PDF) of the random variable X and is defined by

$$f_X(x) = \frac{\mathrm{d}F_X(x)}{\mathrm{d}x}$$

This means that

$$F_X(x) = \int_{-\infty}^{x} f_X(u)du$$

1.2.4 Expectations

If X is a random variable, then the *expectation* (or *expected value* or *mean*) of X, denoted by $E[X]$ or $\overline{X}$, is defined by

$$E[X] = \overline{X} = \begin{cases} \sum_i x_i p_X(x_i) & X \text{ discrete} \\ \int_{-\infty}^{\infty} x f_X(x)dx & X \text{ continuous} \end{cases}$$

Thus, the expected value of X is a weighted average of the possible values that X can take, where each value is weighted by the probability that X takes that value.

1.2.5 Expectation of Nonnegative Random Variables

Some random variables assume only nonnegative values. For example, the time X until a component fails cannot be negative. For a nonnegative random variable X with the PDF $f_X(x)$ and the CDF is $F_X(x)$, the expected value is given by

$$E[X] = \int_0^{\infty} P[X > x]dx = \int_0^{\infty} [1 - F_X(x)]dx$$

For a discrete random variable X that assumes only nonnegative values,

$$E[X] = \sum_{k=0}^{\infty} P[X > k]$$

1.2.6 Moments of Random Variables and the Variance

The nth moment of the random variable X, denoted by $E[X^n] = \overline{X^n}$, is defined by

$$E[X^n] = \overline{X^n} = \begin{cases} \sum_i x_i^n p_X(x_i) & X \text{ discrete} \\ \int_{-\infty}^{\infty} x^n f_X(x)dx & X \text{ continuous} \end{cases}$$

for $n = 1, 2, \ldots$. The first moment, $E[X]$, is the expected value of X.

We can also define the *central moments* (or *moments about the mean*) of a random variable. These are the moments of the difference between a random variable and its expected value. The nth central moment is defined by

$$E[(X-\overline{X})^n] = \overline{(X-\overline{X})^n} = \begin{cases} \sum_i (x_i-\overline{X})^n p_X(x_i) & X \text{ discrete} \\ \int_{-\infty}^{\infty} (x-\overline{X})^n f_X(x)dx & X \text{ continuous} \end{cases}$$

The central moment for the case of $n = 2$ is very important and carries a special name, the *variance*, which is usually denoted by σ_X^2. Thus,

$$\sigma_X^2 = E[(X-\overline{X})^2] = \overline{(X-\overline{X})^2} = \begin{cases} \sum_i (x_i-\overline{X})^2 p_X(x_i) & X \text{ discrete} \\ \int_{-\infty}^{\infty} (x-\overline{X})^2 f_X(x)dx & X \text{ continuous} \end{cases}$$

It can be shown that

$$\sigma_X^2 = E[X^2] - (E[X])^2$$

The square root of the variance is called the *standard deviation* of X and denoted by σ_X.

1.3 Transform Methods

Different types of transforms are used in science and engineering. These include the z-transform, Laplace transform, and Fourier transform. We consider two types of transforms: the z-transform (or moment-generating function) of PMFs and the s-transform (or unilateral Laplace transform) of PDFs.

1.3.1 The s-Transform

Let $f_X(x)$ be the PDF of the continuous random variable X that takes only nonnegative values; that is, $f_X(x) = 0$ for $x < 0$. The s-transform of $f_X(x)$, denoted by $M_X(s)$, is defined by

$$M_X(s) = E[e^{-sX}] = \int_0^{\infty} e^{-sx} f_X(x)dx$$

One important property of an s-transform is that when it is evaluated at the point $s = 0$, its value is equal to 1. That is,

$$M_X(s)|_{s=0} = \int_0^{\infty} f_X(x)dx = 1$$

For example, the value of K for which the function $A(s) = K/s + 5$ is a valid s-transform of a PDF is obtained by setting $A(0) = 1$, which gives

$$K/5 = 1 \Rightarrow K = 5$$

One of the primary reasons for studying the transform methods is to use them to derive the moments of the different probability distributions. By definition

$$M_X(s) = \int_0^\infty e^{-sx} f_X(x) dx$$

Taking different derivatives of $M_X(s)$ and evaluating them at $s = 0$, we obtain the following results:

$$\frac{d}{ds} M_X(s) = \frac{d}{ds} \int_0^\infty e^{-sx} f_X(x) dx = \int_0^\infty \frac{d}{ds} e^{-sx} f_X(x) dx$$

$$= - \int_0^\infty x e^{-sx} f_X(x) dx$$

$$\frac{d}{ds} M_X(s)|_{s=0} = - \int_0^\infty x f_X(x) dx = - E[X]$$

$$\frac{d^2}{ds^2} M_X(s) = \frac{d}{ds} (-1) \int_0^\infty x e^{-sx} f_X(x) dx = \int_0^\infty x^2 e^{-sx} f_X(x) dx$$

$$\frac{d^2}{ds^2} M_X(s)|_{s=0} = \int_0^\infty x^2 f_X(x) dx = E[X^2]$$

In general,

$$\frac{d^n}{ds^n} M_X(s)|_{s=0} = (-1)^n E[X^n]$$

1.3.2 The z-Transform

Let $p_X(x)$ be the PMF of the discrete nonnegative random variable X. The z-transform of $p_X(x)$, denoted by $G_X(z)$, is defined by

$$G_X(z) = E[z^X] = \sum_{x=0}^\infty z^x p_X(x)$$

The sum is guaranteed to converge and, therefore, the z-transform exists, when evaluated on or within the unit circle (where $|z| \le 1$). Note that

$$G_X(1) = \sum_{x=0}^\infty p_X(x) = 1$$

This means that a valid z-transform of a PMF reduces to unity when evaluated at $z = 1$. However, this is a necessary but not sufficient condition for a function to the z-transform of a PMF. By definition,

$$G_X(z) = \sum_{x=0}^{\infty} z^x p_X(x)$$

$$= p_X(0) + z p_X(1) + z^2 p_X(2) + z^3 p_X(3) + \cdots$$

This means that $P[X = k] = p_X(k)$ is the coefficient of z^k in the series expansion. Thus, given the z-transform of a PMF, we can uniquely recover the PMF. The implication of this statement is that not every polynomial that has a value 1 when evaluated at $z = 1$ is a valid z-transform of a PMF. For example, consider the function $A(z) = 2z - 1$. Although $A(1) = 1$, the function contains invalid coefficients in the sense that these coefficients either have negative values or positive values that are greater than 1. Thus, for a function of z to be a valid z-transform of a PMF, it must have a value of 1 when evaluated at $z = 1$, and the coefficients of z must be nonnegative numbers that cannot be greater than 1.

The individual terms of the PMF can also be determined as follows:

$$p_X(x) = \frac{1}{x!} \left[\frac{d^x}{dz^x} G_X(z) \right]_{z=0} \quad x = 0, 1, 2, \ldots$$

This feature of the z-transform is the reason it is sometimes called the *probability-generating function*.

As stated earlier, one of the major motivations for studying transform methods is their usefulness in computing the moments of the different random variables. Unfortunately, the moment-generating capability of the z-transform is not as computationally efficient as that of the s-transform. The moment-generating capability of the z-transform lies in the results obtained from evaluating the derivatives of the transform at $z = 1$. For a discrete random variable X with PMF $p_X(x)$, we have that

$$\frac{d}{dz} G_X(z) = \frac{d}{dz} \sum_{x=0}^{\infty} z^x p_X(x) = \sum_{x=0}^{\infty} \frac{d}{dz} z^x p_X(x) = \sum_{x=0}^{\infty} x z^{x-1} p_X(x)$$

$$= \sum_{x=1}^{\infty} x z^{x-1} p_X(x)$$

$$\frac{d}{dz} G_X(z)|_{z=1} = \sum_{x=1}^{\infty} x p_X(x) = \sum_{x=0}^{\infty} x p_X(x) = E[X]$$

Similarly,

$$\frac{d^2}{dz^2} G_X(z) = \frac{d}{dz} \sum_{x=1}^{\infty} xz^{x-1} p_X(x) = \sum_{x=1}^{\infty} x \frac{d}{dz} z^{x-1} p_X(x)$$

$$= \sum_{x=1}^{\infty} x(x-1) z^{x-2} p_X(x)$$

$$\frac{d^2}{dz^2} G_X(z)|_{z=1} = \sum_{x=1}^{\infty} x(x-1) p_X(x) = \sum_{x=0}^{\infty} x(x-1) p_X(x)$$

$$= \sum_{x=0}^{\infty} x^2 p_X(x) - \sum_{x=0}^{\infty} x p_X(x) = E[X^2] - E[X]$$

$$E[X^2] = \frac{d^2}{dz^2} G_X(z)|_{z=1} + \frac{d}{dz} G_X(z)|_{z=1}$$

We can obtain higher moments in a similar manner.

1.4 Bivariate Random Variables

Consider two random variables X and Y defined on the same sample space. For example, X can denote the grade of a student and Y can denote the height of the same student. The *joint* CDF of X and Y is given by

$$F_{XY}(x, y) = P[X \leq x, Y \leq y]$$

The pair (X, Y) is referred to as a *bivariate* random variable. If we define $F_X(x) = P[X \leq x]$ as the *marginal* CDF of X and $F_Y(y) = P[Y \leq y]$ as the *marginal* CDF of Y, then we define the random variables X and Y to be independent if

$$F_{XY}(x, y) = F_X(x) F_Y(y)$$

for every value of x and y. The marginal CDFs are obtained as follows:

$$F_X(x) = F_{XY}(x, \infty)$$
$$F_Y(y) = F_{XY}(\infty, y)$$

From the above properties, we can answer questions about X and Y.

1.4.1 Discrete Bivariate Random Variables

When both X and Y are discrete random variables, we define their joint PMF as follows:

$$p_{XY}(x, y) = P[X = x, Y = y]$$

The marginal PMFs are obtained as follows:

$$p_X(x) = \sum_y p_{XY}(x, y)$$
$$p_Y(y) = \sum_x p_{XY}(x, y)$$

If X and Y are independent random variables,

$$p_{XY}(x, y) = p_X(x)p_Y(y)$$

1.4.2 Continuous Bivariate Random Variables

If both X and Y are continuous random variables, their joint PDF is given by

$$f_{XY}(x, y) \frac{\partial^2}{\partial x \partial y} F_{XY}(x, y)$$

The joint PDF satisfies the following condition:

$$F_{XY}(x, y) = \int_{u=-\infty}^{x} \int_{v=-\infty}^{y} f_{XY}(u, v) dv\, du$$

The marginal PDFs are given by

$$f_X(x) = \int_{y=-\infty}^{\infty} f_{XY}(x, y) dy$$
$$f_Y(y) = \int_{x=-\infty}^{\infty} f_{XY}(x, y) dx$$

If X and Y are independent random variables, then

$$f_{XY}(x, y) = f_X(x)f_Y(y)$$

1.4.3 Covariance and Correlation Coefficient

Consider two random variables X and Y with expected values $E[X] = \mu_X$ and $E[Y] = \mu_Y$, respectively, and variances σ_X^2 and σ_Y^2, respectively. The *covariance* of X and Y, which is denoted by σ_{XY} or $\mathrm{Cov}(X, Y)$, is defined by

$$\sigma_{XY} = E[(X - \mu_X)(Y - \mu_Y)] = E[XY - X\mu_Y - Y\mu_X + \mu_X\mu_Y]$$
$$= E[XY] - \mu_X\mu_Y$$

If X and Y are independent, then $E[XY] = E[X]E[Y] = \mu_X\mu_Y$ and $\sigma_{XY} = 0$. However, the converse is not true; that is, if the covariance of X and Y is zero, it does not mean that X and Y are independent random variables. If the covariance of two random variables is zero, we define the two random variables to be *uncorrelated*.

We define the *correlation coefficient* of X and Y, denoted by ρ_{XY} or $\rho(X,Y)$, as follows:

$$\rho_{XY} = \frac{\mu_Y}{\sqrt{\sigma_X^2\sigma_Y^2}} = \frac{\mu_Y}{\sigma_X\sigma_Y}$$

where σ_X and σ_Y are the standard deviations of X and Y, respectively. The correlation coefficient has the property that

$$-1 \leq \rho_{XY} \leq 1$$

1.5 Many Random Variables

In the previous sections, we considered a system of two random variables. In this section, we extend the concepts developed for two random variables to systems of more than two random variables.

Let X_1, X_2, ..., X_n be a set of random variables that are defined on the same sample space. Their joint CDF is defined as

$$F_{X_1X_2\cdots X_n}(x_1,x_2,\ldots,x_n) = P[X_1 \leq x_1, X_2 \leq x_2,\ldots,X_n \leq x_n]$$

If all the random variables are discrete random variables, their joint PMF is defined by

$$p_{X_1X_2\cdots X_n}(x_1,x_2,\ldots,x_n) = P[X_1 = x_1, X_2 = x_2,\ldots,X_n = x_n]$$

The marginal PMFs are obtained by summing the joint PMF over the appropriate ranges. For example, the marginal PMF of X_n is given by

$$p_{X_n}(x_n) = \sum_{x_1}\sum_{x_2}\cdots\sum_{x_{n-1}}p_{X_1X_2\cdots X_n}(x_1,x_2,\ldots,x_n)$$

The conditional PMFs are similarly obtained. For example,

$$p_{X_n|X_1X_2\cdots X_{n-1}}(x_n|x_1,x_2,\ldots,x_{n-1}) = P[X_n = x_n|X_1 = x_1, X_2 = x_2,\ldots,X_{n-1} = x_{n-1}]$$

$$= \frac{p_{X_1X_2\cdots X_n}(x_1,x_2,\ldots,x_n)}{p_{X_1X_2\cdots X_{n-1}}(x_1,x_2,\ldots,x_{n-1})}$$

The random variables are defined to be mutually independent if

$$p_{X_1 X_2 \cdots X_n}(x_1, x_2, \ldots, x_n) = \prod_{i=1}^{n} p_{X_i}(x_i)$$

If all the random variables are continuous random variables, their joint PDF can be obtained from the joint CDF as follows:

$$f_{X_1 X_2 \cdots X_n}(x_1, x_2, \ldots, x_n) = \frac{\partial^n}{\partial x_1 \partial x_2 \cdots \partial x_n} F_{X_1 X_2 \cdots X_n}(x_1, x_2, \ldots, x_n)$$

The conditional PDFs can also be defined. For example,

$$f_{X_n | X_1 X_2 \cdots X_{n-1}}(x_n | x_1, x_2, \ldots, x_{n-1}) = \frac{f_{X_1 X_2 \cdots X_n}(x_1, x_2, \ldots, x_n)}{f_{X_1 X_2 \cdots X_{n-1}}(x_1, x_2, \ldots, x_{n-1})}$$

Similarly, if the random variables are mutually independent, then

$$f_{X_1 X_2 \cdots X_n}(x_1, x_2, \ldots, x_n) = \prod_{i=1}^{n} f_{X_i}(x_i)$$

1.6 Fubini's Theorem

The expectation of a function of random vectors (i.e., d-dimensional random variables, where $d > 1$) is obtained as a multidimensional integral. For example, if g is a function defined in the probability space (Ω, F, P) and (X, Y) is a random vector, then

$$E[g(X, Y)] = \int_{\Omega} g(X, Y) dP = \int_{R^2} g(x, y) dF_{XY}(x, y)$$

$$= \begin{cases} \sum_{i=1}^{\infty} \sum_{j=1}^{\infty} g(x, y) p_{XY}(x, y) & \text{discrete case} \\ \int_{R^2} g(x, y) f_{XY}(x, y) dx\, dy & \text{continuous case} \end{cases}$$

Fubini's theorem allows us to compute expectations of functions of random variables in a rather simpler manner when the probability spaces are product spaces. Specifically, in the case of a function of two random variables, the theorem allows us to evaluate the expectation, which involves double integrals, as iterated single integrals. We state the theorem without proof.

Theorem 1.1 (Fubini's Theorem) Let (X,Y) be a two-dimensional random variable where $(X,Y) \in \{(x,y) \in R^2\}$ and assume that g is R^2-measurable and is nonnegative and integrable. Then

$$E[g(X, Y)] = \iint_{R^2} g(x, y) dF_X(x) dF_Y(y)$$

$$= \int_R \left\{ \int_R g(x, y) dF_X(x) \right\} dF_Y(y)$$

$$= \int_R \left\{ \int_R g(x, y) dF_Y(y) \right\} dF_X(x)$$

This means that to compute the double integral with respect to the product measure, we integrate first with respect to one variable and then with respect to the other variable.

1.7   Sums of Independent Random Variables

Consider two independent continuous random variables X and Y. We are interested in computing the CDF and PDF of their sum $g(X,Y) = V = X + Y$. The random variable V can be used to model the reliability of systems with standby connections. In such systems, the component A whose time-to-failure represented by the random variable X is the primary component, and the component B whose time-to-failure represented by the random variable Y is the backup component that is brought into operation when the primary component fails. Thus, V represents the time until the system fails, which is the sum of the lifetimes of both components.

The CDF of V can be obtained as follows:

$$F_V(v) = P[V \leq v] = P[X + Y \leq v] = \int_D \int f_{XY}(x, y) dx\, dy$$

where D is the set $D = \{(x, y) | x + y \leq v\}$. Thus,

$$F_V(v) = \int_{-\infty}^{\infty} \int_{-\infty}^{v-y} f_{XY}(x, y) dx\, dy = \int_{-\infty}^{\infty} \int_{-\infty}^{v-y} f_X(x) f_Y(y) dx\, dy$$

$$= \int_{-\infty}^{\infty} \left\{ \int_{-\infty}^{v-y} f_X(x) dx \right\} f_Y(y) dy$$

$$= \int_{-\infty}^{\infty} F_X(v - y) f_Y(y) dy$$

The PDF of V is obtained by differentiating the CDF as follows:

$$f_V(v) = \frac{d}{dv} F_V(v) = \frac{d}{dv} \int_{-\infty}^{\infty} F_X(v-y)f_Y(y)dy = \int_{-\infty}^{\infty} \frac{d}{dv} F_X(v-y)f_Y(y)dy$$

$$= \int_{-\infty}^{\infty} f_X(v-y)f_Y(y)dy$$

where we have assumed that we can interchange differentiation and integration. The expression on the right-hand side is a well-known result in signal analysis called the *convolution integral*. Thus, we find that the PDF of the sum V of two independent random variables X and Y is the convolution of the PDFs of the two random variables, that is,

$$f_V(v) = f_X(v) * f_Y(v)$$

In general, if V is the sum of n mutually independent random variables X_1, X_2, ..., X_n whose PDFs are $f_{X_i}(x_i), i = 1, 2, \ldots, n$, then we have that

$$V = X_1 + X_2 + \cdots + X_n$$
$$f_V(v) = f_{X_1}(v) * f_{X_2}(v) * \cdots * f_{X_n}(v)$$

Thus, the s-transform of the PDF of V is given by

$$M_V(s) = \prod_{i=1}^{n} M_{X_i}(s)$$

1.8   Some Probability Distributions

Random variables with special probability distributions are encountered in different fields of science and engineering. In this section, we describe some of these distributions, including their expected values, variances, and s-transforms (or z-transforms, as the case may be). These include the Bernoulli distribution, binomial distribution, geometric distribution, Pascal distribution, Poisson distribution, exponential distribution, Erlang distribution, and normal distribution.

1.8.1   *The Bernoulli Distribution*

A Bernoulli trial is an experiment that results in two outcomes: *success* and *failure*. One example of a Bernoulli trial is the coin-tossing experiment, which results in heads or tails. In a Bernoulli trial, we define the probability of success and probability of failure as follows:

$$P[\text{success}] = p \quad 0 \leq p \leq 1$$
$$P[\text{failure}] = 1 - p$$

Let us associate the events of the Bernoulli trial with a random variable X such that when the outcome of the trial is a success, we define $X = 1$, and when the outcome is a failure, we define $X = 0$. The random variable X is called a Bernoulli random variable and its PMF is given by

$$p_X(x) = \begin{cases} 1 - p & x = 0 \\ p & x = 1 \end{cases}$$

An alternative way to define the PMF of X is as follows:

$$p_X(x) = p^x(1-p)^{1-x} \quad x = 0, 1$$

The CDF is given by

$$F_X(x) = \begin{cases} 0 & x < 0 \\ 1 - p & x \leq x < 1 \\ 1 & x \geq 1 \end{cases}$$

The expected value of X is given by

$$E[X] = 0(1 - p) + 1(p) = p$$

Similarly, the second moment of X is given by

$$E[X^2] = 0^2(1 - p) + 1^2(p) = p$$

Thus, the variance of X is given by

$$\sigma_X^2 = E[X^2] - (E[X])^2 = p - p^2 = p(1 - p)$$

The z-transform of the PMF is given by

$$G_X(z) = \sum_{x=0}^{\infty} z^x p_X(x) = \sum_{x=0}^{1} z^x p_X(x) = z^0(1 - p) + z(p)$$

$$= 1 - p + zp$$

1.8.2 The Binomial Distribution

Suppose we conduct n independent Bernoulli trials and we represent the number of successes in those n trials by the random variable $X(n)$. Then $X(n)$ is defined as a binomial random variable with parameters (n, p). The PMF of a random variable $X(n)$ with parameters (n, p) is given by

$$p_{X(n)}(x) = \binom{n}{x} p^x (1-p)^{n-x} \quad x = 0, 1, 2, \ldots, n$$

The binomial coefficient, $\binom{n}{x}$, represents the number of ways of arranging x successes and $n - x$ failures.

Because $X(n)$ is essentially the sum of n independent Bernoulli random variables, its CDF, mean, variance, and the z-transform of its PMF are given by

$$F_{X(n)}(x) = P[X(n) \le x] = \sum_{k=0}^{x} \binom{n}{k} p^k (1-p)^{n-k}$$

$$E[X(n)] = np$$

$$E[X^2(n)] = n(n-1)p^2 + np$$

$$\sigma_{X(n)}^2 = E[X^2(n)] - (E[X(n)])^2 = np(1-p)$$

$$G_{X(n)}(z) = (1 - p + zp)^n$$

1.8.3   The Geometric Distribution

The geometric random variable is used to describe the number of independent Bernoulli trials until the first success occurs. Let X be a random variable that denotes the number of Bernoulli trials until the first success. If the first success occurs on the xth trial, then we know that the first $x - 1$ trials resulted in failures. Thus, since the trials are independent, the PMF of a geometric random variable, X, is given by

$$p(x) = p(1-p)^{x-1} \quad x = 1, 2, \ldots$$

The CDF, mean, and variance of X, and the z-transform of its PMF are given by

$$F_X(x) = P[X \le x] = 1 - (1-p)^x$$

$$E[X] = \frac{1}{p}$$

$$E[X^2] = \frac{2-p}{p^2}$$

$$\sigma_X^2 = E[X^2] - (E[X])^2 = \frac{1-p}{p^2}$$

$$G_X(z) = \frac{zp}{1 - z(1-p)}$$

The geometric distribution possesses the so-called forgetfulness property, which means that given that success has not occurred at the end of the nth trial, the time until the next success is geometrically distributed with mean $1/p$. Thus, the process "starts from scratch," and as shown in Ibe (2005),

$$p_{X|X>n}(x|X>n) = p(1-p)^{x-n-1} \quad x > n$$

1.8.4 The Pascal Distribution

The Pascal random variable is an extension of the geometric random variable. It describes the number of trials until the kth success, which is why it is sometimes called the "kth-order interarrival time for a Bernoulli process." The Pascal distribution is also called the *negative binomial distribution*.

Let X_k be a kth-order Pascal random variable. Then its PMF is given by

$$p_{X_k}(x) = \binom{x-1}{k-1} p^k (1-p)^{x-k} \quad k = 1,2,\dots; \ x = k, \ k+1,\dots$$

Because X_k is essentially the sum of k independent geometric random variables, its CDF, mean, variance, and the z-transform of its PMF are given by

$$F_{X_k}(x) = P[X_k \le x] = \sum_{n=k}^{x} \binom{n-1}{k-1} p^k (1-p)^{n-k}$$

$$E[X_k] = \frac{k}{p}$$

$$E[X_k^2] = \frac{k^2 + k(1-p)}{p^2}$$

$$\sigma_{X_k}^2 = E[X_k^2] - (E[X_k])^2 = \frac{k(1-p)}{p^2}$$

$$G_X(z) = \left[\frac{zp}{1-z(1-p)}\right]^k$$

Note that the geometric random variable is the *first-order* Pascal random variable.

1.8.5 The Poisson Distribution

A discrete random variable K is called a Poisson random variable with parameter λ, where $\lambda > 0$, if its PMF is given by

$$p_K(k) = \frac{\lambda^k}{k!} e^{-\lambda} \quad k = 0,1,2,\dots$$

The CDF, mean, and variance of K, and the z-transform of its PMF are given by

$$F_K(k) = P[K \le k] = \sum_{r=0}^{k} \frac{\lambda^r}{r!} e^{-\lambda}$$

$$E[K] = \lambda$$

$$E[K^2] = \lambda^2 + \lambda$$

$$\sigma_K^2 = E[K^2] - (E[K])^2 = \lambda$$

$$G_K(z) = e^{-\lambda(1-z)}$$

Note that the mean and the variance of the Poisson random variable are equal. This is a special characteristic of the Poisson random variable.

1.8.6 The Exponential Distribution

A continuous random variable X is defined to be an exponential random variable (or X has an exponential distribution) if for some parameter $\lambda > 0$ its PDF is given by

$$f_X(x) = \begin{cases} \lambda e^{-\lambda x} & x \ge 0 \\ 0 & x < 0 \end{cases}$$

The CDF, mean, and variance of X, and the s-transform of its PDF are given by

$$F_X(x) = P[X \le x] = 1 - e^{-\lambda x}$$

$$E[X] = \frac{1}{\lambda}$$

$$E[X^2] = \frac{2}{\lambda^2}$$

$$\sigma_X^2 = E[X^2] - (E[X])^2 = \frac{1}{\lambda^2}$$

$$M_X(s) = \frac{\lambda}{s + \lambda}$$

Like the geometric distribution, the exponential distribution possesses the forgetfulness property. Thus, if we consider the occurrence of an event governed by the exponential distribution as an arrival, then given that no arrival has occurred up to time t, the time until the next arrival is exponentially distributed with mean $1/\lambda$. In particular, it can be shown, as in Ibe (2005), that

$$f_{X|X>t}(x|X>t) = \lambda e^{-\lambda(x-t)} \quad x > t$$

1.8.7 The Erlang Distribution

The Erlang distribution is a generalization of the exponential distribution. While the exponential random variable describes the time between adjacent events, the Erlang random variable describes the time interval between any event and the kth following event. A random variable X_k is referred to as a kth-order Erlang (or Erlang-k) random variable with parameter λ if its PDF is given by

$$f_{X_k}(x) = \begin{cases} \dfrac{\lambda^k x^{k-1} e^{-\lambda x}}{(k-1)!} & k = 1, 2, 3, \ldots; \ x \geq 0 \\ 0 & x < 0 \end{cases}$$

Because X_k is the sum of k independent exponential random variables, its CDF, mean, variance, and the s-transform of its PDF are given by

$$F_{X_k}(x) = P[X_k \leq x] = 1 - \sum_{j=0}^{k-1} \frac{(\lambda x)^j e^{-\lambda x}}{j!}$$

$$E[X_k] = \frac{k}{\lambda}$$

$$E[X_k^2] = \frac{k(k+1)}{\lambda^2}$$

$$\sigma_{X_k}^2 = E[X_k^2] - (E[X_k])^2 = \frac{k}{\lambda^2}$$

$$M_{X_k}(s) = \left[\frac{\lambda}{s+\lambda} \right]^k$$

The exponential random variable is the first-order Erlang random variable.

1.8.8 Normal Distribution

A continuous random variable X is defined to be a normal random variable with parameters μ_X and σ_X^2 if its PDF is given by

$$f_X(x) = \frac{1}{\sqrt{2\pi\sigma_X^2}} e^{-(x-\mu_X)^2/2\sigma_X^2} \quad -\infty < x < \infty$$

The PDF is a bell-shaped curve that is symmetric about μ_X, which is the mean of X. The parameter σ_X^2 is the variance. Figure 1.1 illustrates the shape of the PDF. Because the variance (or more precisely, the standard deviation) is a measure of the spread around the mean, the larger the variance, the lower the peak of the curve and the more spread out it will be.

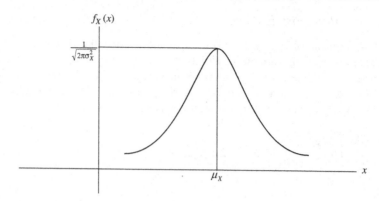

Figure 1.1 PDF of the normal random variable.

The CDF of X is given by

$$F_X(x) = P[X \leq x] = \frac{1}{\sigma_X \sqrt{2\pi}} \int_{-\infty}^{x} e^{-(u-\mu_X)^2/2\sigma_X^2} \, du$$

The normal random variable X with parameters μ_X and σ_X^2 is usually designated $X \sim N(\mu_X, \sigma_X^2)$. The special case of zero mean and unit variance (i.e., $\mu_X = 0$ and $\sigma_X^2 = 1$) is designated $X \sim N(0,1)$ and is called the *standard normal random variable*. Let $y = (u - \mu_X)/\sigma_X$. Then $du = \sigma_X \, dy$, and the CDF of X becomes

$$F_X(x) = \frac{1}{\sqrt{2\pi}} \int_{-\infty}^{(x-\mu_X)/\sigma_X} e^{-y^2/2} \, dy$$

Thus, with the above transformation, X becomes a standard normal random variable. The above integral cannot be evaluated in closed form. It is usually evaluated numerically through the function $\Phi(x)$, which is defined as follows:

$$\Phi(x) = \frac{1}{\sqrt{2\pi}} \int_{-\infty}^{x} e^{-y^2/2} \, dy$$

Thus, the CDF of X is given by

$$F_X(x) = \frac{1}{\sqrt{2\pi}} \int_{-\infty}^{(x-\mu_X)/\sigma_X} e^{-y^2/2} \, dy = \Phi\left(\frac{x - \mu_X}{\sigma_X}\right)$$

The values of $\Phi(x)$ are usually given for nonnegative values of x. For negative values of x, $\Phi(x)$ can be obtained from the following relationship:

$$\Phi(-x) = 1 - \Phi(x)$$

Values of $\Phi(x)$ are usually available in standard probability and statistics textbooks.

1.9 Limit Theorems

In this section, we discuss two fundamental theorems in probability. These are the law of large numbers, which is regarded as the first fundamental theorem, and the central limit theorem, which is regarded as the second fundamental theorem. We begin the discussion with the Markov and Chebyshev inequalities that enable us to prove these theorems.

1.9.1 Markov Inequality

The Markov inequality applies to random variables that take only nonnegative values. It can be stated as follows:

Proposition 1.1 If X is a random variable that takes only nonnegative values, then for any $a > 0$,

$$P[X \geq a] \leq \frac{E[X]}{a}$$

Proof We consider only the case when X is a continuous random variable. Thus,

$$E[X] = \int_0^\infty x f_X(x)\mathrm{d}x = \int_0^a x f_X(x)\mathrm{d}x + \int_a^\infty x f_X(x)\mathrm{d}x \geq \int_a^\infty x f_X(x)\mathrm{d}x$$
$$\geq \int_a^\infty a f_X(x)\mathrm{d}x = a \int_a^\infty f_X(x)\mathrm{d}x = aP[X \geq a]$$

and the result follows.

1.9.2 Chebyshev Inequality

The Chebyshev inequality enables us to obtain bounds on probability when both the mean and variance of a random variable are known. The inequality can be stated as follows:

Proposition 1.2 Let X be a random variable with mean μ and variance σ^2. Then, for any $b > 0$,

$$P[|X - \mu| \geq b] \leq \frac{\sigma^2}{b^2}$$

Proof Since $(X - \mu)^2$ is a nonnegative random variable, we can invoke the Markov inequality, with $a = b^2$, to obtain

$$P\left[(X-\mu)^2 \geq b^2\right] \leq \frac{E[(X-\mu)^2]}{b^2}$$

Since $(X - \mu)^2 \geq b^2$ if and only if $|X - \mu| \geq b$, the preceding inequality is equivalent to

$$P[|X - \mu| \geq b] \leq \frac{E[(X-\mu)^2]}{b^2} = \frac{\sigma^2}{b^2}$$

which completes the proof.

1.9.3 Laws of Large Numbers

There are two laws of large numbers that deal with the limiting behavior of random sequences. One is called the "weak" law of large numbers, and the other is called the "strong" law of large numbers. We will discuss only the weak law of large numbers.

Proposition 1.3 Let X_1, X_2, ..., X_n be a sequence of mutually independent and identically distributed random variables, and let their mean be $E[X_k] = \mu < \infty$. Similarly, let their variance be $\sigma_{X_k}^2 = \sigma^2 < \infty$. Let S_n denote the sum of the n random variables, that is,

$$S_n = X_1 + X_2 + \cdots + X_n$$

Then the weak law of large numbers states that for any $\varepsilon > 0$,

$$\lim_{n \to \infty} P\left[\left|\frac{S_n}{n} - \mu\right| \geq \varepsilon\right] \to 0$$

Equivalently,

$$\lim_{n \to \infty} P\left[\left|\frac{S_n}{n} - \mu\right| < \varepsilon\right] \to 1$$

Proof Since X_1, X_2, ..., X_n are independent and have the same distribution, we have that

$$\text{Var}(S_n) = n\sigma^2$$

$$\text{Var}\left(\frac{S_n}{n}\right) = \frac{n\sigma^2}{n^2} = \frac{\sigma^2}{n}$$

$$E\left[\frac{S_n}{n}\right] = \frac{n\mu}{n} = \mu$$

From Chebyshev inequality, for $\varepsilon > 0$, we have that

$$P\left[\left|\frac{S_n}{n} - \mu\right| \geq \varepsilon\right] \leq \frac{\sigma^2}{n\varepsilon^2}$$

Thus, for a fixed ε,

$$P\left[\left|\frac{S_n}{n} - \mu\right| \geq \varepsilon\right] \to 0$$

as $n \to \infty$, which completes the proof.

1.9.4 The Central Limit Theorem

The central limit theorem provides an approximation to the behavior of sums of random variables. The theorem states that as the number of independent and identically distributed random variables with finite mean and finite variance increases, the distribution of their sum becomes increasingly normal regardless of the form of the distribution of the random variables. More formally, let X_1, X_2, ..., X_n be a sequence of mutually independent and identically distributed random variables each of which has a finite mean μ_X and a finite variance σ_X^2. Let S_n be defined as follows:

$$S_n = X_1 + X_2 + \cdots + X_n$$

Now, $E[S_n] = n\mu_X$ and $\sigma_{S_n}^2 = n\sigma_X^2$. Converting S_n to a standard normal random variable (i.e., zero mean and unit variance), we obtain

$$Y_n = \frac{S_n - E[S_n]}{\sigma_{S_n}} = \frac{S_n - n\mu_X}{\sqrt{n\sigma_X^2}} = \frac{S_n - n\mu_X}{\sigma_X\sqrt{n}}$$

The central limit theorem states that if $F_{Y_n}(y)$ is the CDF of Y_n, then

$$\lim_{n \to \infty} F_{Y_n}(y) = \lim_{n \to \infty} P[Y_n \le y] = \frac{1}{\sqrt{2\pi}} \int_{-\infty}^{y} e^{-u^2/2} \, du = \Phi(y)$$

This means that $\lim_{n \to \infty} Y_n \sim N(0, 1)$. Thus, one of the important roles that the normal distribution plays in statistics is its usefulness as an approximation of other probability distribution functions.

An alternate statement of the theorem is that in the limit as n becomes very large,

$$Z_n = \frac{X_1 + X_2 + \cdots + X_n}{\sigma_X \sqrt{n}}$$

is a normal random variable with unit variance.

1.10 Problems

1.1 A sequence of Bernoulli trials consists of choosing seven components at random from a batch of components. A selected component is classified as either defective or nondefective. A nondefective component is considered to be a success, while a defective component is considered to be a failure. If the probability that a selected component is nondefective is 0.8, then what is the probability of three successes?

1.2 The probability that a patient recovers from a rare blood disease is 0.3. If 15 people are known to have contracted this disease, find the following probabilities:
a. At least 10 survive.
b. From 3 to 8 survive.
c. Exactly 6 survive.

1.3 A sequence of Bernoulli trials consists of choosing components at random from a batch of components. A selected component is classified as either defective or nondefective. A nondefective component is considered to be a success, while a defective component is considered to be a failure. If the probability that a selected component is nondefective is 0.8, determine the probabilities of the following events:
a. The first success occurs on the 5th trial.
b. The third success occurs on the 8th trial.
c. There are 2 successes by the 4th trial, there are 4 successes by the 10th trial, and there are 10 successes by the 18th trial.

1.4 A lady invites 12 people for dinner at her house. Unfortunately, the dining table can only seat six people. Her plan is that if six or fewer guests come, then they will be seated at the table (i.e., they will have a sit-down dinner); otherwise she will set up a buffet-style meal. The probability that each invited guest will come to dinner is 0.4, and each guest's decision is independent of other guests' decisions. Determine the following:
a. The probability that she has a sit-down dinner.
b. The probability that she has a buffet-style dinner.
c. The probability that there are at most three guests.

1.5 A Girl Scout troop sells cookies from house to house. One of the parents of the girls figured out that the probability that they sell a set of packs of cookies at any house they visit is 0.4, where it is assumed that they sell exactly one set to each house that buys their cookies.

    a. What is the probability that the house where they make their first sale is the fifth house they visit?

    b. Given that they visited 10 houses on a particular day, what is the probability that they sold exactly six sets of cookie packs?

    c. What is the probability that on a particular day the third set of cookie packs is sold at the seventh house that the girls visit?

1.6 Students arrive for a lab experiment according to a Poisson process with a rate of 12 students per hour. However, the lab attendant opens the door to the lab when at least four students are waiting at the door. What is the probability that the waiting time of the first student to arrive exceeds 20 min? (By waiting time we mean the time that elapses from when a student arrives until the door is opened by the lab attendant.)

1.7 Cars arrive at a gas station according to a Poisson process at an average rate of 12 cars per hour. The station has only one attendant. If the attendant decides to take a 2-min coffee break when there were no cars at the station, what is the probability that one or more cars will be waiting when he comes back from the break, given that any car that arrives when he is on coffee break waits for him to get back?

1.8 Customers arrive at the neighborhood bookstore according to a Poisson process with an average rate of 10 customers per hour. Independently of other customers, each arriving customer buys a book with probability 1/8.

    a. What is the probability that the bookstore sells no book during a particular hour?

    b. What is the PDF of the time until the first book is sold?

1.9 Joe is a student who is conducting experiments with a series of light bulbs. He started with 10 identical light bulbs each of which has an exponentially distributed lifetime with a mean of 200 h. He wants to know how long it will take until the last bulb burns out (or fails). At noontime he stepped out to get some lunch with six bulbs still on. Assume that he came back and found that none of the six bulbs has failed.

    a. After Joe came back, what is the expected time until the next bulb failure?

    b. What is the expected length of time between the fourth bulb failure and the fifth bulb failure?

1.10 Students arrive at the professor's office for extra help according to a Poisson process with an average rate of four students per hour. The professor does not start the tutorial until at least three students are available. Students who arrive while the tutorial is going on will have to wait for the next session.

    a. Given that a tutorial has just ended and there are no students currently waiting for the professor, what is the mean time until another tutorial can start?

    b. Given that one student was waiting when the tutorial ended, what is the probability that the next tutorial does not start within the first 2 h?

2 Basic Concepts in Stochastic Processes

2.1 Introduction

Stochastic processes deal with the dynamics of probability theory. The concept of stochastic processes enlarges the random variable concept to include time. Thus, instead of thinking of a random variable X that maps an event $w \in \Omega$, where Ω is the sample space, to some number $X(w)$, we think of how the random variable maps the event to different numbers at different times. This implies that instead of the number $X(w)$ we deal with $X(t, w)$, where $t \in T$ and T is called the *parameter set* of the process and is usually a set of times.

If we fix the sample point w, $X(t)$ is some real function of time. For each w, we have a function $X(t)$. Thus, $X(t, w)$ can be viewed as a collection of time functions, one for each sample point w. On the other hand, if we fix t, we have a function $X(w)$ that depends only on w and thus is a random variable. Therefore, a stochastic process becomes a random variable when time is fixed at some particular value. With many values of t, we obtain a collection of random variables. Thus, we can define a stochastic process as a family of random variables $\{X(t, w)|t \in T, w \in \Omega\}$ defined over a given probability space and indexed by the time parameter t. A stochastic process is also called a *random process*. Thus, we will use the terms "random process" and "stochastic process" interchangeably.

2.2 Classification of Stochastic Processes

A stochastic process can be classified according to the nature of the time parameter and the values that $X(t, w)$ can assume. As discussed earlier, T is called the parameter set of the stochastic process and is usually a set of times. If T is an interval of real numbers and hence is continuous, the process is called a *continuous-time* stochastic process. Similarly, if T is a countable set and hence is discrete, the process is called a *discrete-time* stochastic process. A discrete-time stochastic process is also called a *random sequence*, which is denoted by $\{X[n]: n = 1, 2, \ldots\}$.

The values that $X(t, w)$ assumes are called the *states* of the stochastic process. The set of all possible values of $X(t, w)$ forms the *state space*, E, of the stochastic process. If E is continuous, the process is called a *continuous-state* stochastic process. Similarly, if E is discrete, the process is called a *discrete-state* stochastic process.

Markov Processes for Stochastic Modeling. DOI: http://dx.doi.org/10.1016/B978-0-12-407795-9.00002-5

2.3   Characterizing a Stochastic Process

In the remainder of the discussion, we will represent the stochastic process $X(t, w)$ by $X(t)$; that is, we will suppress w, the sample space parameter. A stochastic process is completely described or characterized by the joint cumulative distribution function (CDF). Because the value of a stochastic process $X(t)$ at time $t_i, X(t_i)$, is a random variable, let

$$F_X(x_1, t_1) = F_X(x_1) = P[X(t_1) \leq x_1]$$
$$F_X(x_2, t_2) = F_X(x_2) = P[X(t_2) \leq x_2]$$
$$\vdots$$
$$F_X(x_n, t_n) = F_X(x_n) = P[X(t_n) \leq x_n]$$

where $0 < t_1 < t_2 < \cdots < t_n$. Then the joint CDF, which is defined by

$$F_X(x_1, \ldots, x_n; t_1, \ldots, t_n) = P[X(t_1) \leq x_1, \ldots, X(t_n) \leq x_n] \quad \text{for all } n$$

completely characterizes the stochastic process. If $X(t)$ is a continuous-time stochastic process, then it is specified by a collection of probability density functions (PDFs):

$$f_X(x_1, \ldots, x_n; t_1, \ldots, t_n) = \frac{\partial^n}{\partial x_1 \cdots \partial x_n} F_X(x_1, \ldots, x_n; t_1, \ldots, t_n)$$

Similarly, if $X(t)$ is a discrete-time stochastic process, then it is specified by a collection of probability mass functions (PMFs):

$$p_X(x_1, \ldots, x_n; t_1, \ldots, t_n) = P[X(t_1) = x_1, \ldots, X(t_n) = x_n]$$

2.4   Mean and Autocorrelation Function of a Stochastic Process

The mean of $X(t)$ is a function of time called the *ensemble average* and is denoted by

$$\mu_{X(t)} = E[X(t)]$$

The autocorrelation function provides a measure of similarity between two observations of the stochastic process $X(t)$ at different points in time t and s. The autocorrelation function of $X(t)$ and $X(s)$ is denoted by $R_{XX}(t, s)$ and is defined as follows:

$$R_{XX}(t, s) = E[X(t)X(s)] = E[X(s)X(t)] = R_{XX}(s, t)$$
$$R_{XX}(t, t) = E[X^2(t)]$$

It is a common practice to define $s = t + \tau$, which gives the autocorrelation function as

$$R_{XX}(t, t + \tau) = E[X(t)X(t + \tau)]$$

The parameter τ is sometimes called the *delay time* (or *lag time*). The autocorrelation function of a deterministic periodic function of period T is given by

$$R_{XX}(t, t + \tau) = \frac{1}{2T} \int_{-T}^{T} f_X(t) f_X(t + \tau) dt$$

Similarly, for an aperiodic function, the autocorrelation function is given by

$$R_{XX}(t, t + \tau) = \int_{-\infty}^{\infty} f_X(t) f_X(t + \tau) dt$$

Basically the autocorrelation function defines how much a signal is similar to a time-shifted version of itself. A random process $X(t)$ is called a *second-order process* if $E[X^2(t)] < \infty$ for each $t \in T$.

2.5 Stationary Stochastic Processes

There are several ways to define a stationary stochastic process. At a high level, it is a process whose statistical properties do not vary with time. In this book, we consider only two types of stationary processes: *strict-sense stationary* processes and the *wide-sense stationary* (WSS) processes.

2.5.1 *Strict-Sense Stationary Processes*

A random process is defined to be a strict-sense stationary process if its CDF is invariant to a shift in the time origin. This means that the process $X(t)$ with the CDF $F_X(x_1, x_2, \ldots, x_n; t_1, t_2, \ldots, t_n)$ is a strict-sense stationary process if its CDF is identical to that of $X(t + \varepsilon)$ for any arbitrary ε. Thus, being a strict-sense stationary process implies that for any arbitrary ε and for all n,

$$F_X(x_1, x_2, \ldots, x_n; t_1, t_2, \ldots, t_n) = F_X(x_1, x_2, \ldots, x_n; t_1 + \varepsilon, t_2 + \varepsilon, \ldots, t_n + \varepsilon)$$

When the CDF is differentiable, the equivalent condition for strict-sense stationarity is that the PDF is invariant to a shift in the time origin; that is, for all n,

$$f_X(x_1, x_2, \ldots, x_n; t_1, t_2, \ldots, t_n) = f_X(x_1, x_2, \ldots, x_n; t_1 + \varepsilon, t_2 + \varepsilon, \ldots, t_n + \varepsilon)$$

If $X(t)$ is a strict-sense stationary process, then the CDF $F_{X_1 X_2}(x_1, x_2; t_1, t_1 + \tau)$ does not depend on t, but it might depend on τ. Thus, if $t_2 = t_1 + \tau$, then

$F_{X_1X_2}(x_1, x_2; t_1, t_2)$ might depend on $t_2 - t_1$, but not on t_1 and t_2 individually. This means that if $X(t)$ is a strict-sense stationary process, then the autocorrelation and autocovariance functions do not depend on t. Thus, we have that for all $\tau \in T$:

$$\mu_X(t) = \mu_X(0)$$
$$R_{XX}(t, t + \tau) = R_{XX}(0, \tau)$$
$$C_{XX}(t, t + \tau) = C_{XX}(0, \tau)$$

where $C_{XX}(t_1, t_2) = E[\{X(t_1) - \mu_X(t_1)\}\{X(t_2) - \mu_X(t_2)\}]$ is the autocovariance function. If the condition $\mu_X(t) = \mu_X(0)$ holds for all t, the mean is constant and denoted by μ_X. Similarly, if the function $R_{XX}(t, t + \tau)$ does not depend on t but is a function of τ, we write $R_{XX}(0, \tau) = R_{XX}(\tau)$. Finally, whenever the condition $C_{XX}(t, t + \tau) = C_{XX}(0, \tau)$ holds for all t, we write $C_{XX}(0, \tau) = C_{XX}(\tau)$.

2.5.2 Wide-Sense Stationary Processes

Many practical problems that we encounter require that we deal with only the mean and autocorrelation function of a random process. Solutions to these problems are simplified if these quantities do not depend on absolute time. Random processes in which the mean and autocorrelation function do not depend on absolute time are called WSS processes. Thus, for a WSS process $X(t)$,

$$E[X(t)] = \mu_X \text{ (constant)}$$
$$R_{XX}(t, t + \tau) = R_{XX}(\tau)$$

Note that a strict-sense stationary process is also a WSS process. However, in general, the converse is not true; that is, a WSS process is not necessarily stationary in the strict sense.

2.6 Ergodic Stochastic Processes

One desirable property of a stochastic process is the ability to estimate its parameters from measurement data. Consider a random process $X(t)$ whose observed samples are $x(t)$. The time average of a function of $x(t)$ is defined by

$$\bar{x} = \lim_{T \to \infty} \frac{1}{2T} \int_{-T}^{T} x(t) \mathrm{d}t$$

The statistical average of the random process $X(t)$ is the expected value $E[X(t)]$ of the process. The expected value is also called the *ensemble average*. An ergodic stochastic process is a stationary process in which every member of the ensemble exhibits the same statistical behavior as the ensemble. This implies that it is possible to determine the statistical behavior of the ensemble by examining only one typical sample function. Thus, for an ergodic stochastic process, the mean values and moments can be determined by time averages as well as by ensemble averages (or expected values), which are equal. That is,

$$E[X^n] = \overline{X^n} = \int_{-\infty}^{\infty} x^n f_X(x)\mathrm{d}x = \lim_{T \to \infty} \frac{1}{2T} \int_{-T}^{T} x^n(t)\mathrm{d}t$$

A stochastic process $X(t)$ is defined to be *mean-ergodic* (or *ergodic in the mean*) if $E[X(t)] = \overline{x}$.

2.7 Some Models of Stochastic Processes

In this section, we consider some examples of random processes that we will encounter in the remainder of the book.

2.7.1 Martingales

A stochastic process $\{X_n, n = 1, 2, \ldots\}$ is defined to be a martingale process (or a martingale) if it has the following properties:

- $E[|X_n|] < \infty$ for all n; that is, it has finite means.
- $E[X_{n+1}|X_1, X_2, \ldots, X_n] = X_n$; that is, the best prediction of its future values is its present value.

If $E[X_{n+1}|X_1, X_2, \ldots, X_n] \leq X_n$, then $\{X_n, n = 1, 2, \ldots\}$ is called a *supermartingale*. Similarly, if $E[X_{n+1}|X_1, X_2, \ldots, X_n] \geq X_n$, then $\{X_n, n = 1, 2, \ldots\}$ is called a *submartingale*. Thus, a martingale satisfies the conditions for both a supermartingale and a submartingale.

Sometimes the martingale property is defined with respect to another stochastic process. Specifically, let $\{X_n, n = 1, 2, \ldots\}$ and $\{Y_n, n = 1, 2, \ldots\}$ be stochastic processes. $\{X_n\}$ is defined to be a martingale with respect to $\{Y_n\}$ if, for $n = 1, 2, \ldots$, the following conditions hold:

- $E[|X_n|] < \infty$
- $E[X_{n+1}|Y_1, Y_2, \ldots, Y_n] = X_n$.

A martingale captures the essence of a fair game in the sense that regardless of a player's current and past fortunes, his expected fortune at any time in the future is the same as his current fortune. Thus, on the average, he neither wins nor loses any money. Also, martingales fundamentally deal with conditional expectation. If we define $\mathfrak{I}_n = \{Y_1, Y_2, \ldots, Y_n\}$, then $\mathfrak{I}_n$ can be considered the potential information that is being revealed as time progresses. Therefore, we can consider a martingale as a process whose expected value, conditional on some potential information, is equal to the value revealed by the last available information. Similarly, a submartingale represents a favorable game because the expected fortune increases in the future, while a supermartingale represents an unfavorable game because the expected fortune decreases in the future.

Martingales occur in many stochastic processes. They have also become an important tool in modern financial mathematics because martingales provide one idea of fair value in financial markets.

Theorem 2.1 If $\{X_n, n \geq 0\}$ is a martingale, then $E[X_n] = E[X_0]$ for all $n \geq 0$.

Proof Let $\Im_n$ be as defined earlier. We know that $E[E[X|Y]] = E[X]$. Also, because $\{X_n, n \geq 0\}$ is a martingale, $E[X_n|\Im_0] = X_0$. Thus, we have that

$$E[X_n] = E[E[X_n|\Im_0]] = E[X_0]$$

Example 2.1

Let $X_1, X_2, \ldots$ be independent random variables with mean 0, and let $Y_n = \sum_{k=1}^{n} X_k$. We show that the process $\{Y_n, n \geq 1\}$ is a martingale as follows:

$$
\begin{aligned}
E[Y_{n+1}|Y_1, Y_2, \ldots, Y_n] &= E[Y_n + X_{n+1}|Y_1, Y_2, \ldots, Y_n] \\
&= E[Y_n|Y_1, Y_2, \ldots, Y_n] + E[X_{n+1}|Y_1, Y_2, \ldots, Y_n] \\
&= Y_n + E[X_{n+1}] \\
&= Y_n
\end{aligned}
$$

Example 2.2

Consider the variance of a sum of random variables. Specifically, let $X_0 = 0$ and $X_1, X_2, \ldots$ be independent and identically distributed random variables with mean $E[X_k] = 0$ and finite variance $E[X_k^2] = \sigma^2, k \geq 1$. If we let $Y_0 = 0$ and define

$$Y_n = \left(\sum_{k=1}^{n} X_k \right)^2 - n\sigma^2$$

we show that $\{Y_n\}$ is a martingale with respect to $\{X_n\}$ as follows:

$$
\begin{aligned}
E[Y_{n+1}|X_0, X_1, \ldots, X_n] &= E\left[\left(X_{n+1} + \sum_{k=1}^{n} X_k \right)^2 - (n+1)\sigma^2 |X_0, X_1, \ldots, X_n \right] \\
&= E\left[X_{n+1}^2 + 2X_{n+1}\sum_{k=1}^{n} X_k + \left(\sum_{k=1}^{n} X_k \right)^2 - (n+1)\sigma^2 |X_0, X_1, \ldots, X_n \right] \\
&= E\left[\left\{ \left(\sum_{k=1}^{n} X_k \right)^2 - n\sigma^2 \right\} + X_{n+1}^2 + 2X_{n+1}\sum_{k=1}^{n} X_k - \sigma^2 |X_0, X_1, \ldots, X_n \right] \\
&= Y_n + E[X_{n+1}^2|X_0, X_1, \ldots, X_n] + 2\left(\sum_{k=1}^{n} X_k \right) E[X_{n+1}|X_0, X_1, \ldots, X_n] - \sigma^2 \\
&= Y_n + \sigma^2 + 0 - \sigma^2 = Y_n
\end{aligned}
$$

Example 2.3

Let $X_1, X_2, \ldots$ be independent random variables with mean $E[X_k] = 1$, $k \geq 1$, and let $Y_n = \prod_{k=1}^{n} X_k$. We show that the process $\{Y_n, n \geq 1\}$ is a martingale as follows:

$$
\begin{aligned}
E[Y_{n+1} | Y_1, \ldots, Y_n] &= E[Y_n X_{n+1} | Y_1, \ldots, Y_n] \\
&= Y_n E[X_{n+1} | Y_1, \ldots, Y_n] = Y_n E[X_{n+1}] \\
&= Y_n
\end{aligned}
$$

Stopping Times

Consider a stochastic process $\{X_n, n \geq 0\}$. The nonnegative integer-valued random variable T is called a stopping time for X if the event $\{T = n\}$ depends only on $\{X_1, X_2, \ldots, X_n\}$ and does not depend on $\{X_{n+k}, k \geq 1\}$. If T_k is a stopping time, then we have that

$$\{T_k = n\} = \{X_1 \neq k, \ldots, X_{n-1} \neq k, X_n = k\}$$

The use of stopping times in martingales is given by the following proposition, which is stated without proof.

Proposition 2.1 Let T be a stopping time for a stochastic process $\{X_n\}$, and let $a \wedge b = \min(a, b)$.

1. If $\{X_n\}$ is a martingale, then so is $\{X_{T \wedge n}\}$.
2. If $\{X_n\}$ is a supermartingale, then so is $\{X_{T \wedge n}\}$.
3. If $\{X_n\}$ is a submartingale, then so is $\{X_{T \wedge n}\}$.

Stopping times can be thought of as the time when a given event occurs. If it has the value $T = \infty$, then the event never occurs. For example, we might be interested in the first time the value of a random sequence that is known to be a martingale is 6. Then we consider the martingale $\{X_n, n \geq 0\}$ and a random variable T that is defined by

$$
T = \begin{cases} \inf_{n \geq 0} \{n | X_n = 6\} & \text{if } X_n = 6 \quad \text{for some } n \in \aleph \\ \infty & \text{otherwise} \end{cases}
$$

where $\aleph$ is the set of positive integers.

Theorem 2.2 (Optional Stopping Theorem) Let T be a stopping time for the martingale $\{X_n\}$. Then $E[X_T] = E[X_0]$ if at least one of the following conditions holds:

a. T is finite (i.e., $P[T < \infty] = 1$) and there exists a finite constant C_1 such that $|X_n| \leq C_1$ for all $n \leq T$.
b. T is bounded; that is, there exists a finite constant C_2 so that with probability 1, $T \leq C_2$.
c. $E[T]$ is finite and there exists a finite constant C_3 such that $E[|X_{n+1} - X_n| | X_1, \ldots, X_n] < C_3$ for $n = 0, 1, \ldots$.

This theorem is also called the *stopping time theorem*.

Proof The proof of this theorem can be found in any standard stochastic processes book, such as Grimmett and Stirzaker (2001).

2.7.2 Counting Processes

A stochastic process $\{X(t), t \geq 0\}$ is called a counting process if $X(t)$ represents the total number of "events" that have occurred in the interval $[0, t)$. An example of a counting process is the number of customers that arrive at a bank from the time the bank opens its doors for business until some time t. A counting process satisfies the following conditions:

1. $X(t) \geq 0$, which means that it has nonnegative values.
2. $X(0) = 0$, which means that the counting of events begins at time 0.
3. $X(t)$ is integer-valued.
4. If $s < t$, then $X(s) \leq X(t)$, which means that it is a nondecreasing function of time.
5. $X(t) - X(s)$ represents the number of events that have occurred in the interval $[s, t]$.

Figure 2.1 represents a sample path of a counting process. The first event occurs at time t_1, and subsequent events occur at times t_2, t_3, and t_4. Thus, the number of events that occur in the interval $[0, t_4]$ is 4.

2.7.3 Independent Increment Processes

A counting process is defined to be an independent increment process if the number of events that occur in disjoint time intervals is an independent random variable. For example, in Figure 2.1, consider the two nonoverlapping (i.e., disjoint) time intervals $[0, t_1]$ and $[t_2, t_4]$. If the number of events occurring in one interval is independent of the number of events that occur in the other, then the process is an independent increment process. Thus, $X(t)$ is an independent increment process if for every set of time instants $t_0 = 0 < t_1 < t_2 < \cdots < t_n$, the increments $X(t_1) - X(t_0), X(t_2) - X(t_1), \ldots, X(t_n) - X(t_{n-1})$ are mutually independent random variables.

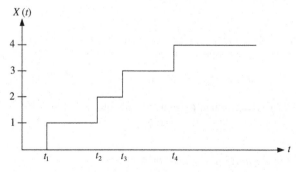

Figure 2.1 Sample function of a counting process.

2.7.4 Stationary Increment Process

A counting process $X(t)$ is defined to possess stationary increments if for every set of time instants $t_0 = 0 < t_1 < t_2 < \cdots < t_n$, the increments $X(t_1) - X(t_0), X(t_2) - X(t_1), \ldots, X(t_n) - X(t_{n-1})$ are identically distributed. In general, the mean of an independent increment process $X(t)$ with stationary increments has the form

$$E[X(t)] = mt$$

where the constant m is the value of the mean at time $t = 1$. That is, $m = E[X(1)]$. Similarly, the variance of an independent increment process $X(t)$ with stationary increments has the form

$$\mathrm{Var}[X(t)] = \sigma^2 t$$

where σ^2 the constant is the value of the variance at time $t = 1$; that is, $\sigma^2 = \mathrm{Var}[X(1)]$.

2.7.5 Poisson Processes

Poisson processes are widely used to model arrivals (or occurrence of events) in a system. For example, they are used to model the arrival of telephone calls at a switchboard, the arrival of customers' orders at a service facility, and the random failures of equipment.

There are two ways to define a Poisson process. The first definition of the process is that it is a counting process $X(t)$ in which the number of events in any interval of length t has a Poisson distribution with mean λt, $\lambda > 0$. Thus, for all $s, t > 0$,

$$P[X(s + t) - X(s) = n] = \frac{(\lambda t)^n}{n!} e^{-\lambda t} \quad n = 0, 1, 2, \ldots$$

The second way to define the Poisson process $X(t)$ is that it is a counting process with stationary and independent increments such that for a rate $\lambda > 0$, the following conditions hold:

1. $P[X(t + \Delta t) - X(t) = 1] = \lambda \, \Delta t + o(\Delta t)$, which means that the probability of one event within a small time interval Δt is approximately $\lambda \, \Delta t$, where $o(\Delta t)$ is a function of Δt that goes to zero faster than Δt docs. That is,

$$\lim_{\Delta t \to 0} \frac{o(\Delta t)}{\Delta t} = 0$$

2. $P[X(t + \Delta t) - X(t) \geq 2] = o(\Delta t)$, which means that the probability of two or more events within a small time interval Δt is $o(\Delta t)$. This implies that the probability of two or more events within a small time interval Δt is negligibly small.
3. $P[X(t + \Delta t) - X(t) = 0] = 1 - \lambda \, \Delta t + o(\Delta t)$.

These three properties enable us to derive the PMF of the number of events in a time interval of length t as follows:

$$
\begin{aligned}
P[X(t + \Delta t) = n] &= P[X(t) = n]P[X(\Delta t) = 0] \\
&\quad + P[X(t) = n - 1]P[X(\Delta t) = 1] \\
&= P[X(t) = n](1 - \lambda \Delta t) \\
&\quad + P[X(t) = n - 1]\lambda \Delta t
\end{aligned}
$$

$$
\begin{aligned}
P[X(t + \Delta t) = n] - P[X(t) = n] &= -\lambda P[X(t) = n]\Delta t \\
&\quad + \lambda P[X(t) = n - 1]\Delta t
\end{aligned}
$$

$$
\frac{P[X(t + \Delta t) = n] - P[X(t) = n]}{\Delta t} = -\lambda P[X(t) = n] + \lambda P[X(t) = n - 1]
$$

$$
\lim_{\Delta t \to 0} \left\{ \frac{P[X(t + \Delta t) = n] - P[X(t) = n]}{\Delta t} \right\} = \frac{\mathrm{d}}{\mathrm{d}t} P[X(t) = n] = -\lambda P[X(t) = n]
$$

$$
+ \lambda P[X(t) = n - 1]
$$

$$
\frac{\mathrm{d}}{\mathrm{d}t} P[X(t) = n] + \lambda P[X(t) = n] = \lambda P[X(t) = n - 1]
$$

The last equation can be solved iteratively for $n = 0, 1, 2, \ldots$, subject to the initial conditions

$$
P[X(0) = n] = \begin{cases} 1 & n = 0 \\ 0 & n \neq 0 \end{cases}
$$

This gives the PMF of the number of events (or "arrivals") in an interval of length t as

$$
p_{X(t)}(n, t) = \frac{(\lambda t)^n}{n!} e^{-\lambda t} \quad t \geq 0, \quad n = 0, 1, 2, \ldots
$$

From the results obtained for Poisson random variables earlier in the chapter, we have that

$$
\begin{aligned}
G_{X(t)}(z) &= e^{-\lambda t(1-z)} \\
E[X(t)] &= \lambda t \\
\sigma^2_{X(t)} &= \lambda t
\end{aligned}
$$

The fact that the mean $E[X(t)] = \lambda t$ indicates that λ is the expected number of arrivals per unit time in the Poisson process. Thus, the parameter λ is called the *arrival rate* for the process. If λ is independent of time, the Poisson process is called a *homogeneous Poisson process*. Sometimes the arrival rate is a function of time, and we represent it as $\lambda(t)$. Such processes are called *nonhomogeneous Poisson processes*. In this book, we are concerned mainly with homogeneous Poisson processes.

Interarrival Times for the Poisson Process

Let L_r be a continuous random variable that is defined to be the interval between any event in a Poisson process and the rth event after it. Then L_r is called the rth-order interarrival time. Let $f_{L_r}(l)$ be the PDF of L_r. To derive the expression for $f_{L_r}(l)$, we consider time of length l over which we know that $r - 1$ events have occurred. Assume that the next event (i.e., the rth event) occurs during the next time of length Δl, as shown in Figure 2.2.

Because the intervals l and Δl are nonoverlapping, the number of events that occur within one interval is independent of the number of events that occur within the other interval. Thus, the PDF of L_r can be obtained as follows:

$$
\begin{aligned}
f_{L_r}(l)\Delta l &= P[l < L_r \leq l + \Delta l] = P[\{X(l) = r - 1\} \cap \{X(\Delta l) = 1\}] \\
&= P[X(l) = r - 1]P[X(\Delta l) = 1] \\
&= \left\{ \frac{(\lambda l)^{r-1}}{(r-1)!} e^{-\lambda l} \right\} \{\lambda \, \Delta l\} = \frac{\lambda^r l^{r-1}}{(r-1)!} e^{-\lambda l} \, \Delta l
\end{aligned}
$$

Thus, $f_{L_r}(l)$ is given by

$$
f_{L_r}(l) = \frac{\lambda^r l^{r-1}}{(r-1)!} e^{-\lambda l} \quad l \geq 0; \quad r = 1, 2, \ldots
$$

which is the Erlang-r (or rth-order Erlang) distribution. The special case of $r = 1$ is the exponential distribution. That is,

$$
f_{L_1}(l) = \lambda e^{-\lambda l} \quad l \geq 0
$$

This result provides another definition of a Poisson process: It is a counting process with stationary and independent increments in which the intervals between consecutive events are exponentially distributed. The sample path for the Poisson process is similar to that shown in Figure 2.1 with the added constraint that the times between arrivals are exponentially distributed.

Figure 2.2 Definition of event intervals.

Compound Poisson Process

Let $\{N(t), t \geq 0\}$ be a Poisson process with arrival rate λ. Let $\{Y_i, i = 1, 2, \ldots\}$ be a family of independent and identically distributed random variables. Assume that the Poisson process $\{N(t), t \geq 0\}$ and the sequence $\{Y_i, i = 1, 2, \ldots\}$ are independent. We define a random process $\{X(t), t \geq 0\}$ to be a *compound Poisson process* if, for $t \geq 0$, it can be represented by

$$X(t) = \sum_{i=1}^{N(t)} Y_i$$

Thus, $X(t)$ is a Poisson sum of random variables. One example of the concept of compound Poisson process is the following. Assume students arrive at the university bookstore to buy books in a Poisson manner. If the number of books that each of these students buys is an independent and identically distributed random variable, then the number of books bought by time t is a compound Poisson process.

Figure 2.3 illustrates the sample path for a compound Poisson process. Here we have a batch arrival in which the number of elements in each batch is a random variable. In the example shown in Figure 2.3, the number in each batch can be positive or negative.

Because the compound Poisson process has a rate that takes on a stochastic nature, it is also called a *doubly stochastic Poisson process*. This term is used to emphasize the fact that the process involves two kinds of randomness: There is a randomness that is associated with the main process that is sometimes called the *Poisson point process*, and there is another independent randomness that is associated with its rate.

Assume that the Y_i are discrete random variables with the PMF $p_Y(y)$. The value of $X(t)$, given that $N(t) = n$, is $X(t) = Y_1 + Y_2 + \cdots + Y_n$. Thus, the conditional z-transform of the PMF of $X(t)$, given that $N(t) = n$, is given by

$$G_{X(t)|N(t)}(z|n) = E[z^{Y_1 + Y_2 + \cdots + Y_n}] = (E[z^Y])^n = \{G_Y(z)\}^n$$

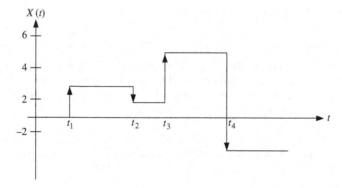

Figure 2.3 Example of sample path of a compound Poisson process.

where the last two equalities follow from the fact that the Y_i are independent. Thus, the unconditional z-transform of the PMF of $X(t)$ is given by

$$G_{X(t)}(z) = \sum_{n=0}^{\infty} G_{X(t)N(t)}(z|n)p_{N(t)}(n) = \sum_{n=0}^{\infty} \{G_Y(z)\}^n p_{N(t)}(n)$$

$$= \sum_{n=0}^{\infty} \{G_Y(z)\}^n \frac{(\lambda t)^n}{n!} e^{-\lambda t} = e^{-\lambda t} \sum_{n=0}^{\infty} \frac{\{\lambda t G_Y(z)\}^n}{n!} = e^{-\lambda t} e^{\lambda t G_Y(z)}$$

$$= e^{-\lambda t[1 - G_Y(z)]}$$

The mean and variance of $X(t)$ can be obtained through differentiating the above function. These are given by

$$E[X(t)] = \frac{d}{dz} G_{X(t)}(z)|_{z=1} = \lambda t\, E[Y]$$

$$E[X^2(t)] = \frac{d^2}{dz^2} G_{X(t)}(z)|_{z=1} + \frac{d}{dz} G_{X(t)}(z)|_{z=1} = \lambda t\, E[Y^2] + (\lambda t\, E[Y])^2$$

$$\sigma_{X(t)}^2 = E[X^2(t)] - (E[X(t)])^2 = \lambda t\, E[Y^2]$$

Note that in the case when the Y_i are continuous random variables, we would use the s-transform of the PDF of $X(t)$; the result would be $M_{X(t)}(s) = e^{-\lambda t[1 - M_Y(s)]}$ and the above results still hold.

Example 2.4

Customers arrive at a grocery store in a Poisson manner at an average rate of 10 customers per hour. The amount of money that each customer spends is uniformly distributed between $8.00 and $20.00. What is the average total amount of money that customers who arrive over a 2 h interval spend in the store? What is the variance of this total amount?

Solution
This is a compound Poisson process with $\lambda = 10$ customers per hour. Let Y be the random variable that represents the amount of money a customer spends in the store. Since Y is uniformly distributed over the interval $(8,20)$, we have that

$$E[Y] = \frac{8+20}{2} = 14$$

$$\sigma_Y^2 = \frac{(20-8)^2}{12} = 12$$

$$E[Y^2] = \sigma_Y^2 + (E[Y])^2 = 12 + 196 = 208$$

Therefore, the mean and the variance of the total amount of money that customers arriving over a 2 h time interval $(t = 2)$ spend in the store are given by

$$E[X(t)]_{t=2} = 2\lambda\, E[Y] = 2(10)(14) = 280$$

$$\sigma_{X(t)}^2|_{t=2} = 2\lambda\, E[Y^2] = 2(10)(208) = 4160$$

Combinations of Independent Poisson Processes

Consider two independent Poisson processes $\{X(t), t \geq 0\}$ and $\{Y(t), t \geq 0\}$ with arrival rates λ_X and λ_Y, respectively. Consider a random process $\{N(t), t \geq 0\}$ that is the sum of the two Poisson processes; that is, $N(t) = X(t) + Y(t)$. Thus, $\{N(t)\}$ is the process consisting of arrivals from the two Poisson processes. We want to show that $\{N(t)\}$ is also a Poisson process with arrival rate $\lambda_X + \lambda_Y$.

To do this, we note that $\{X(t), t \geq 0\}$ and $\{Y(t), t \geq 0\}$ are specified as being independent. Therefore,

$$
\begin{aligned}
P[N(t + \Delta t) - N(t) = 0] &= P[X(t + \Delta t) - X(t) = 0]P[Y(t + \Delta t) - Y(t) = 0] \\
&= [1 - \lambda_X \Delta t + o(\Delta t)][1 - \lambda_Y \Delta t + o(\Delta t)] \\
&= 1 - \lambda_X \Delta t - \lambda_Y \Delta t + o(\Delta t) = 1 - (\lambda_X + \lambda_Y)\Delta t + o(\Delta t) \\
&= 1 - \lambda \Delta t + o(\Delta t)
\end{aligned}
$$

where $\lambda = \lambda_X + \lambda_Y$. Since the last equation is the probability that there is no arrival within an interval of length Δt, it follows that $\{N(t)\}$ is a Poisson process.

Another way to prove this is to note that when t is fixed, $N(t)$ is a random variable that is a sum of two independent random variables. Thus, the z-transform of the PMF of $N(t)$ is the product of the z-transform of the PMF of $X(t)$ and the z-transform of the PMF of $Y(t)$. That is,

$$
\begin{aligned}
G_{N(t)}(z) &= G_{X(t)}(z)G_{Y(t)}(z) = e^{-\lambda_X t(1-z)}e^{-\lambda_Y t(1-z)} = e^{-(\lambda_X + \lambda_Y)t(1-z)} \\
&= e^{-\lambda t(1-z)}
\end{aligned}
$$

Thus, $N(t)$ is a Poisson random variable. Since for each t, $N(t)$ is a random variable, the collection of these random variables over time (i.e., $\{N(t), t \geq 0\}$) constitutes a Poisson random process.

The third way to show this is via the interarrival time. Let L denote the time until the first arrival in the process $\{N(t), t \geq 0\}$, let L_X denote the time until the first arrival in the process $\{X(t), t \geq 0\}$, and let L_Y denote the time until the first arrival in the process $\{Y(t), t \geq 0\}$. Then, since the two Poisson processes are independent,

$$
\begin{aligned}
P[L > t] &= P[L_X > t]P[L_X > t] = e^{-\lambda_X t}\, e^{-\lambda_Y t} = e^{-(\lambda_X + \lambda_X)t} \\
&= e^{-\lambda t}
\end{aligned}
$$

which shows that $\{N(t), t \geq 0\}$ exhibits the same memoryless property as $\{X(t), t \geq 0\}$ and $\{Y(t), t \geq 0\}$. Therefore, $\{N(t), t \geq 0\}$ must be a Poisson process.

Example 2.5

Two light bulbs, labeled A and B, have exponentially distributed lifetimes. If the two lifetimes of the two bulbs are independent and the mean lifetime of bulb A is 500 h, while the mean lifetime of bulb B is 200 h, what is the mean time to a bulb failure?

Solution

Let λ_A denote the burnout rate of bulb A and λ_B the burnout rate of bulb B. Since $1/\lambda_A = 500$ and $1/\lambda_B = 200$, the rates are $\lambda_A = 1/500$ and $\lambda_B = 1/200$. From the results obtained above, the two bulbs behave like a single system with exponentially distributed lifetime with a mean of $1/\lambda$, where $\lambda = \lambda_A + \lambda_B$. Thus, the mean time until a bulb fails in hours is

$$\frac{1}{\lambda} = \frac{1}{\lambda_A + \lambda_B} = \frac{1}{(1/500) + (1/200)} = \frac{1000}{7} = 142.86$$

Competing Independent Poisson Processes

In this section, we extend the combination problem discussed in the previous section. Thus, we consider two independent Poisson processes $\{X(t), t \geq 0\}$ and $\{Y(t), t \geq 0\}$ with arrival rates λ_X and λ_Y, respectively. The question we are interested in is this: What is the probability that an arrival from $\{X(t), t \geq 0\}$ occurs before an arrival from $\{Y(t), t \geq 0\}$? Since the interarrival times in a Poisson process are exponentially distributed, let T_X be the random variable that denotes the interarrival time in the $\{X(t), t \geq 0\}$ process and let T_Y be the random variable that denotes the interarrival time in the $\{Y(t), t \geq 0\}$ process. Thus, we are interested in the computing $P[T_X < T_Y]$, where $f_{T_X}(x) = \lambda_X \, e^{-\lambda_X x}, x \geq 0$ and $f_{T_Y}(y) = \lambda_Y \, e^{-\lambda_Y y}, y \geq 0$. Because the two processes are independent, the joint PDF of T_X and T_Y is given by

$$f_{T_X T_Y}(x, y) = \lambda_X \lambda_Y \, e^{-\lambda_X x} \, e^{-\lambda_Y y} \quad x \geq 0, y \geq 0$$

In order to evaluate the probability, we consider the limits of integration by observing Figure 2.4.

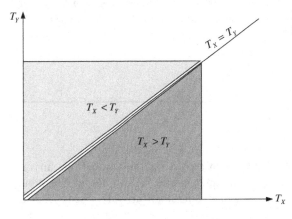

Figure 2.4 Partitioning the regions around the line $T_X = T_Y$.

Thus,

$$P[T_X < T_Y] = \int_{x=0}^{\infty} \int_{y=x}^{\infty} f_{T_X T_Y}(x,y) dy\, dx = \int_{x=0}^{\infty} \int_{y=x}^{\infty} \lambda_X \lambda_Y\, e^{-\lambda_X x}\, e^{-\lambda_Y y} dy\, dx$$

$$= \int_{x=0}^{\infty} \lambda_X\, e^{-\lambda_X x}\, e^{-\lambda_Y x}\, dx = \int_{x=0}^{\infty} \lambda_X\, e^{-(\lambda_X - \lambda_Y)x}\, dx$$

$$= \frac{\lambda_X}{\lambda_X + \lambda_Y}$$

Another way to derive this result is by considering events that occur within the small time interval $[t, t + \Delta t]$. Then, since the probability of an arrival from $X(t)$ within the interval $[t, t + \Delta t]$ is approximately $\lambda_X \Delta t$ and the probability of an arrival (from either $X(t)$ or $Y(t)$) is approximately $(\lambda_X + \lambda_Y)\Delta t$, the probability that the $X(t)$ process occurs in the interval $[t, t + \Delta t]$, given an arrival in that interval, is $\lambda_X \Delta t/(\lambda_X + \lambda_Y)\Delta t = \lambda_X/(\lambda_X + \lambda_Y)$.

The third way to solve the problem is to consider a time interval T. Within this interval, the average number of arrivals from the $\{X(t), t \geq 0\}$ process is $\lambda_X T$. Since the two processes form a combination of independent Poisson processes with rate $\lambda_X + \lambda_Y$, the average total number of arrivals from both processes is $(\lambda_X + \lambda_Y)T$. Thus, the probability that an $\{X(t), t \geq 0\}$ process occurs is $\lambda_X T/(\lambda_X + \lambda_Y)T = \lambda_X/(\lambda_X + \lambda_Y)$.

Example 2.6

Two light bulbs, labeled A and B, have exponentially distributed lifetimes. If the two lifetimes of the two bulbs are independent and the mean lifetime of bulb A is 500 h, while the mean lifetime of bulb B is 200 h, what is the probability that bulb A fails before bulb B?

Solution

Let λ_A denote the burnout rate of bulb A and λ_B the burnout rate of bulb B. Since $1/\lambda_A = 500$ and $1/\lambda_B = 200$, the rates are $\lambda_A = 1/500$ and $\lambda_B = 1/200$. Thus, the probability that bulb A fails before bulb B is

$$\frac{\lambda_A}{\lambda_A + \lambda_B} = \frac{1/500}{(1/500) + (1/200)} = \frac{2}{7}$$

Subdivision of a Poisson Process

Consider a Poisson process $\{X(t), t \geq 0\}$ with arrival rate λ. Assume that arrivals in $\{X(t), t \geq 0\}$ can be sent to one of two outputs, which we call output A and output B. Assume that the decision on which output an arrival is sent is made

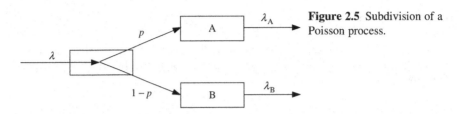

Figure 2.5 Subdivision of a Poisson process.

independently of other arrivals. Furthermore, assume that each arrival is sent to output A with probability p and to output B with probability $1 - p$, as shown in Figure 2.5.

The arrival rate at output A is $\lambda_A = \lambda p$, and the arrival rate at output B is $\lambda_B = \lambda(1 - p)$. The two outputs are independent. Consider a small time interval $(t, t + \Delta t)$. The probability that there is an arrival in the original process over this interval is approximately $\lambda \Delta t$, if we ignore higher-order terms of Δt. Thus, the probability that there is an arrival in output A over this interval is approximately $\lambda p \Delta t$, and the probability that there is an arrival in output B over this interval is $\lambda(1 - p)\Delta t$. Since the original process is a stationary and independent increment process and the two outputs are independent, each output is a stationary and independent increment process. Thus, each output is a Poisson process. We can then refer to output A as the Poisson process $\{X_A(t), t \geq 0\}$ with arrival rate λp. Similarly, we can refer to output B as the Poisson process $\{X_B(t), t \geq 0\}$ with arrival rate $\lambda(1 - p)$.

A *filtered Poisson process* $Y(t)$ is a process in which events occur according to a Poisson process $X(t)$ with rate λ, but each event is independently recorded with a probability p. From the discussion above, we observe that $Y(t)$ is a Poisson process with rate λp.

Example 2.7

A gas station is located next to a fast-food restaurant along a highway. Cars arrive at the restaurant according to a Poisson process at an average rate of 12 per hour. Independently of other cars, each car that stops at the restaurant will go to refuel at the gas station before going back to the highway with a probability of 0.25. What is the probability that exactly 10 cars have been refueled at the gas station within a particular 2 h period?

Solution

The process that governs car arrivals at the gas station is Poisson with a rate of $\lambda_G = \lambda p = (12)(0.25) = 3$ cars per hour. Thus, if K represents that number of cars that arrive at the gas within 2 h, the probability that $K = 10$ cars is given by

$$P[K = 10] = \frac{(2\lambda_G)^{10}}{10!} e^{-2\lambda_G} = \frac{6^{10}}{10!} e^{-6} = 0.0413$$

2.8 Problems

2.1 Suppose $X(t)$ is a Gaussian random process with a mean $E[X(t)] = 0$ and autocorrelation function $R_{XX}(\tau) = e^{-|\tau|}$. Assume that the random variable A is defined as follows:

$$A = \int_0^1 X(t)dt$$

Determine the following:
a. $E[A]$
b. σ_A^2

2.2 Suppose $X(t)$ is a Gaussian random process with a mean $E[X(t)] = 0$ and autocorrelation function $R_{XX}(\tau) = e^{-|\tau|}$. Assume that the random variable A is defined as follows:

$$A = \int_0^B X(t)dt$$

where B is a uniformly distributed random variable with values between 1 and 5 and is independent of the random process $X(t)$. Determine the following:
a. $E[A]$
b. σ_A^2

2.3 Three customers A, B, and C simultaneously arrive at a bank with two tellers on duty. The two tellers were idle when the three customers arrived, and A goes directly to one teller, B goes to the other teller, and C waits until either A or B leaves before she can begin receiving service. If the service times provided by the tellers are exponentially distributed with a mean of 4 min, what is the probability that customer A is still in the bank after the other two customers leave?

2.4 Customers arrive at a bank according to a Poisson process with an average rate of 6 customers per hour. Each arriving customer is either a man with probability p or a woman with probability $1 - p$. It was found that in the first 2 h, the average number of men who arrived at the bank was 8. What is the average number of women who arrived over the same period?

2.5 Chris is conducting an experiment to test the mean lifetimes of two sets of electric bulbs labeled A and B. The manufacturer claims that the mean lifetime of bulbs in set A is 200 h, while the mean lifetime of the bulbs in set B is 400 h. The lifetimes for both sets are exponentially distributed. Chris' experimental procedure is as follows. He started with one bulb from each set. As soon as a bulb from a given set fails (or burns out), he immediately replaces it with a new bulb from the same set and writes down the lifetime of the burnt-out bulb. Thus, at any point in time he has two bulbs on, one from each set. If at the end of the week Chris tells you that eight bulbs have failed, determine the following:
a. The probability that exactly five of those eight bulbs are from set B.
b. The probability that no bulb will fail in the first 100 h.
c. The mean time between two consecutive bulb failures.

2.6 Bob has a pet that requires the light in his apartment to always be on. To achieve this, Bob keeps three light bulbs on with the hope that at least one bulb will be operational when he is not at the apartment. The light bulbs have independent and identically distributed lifetimes T with PDF $f_T(t) = \lambda e^{-\lambda t}, \lambda > 0, t \geq 0$.

a. Probabilistically speaking, given that Bob is about to leave the apartment and all three bulbs are working fine, what does he gain by replacing all three bulbs with new ones before he leaves?

b. Suppose X is the random variable that denotes the time until the first bulb fails. What is the PDF of X?

c. Given that Bob is going away for an indefinite period of time and all three bulbs are working fine before he leaves, what is the PDF of Y, the time until the third bulb failure after he leaves?

d. What is the expected value of Y?

2.7 Joe replaced two light bulbs, one of which is rated 60 W with an exponentially distributed lifetime whose mean is 200 h, and the other is rated 100 W with an exponentially distributed lifetime whose mean is 100 h.

a. What is the probability that the 60 W bulb fails before the 100 W bulb?

b. What is the mean time until the first of the two bulbs fails?

c. Given that the 60 W bulb has not failed after 300 h, what is the probability that it will last at least another 100 h?

2.8 A five-motor machine can operate properly if at least three of the five motors are functioning. If the lifetime X of each motor has the PDF $f_X(x) = \lambda e^{-\lambda x}, \lambda > 0, x \geq 0$, and if the lifetimes of the motors are independent, what is the mean of the random variable Y, the time until the machine fails?

2.9 Suzie has two identical personal computers, which she never uses at the same time. She uses one PC at a time, and the other is a backup. If the one she is currently using fails, she turns it off, calls the PC repairman, and turns on the backup PC. The time until either PC fails when it is in use is exponentially distributed with a mean of 50 h. The time between the moment a PC fails until the repairman comes and finishes repairing it is also exponentially distributed with a mean of 3 h. What is the probability that Suzie is idle because neither PC is operational?

2.10 Cars arrive from the northbound section of an intersection in a Poisson manner at the rate of λ_N cars per minute and from the eastbound section in a Poisson manner at the rate of λ_E cars per minute.

a. Given that there is currently no car at the intersection, what is the probability that a northbound car arrives before an eastbound car?

b. Given that there is currently no car at the intersection, what is the probability that the fourth northbound car arrives before the second eastbound car?

2.11 A one-way street has a fork in it, and cars arriving at the fork can either bear right or left. A car arriving at the fork will bear right with probability 0.6 and will bear left with probability 0.4. Cars arrive at the fork in a Poisson manner with a rate of 8 cars per minute.

a. What is the probability that at least 4 cars bear right at the fork in 3 min?

b. Given that 3 cars bear right at the fork in 3 min, what is the probability that 2 cars bear left at the fork in 3 min?

c. Given that 10 cars arrive at the fork in 3 min, what is the probability that 4 of the cars bear right at the fork?

2.12 Let the random variable S_n be defined as follows:

$$S_n = \begin{cases} 0 & n=0 \\ \sum_{k=1}^{n} X_k & n \geq 1 \end{cases}$$

where X_k is the kth outcome of a Bernoulli trial such that $P[X_k = 1] = p$ and $P[X_k = -1] = q = 1 - p$, and the X_k are independent and identically distributed. Consider the process $\{S_n | n = 1, 2, \ldots\}$.

a. For what values of p (relative to q) is $\{S_n\}$ a martingale?
b. For what values of p is $\{S_n\}$ a submartingale?
c. For what values of p is $\{S_n\}$ a supermartingale?

2.13 Let $X_1, X_2, \ldots$ be independent and identically distributed Bernoulli random variables with values ± 1 that have equal probability of $1/2$. Show that the partial sums

$$S_n = \sum_{k=1}^{n} \frac{X_k}{k} \quad n = 1, 2, \ldots$$

form a martingale with respect to $\{X_n\}$.

2.14 Let $X_1, X_2, \ldots$ be independent and identically distributed Bernoulli random variables with values ± 1 that have equal probability of $1/2$. Let K_1 and K_2 be positive integers, and define N as follows:

$$N = \min\{n | S_n = K_1 \quad \text{or} \quad S_n = -K_2\}$$

where

$$S_n = \sum_{k=1}^{n} X_k \quad n = 1, 2, \ldots$$

is called a symmetric random walk.

a. Show that $E[N] < \infty$.
b. Show that $P[S_n = K_1] = \frac{K_2}{K_1 + K_2}$.

2.15 A symmetric random walk $\{S_n | n = 0, 1, 2, \ldots\}$ starts at the position $S_0 = k$ and ends when the walk first reaches either the origin or the position m, where $0 < k < m$. Let T be defined by

$$T = \min\{n | S_n = 0 \quad \text{or} \quad m\}$$

That is, T is the stopping time.

a. Show that $E[S_T] = k$.
b. Define $Y_n = S_n^2 - n$ and show that $\{Y_n\}$ is a martingale with respect to $\{S_n\}$.

3 Introduction to Markov Processes

3.1 Introduction

The focus of this book is on Markov processes and their applications. In this chapter, we define these processes and discuss some of their applications. The chapters that follow are devoted to the different types of Markov processes.

A stochastic process $\{X(t), t \in T\}$ is called a first-order Markov process if for any $t_0 < t_1 < \cdots < t_n$, the conditional CDF of $X(t_n)$ for given values of $X(t_0), X(t_1), \ldots, X(t_{n-1})$ depends only on $X(t_{n-1})$. That is,

$$P[X(t_n) \le x_n | X(t_{n-1}) = x_{n-1}, X(t_{n-2}) = x_{n-2}, \ldots, X(t_0) = x_0]$$
$$= P[X(t_n) \le x_n | X(t_{n-1}) = x_{n-1}]$$

This means that, given the present state of the process, the future state is independent of the past. This property is usually referred to as the *Markov property*. In second-order Markov processes, the future state depends on both the current state and the immediate past state, and so on for higher-order Markov processes. In this chapter, we consider only first-order Markov processes.

Markov processes are classified according to the nature of the time parameter and the nature of the state space. With respect to state space, a Markov process can be either a discrete-state Markov process or a continuous-state Markov process. A discrete-state Markov process is called a *Markov chain*. Similarly, with respect to time, a Markov process can be either a discrete-time Markov process or a continuous-time Markov process. Thus, there are four basic types of Markov processes:

1. Discrete-time Markov chain (or discrete-time discrete-state Markov process)
2. Continuous-time Markov chain (or continuous-time discrete-state Markov process)
3. Discrete-time Markov process (or discrete-time continuous-state Markov process)
4. Continuous-time Markov process (or continuous-time continuous-state Markov process).

This classification of Markov processes is illustrated in Figure 3.1.

In the remainder of this chapter, we discuss the structure and properties of Markov processes as well as the applications of these processes. A detailed discussion on each type of Markov process is presented in different chapters of the book. Specifically, discrete-time Markov chains are discussed in Chapter 4, continuous-time Markov chains are discussed in Chapter 5, and continuous-state Markov processes are discussed in Chapters 9 and 10. Continuous-state processes include the Brownian motion and diffusion processes.

Markov Processes for Stochastic Modeling. DOI: http://dx.doi.org/10.1016/B978-0-12-407795-9.00003-7

State space		
	Discrete	Continuous
Time	Discrete	Discrete-time Markov chain	Discrete-time Markov process
	Continuous	Continuous-time Markov chain	Continuous-time Markov process

Figure 3.1 Classification of Markov processes.

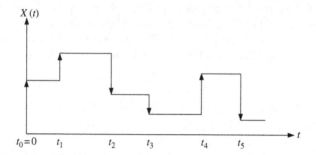

Figure 3.2 Realization of a pure jump process.

3.2 Structure of Markov Processes

A jump process is a stochastic process that makes transitions between discrete states at times that can be fixed or random. In such a process, the system enters a state, spends an amount of time called the *holding time* (or *sojourn time*), and then jumps to another state where it spends another holding time, and so on. If the jump times are $t_0 = 0 < t_1 < t_2 < \cdots$, then the sample path of the process is constant between t_i and t_{i+1}. If the jump times are discrete, the jump process is called a *jump chain*.

There are two types of jump processes: *pure* (or *nonexplosive*) and *explosive*. In an explosive jump process, the process makes an infinitely many jumps within a finite time interval. In a pure jump process, there are a finite number of jumps in a finite interval. Figure 3.2 illustrates a realization of a pure jump process.

If the holding times of a continuous-time jump process are exponentially distributed, the process is called a Markov jump process. A Markov jump process is a continuous-time Markov chain if the holding time depends only on the current state. If the holding times of a discrete-time jump process are geometrically distributed, the process is called a *Markov jump chain*. However, not all discrete-time Markov chains are Markov jump chains. For many discrete-time Markov chains, transitions occur in equally spaced intervals, such as every day, every week, and every year. Such Markov chains are not Markov jump chains.

Unfortunately, not every physical system can be modeled by a jump process. Such systems can be modeled by processes that move continuously between all possible states that lie in some interval of the real line. Thus, such processes have continuous space and continuous time. One example of a continuous-time continuous-space process is the Brownian motion, which was first described in 1828 by the botanist Robert Brown, who observed that pollen particles suspended in a fluid moved in an irregular random manner. In his mathematical theory of speculation, Bachelier (1900) used the Brownian motion to model the movement of stock prices. Arguing that the Brownian motion is caused by the bombardment of particles by the molecules of the fluid, Einstein (1905) obtained the equation for Brownian motion. Finally, Wiener (1923) established the mathematical foundation of the Brownian motion as a stochastic process. Consequently, the Brownian motion is also called the *Wiener process* and is discussed in great detail in Chapter 9. The Brownian motion has been successfully used to describe thermal noise in electric circuits, limiting behavior of queueing networks under heavy traffic, population dynamics in biological systems, and in modeling various economic processes.

A related process is the diffusion process. Diffusion is the process by which particles are transported from one part of a system to another as a result of random molecular motion. The direction of the motion of particles is from a region of higher concentration to a region of lower concentration of the particle. The laws of diffusion were first formulated by Fick, and Fick's first law of diffusion states that the diffusion flux (or amount of substance per unit area per unit time, or the rate of mass transfer per unit area) between two points of different concentrations in a fluid is proportional to the concentration gradient between these points. The constant of proportionality is called the diffusion gradient and is measured in units of area per unit time. Fick's second law, which is a consequence of his first law and the principle of conservation of mass, states that the rate of change of the concentration of a solute diffusing in a solvent is equal to the negative of the divergence of the diffusion flux. In 1905 Einstein, and independently in 1906, Smoluchowski demonstrated theoretically that the phenomenon of diffusion is the result of Brownian motion.

There is a subtle difference between Brownian motion and diffusion process. Brownian motion is the random motion of molecules, and the direction of motion of these molecules is random. Diffusion is the movement of particles from areas of high concentration to areas of low concentration. Thus, while Brownian motion is completely random, diffusion is not exactly as random as Brownian motion. For example, diffusion does not occur in a homogeneous medium where there is no concentration gradient. Thus, Brownian motion may be considered a probabilistic model of diffusion in a homogeneous medium.

Consider a physical system with state $x(t), t \geq 0$. The behavior of the system when an input $w(t), t \geq 0$, is presented to it is governed by a differential equation of the following form that gives the rate of change of the state:

$$\frac{dx(t)}{dt} = a(x(t), t) + b(x(t), t)w(t) \quad t \geq 0 \tag{3.1}$$

where the functions a and b depend on the system properties. Equation (3.1) assumes that the system properties and the input are perfectly known and deterministic. However, when the input is a random function, the state function will be a stochastic process. Under this condition, it is a common practice to assume that the input is a white noise process. Also, instead of dealing with a differential equation, we deal with increments in the system state. Thus, the evolution of the state $X(t)$ is given by the following *stochastic differential equation*:

$$dX(t) = a(X(t), t)dt + b(X(t), t)dW(t) \quad t \geq 0 \tag{3.2}$$

For a diffusion process, the function a is called the *drift coefficient*, the function b is called the *diffusion coefficient*, and $W(t)$ is the Brownian motion. Thus, a stochastic differential equation can be regarded as a mathematical description of the motion of a particle in a moving fluid. The Markov property of the diffusion process is discussed in Chapter 10. The solution to the stochastic differentiation is obtained via the following stochastic integral equation:

$$X(t) = X(0) + \int_0^t a(X(u), u)du + \int_0^t b(X(u), u)dW(u) \quad t \geq 0 \tag{3.3}$$

Different types of diffusion processes are discussed in Chapter 10, and they differ in the way the drift and diffusion coefficients are defined.

3.3   Strong Markov Property

The Markov property implies that for all t, the process $\{X(t + s) - X(t), s \geq 0\}$ has the same distribution as the process $\{X(s), s \geq 0\}$ and is independent of $\{X(s), 0 \leq s \leq t\}$. Thus, once the state of the process is known at time t, the probability law of the future change of state of the process will be determined as if the process started at time t, independently of the history of the process up to time t. While the time t is arbitrary, it is constant. The strong Markov property allows us to replace the fixed time t with a nonconstant random time. Before we state the strong Markov property, we first revisit the concept of stopping time that was discussed in Chapter 2.

Consider a stochastic process $\{X_k, k \geq 0\}$. The nonnegative integer-valued random variable T is called a stopping time for $\{X_k\}$ if, for all n, the event $\{T = n\}$ depends only on $\{X_0, X_1, \ldots, X_n\}$ and does not depend on $\{X_{n+m}, m \geq 1\}$. Thus, the event $\{T = n\}$ is nonanticipating in the sense that it is required to be independent of the future; it does not depend on $\{X_{n+1}, X_{n+2}, \ldots\}$. If T_r is a stopping time, then we have

$$\{T_r = n\} = \{X_1 \neq r, \ldots, X_{n-1} \neq r, X_n = r\}$$

For example, if we define the *recurrence time* of state i, $f_i(n)$, as the conditional probability that given that the process is presently in state i, the first time it will return to state i occurs in exactly n transitions, then we have that

$$\{T_i = n\} = \{X_0 = i, X_1 \neq i, \ldots, X_{n-1} \neq i, X_n = i\}$$

Similarly, if we define the *first passage time* between state i and state j, $f_{ij}(n)$, as the conditional probability that given that the process is presently in state i, the first time it will enter state j occurs in exactly n transitions, then we have that

$$\{T_{ij} = n\} = \{X_0 = i, X_1 \neq j, \ldots, X_{n-1} \neq j, X_n = j\}$$

Thus, both the recurrence time and the first passage time are stopping times.

With respect to the Markov process, the strong Markov property is stated as follows. Let $T < \infty$ be a stopping time with respect to the Markov chain $X = \{X_k, k \geq 0\}$. Given that $X_T = m$, the sequence $X_{T+1}, X_{T+2}, \ldots$ is a Markov chain that behaves as if X started at m, independently of $X_0, X_1, \ldots, X_T$. More specifically, we state the following theorem without proof. The proof can be found in Iosifescu (1980), Norris (1997), and Stirzaker (2005), among other books.

Theorem 3.1 Let T be a stopping time for a Markov chain $\{X_k, k \geq 0\}$. If $T < \infty$ and E is a random event prior to T, then

$$P[X_{T+1} = j | X_T = i, E] = P[X_{T+1} = j | X_T = i]$$

3.4   Applications of Discrete-Time Markov Processes

Discrete-time Markov chains have applications in many systems. Some of these applications have already been identified in the preceding discussion. In this section, we discuss other areas.

3.4.1   Branching Processes

Consider a system that initially consists of a finite set of elements. As time passes, each element can independently disappear with probability p_0 or produce k other elements with probability p_k, where $k = 1, 2, \ldots$. The behavior of each of these k elements is similar to that of their parents. Let X_n denote the size of the population after n such events. The process $\{X_n, n \geq 0\}$ is a Markov chain called a branching process.

Branching processes are used to model many problems in science and engineering. These problems include population growth, the spread of epidemics, and nuclear fission. A good discussion on the application of Markov chains in biology can be found in Norris (1997).

3.4.2 Social Mobility

Prais (1955) discusses how sociologists have used Markov chains to determine how the social class of the father, grandfather, and so on, affects the social class of a son. Such a determination is based on the fact that people can be classified into three social classes: upper class, middle class, and lower class. Thus, when the conditional probabilities are known, they can be used to represent the transition probabilities between social classes of the successive generations in a family, thereby modeling the social mobility between classes by a Markov chain.

3.4.3 Markov Decision Processes

A Markov decision process is used to model an uncertain dynamic system whose states change with time. In such a system, a decision maker is required to make a sequence of decisions over time with uncertain outcomes. Each action taken by the decision maker can either yield a reward or incur a cost. Thus, the goal is to find an optimal sequence of actions that the decision maker must take to maximize the expected reward over a given time interval, which can be finite or infinite.

3.5 Applications of Continuous-Time Markov Processes

Similar to their discrete-time counterpart, continuous-time Markov chains have applications in many systems. Some of these applications have already been identified in the preceding discussion. In this section, we discuss other areas.

3.5.1 Queueing Systems

A queue is a waiting line, and queueing systems are encountered almost everywhere including checkout counters in grocery stores and people waiting for service at banks, post offices, movie theaters, and cafeterias. A queueing system consists of one or more servers who attend to customers that arrive according to a well-defined stochastic process. Any customer who arrives when at least one server is idle goes and receives service from a server without having to wait. Customers that arrive when all the servers are busy wait to be served according to a specified service policy, such as first-come, first-served. Let n denote the number of customers in the system. If the arrival process is a Poisson process and the service time is exponentially distributed, then the process $\{n|n = 0, 1, \ldots\}$ is a Markov chain.

Sometimes, the service center is organized in stages such that after a customer has finished receiving service at one stage, he/she can proceed to receive additional service at other stages or exit the system. In this case, we have a network of queues. The basic Markovian queueing systems are discussed in Chapter 6.

3.5.2   Continuous-Time Markov Decision Processes

As discussed in the previous section, the Markov decision process is used to model an uncertain dynamic system whose states change with time. A decision maker is required to make a sequence of decisions over time with uncertain outcomes, and an action can either yield a reward or incur a cost. Thus, the goal is to find an optimal sequence of actions that the decision maker must take to maximize the expected reward over a given time interval, which can be finite or infinite. In discrete-time Markov decision processes, actions are taken at discrete intervals. However, in continuous-time Markov decision processes, actions are taken over exponentially distributed time intervals.

3.5.3   Stochastic Storage Systems

Inventory systems, dams, and insurance claims involve activities where some resource is kept in storage until it is used. Because the demand for these resources is generally random, they are generically referred to as stochastic storage systems.

In an inventory system, the goal is to coordinate the production rate and the inventory to cope with the random fluctuations in demand. There are different types of stochastic inventory control models, including the continuous review model, the single-period model, and the multiperiod model. In the continuous review model, the stock is continuously reviewed and an order of a fixed size is placed whenever the stock level reaches a certain reorder point. In the single-period model, an item is ordered only once to satisfy the demand of a specific period. In the multiperiod model, an order is placed whenever the quantity in stock cannot satisfy the current demand.

We consider the single-period model. Let c denote the capacity of the warehouse. Assume that in each period n, there is a demand for D_n units. Let Y_n denote the residual stock at the end of period n. Consider the policy that requires the warehouse manager to restock to capacity for the beginning of period $n + 1$ whenever $Y_n \leq m$, where m is a fixed threshold value. Then we have that

$$Y_{n+1} = \begin{cases} \max\{0, c - D_{n+1}\} & \text{if } Y_n \leq m \\ \max\{0, Y_n - D_{n+1}\} & \text{if } Y_n > m \end{cases}$$

If $D_i, i \geq 1$, are independent and identically distributed, then $\{Y_n, n \geq 0\}$ is a Markov chain.

We can extend the same discussion to dams and insurance risks, which Tijms (1986) and Norris (1997) have shown to belong to a class of queueing systems called M/G/1 queue, which is discussed in Chapter 6.

3.6   Applications of Continuous-State Markov Processes

Similar to their discrete-state counterparts, continuous-state Markov processes are used to model many physical systems.

3.6.1   Application of Diffusion Processes to Financial Options

A financial option is a contract that gives a person the right, but not the obligation, to buy (or what is known as a *call option*) or sell (or what is known as a *put option*) an *underlying asset*, such as a stock or commodities, at a given price (known as the *strike price*) at a future date, in the case of the so-called *European option*, or before a future date, in the case of the so-called *American option*.

For a financial market that consists of stocks, which are risky investments, and bonds, which are risk-free investments, the celebrated Black−Scholes model of the price process of stocks assumes that the a and b functions of the stochastic differential equation

$$dX(t) = a(X(t), t)dt + b(X(t), t)dW(t) \quad t \geq 0$$

are given by $a(X(t), t) = \alpha X(t)$ and $b(X(t), t) = \beta X(t)$, where α and β are constants. Thus, the price dynamics of stocks for the European option are given by

$$dX(t) = \alpha X(t)dt + \beta X(t)dW(t) \quad t \geq 0$$

While different refinements of the model have been suggested, a basic tool used in financial mathematics is the diffusion process. The Black−Scholes model is discussed in Chapter 9.

3.6.2   Applications of Brownian Motion

Apart from being an integral part of the diffusion process, the Brownian motion is used to model many physical systems. In this section, we review some of these applications.

Fractal geometry enables fractal image models to be used in medical image processing. Medical images have a degree of randomness associated with both the natural random nature of the underlying structure and the random noise superimposed on the image. The fractional Brownian motion model, which was developed by Mandelbrot (1968), regards naturally occurring surfaces as the result of random walks. This has permitted the intensity of a medical image to be treated fractionally by the Brownian motion model.

An application of the Brownian motion in robotics was reported by Arakawa and Krotkov (1994). In Wein (1990), the Brownian motion was used to model a flexible manufacturing system (FMS) as a network of queues. This enables the FMS scheduling problem to be modeled as a problem of controlling the flow in a queueing network. Another application is in decision making, which was reported by Brekke and Oksendal (1991) and Romanow (1984).

3.7 Summary

In this chapter, we have discussed the structure and general property of Markov processes. We have also discussed different applications of these processes in science, engineering, and economics. Each of the following chapters will focus on the different types of Markov processes and their applications.

4 Discrete-Time Markov Chains

4.1 Introduction

The discrete-time process $\{X_k, k = 0, 1, 2, \ldots\}$ is called a Markov chain if for all $i, j, k, \ldots, m$, the following is true:

$$P[X_k = j | X_{k-1} = i, X_{k-2} = n, \ldots, X_0 = m] = P[X_k = j | X_{k-1} = i] = p_{ijk} \qquad (4.1)$$

The quantity p_{ijk} is called the *state-transition probability*, which is the conditional probability that the process will be in state j at time k immediately after the next transition, given that it is in state i at time $k - 1$. A Markov chain that obeys the preceding rule is called a *nonhomogeneous Markov chain*. In this book, we will consider only *homogeneous Markov chains*, which are Markov chains in which $p_{ijk} = p_{ij}$. This means that homogeneous Markov chains do not depend on the time unit, which implies that

$$P[X_k = j | X_{k-1} = i, X_{k-2} = n, \ldots, X_0 = m] = P[X_k = j | X_{k-1} = i] = p_{ij} \qquad (4.2)$$

This is the so-called Markov property mentioned in Chapter 3. The *homogeneous state-transition probability* p_{ij} satisfies the following conditions:

1. $0 \leq p_{ij} \leq 1$
2. $\sum_j p_{ij} = 1$, $i = 1, 2, \ldots, n$, which follows from the fact that the states are mutually exclusive and collectively exhaustive.

From the preceding definition, we obtain the following *Markov chain rule*:

$$\begin{aligned}
P[X_k &= j, X_{k-1} = i_1, X_{k-2} = i_2, \ldots, X_0 = i_k] \\
&= P[X_k = j | X_{k-1} = i_1, X_{k-2} = i_2, \ldots, X_0 = i_k] P[X_{k-1} = i_1, X_{k-2} = i_2, \ldots, X_0 = i_k] \\
&= P[X_k = j | X_{k-1} = i_1] P[X_{k-1} = i_1, X_{k-2} = i_2, \ldots, X_0 = i_k] \\
&= P[X_k = j | X_{k-1} = i_1] P[X_{k-1} = i_1 | X_{k-2} = i_2, \ldots, X_0 = i_k] P[X_{k-2} = i_2, \ldots, X_0 = i_k] \\
&= P[X_k = j | X_{k-1} = i_1] P[X_{k-1} = i_1 | X_{k-2} = i_2] P[X_{k-2} = i_2, \ldots, X_0 = i_0] \\
&= P[X_k = j | X_{k-1} = i_1] P[X_{k-1} = i_1 | X_{k-2} = i_2] \cdots P[X_1 = i_{k-1} | X_0 = i_k] P[X_0 = i_k] \\
&= p_{i_1 j} p_{i_2 i_1} p_{i_3 i_2} \cdots p_{i_k i_{k-1}} P[X_0 = i_k]
\end{aligned}$$

Markov Processes for Stochastic Modeling. DOI: http://dx.doi.org/10.1016/B978-0-12-407795-9.00004-9

Thus, if we know the probability of being in the initial state X_0 and the state-transition probabilities, we can evaluate the joint probability $P[X_k, X_{k-1}, \ldots, X_0]$.

4.2   State-Transition Probability Matrix

It is customary to display the state-transition probabilities as the entries of an $n \times n$ matrix P, where p_{ij} is the entry in the ith row and jth column:

$$P = \begin{bmatrix} p_{11} & p_{12} & \cdots & p_{1n} \\ p_{21} & p_{22} & \cdots & p_{2n} \\ \vdots & \vdots & \ddots & \vdots \\ p_{n1} & p_{n2} & \cdots & p_{nn} \end{bmatrix}$$

P is called the *transition probability matrix*. It is a *stochastic matrix* because for any row i, $\sum_j p_{ij} = 1$; that is, the sum of the entries in each row is 1.

4.2.1   The n-Step State-Transition Probability

Let $p_{ij}(n)$ denote the conditional probability that the system will be in state j after exactly n transitions, given that it is presently in state i. That is,

$$p_{ij}(n) = P[X_{m+n} = j | X_m = i]$$

$$p_{ij}(0) = \begin{cases} 1 & i = j \\ 0 & i \neq j \end{cases}$$

$$p_{ij}(1) = p_{ij}$$

Consider the two-step transition probability $p_{ij}(2)$, which is defined by

$$p_{ij}(2) = P[X_{m+2} = j | X_m = i]$$

Assume that $m = 0$, then

$$\begin{aligned} p_{ij}(2) &= P[X_2 = j | X_0 = i] = \sum_k P[X_2 = j, X_1 = k | X_0 = i] \\ &= \sum_k P[X_2 = j | X_1 = k, X_0 = i] P[X_1 = k | X_0 = i] \\ &= \sum_k P[X_2 = j | X_1 = k] P[X_1 = k | X_0 = i] = \sum_k p_{kj} p_{ik} \\ &= \sum_k p_{ik} p_{kj} \end{aligned}$$

where the third to the last equality is due to the Markov property. The final equation states that the probability of starting in state i and being in state j at the end of the second transition is the probability that we first go immediately from state i to an intermediate state k and then immediately from state k to state j; the summation is taken over all possible intermediate states k.

The following proposition deals with a class of equations called the *Chapman–Kolmogorov equations*, which provide a generalization of the preceding results obtained for the two-step transition probability.

Proposition 4.1 For all $0 < r < n$,

$$p_{ij}(n) = \sum_k p_{ik}(r)p_{kj}(n - r)$$

This proposition states that the probability that the process starts in state i and finds itself in state j at the end of the nth transition is the product of the probability that the process starts in state i and finds itself in an intermediate state k after r transitions and the probability that it goes from state k to state j after additional $n - r$ transitions.

Proof The proof is a generalization of the proof for the case of $n = 2$ and is as follows:

$$
\begin{aligned}
p_{ij}(n) &= P[X_n = j | X_0 = i] = \sum_k P[X_n = j, X_r = k | X_0 = i] \\
&= \sum_k P[X_n = j | X_r = k, X_0 = i] P[X_r = k | X_0 = i] \\
&= \sum_k P[X_n = j | X_r = k] P[X_r = k | X_0 = i] = \sum_k p_{kj}(n - r) p_{ik}(r) \\
&= \sum_k p_{ik}(r) p_{kj}(n - r)
\end{aligned}
$$

This completes the proof.

It can be shown that $p_{ij}(n)$ is the ijth entry (i.e., entry in the ith row and jth column) in the matrix P^n. That is, for an N-state Markov chain, P^n is the matrix

$$
P^n = \begin{bmatrix}
p_{11}(n) & p_{12}(n) & \cdots & p_{1N}(n) \\
p_{21}(n) & p_{22}(n) & \cdots & p_{2N}(n) \\
\vdots & \vdots & \ddots & \vdots \\
p_{N1}(n) & p_{N2}(n) & \cdots & p_{NN}(n)
\end{bmatrix}
$$

4.3 State-Transition Diagrams

Consider the following problem. It has been observed via a series of tosses of a particular biased coin that the outcome of the next toss depends on the outcome of the current toss. In particular, given that the current toss comes up heads, the next toss will come up heads with probability 0.6 and tails with probability 0.4. Similarly, given that the current toss comes up tails, the next toss will come up heads with probability 0.35 and tails with probability 0.65.

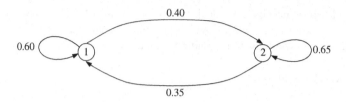

Figure 4.1 Example of state-transition diagram.

If we define state 1 to represent heads and state 2 to represent tails, then the transition probability matrix for this problem is the following:

$$P = \begin{bmatrix} 0.60 & 0.40 \\ 0.35 & 0.65 \end{bmatrix}$$

All the properties of the Markov chain can be determined from this matrix. However, the analysis of the problem can be simplified by the use of the *state-transition diagram* in which the states are represented by circles, and directed arcs represent transitions between states. The state-transition probabilities are labeled on the appropriate arcs. Thus, with respect to the preceding problem, we obtain the state-transition diagram as shown in Figure 4.1.

4.4 Classification of States

A state j is said to be *accessible* (or *can be reached*) from state i if, starting from state i, it is possible that the process will ever enter state j. This implies that $p_{ij}(n) > 0$ for some $n > 0$. Thus, the n-step probability enables us to obtain reachability information between any two states of the process.

Two states that are accessible from each other are said to *communicate* with each other. The concept of communication divides the state space into different classes. Two states that communicate are said to be in the same *class*. All members of one class communicate with one another. If a class is not accessible from any state outside the class, we define the class to be a *closed communicating class*. A Markov chain in which all states communicate, which means that there is only one class, is called an *irreducible* Markov chain. For example, the Markov chain shown in Figure 4.1 is an irreducible chain.

The states of a Markov chain can be classified into two broad groups: those that the process enters infinitely often and those that it enters finitely often. In the long run, the process will be found to be in only those states that it enters infinitely often. Let $f_{ij}(n)$ denote the conditional probability that given that the process is presently in state i, the first time it will enter state j occurs in exactly n transitions (or steps). We call $f_{ij}(n)$ the probability of *first passage* from state i to state j in n transitions. The parameter f_{ij}, which is given by

$$f_{ij} = \sum_{n=1}^{\infty} f_{ij}(n)$$

is the probability of first passage from state i to state j. It is the conditional probability that the process will ever enter state j, given that it was initially in state i. Obviously $f_{ij}(1) = p_{ij}$, and a recursive method of computing $f_{ij}(n)$ is

$$f_{ij}(n) = \sum_{l \neq j} p_{il} f_{lj}(n - 1)$$

The *recurrence time* T_i of state i is the random variable that is defined as follows:

$$T_i = \min \{n \geq 1: X_n = i \text{ given that } X_0 = i\}$$

That is, T_i is the time until the process returns to state i, given that it was initially in state i. The probability that the first return to state i occurs after n transitions, $f_{ii}(n)$, is defined by

$$f_{ii}(n) = P[T_i = n]$$

A state i is said to be *transient* if, given that we start in state i, there is a nonzero probability that we will never return to i. For a transient state i,

$$P[T_i < \infty] = \sum_{n=1}^{\infty} f_{ii}(n) = f_{ii} < 1$$

A state i in which $f_{ii} = P[T_i < \infty] = 1$ is called a *recurrent* state. The *mean recurrence time* of state i, m_i, is given by

$$m_i = E[T_i] = \sum_{n=1}^{\infty} n f_{ii}(n)$$

Note that f_{ii} denotes the probability that a process that starts at state i will ever return to state i. The mean recurrence time is used to classify that states of a Markov chain. A state i is defined to be a *positive recurrent* (or *nonnull persistent*) state if m_i is finite; otherwise, state i is defined to be a *null recurrent* (or *null persistent*) state. A Markov chain is called positive recurrent if all of its states are positive recurrent.

Let V_i denote the total number of visits to i. That is,

$$V_i = \sum_{n=0}^{\infty} I\{X_n = i\}$$

where

$$I(a) = \begin{cases} 1 & \text{if } a \text{ is true} \\ 0 & \text{otherwise} \end{cases}$$

Let $X_0 = j$. Then

$$E[V_i | X_0 = j] = E\left[\sum_{n=0}^{\infty} I\{X_n = i\} | X_0 = j\right] = \sum_{n=0}^{\infty} P[X_n = i | X_0 = j]$$

$$= \sum_{n=0}^{\infty} p_{ji}(n)$$

Thus, $E[V_i | X_0 = i] = \sum_{n=0}^{\infty} p_{ii}(n)$. It can be shown that a state i is a recurrent state if and only if the expected number of visits to this state from itself is infinite; that is, if $\sum_{n=0}^{\infty} p_{ii}(n) = \infty$.

A state i is called *absorbing* or *trapping* state if it is impossible to leave this state once the process enters the state; that is, when the process enters an absorbing state, it is "trapped" and cannot leave the state. Therefore, the state i is absorbing if and only if

$$p_{ii} = 1, \quad \text{and} \quad p_{ij} = 0 \quad \text{for} \quad j \neq i$$

If every state can reach an absorbing state, then the Markov chain is an absorbing Markov chain.

To summarize, we define these states as follows:

a. A state j is called a *transient* (or *nonrecurrent*) state if there is a positive probability that the process will never return to j again after it leaves j.

b. A state j is called a *recurrent* (or *persistent*) state if, with probability 1, the process will eventually return to j after it leaves j. A set of recurrent states forms a *single chain* if every member of the set communicates with all other members of the set.

c. A recurrent state j is called a *periodic* state if there exists an integer $d > 1$ such that $p_{jj}(n)$ is zero for all values of n other than $d, 2d, 3d, \ldots$; d is called the period. If $d = 1$, the recurrent state j is said to be *aperiodic*.

d. A recurrent state j is called a *positive recurrent* state if, starting at state j, the expected time until the process returns to state j is finite. Otherwise, the recurrent state is called a *null recurrent* state.

e. Positive recurrent, aperiodic states are called *ergodic* states.

f. A chain consisting of ergodic states is called an *ergodic chain*.

Example 4.1

Consider the Markov chain with the state-transition diagram shown in Figure 4.2. State 4 is the only transient state; states 1, 2, and 3 are recurrent states; and there is no periodic state. There is a single chain, which is $\{1, 2, 3\}$.

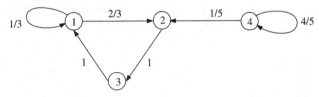

Figure 4.2 State-transition diagram for Example 4.1.

Example 4.2

Consider the state-transition diagram of Figure 4.3. The transition is now from state 2 to state 4 instead of from state 4 to state 2. For this case, states 1, 2, and 3 are now transient states because when the process enters state 2 and makes a transition to state 4, it does not return to these states again. Also, state 4 is a trapping (or absorbing) state because when the process enters the state, the process never leaves the state. As stated in the definition, we identify a trapping state from the fact that, as in this example, $p_{44} = 1$ and $p_{4k} = 0$ for k not equal to 4.

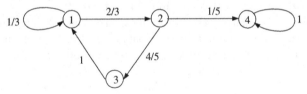

Figure 4.3 State-transition diagram for Example 4.2.

Example 4.3

The Markov chain whose state-transition diagram is shown in Figure 4.4 has a single chain $\{1, 2, 3\}$, and the three states are periodic with a period of 3.

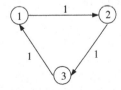

Figure 4.4 State-transition diagram for Example 4.3.

4.5   Limiting-State Probabilities

Recall that the n-step state-transition probability $p_{ij}(n)$ is the conditional probability that the system will be in state j after exactly n transitions, given that it is presently in state i. The n-step transition probabilities can be obtained by multiplying the transition probability matrix by itself n times. For example, consider the following transition probability matrix:

$$P = \begin{bmatrix} 0.4 & 0.5 & 0.1 \\ 0.3 & 0.3 & 0.4 \\ 0.3 & 0.2 & 0.5 \end{bmatrix}$$

$$P^2 = P \times P = \begin{bmatrix} 0.4 & 0.5 & 0.1 \\ 0.3 & 0.3 & 0.4 \\ 0.3 & 0.2 & 0.5 \end{bmatrix} \times \begin{bmatrix} 0.4 & 0.5 & 0.1 \\ 0.3 & 0.3 & 0.4 \\ 0.3 & 0.2 & 0.5 \end{bmatrix} = \begin{bmatrix} 0.34 & 0.37 & 0.29 \\ 0.33 & 0.32 & 0.35 \\ 0.33 & 0.31 & 0.36 \end{bmatrix}$$

$$P^3 = P^2 \times P = \begin{bmatrix} 0.34 & 0.37 & 0.29 \\ 0.33 & 0.32 & 0.35 \\ 0.33 & 0.31 & 0.36 \end{bmatrix} \times \begin{bmatrix} 0.4 & 0.5 & 0.1 \\ 0.3 & 0.3 & 0.4 \\ 0.3 & 0.2 & 0.5 \end{bmatrix} = \begin{bmatrix} 0.334 & 0.339 & 0.327 \\ 0.333 & 0.331 & 0.336 \\ 0.333 & 0.330 & 0.337 \end{bmatrix}$$

From the matrix P^2, we obtain the $p_{ij}(2)$. For example, $p_{23}(2) = 0.35$, which is the entry in the second row and third column of the matrix P^2. Similarly, the entries of the matrix P^3 are the $p_{ij}(3)$; for example, $p_{23}(3) = 0.336$.

For this particular matrix and matrices for a large number of Markov chains, we find that as we multiply the transition probability matrix by itself many times, the entries remain constant. More importantly, all the members of one column will tend to converge to the same value.

If we define $P[X(0) = i]$ as the probability that the process is in state i before it makes the first transition, then the set $\{P[X(0) = i]\}$ defines the initial condition for the process, and for an N-state process,

$$\sum_{i=1}^{N} P[X(0) = i] = 1$$

Let $P[X(n) = j]$ denote the probability that it is in state j at the end of the first n transitions, then for the N-state process,

$$P[X(n) = j] = \sum_{i=1}^{N} P[X(0) = i] p_{ij}(n)$$

For the class of Markov chains previously referenced, it can be shown that as $n \to \infty$ the n-step transition probability $p_{ij}(n)$ does not depend on i, which means that $P[X(n) = j]$ approaches a constant as $n \to \infty$ for this class of Markov chains. That is, the constant is independent of the initial conditions. Thus, for the class of Markov chains in which the limit exists, we define the *limiting-state probabilities* as follows:

$$\lim_{n \to \infty} P[X(n) = j] = \pi_j \quad n = 1, 2, \ldots, N$$

Because the n-step transition probability can be written in the form

$$p_{ij}(n) = \sum_k p_{ik}(n-1)p_{kj}$$

then if the limiting-state probabilities exist and do not depend on the initial state, we have that

$$\lim_{n \to \infty} P[X(n) = j] = \pi_j = \lim_{n \to \infty} \sum_k p_{ik}(n-1)p_{kj} = \sum_k \pi_k p_{kj}$$

If we define the limiting-state probability vector $\pi = [\pi_1, \pi_2, \ldots, \pi_N]$, then we have that

$$\pi_j = \sum_k \pi_k p_{kj}$$

$$\pi = \pi P$$

$$1 = \sum_j \pi_j$$

where the last equation is due to the law of total probability. Each of the first two equations, together with the last equation, gives a system of linear equations that the π_j must satisfy. The following propositions specify the conditions for the existence of the limiting-state probabilities:

a. In any irreducible, aperiodic Markov chain, the limits $\pi_j = \lim_{n \to \infty} p_{ij}(n)$ exist and are independent of the initial distribution.
b. In any irreducible, periodic Markov chain, the limits $\pi_j = \lim_{n \to \infty} p_{ij}(n)$ exist and are independent of the initial distribution. However, they must be interpreted as the long-run probability that the process is in state j.

Example 4.4

Recall the biased coin problem whose state-transition diagram is shown in Figure 4.1 and reproduced in Figure 4.5. Find the limiting-state probabilities.

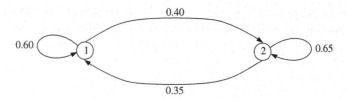

Figure 4.5 State-transition diagram for Example 4.4.

Solution
There are three equations associated with the preceding Markov chain, and they are

$$\pi_1 = 0.6\pi_1 + 0.35\pi_2$$
$$\pi_2 = 0.4\pi_1 + 0.65\pi_2$$
$$1 = \pi_1 + \pi_2$$

Because there are three equations and two unknowns, one of the equations is redundant. Thus, the rule of thumb is that for an N-state Markov chain, we use the first $N-1$ linear equations from the relation $\pi_j = \sum_k \pi_k p_{kj}$ and the total probability, $1 = \sum_j \pi_j$. Applying this rule to the given problem, we have that

$$\pi_1 = 0.6\pi_1 + 0.35\pi_2$$
$$1 = \pi_1 + \pi_2$$

From the first equation, we obtain $\pi_1 = (0.35/0.4)\pi_2$. Substituting for π_1 and solving for π_2 in the second equation, we obtain the result $\pi = [\pi_1, \pi_2] = [7/15, 8/15]$.

Suppose we are also required to compute $p_{12}(3)$, which is the probability that the process will be in state 2 at the end of the third transition, given that it is presently in state 1. We can proceed in two ways: the direct method and the matrix method. We consider both methods:

a. *Direct method*: Under this method, we exhaustively enumerate all the possible ways of a state 1 to state 2 transition in three steps. If we use the notation $a \to b \to c$ to denote a transition from state a to state b and then from state b to state c, the desired result is the following:

$$p_{12}(3) = P[\{1 \to 1 \to 1 \to 2\} \cup \{1 \to 1 \to 2 \to 2\} \cup \{1 \to 2 \to 1 \to 2\} \cup \{1 \to 2 \to 2 \to 2\}]$$

Because the different events are mutually exclusive, we obtain

$$\begin{aligned} p_{12}(3) &= P[\{1 \to 1 \to 1 \to 2\}] + P[\{1 \to 1 \to 2 \to 2\}] + P[\{\}] \\ &\quad + P[\{1 \to 2 \to 2 \to 2\}] \\ &= (0.6)(0.6)(0.4) + (0.6)(0.4)(0.65) + (0.4)(0.35)(0.4) + (0.4)(0.65)(0.65) \\ &= 0.525 \end{aligned}$$

b. *Matrix method*: One of the limitations of the direct method is that it is difficult to exhaustively enumerate the different ways of going from state 1 to state 2 in n steps, especially when n is large. This is where the matrix method becomes very useful. As discussed earlier, $p_{ij}(n)$ is the ijth entry in the matrix P^n. Thus, for the current problem, we are looking for the entry in the first row and second column of the matrix P^3. Therefore, we have

$$P = \begin{bmatrix} 0.60 & 0.40 \\ 0.35 & 0.65 \end{bmatrix}$$

$$P^2 = P \times P = \begin{bmatrix} 0.60 & 0.40 \\ 0.35 & 0.65 \end{bmatrix} \times \begin{bmatrix} 0.60 & 0.40 \\ 0.35 & 0.65 \end{bmatrix} = \begin{bmatrix} 0.50 & 0.50 \\ 0.4375 & 0.5625 \end{bmatrix}$$

$$P^3 = P \times P^2 = \begin{bmatrix} 0.60 & 0.40 \\ 0.35 & 0.65 \end{bmatrix} \times \begin{bmatrix} 0.50 & 0.50 \\ 0.4375 & 0.5625 \end{bmatrix} = \begin{bmatrix} 0.475 & 0.525 \\ 0.459375 & 0.540625 \end{bmatrix}$$

The required result (first row, second column) is 0.525, which is the result obtained via the direct method.

Proposition 4.2 If state i of a Markov chain is aperiodic, $\pi_i = 1/m_i$, where m_i is mean recurrence time of state i. Thus, even if i is a transient state, this holds because in this case $m_i = \infty$ and $\pi_i = 0$. In general, if an ergodic Markov chain is positive recurrent, then $\pi_i = 1/m_i$ for all i.

4.5.1 Doubly Stochastic Matrix

A transition probability matrix P is defined to be a doubly stochastic matrix if each of its columns sums to 1. That is, not only does each row sum to 1, each column also sums to 1. Thus, for every column j of a doubly stochastic matrix, we have that $\sum_i p_{ij} = 1$.

Doubly stochastic matrices have interesting limiting-state probabilities, as the following theorem shows.

Theorem 4.1 If P is a doubly stochastic matrix associated with the transition probabilities of a Markov chain with N states, then the limiting-state probabilities are given by $\pi_i = 1/N$, $i = 1, 2, \ldots, N$.

Proof We know that the limiting-state probabilities satisfy the condition

$$\pi_j = \sum_k \pi_k p_{kj} \tag{4.3}$$

To check the validity of the theorem, we observe that when we substitute $\pi_i = 1/N$, $i = 1, 2, \ldots, N$, in Eq. (4.3), we obtain

$$\frac{1}{N} = \frac{1}{N} \sum_k p_{kj} \Rightarrow 1 = \sum_k p_{kj}$$

This shows that $\pi_i = 1/N$ satisfies the condition $\pi = \pi P$, which the limiting-state probabilities are required to satisfy. Conversely, from Eq. (4.3), we see that if the limiting-state probabilities are given by $\pi_i = 1/N$, then each column j of P sums to 1; that is, P is doubly stochastic. This completes the proof.

Example 4.5

Find the transition probability matrix and the limiting-state probabilities of the process represented by the state-transition diagram shown in Figure 4.6.

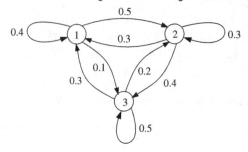

Figure 4.6 State-transition diagram for Example 4.5.

Solution

The transition probability matrix is given by

$$P = \begin{bmatrix} 0.4 & 0.5 & 0.1 \\ 0.3 & 0.3 & 0.4 \\ 0.3 & 0.2 & 0.5 \end{bmatrix}$$

It can be seen that each row of the matrix sums to 1, and each column also sums to 1; that is, it is a doubly stochastic matrix. Because the process is an irreducible, aperiodic Markov chain, the limiting-state probabilities exist and are given by $\pi_1 = \pi_2 = \pi_3 = 1/3$.

4.6 Sojourn Time

Consider a state i for which $p_{ii} > 0$. We are interested in the probability that the process remains in the state for exactly d time units. Thus, if we let the random variable D_i denote the number of time units that the process remains in state i before leaving the state, given that it enters the state, then the probability mass function (PMF) of D_i is given by

$$p_{D_i}(d) = P[D_i = d] = P[X_0 = i, X_1 = i, X_2 = i, \ldots, X_{d-1} = i, X_d \neq i]$$
$$= p_{ii}^{d-1}(1 - p_{ii})$$

where we have used the Markov chain rule. If the state of the process denotes the members of an observation sequence, then $p_{D_i}(d)$ represents the probability that the sequence remains unchanged exactly $d-1$ times before changing. Because D_i is a geometrically distributed random variable, the mean sojourn time in state i is given by

$$E[D_i] = \frac{1}{1 - p_{ii}}$$

Note that if i is a trapping state, then $p_{ii} = 1$ and $E[D_i] = \infty$, which is true because the process remains in the state indefinitely. For $p_{ii} \neq 1, E[D_i]$ is finite.

4.7 Transient Analysis of Discrete-Time Markov Chains

Recall that the n-step transition probability $p_{ij}(n)$, which is the conditional probability that the system will be in state j after exactly n transitions given that it is presently in state i, is given by

$$p_{ij}(n) = \sum_{k} p_{ik}(r)p_{kj}(n-r)$$

In particular, for an N-state Markov chain, we have that

$$p_{ij}(n+1) = \sum_{k=1}^{N} p_{ik}(n)p_{kj} \quad n = 0, 1, 2, \ldots \tag{4.4}$$

Let $g_{ij}(z)$ denote the z-transform of $p_{ij}(n)$, $n = 0, 1, \ldots$. That is,

$$g_{ij}(z) = \sum_{n=0}^{\infty} p_{ij}(n)z^n$$

Then, taking the z-transform on both sides of Eq. (4.4), we obtain

$$z^{-1}[g_{ij}(z) - p_{ij}(0)] = \sum_{k=1}^{N} g_{ik}(z)p_{kj} \tag{4.5}$$

Let $G(z)$ denote the matrix of the $g_{ij}(z)$. Recall that $p_{ij}(0) = 1$ if $i = j$, and $p_{ij}(0) = 0$ otherwise. Thus, if $P(0)$ is the matrix of the $p_{ij}(0)$, then $P(0) = I$, where I is the identity matrix, and we have that

$$z^{-1}[G(z) - I] = G(z)P \Rightarrow G(z) - I = zG(z)P$$

This gives

$$G(z) = [I - Pz]^{-1} \tag{4.6}$$

We obtain P^n as the inverse of $G(z)$. In general P^n obtained via this operation consists of two sets of components: a constant term C and a transient term $T(n)$ that is a function of n. That is,

$$P^n = C + T(n) \tag{4.7}$$

The constant term C has the characteristic that all the n rows are identical, and the elements of the rows are the limiting-state probabilities. This means that $G(z)$ can be written as follows:

$$G(z) = [I - Pz]^{-1} = \frac{1}{1 - z} C + T(z) \tag{4.8}$$

where $T(z)$ is the z-transform of $T(n)$.

Example 4.6

Consider the transition probability matrix given by

$$P = \begin{bmatrix} 0.4 & 0.5 & 0.1 \\ 0.3 & 0.3 & 0.4 \\ 0.3 & 0.2 & 0.5 \end{bmatrix}$$

We would like to obtain P^n.

Solution

We proceed as follows:

$$I - Pz = \begin{bmatrix} 1 & 0 & 0 \\ 0 & 1 & 0 \\ 0 & 0 & 1 \end{bmatrix} - z \begin{bmatrix} 0.4 & 0.5 & 0.1 \\ 0.3 & 0.3 & 0.4 \\ 0.3 & 0.2 & 0.5 \end{bmatrix} = \begin{bmatrix} 1 - 0.4z & -0.5z & -0.1z \\ -0.3z & 1 - 0.3z & -0.4z \\ -0.3z & -0.2z & 1 - 0.5z \end{bmatrix}$$

The determinant of $I - Pz$ is

$$|I - Pz| = 1 - 1.2z + 0.21z^2 - 0.001z^3 = (1 - z)(1 - 0.2z + 0.01z^2) = (1 - z)(1 - 0.1z)^2$$

From this we have that

$$
[I-Pz]^{-1} = \frac{1}{(1-z)(1-0.1z)^2}\begin{bmatrix} 1-0.8z+0.07z^2 & 0.5z-0.23z^2 & 0.1z+0.17z^2 \\ 0.3z-0.03z^2 & 1-0.9z+0.17z^2 & 0.4z-0.13z^2 \\ 0.3z-0.03z^2 & 0.2z+0.07z^2 & 1-0.7z-0.03z^2 \end{bmatrix}
$$

$$
= \frac{1}{1-z}\begin{bmatrix} 1/3 & 1/3 & 1/3 \\ 1/3 & 1/3 & 1/3 \\ 1/3 & 1/3 & 1/3 \end{bmatrix} + \frac{1}{1-0.1z}\begin{bmatrix} 2/3 & -7/3 & 5/3 \\ -1/3 & 5/3 & -4/3 \\ -1/3 & 2/3 & -1/3 \end{bmatrix}
$$

$$
+ \frac{1}{(1-0.1z)^2}\begin{bmatrix} 0 & 2 & -2 \\ 0 & -1 & 1 \\ 0 & -1 & 1 \end{bmatrix}
$$

Thus, we obtain

$$
P^n = \begin{bmatrix} 1/3 & 1/3 & 1/3 \\ 1/3 & 1/3 & 1/3 \\ 1/3 & 1/3 & 1/3 \end{bmatrix} + (0.1)^n \begin{bmatrix} 2/3 & -7/3 & 5/3 \\ -1/3 & 5/3 & -4/3 \\ -1/3 & 2/3 & -1/3 \end{bmatrix}
$$

$$
+ (n+1)(0.1)^n \begin{bmatrix} 0 & 2 & -2 \\ 0 & -1 & 1 \\ 0 & -1 & 1 \end{bmatrix}
$$

Note that the matrix associated with the root $1-z$ gives the limiting-state probabilities, which can be seen to be $\pi = [1/3, 1/3, 1/3]$. The reason why the limiting-state probabilities are equal is because P is a doubly stochastic matrix. Also, each row in the two other matrices sums to zero. Finally, it must be observed that when $n=0$, we obtain the identity matrix, and when $n=1$ we obtain P.

4.8 First Passage and Recurrence Times

We have earlier defined the first passage time from state i to state j, T_{ij}, as

$$
T_{ij} = \min\{n \geq 1: X_n = j | X_0 = i\}
$$

Thus, the probability of first passage time from state i to state j in n transitions, $f_{ij}(n)$, is the conditional probability that given that the process is presently in state i, the first time it will enter state j occurs in exactly n transitions (or steps). The probability of first passage from state i to state j, f_{ij}, is also defined as follows:

$$
f_{ij} = \sum_{n=1}^{\infty} f_{ij}(n)
$$

This means that f_{ij} is the conditional probability that the process will ever enter state j, given that it was initially in state i. Obviously $f_{ij}(1) = p_{ij}$ and a recursive method of computing $f_{ij}(n)$ is

$$f_{ij}(n) = \sum_{l \neq j} p_{il} f_{lj}(n-1)$$

When $i = j$, the first passage time becomes the *recurrence time* for state i. That is, $f_{ii}(n)$ is the conditional probability that given that the process is presently in state i, the first time it will return to state i occurs in exactly n transitions. Thus, f_{ii} is the conditional probability that a process that starts in state i will ever return to state i. The relationship between the n-step transition probability $p_{ij}(n)$ and the first passage time probability $f_{ij}(n)$ can be obtained as follows. A process that starts in state i can be in state j in n transitions if it entered state j for the first time after $m \leq n$ transitions and reached state j again after another $n - m$ transitions. Thus, we have that

$$p_{ij}(n) = \sum_{m=1}^{n} f_{ij}(m) p_{jj}(n-m)$$

This expression can also be written in the following form:

$$p_{ij}(n) = \sum_{m=1}^{n-1} f_{ij}(m) p_{jj}(n-m) + f_{ij}(n)$$

where the last equation follows from the fact that $p_{jj}(0) = 1$. From this, we have that

$$f_{ij}(n) = \begin{cases} 0 & n = 0 \\ p_{ij}(n) & n = 1 \\ p_{ij}(n) - \sum_{m=1}^{n-1} f_{ij}(m) p_{jj}(n-m) & n = 2, 3, \ldots \end{cases}$$

If we define m_{ij} as the mean first passage time from state i to state j, then it can be shown that

$$m_{ij} = \sum_{n=1}^{\infty} n f_{ij}(n) = 1 + \sum_{k \neq j} p_{ik} m_{kj} \quad i \neq j \tag{4.9}$$

The meaning of the second equation is as follows. Because the time the process spends in each state (called the *holding time*) is 1, the equation says that the mean first passage time from state i to state j is the holding time in state i plus the mean first passage time from state k to state j, $k \neq j$, given that the next state the process visits when it leaves state i is state k. The probability of this transition is, of course, p_{ik}, and we sum over all $k \neq j$. Similarly, the mean recurrence time is given by

$$m_{ii} = \sum_{n=1}^{\infty} n f_{ii}(n) = 1 + \sum_{k \neq i} p_{ik} m_{ki} \qquad (4.10)$$

Example 4.7

Consider the transition probability matrix associated with Example 4.5. We would like to obtain the mean first passage time m_{13}.

Solution
The transition probability matrix associated with Example 4.5 is

$$P = \begin{bmatrix} 0.4 & 0.5 & 0.1 \\ 0.3 & 0.3 & 0.4 \\ 0.3 & 0.2 & 0.5 \end{bmatrix}$$

We have that $m_{13} = 1 + p_{11} m_{13} + p_{12} m_{23}$. Thus, to compute m_{13}, we must obtain m_{23}. Therefore, we must solve the following system of equations:

$$m_{13} = 1 + p_{11} m_{13} + p_{12} m_{23}$$
$$= 1 + 0.4 m_{13} + 0.5 m_{23} \Rightarrow 0.6 m_{13} = 1 + 0.5 m_{23}$$
$$m_{23} = 1 + p_{21} m_{13} + p_{22} m_{23}$$
$$= 1 + 0.3 m_{13} + 0.2 m_{23} \Rightarrow 0.8 m_{23} = 1 + 0.3 m_{13}$$

From this we obtain

$$m_{13} = 3.939$$
$$m_{23} = 2.737$$

We can also obtain the mean recurrence time m_{11} as follows:

$$m_{11} = 1 + p_{12} m_{21} + p_{13} m_{31} = 1 + 0.5 m_{21} + 0.1 m_{31}$$
$$m_{21} = 1 + p_{22} m_{21} + p_{23} m_{31}$$
$$= 1 + 0.3 m_{21} + 0.4 m_{31} \Rightarrow 0.7 m_{21} = 1 + 0.4 m_{31}$$
$$m_{31} = 1 + p_{32} m_{21} + p_{33} m_{31}$$
$$= 1 + 0.2 m_{21} + 0.5 m_{31} \Rightarrow 0.5 m_{31} = 1 + 0.2 m_{21}$$

The solution to the system of equations is

$$m_{11} = 1.7187$$
$$m_{21} = 2.8125$$
$$m_{31} = 3.1250$$

4.9   Occupancy Times

Consider a discrete-time Markov chain $\{X_n, n \geq 0\}$. Let $N_i(n)$ denote the number of times that the process visits state i in the first n transitions, $n = 1, 2, \ldots$. Let $\phi_{ik}(n)$ be defined by

$$\phi_{ik}(n) = E[N_k(n)|X_0 = i]$$

That is, $\phi_{ik}(n)$ is the expected number of times that the process visits state k in the first n transitions, given that it starts in state i, and is called the *mean occupancy time* of state k up to n transitions given that the process started from state i. It can be shown that $\phi_{ik}(n)$ is given by

$$\phi_{ik}(n) = \sum_{r=0}^{n} p_{ik}(r) \tag{4.11}$$

where $p_{ik}(r)$ is the r-step transition probability from state i to state k. Because $p_{ik}(r)$ is the ikth entry in the matrix P^r, we can define the matrix for an N-state Markov chain

$$\Phi(n) = \begin{bmatrix} \phi_{11}(n) & \phi_{12}(n) & \cdots & \phi_{1N}(n) \\ \phi_{21}(n) & \phi_{22}(n) & \cdots & \phi_{2N}(n) \\ \vdots & \vdots & \ddots & \vdots \\ \phi_{N1}(n) & \phi_{N2}(n) & \cdots & \phi_{NN}(n) \end{bmatrix}$$

Then we have that

$$\Phi(n) = \sum_{r=0}^{n} P^r \tag{4.12}$$

Example 4.8

Consider the transition probability matrix associated with Example 4.6. We would like to obtain the mean first passage time $\phi_{13}(5)$.

Solution
The matrix $\Phi(5)$ is given by

$$\Phi(5) = \sum_{r=0}^{5} P^r = \sum_{r=0}^{5} \begin{bmatrix} 0.4 & 0.5 & 0.1 \\ 0.3 & 0.3 & 0.4 \\ 0.3 & 0.2 & 0.5 \end{bmatrix}^r$$

From Example 4.6, we have that

$$P^r = \frac{1}{3}\begin{bmatrix} 1 & 1 & 1 \\ 1 & 1 & 1 \\ 1 & 1 & 1 \end{bmatrix} + (0.1)^r \begin{bmatrix} 2/3 & -7/3 & 5/3 \\ -1/3 & 5/3 & -4/3 \\ -1/3 & 2/3 & -1/3 \end{bmatrix} + (r+1)(0.1)^r \begin{bmatrix} 0 & 2 & -2 \\ 0 & -1 & 1 \\ 0 & -1 & 1 \end{bmatrix}$$

Thus,

$$\phi_{13}(5) = \sum_{r=0}^{5} p_{13}(r) = \sum_{r=0}^{5} \left\{ \frac{1}{3} + \frac{5}{3}(0.1)^r - 2(r+1)(0.1)^r \right\}$$

$$= 2 - \frac{1}{3} \left\{ \frac{1-(0.1)^6}{1-0.1} \right\} - 2\sum_{r=0}^{5} r(0.1)^r = 2 - 0.37037 - 0.24690$$

$$= 1.38273$$

4.10 Absorbing Markov Chains and the Fundamental Matrix

As we defined earlier, an absorbing Markov chain is a Markov chain with at least one absorbing state that is reachable from other states. A problem of interest in the study of absorbing Markov chains is the probability that the process eventually reaches an absorbing state. We might also be interested in how long it will take for the process to be absorbed. To study this class of Markov chains, we use the canonical form used in Kemeny and Snell (1976) and Doyle and Snell (1984).

Let P be an $N \times N$ transition probability matrix of an absorbing Markov chain, and assume that there are k absorbing states and $m = N - k$ nonabsorbing (or transient) states. If we reorder the states so that the absorbing states (A) come first and the transient states (T) come last, we obtain the following canonical form:

$$P = \begin{bmatrix} I & 0 \\ R & Q \end{bmatrix}$$

Here I is a $k \times k$ identity matrix associated with the absorbing states, 0 is a $k \times m$ matrix whose entries are all 0, R is an $m \times k$ matrix associated with the transitions from the transient states to the absorbing states, and Q is an $m \times m$ matrix associated with transitions among the transient states. We observe that

$$P^2 = \begin{bmatrix} I & 0 \\ R+QR & Q^2 \end{bmatrix} = \begin{bmatrix} I & 0 \\ (I+Q)R & Q^2 \end{bmatrix}$$

$$P^3 = \begin{bmatrix} I & 0 \\ R+QR+Q^2R & Q^3 \end{bmatrix} = \begin{bmatrix} I & 0 \\ (I+Q+Q^2)R & Q^3 \end{bmatrix}$$

$$\vdots$$

$$P^n = \begin{bmatrix} I & 0 \\ (I+Q+\cdots+Q^{n-1})R & Q^n \end{bmatrix}$$

A primary parameter of interest is the quantity N_{ij}, which is the mean number of times the process is in transient state j before hitting an absorbing state, given that it starts in transient state i. Note that the emphasis is on both state i and j being transient states. If, for example, state i is an absorbing state and $i \neq j$, then the quantity is zero. Similarly, if state j is an absorbing state and $i \neq j$, then the quantity is infinity if state j is accessible from state i. The following theorem, which is proved in Grinstead and Snell (1997), establishes the relationship between $N = [N_{ij}]$ (i.e., N is the matrix of the N_{ij}) and Q.

Theorem 4.2

$$N = \sum_{k=0}^{\infty} Q^k = [I-Q]^{-1} \tag{4.13}$$

The matrix $N = [I-Q]^{-1}$ is called the *fundamental matrix* for P, the transition probability matrix for the absorbing Markov chain.

4.10.1 Time to Absorption

The mean time to absorption is defined as the expected number of transitions until the process reaches a nontransient state, given that it starts from a particular transient state. The following recursive equation gives the expected absorption time μ_i for a process that starts in state $i \in T$:

$$\mu_i = \begin{cases} 0 & i \in R \\ 1 + \sum_{j \in T} p_{ij} \mu_j & i \in T \end{cases}$$

This equation can be explained as follows. Given that the process is currently in state i, the mean number of transitions until absorption is 1 plus the mean number of transitions until absorption from state j given that the next transition from state i is state j. This is true for all j and so we sum over all $j \in T$.

The mean time to absorption can also be computed directly from the fundamental matrix. The following theorem, which is proved in Grinstead and Snell (1997), defines how to compute the mean times to absorption for the different transient states.

Theorem 4.3 Let μ_i denote the number of transitions before the process hits an absorption state, given that the chain starts in state i, and let M be the column vector whose ith entry is μ_i. Then

$$M = N\Gamma = [I-Q]^{-1}\Gamma$$

where Γ is a column vector whose entries are all 1.

Example 4.9

Consider the Markov chain whose state-transition diagram is shown in Figure 4.7. Find μ_3.

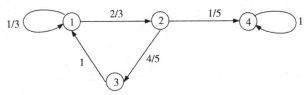

Figure 4.7 State-transition diagram for Example 4.9.

Solution

The sets of nontransient and transient states are as follows:

$$A = \{4\}$$
$$T = \{1, 2, 3\}$$

Thus, using the direct method, we obtain

$$\mu_3 = 1 + p_{31}\mu_1 + p_{32}\mu_2 = 1 + \mu_1$$

$$\mu_2 = 1 + p_{21}\mu_1 + p_{23}\mu_3 = 1 + \frac{4}{5}\mu_3$$

$$\mu_1 = 1 + p_{11}\mu_1 + p_{12}\mu_2 = 1 + \frac{1}{3}\mu_1 + \frac{2}{3}\mu_2$$

From this system of equations, we obtain

$$\mu_3 = 17.5$$
$$\mu_2 = 15.0$$
$$\mu_1 = 16.5$$

Alternatively, the matrix Q associated with the transient states and the fundamental matrix are given by

$$Q = \begin{bmatrix} 1/3 & 2/3 & 0 \\ 0 & 0 & 4/5 \\ 1 & 0 & 0 \end{bmatrix} \Rightarrow I - Q = \begin{bmatrix} 2/3 & -2/3 & 0 \\ 0 & 1 & -4/5 \\ -1 & 0 & 1 \end{bmatrix} \Rightarrow |I - Q| = \frac{2}{15}$$

$$[I - Q]^{-1} = \frac{15}{2}\begin{bmatrix} 1 & 2/3 & 8/15 \\ 4/5 & 2/3 & 8/15 \\ 1 & 2/3 & 2/3 \end{bmatrix} = \begin{bmatrix} 15/2 & 5 & 4 \\ 6 & 5 & 4 \\ 15/2 & 5 & 5 \end{bmatrix} = N$$

Thus,

$$M = N\Gamma = \begin{bmatrix} 15/2 & 5 & 4 \\ 6 & 5 & 4 \\ 15/2 & 5 & 5 \end{bmatrix}\begin{bmatrix} 1 \\ 1 \\ 1 \end{bmatrix} = \begin{bmatrix} 33/2 \\ 15 \\ 35/2 \end{bmatrix} = \begin{bmatrix} 16.5 \\ 15 \\ 17.5 \end{bmatrix}$$

Example 4.10

Consider the Markov chain whose state-transition diagram is shown in Figure 4.8. Find the μ_i for $i = 2, 3, 4$.

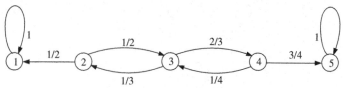

Figure 4.8 State-transition diagram for Example 4.10.

Solution

Because the transient states are $T = \{2, 3, 4\}$ and the absorbing states are $A = \{1, 5\}$, we reorder the states as $\{1, 5, 2, 3, 4\}$ and the P, Q, and R matrices are as follows:

$$P = \begin{bmatrix} I & 0 \\ R & Q \end{bmatrix} = \begin{bmatrix} 1 & 0 & 0 & 0 & 0 \\ 0 & 1 & 0 & 0 & 0 \\ 1/2 & 0 & 0 & 1/2 & 0 \\ 0 & 0 & 1/3 & 0 & 2/3 \\ 0 & 3/4 & 0 & 1/4 & 0 \end{bmatrix} \Rightarrow I = \begin{bmatrix} 1 & 0 \\ 0 & 1 \end{bmatrix}$$

$$Q = \begin{bmatrix} 0 & 1/2 & 0 \\ 1/3 & 0 & 2/3 \\ 0 & 1/4 & 0 \end{bmatrix} \quad R = \begin{bmatrix} 1/2 & 0 \\ 0 & 0 \\ 0 & 3/4 \end{bmatrix}$$

Thus,

$$I - Q = \begin{bmatrix} 1 & -1/2 & 0 \\ -1/3 & 1 & -2/3 \\ 0 & -1/4 & 1 \end{bmatrix} \Rightarrow |I - Q| = \frac{2}{3}$$

$$N = [I - Q]^{-1} = \frac{3}{2} \begin{bmatrix} 5/6 & 1/2 & 1/3 \\ 1/3 & 1 & 2/3 \\ 1/12 & 1/4 & 5/6 \end{bmatrix} = \begin{bmatrix} 5/4 & 3 & 1/2 \\ 1/2 & 3/2 & 1 \\ 1/8 & 3/8 & 5/4 \end{bmatrix}$$

From this we obtain

$$M = N\Gamma = \begin{bmatrix} 5/4 & 3 & 1/2 \\ 1/2 & 3/2 & 1 \\ 1/8 & 3/8 & 5/4 \end{bmatrix} \begin{bmatrix} 1 \\ 1 \\ 1 \end{bmatrix} = \begin{bmatrix} 19/4 \\ 3 \\ 7/4 \end{bmatrix} = \begin{bmatrix} 4.75 \\ 3 \\ 1.75 \end{bmatrix}$$

That is, $\mu_2 = 4.75$, $\mu_3 = 3$, $\mu_4 = 1.75$.

4.10.2 Absorption Probabilities

For an absorbing Markov chain, the probability that the chain that starts in a transient state i will be absorbed in state j is denoted by b_{ij}. Let B be the $m \times k$ matrix whose entries are b_{ij}. Then B is given by

$$B = [I - Q]^{-1} R = NR$$

where N is the fundamental matrix and R is the $m \times k$ matrix whose entries are transition probabilities from the transient states to the absorbing states.

Example 4.11

For the Markov chain whose state-transition diagram is shown in Figure 4.8, find the absorption probabilities b_{ij} for $i = 2, 3, 4$ and $j = 1, 5$.

Solution

Using $B = [I - Q]^{-1}R = NR$, where N and R are as defined in Example 4.10, we obtain

$$B = \begin{bmatrix} 5/4 & 3 & 1/2 \\ 1/2 & 3/2 & 1 \\ 1/8 & 3/8 & 5/4 \end{bmatrix} \begin{bmatrix} 1/2 & 0 \\ 0 & 0 \\ 0 & 3/4 \end{bmatrix} = \begin{bmatrix} 5/8 & 3/8 \\ 1/4 & 3/4 \\ 1/16 & 15/16 \end{bmatrix}$$

That is,

$$b_{21} = 5/8, b_{25} = 3/8; \quad b_{31} = 1/4, b_{35} = 3/4; \quad b_{41} = 1/16, b_{45} = 15/16$$

4.11   Reversible Markov Chains

A Markov chain $\{X_n\}$ is defined to be a reversible Markov chain if the sequence of states $\ldots, X_{n+1}, X_n, X_{n-1}, \ldots$ has the same probabilistic structure as the sequence $\ldots, X_{n-1}, X_n, X_{n+1}, \ldots$. That is, the sequence of states looked at backward in time has the same structure as the sequence running forward in time. Consider a Markov chain $\{X_n\}$ with limiting-state probabilities $\{\pi_1, \pi_2, \pi_3, \ldots\}$ and transition probabilities p_{ij}. Suppose that starting at time n we consider the sequence $X_n, X_{n-1}, \ldots$, and let $\hat{p}_{ij}$ be the transition probabilities of the reversed process. That is,

$$\begin{aligned}
\hat{p}_{ij} &= P[X_n = j | X_{n+1} = i, X_{n+2} = i_2, \ldots, X_{n+k} = i_k] \\
&= \frac{P[X_n = j, X_{n+1} = i, X_{n+2} = i_2, \ldots, X_{n+k} = i_k]}{P[X_{n+1} = i, X_{n+2} = i_2, \ldots, X_{n+k} = i_k]} \\
&= \frac{P[X_n = j]P[X_{n+1} = i | X_n = j]P[X_{n+2} = i_2, \ldots, X_{n+k} = i_k | X_n = j, X_{n+1} = i]}{P[X_{n+1} = i]P[X_{n+2} = i_2, \ldots, X_{n+k} = i_k | X_{n+1} = i]} \\
&= \frac{P[X_n = j]P[X_{n+1} = i | X_n = j]P[X_{n+2} = i_2, \ldots, X_{n+k} = i_k | X_{n+1} = i]}{P[X_{n+1} = i]P[X_{n+2} = i_2, \ldots, X_{n+k} = i_k | X_{n+1} = i]} \\
&= \frac{P[X_n = j]P[X_{n+1} = i | X_n = j]}{P[X_{n+1} = i]} \\
&= \frac{\pi_j p_{ji}}{\pi_i}
\end{aligned}$$

Thus, the backward chain is homogeneous if the forward process is in steady state, and the backward transition probabilities $\hat{p}_{ij}$ are given by

$$\hat{p}_{ij} = \frac{\pi_j p_{ji}}{\pi_i}$$

A Markov chain is said to be *reversible* if $\hat{p}_{ij} = p_{ij}$ for all i and j. Thus, for a reversible Markov chain,

$$\pi_i p_{ij} = \pi_j p_{ji} \quad \forall\, i, j$$

This condition simply states that for states i and j, the rate $\pi_i p_{ij}$ at which the process goes from state i to state j is equal to the rate $\pi_j p_{ji}$ at which the process goes from state j to state i. This "local balance" is particularly used in the steady-state analysis of birth-and-death processes that are discussed in Chapter 5.

4.12   Problems

4.1 Consider the following transition probability matrix:

$$P = \begin{bmatrix} 0.6 & 0.2 & 0.2 \\ 0.3 & 0.4 & 0.3 \\ 0.0 & 0.3 & 0.7 \end{bmatrix}$$

a. Give the state-transition diagram.
b. Given that the process is currently in state 1, what is the probability that it will be in state 2 at the end of the third transition?
c. Given that the process is currently in state 1, what is the probability that the first time it enters state 3 is the fourth transition?

4.2 Consider the following social mobility problem. Studies indicate that people in a society can be classified as belonging to the upper class (state 1), middle class (state 2), and lower class (state 3). Membership in any class is inherited in the following probabilistic manner. Given that a person is raised in an upper-class family, he will have an upper-class family with probability 0.45, a middle-class family with probability 0.48, and a lower-class family with probability 0.07. Given that a person is raised in a middle-class family, he will have an upper-class family with probability 0.05, a middle-class family with probability 0.70, and a lower-class family with probability 0.25. Finally, given that a person is raised in a lower-class family, he will have an upper-class family with probability 0.01, a middle-class family with probability 0.50, and a lower-class family with probability 0.49. Determine the following:

a. The state-transition diagram of the process.
b. The transition probability matrix of the process.
c. The limiting-state probabilities. Interpret what they mean to the layperson.

4.3 A taxi driver conducts his business in three different towns 1, 2, and 3. On any given day, when he is in town 1, the probability that the next passenger he picks up is going to a place in town 1 is 0.3, the probability that the next passenger he picks up is going to town 2 is 0.2, and the probability that the next passenger he picks up is going to town 3 is 0.5. When he is in town 2, the probability that the next passenger he picks up is going to town 1 is 0.1, the probability that the next passenger he picks up is going to town 2 is 0.8, and the probability that the next passenger he picks up is going to town 3 is 0.1. When he is in town 3, the probability that the next passenger he picks up is going to town 1 is 0.4, the probability that the next passenger he picks up is going to town 2 is 0.4, and the probability that the next passenger he picks up is going to town 3 is 0.2.

a. Determine the state-transition diagram for the process.
b. Give the transition probability matrix for the process.
c. What are the limiting-state probabilities?
d. Given that the taxi driver is currently in town 2 and is waiting to pick up his first customer for the day, what is the probability that the first time he picks up a passenger to town 2 is when he picks up his third passenger for the day?
e. Given that he is currently in town 2, what is the probability that his third passenger from now will be going to town 1?

4.4 The New England fall weather can be classified as sunny, cloudy, or rainy. A student conducted a detailed study of the weather conditions and came up with the following conclusion: Given that it is sunny on any given day, then on the following day it will be sunny again with probability 0.5, cloudy with probability 0.3 and rainy with probability 0.2. Given that it is cloudy on any given day, then on the following day it will be sunny with probability 0.4, cloudy again with probability 0.3 and rainy with probability 0.3. Finally, given that it is rainy on any given day, then on the following day it will be sunny with probability 0.2, cloudy with probability 0.5 and rainy again with probability 0.3.

a. Give the state-transition diagram of New England fall weather with the state "sunny" as state 1, the state "cloudy" as state 2, and the state "rainy" as state 3.
b. Using the same convention as in part (a), give the transition probability matrix of the New England fall weather.
c. Given that it is sunny today, what is the probability that it will be sunny four days from now?
d. Determine the limiting-state probabilities of the weather.

4.5 Consider the following transition probability matrix:

$$P = \begin{bmatrix} 0.3 & 0.2 & 0.5 \\ 0.1 & 0.8 & 0.1 \\ 0.4 & 0.4 & 0.2 \end{bmatrix}$$

a. What is P^n?
b. Obtain $\phi_{13}(5)$, the mean occupancy time of state 3 up to five transitions given that the process started from state 1.

4.6 Consider the following transition probability matrix:

$$P = \begin{bmatrix} 0.5 & 0.25 & 0.25 \\ 0.3 & 0.3 & 0.4 \\ 0.25 & 0.5 & 0.25 \end{bmatrix}$$

a. Calculate $p_{13}(3), p_{22}(2)$, and $p_{32}(4)$.
b. Calculate $p_{32}(\infty)$.

4.7 Consider the following transition probability matrix:

$$P = \begin{bmatrix} 1 & 0 & 0 & 0 \\ 0.75 & 0 & 0.25 & 0 \\ 0 & 0.25 & 0 & 0.75 \\ 0 & 0 & 0 & 1 \end{bmatrix}$$

    a. Put the matrix in the canonical form $P = \begin{bmatrix} I & 0 \\ R & Q \end{bmatrix}$.

    b. Calculate the expected absorption times μ_2 and μ_3.

4.8 Consider the following transition probability matrix:

$$P = \begin{bmatrix} 0.5 & 0.25 & 0.25 \\ 0.3 & 0.3 & 0.4 \\ 0.25 & 0.5 & 0.25 \end{bmatrix}$$

    a. Calculate $f_{13}(4)$ the probability of first passage from state 1 to state 3 in four transitions.

    b. Calculate the mean sojourn time in state 2.

4.9 Let $\{X_n\}$ be a Markov chain with the state space $\{1, 2, 3\}$ and transition probability matrix

$$P = \begin{bmatrix} 0 & 0.4 & 0.6 \\ 0.25 & 0.75 & 0 \\ 0.4 & 0 & 0.6 \end{bmatrix}$$

    Let the initial distribution be $p(0) = [p_1(0), p_2(0), p_3(0)] = [0.4, 0.2, 0.4]$. Calculate the following probabilities:

    a. $P[X_1 = 2, X_3 = 2, X_3 = 1 | X_0 = 1]$

    b. $P[X_1 = 2, X_3 = 2, X_3 = 1]$

    c. $P[X_1 = 2, X_4 = 2, X_6 = 2]$

4.10 On a given day Mark is cheerful, so-so, or glum. Given that he is cheerful on a given day, then he will be cheerful again the next day with probability 0.6, so-so with probability 0.2, and glum with probability 0.2. Given that he is so-so on a given day, then he will be cheerful the next day with probability 0.3, so-so again with probability 0.5, and glum with probability 0.2. Given that he is glum on a given day, then he will be so-so the next day with probability 0.5, and glum again with probability 0.5. Let state 1 denote the cheerful state, state 2 denote the so-so state, and state 3 denote the glum state. Let X_n denote Mark's mood on the nth day, then $\{X_n, n = 0, 1, 2, \ldots\}$ is a three-state Markov chain.

    a. Draw the state-transition diagram of the process.

    b. Give the state-transition probability matrix.

    c. Given that Mark was so-so on Monday, what is the probability that he will be cheerful on Wednesday and Friday and glum on Sunday?

    d. On the long run, what proportion of time is Mark in each of his three moods?

5 Continuous-Time Markov Chains

5.1 Introduction

A stochastic process $\{X(t), t \geq 0\}$ is a continuous-time Markov chain (CTMC) if, for all $s, t \geq 0$ and nonnegative integers i, j, k:

$$P[X(t+s) = j | X(s) = i, X(u) = k, 0 \leq u \leq s] = P[X(t+s) = j | X(s) = i]$$

This means that in a CTMC, the conditional probability of the future state at time $t + s$ given the present state at s and all past states depends only on the present state and is independent of the past. If in addition $P[X(t+s) = j | X(s) = i]$ is independent of s, then the process $\{X(t), t \geq 0\}$ is said to be *time homogeneous* or have the *time homogeneity property*. Time homogeneous Markov chains have stationary (or homogeneous) transition probabilities. Let

$$p_{ij}(t) = P[X(t+s) = j | X(s) = i]$$
$$p_j(t) = P[X(t) = j]$$

That is, $p_{ij}(t)$ is the probability that a Markov chain that is presently in state i will be in state j after an additional time t, and $p_j(t)$ is the probability that a Markov chain is in state j at time t. Thus, the $p_{ij}(t)$ are the *transition probability functions* that satisfy the condition $0 \leq p_{ij}(t) \leq 1$. Also,

$$\sum_j p_{ij}(t) = 1$$

$$\sum_j p_j(t) = 1$$

The last equation follows from the fact that at any given time the process must be in some state. Furthermore,

$$
\begin{aligned}
p_{ij}(t+s) &= \sum_k P[X(t+s) = j, X(t) = k | X(0) = i] \\
&= \sum_k \left\{ \frac{P[X(0) = i, X(t) = k, X(t+s) = j]}{P[X(0) = i]} \right\} \\
&= \sum_k \left\{ \frac{P[X(0) = i, X(t) = k]}{P[X(0) = i]} \right\} \left\{ \frac{P[X(0) = i, X(t) = k, X(t+s) = j]}{P[X(0) = i, X(t) = k]} \right\} \\
&= \sum_k P[X(t) = k | X(0) = i] P[X(t+s) = j | X(0) = i, X(t) = k] \\
&= \sum_k P[X(t) = k | X(0) = i] P[X(t+s) = j | X(t) = k] \\
&= \sum_k p_{ik}(t) p_{kj}(s)
\end{aligned}
$$

Markov Processes for Stochastic Modeling. DOI: http://dx.doi.org/10.1016/B978-0-12-407795-9.00005-0

This equation is called the Chapman–Kolmogorov equation for the CTMC. Note that the second to last equation is due to the Markov property. If we define $P(t)$ as the matrix of the $p_{ij}(t)$, that is,

$$P(t) = \begin{bmatrix} p_{11}(t) & p_{12}(t) & p_{13}(t) & \cdots \\ p_{21}(t) & p_{22}(t) & p_{23}(t) & \cdots \\ p_{31}(t) & p_{32}(t) & p_{33}(t) & \cdots \\ \cdots & \cdots & \cdots & \cdots \end{bmatrix}$$

then the Chapman–Kolmogorov equation becomes

$$P(t + s) = P(t)P(s)$$

Whenever a CTMC enters a state i, it spends an amount of time called the *dwell time* (or *holding time*) in that state. The holding time in state i is exponentially distributed with mean $1/v_i$. At the expiration of the holding time the process makes a transition to another state j with probability p_{ij}, where

$$\sum_j p_{ij} = 1$$

Because the mean holding time in state i is $1/v_i$, v_i represents the rate at which the process leaves state i, and $v_i p_{ij}$ represents the rate when in state i that the process makes a transition to state j. Also, because the holding times are exponentially distributed, the probability that when the process is in state i a transition to state $j \neq i$ will take place in the next small time Δt is $v_i p_{ij} \Delta t$. The probability that no transition out of state i will take place in Δt given that the process is presently in state i is $1 - \sum_{j \neq i} p_{ij} v_i \Delta t$, and $\sum_{j \neq i} p_{ij} v_i \Delta t$ is the probability that it leaves state i in Δt.

With these definitions we consider the state-transition diagram for the process, which is shown in Figure 5.1 for state i. We consider the transition equations for state i for the small time interval Δt.

From Figure 5.1, we obtain the following equation:

$$p_i(t + \Delta t) = p_i(t) \left\{ 1 - \sum_{j \neq i} p_{ij} v_i \Delta t \right\} + \sum_{j \neq i} p_j(t) p_{ji} v_j \Delta t$$

Thus,

$$p_i(t + \Delta t) - p_i(t) = -p_i(t) \sum_{j \neq i} p_{ij} v_i \Delta t + \sum_{j \neq i} p_j(t) p_{ji} v_j \Delta t$$

$$\frac{p_i(t + \Delta t) - p_i(t)}{\Delta t} = -v_i p_i(t) \sum_{j \neq i} p_{ij} + \sum_{j \neq i} p_j(t) p_{ji} v_j \tag{5.1}$$

$$\lim_{\Delta t \to 0} \left\{ \frac{p_i(t + \Delta t) - p_i(t)}{\Delta t} \right\} = \frac{dp_i(t)}{dt} = -v_i p_i(t) \sum_{j \neq i} p_{ij} + \sum_{j \neq i} p_j(t) p_{ji} v_j$$

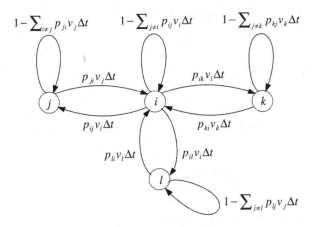

Figure 5.1 State-transition diagram for state i over small time Δt.

In the steady state, $p_i(t) \to p_j$ and

$$\lim_{t \to \infty} \left\{ \frac{dp_i(t)}{dt} \right\} = 0$$

Thus, we obtain

$$0 = -v_i p_i(t) \sum_{j \neq i} p_{ij} + \sum_{j \neq i} p_j(t) p_{ji} v_j$$

$$1 = \sum_i p_i$$

Alternatively, we can write

$$v_i p_i(t) \sum_{j \neq i} p_{ij} = \sum_{j \neq i} p_j(t) p_{ji} v_j$$

$$1 = \sum_i p_i \tag{5.2}$$

The left side of the first line of Eq. (5.2) is the rate of transition out of state i, while the right hand is the rate of transition into state i. This "balance" equation states that in the steady state the two rates are equal for any state in the Markov chain.

5.2 Transient Analysis

Recall from Eq. (5.1) that

$$\frac{dp_i(t)}{dt} = -v_i p_i(t) \sum_{j \neq i} p_{ij} + \sum_{j \neq i} p_j(t) p_{ji} v_j$$

We define the following parameters:

$$q_{ji} = p_{ji}v_j$$

$$q_i = v_i \sum_{j \neq i} p_{ij} = \sum_{j \neq i} q_{ij}$$

Then Eq. (5.1) becomes

$$\frac{dp_i(t)}{dt} = -q_i p_i(t) + \sum_{j \neq i} p_j(t) q_{ji} \tag{5.3}$$

We further define the following vectors and matrix:

$$p(t) = [p_1(t), p_2(t), p_3(t), \ldots]$$

$$\frac{dp(t)}{dt} = \left[\frac{dp_1(t)}{dt}, \frac{dp_2(t)}{dt}, \frac{dp_3(t)}{dt}, \ldots\right]$$

$$Q = \begin{bmatrix} -q_1 & q_{12} & q_{13} & q_{14} & \cdots \\ q_{21} & -q_2 & q_{23} & q_{24} & \cdots \\ q_{31} & q_{32} & -q_3 & q_{34} & \cdots \\ \cdots & \cdots & \cdots & \cdots & \cdots \\ \cdots & \cdots & \cdots & \cdots & \cdots \end{bmatrix}$$

Then Eq. (5.3) becomes

$$\frac{dp(t)}{dt} = p(t)Q \tag{5.4}$$

Q is usually called the *infinitesimal generator matrix* (or the *intensity matrix*). Under the initial condition that $p(0) = I$, where I is the identity matrix, the solution to this matrix equation is

$$p(t) = e^{Qt} = I + \sum_{k=1}^{\infty} \frac{Q^k t^k}{k!} \tag{5.5}$$

Example 5.1

Find the transition probability functions for the two-state Markov chain shown in Figure 5.2.

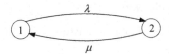

Figure 5.2 State-transition-rate diagram for Example 5.1.

Solution

In this example, $v_1 = \lambda$ and $v_2 = \mu$. Also, the Q-matrix is given by

$$Q = \begin{bmatrix} -\lambda & \lambda \\ \mu & -\mu \end{bmatrix}$$

Thus,

$$
\begin{aligned}
Q^2 &= \begin{bmatrix} -\lambda & \lambda \\ \mu & -\mu \end{bmatrix} \times \begin{bmatrix} -\lambda & \lambda \\ \mu & -\mu \end{bmatrix} = \begin{bmatrix} \lambda^2 + \lambda\mu & -\lambda^2 - \lambda\mu \\ -\mu^2 - \lambda\mu & \mu^2 + \lambda\mu \end{bmatrix} = -(\lambda + \mu)\begin{bmatrix} -\lambda & \lambda \\ \mu & -\mu \end{bmatrix} \\
&= -(\lambda + \mu)Q \\
Q^3 &= Q \times Q^2 = -(\lambda + \mu)Q \times Q = -(\lambda + \mu)Q^2 = (\lambda + \mu)^2 Q \\
&\cdots \\
Q^k &= [-(\lambda + \mu)]^{k-1} Q
\end{aligned}
$$

With this we obtain

$$
\begin{aligned}
p(t) &= I + \sum_{k=1}^{\infty} \frac{[-(\lambda + \mu)]^{k-1} Q t^k}{k!} = I - \frac{1}{\lambda + \mu}\sum_{k=1}^{\infty} \frac{[-(\lambda + \mu)t]^k}{k!} Q \\
&= I - \frac{1}{\lambda + \mu}\left\{ e^{-(\lambda + \mu)t} - 1 \right\} Q = I + \frac{1}{\lambda + \mu} Q - \frac{1}{\lambda + \mu} Q e^{-(\lambda + \mu)t} \\
&= \begin{bmatrix} 1 & 0 \\ 0 & 1 \end{bmatrix} + \begin{bmatrix} \dfrac{-\lambda}{\lambda + \mu} & \dfrac{\lambda}{\lambda + \mu} \\ \dfrac{\mu}{\lambda + \mu} & \dfrac{-\mu}{\lambda + \mu} \end{bmatrix} - \begin{bmatrix} \dfrac{-\lambda}{\lambda + \mu} & \dfrac{\lambda}{\lambda + \mu} \\ \dfrac{\mu}{\lambda + \mu} & \dfrac{-\mu}{\lambda + \mu} \end{bmatrix} e^{-(\lambda + \mu)t} \\
&= \begin{bmatrix} \dfrac{\mu}{\lambda + \mu} & \dfrac{\lambda}{\lambda + \mu} \\ \dfrac{\mu}{\lambda + \mu} & \dfrac{\lambda}{\lambda + \mu} \end{bmatrix} + \begin{bmatrix} \dfrac{\lambda}{\lambda + \mu} & \dfrac{-\lambda}{\lambda + \mu} \\ \dfrac{-\mu}{\lambda + \mu} & \dfrac{\mu}{\lambda + \mu} \end{bmatrix} e^{-(\lambda + \mu)t}
\end{aligned}
$$

Continuing the discussion, in the limit as t becomes very large, we have that

$$
\begin{aligned}
\lim_{t \to \infty} p(t) &= p \\
\lim_{t \to \infty} \frac{dp(t)}{dt} &= 0 \\
pQ &= 0 \\
pe^T &= 1
\end{aligned}
$$

where $e = [1, 1, \ldots, 1]$ and p is the vector of the limiting-state probabilities.

5.2.1 The s-Transform Method

Another method of analysis is via the s-transform. Consider again the equation:

$$\frac{dp(t)}{dt} = p(t)Q \tag{5.6}$$

Let $M_{P(t)}(s)$ denote the s-transform of $p(t)$. Then taking the s-transform of both sides of Eq. (5.6) we obtain

$$sM_{P(t)}(s) - p(0) = M_{P(t)}(s)Q \Rightarrow M_{P(t)}(s)[sI - Q] = p(0)$$

From this we obtain

$$M_{P(t)}(s) = p(0)[sI - Q]^{-1}$$

The matrix $[sI - Q]^{-1}$ is generally of the form

$$[sI - Q]^{-1} = \frac{1}{s}p + T(s)$$

where p is the matrix of the limiting-state probabilities and $T(s)$ represents transient components of the form $e^{-qt}, te^{-qt}, t^2 e^{-qt}$, and so on. These transient components vanish as t goes to infinity.

For example, recalling Example 5.1 we have that

$$Q = \begin{bmatrix} -\lambda & \lambda \\ \mu & -\mu \end{bmatrix}$$

$$sI - Q = \begin{bmatrix} s+\lambda & -\lambda \\ -\mu & s+\mu \end{bmatrix} \Rightarrow |sI - Q| = s(s + \lambda + \mu)$$

$$[sI-Q]^{-1} = \frac{1}{s(s+\lambda+\mu)}\begin{bmatrix} s+\mu & \lambda \\ \mu & s+\lambda \end{bmatrix}$$

$$= \begin{bmatrix} \dfrac{\mu/(\lambda+\mu)}{s} + \dfrac{\lambda/(\lambda+\mu)}{s+\lambda+\mu} & \dfrac{\lambda/(\lambda+\mu)}{s} - \dfrac{\lambda/(\lambda+\mu)}{s+\lambda+\mu} \\ \dfrac{\mu/(\lambda+\mu)}{s} - \dfrac{\mu/(\lambda+\mu)}{s+\lambda+\mu} & \dfrac{\lambda/(\lambda+\mu)}{s} + \dfrac{\mu/(\lambda+\mu)}{s+\lambda+\mu} \end{bmatrix}$$

$$= \frac{1}{s}\begin{bmatrix} \dfrac{\mu}{\lambda+\mu} & \dfrac{\lambda}{\lambda+\mu} \\ \dfrac{\mu}{\lambda+\mu} & \dfrac{\lambda}{\lambda+\mu} \end{bmatrix} + \frac{1}{s+\lambda+\mu}\begin{bmatrix} \dfrac{\lambda}{\lambda+\mu} & -\dfrac{\lambda}{\lambda+\mu} \\ -\dfrac{\mu}{\lambda+\mu} & \dfrac{\mu}{\lambda+\mu} \end{bmatrix}$$

If, as before, we assume that $p(0) = I$, we obtain

$$p(t) = \begin{bmatrix} \dfrac{\mu}{\lambda+\mu} & \dfrac{\lambda}{\lambda+\mu} \\[2ex] \dfrac{\mu}{\lambda+\mu} & \dfrac{\lambda}{\lambda+\mu} \end{bmatrix} + \begin{bmatrix} \dfrac{\lambda}{\lambda+\mu} & -\dfrac{\lambda}{\lambda+\mu} \\[2ex] -\dfrac{\mu}{\lambda+\mu} & \dfrac{\mu}{\lambda+\mu} \end{bmatrix} e^{-(\lambda+\mu)t}$$

which is the same result we obtained earlier.

5.3 Birth and Death Processes

Birth and death processes are a special type of CTMC. Consider a CTMC with states $0, 1, 2, \ldots$. If $p_{ij} = 0$ whenever $j \neq i - 1$ or $j \neq i + 1$, then the Markov chain is called a birth and death process. Thus, a birth and death process is a CTMC with states $0, 1, 2, \ldots$, in which transitions from state i can only go to either state $i + 1$ or state $i - 1$. That is, a transition either causes an increase in state by one or a decrease in state by one. A birth is said to occur when the state increases by one, and a death is said to occur when the state decreases by one. For a birth and death process, we define the following *transition rates* from state i:

$$\lambda_i = v_i p_{i(i+1)}$$

$$\mu_i = v_i p_{i(i-1)}$$

Thus, λ_i is the rate at which a birth occurs when the process is in state i and μ_i is the rate at which a death occurs when the process is in state i. The sum of these two rates is $\lambda_i + \mu_i = v_i$, which is the rate of transition out of state i. The *state-transition-rate diagram* of a birth and death process is shown in Figure 5.3. It is called a state-transition-rate diagram as opposed to a state-transition diagram because it shows the rate at which the process moves from state to state and not the probability of moving from one state to another. Note that $\mu_0 = 0$, because there can be no death when the process is in empty state.

The actual state-transition probabilities when the process is in state i are $p_{i(i+1)}$ and $p_{i(i-1)}$. By definition, $p_{i(i+1)} = \lambda_i / (\lambda_i + \mu_i)$ is the probability that a birth occurs before a death when the process is in state i. Similarly, $p_{i(i-1)} = \mu_i / (\lambda_i + \mu_i)$ is the probability that a death occurs before a birth when the process is in state i.

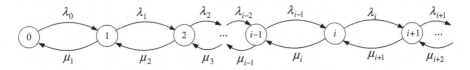

Figure 5.3 State-transition-rate diagram for birth and death process.

Recall from Eq. (5.1) that the rate at which the probability of the process being in state i changes with time is given by

$$\frac{dp_i(t)}{dt} = -v_i p_i(t) \sum_{j \neq i} p_{ij} + \sum_{j \neq i} p_j(t) p_{ji} v_j$$

$$= -(\lambda_i + \mu_i) p_i(t) + \mu_{i+1} p_{i+1}(t) + \lambda_{i-1} p_{i-1}(t)$$

Thus, for the birth and death process we have that

$$\frac{dp_0(t)}{dt} = -\lambda_0 p_0(t) + \mu_1 p_1(t)$$

$$\frac{dp_i(t)}{dt} = -(\lambda_i + \mu_i) p_i(t) + \mu_{i+1} p_{i+1}(t) + \lambda_{i-1} p_{i-1}(t) \quad i > 0$$

In the steady state,

$$\lim_{t \to \infty} \left\{ \frac{dp_i(t)}{dt} \right\} = 0$$

If we assume that the limiting probabilities $\lim_{t \to \infty} p_{ij}(t) = p_j$ exist, then from the preceding equation we obtain the following:

$$\lambda_0 p_0 = \mu_1 p_1$$
$$(\lambda_i + \mu_i) p_i = \mu_{i+1} p_{i+1} + \lambda_{i-1} p_{i-1} \quad i = 1, 2, \ldots$$
$$\sum_i p_i = 1$$

The equation states that the rate at which the process leaves state i either through a birth or a death is equal to the rate at which it enters the state through a birth when the process is in state $i - 1$ or through a death when the process is in state $i + 1$. This is called the *balance equation* because it balances (or equates) the rate at which the process enters state i with the rate at which it leaves state i.

Example 5.2

A machine is operational for an exponentially distributed time with mean $1/\lambda$ before breaking down. When it breaks down, it takes a time that is exponentially distributed with mean $1/\mu$ to repair it. What is the fraction of time that the machine is operational (or available)?

Solution

This is a two-state birth and death process. Let U denote the up state and D the down state. Then, the state-transition-rate diagram is shown in Figure 5.4.

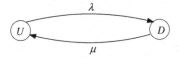

Figure 5.4 State-transition-rate diagram for Example 5.2.

Let p_U denote the steady-state probability that the process is in the operational state, and let p_D denote the steady-state probability that the process is in the down state. Then the balance equations become

$$\lambda p_U = \mu p_D$$

$$p_U + p_D = 1 \Rightarrow p_D = 1 - p_U$$

Substituting $p_D = 1 - p_U$ in the first equation gives $p_U = \mu/(\lambda + \mu)$.

Example 5.3

Customers arrive at a bank according to a Poisson process with rate λ. The time to serve each customer is exponentially distributed with mean $1/\mu$. There is only one teller at the bank, and an arriving customer who finds the teller busy when she arrives will join a single queue that operates on a first-come-first-served basis. Determine the limiting-state probabilities given that $\mu > \lambda$.

Solution

This is a CTMC in which arrivals constitute births and service completions constitute deaths. Also, for all i, $\mu_i = \mu$ and $\lambda_i = \lambda$. Thus, if p_k denotes the steady-state probability that there are k customers in the system, the balance equations are as follows:

$$\lambda p_0 = \mu p_1 \Rightarrow p_1 = \left(\frac{\lambda}{\mu}\right) p_0$$

$$(\lambda + \mu)p_1 = \lambda p_0 + \mu p_2 \Rightarrow p_2 = \left(\frac{\lambda}{\mu}\right) p_1 = \left(\frac{\lambda}{\mu}\right)^2 p_0$$

$$(\lambda + \mu)p_2 = \lambda p_1 + \mu p_3 \Rightarrow p_3 = \left(\frac{\lambda}{\mu}\right) p_2 = \left(\frac{\lambda}{\mu}\right)^3 p_0$$

Similarly, it can be shown that

$$p_k = \left(\frac{\lambda}{\mu}\right)^k p_0 \quad k = 0, 1, 2, \ldots$$

Now,

$$\sum_{k=0}^{\infty} p_k = 1 = p_0 \sum_{k=0}^{\infty} \left(\frac{\lambda}{\mu}\right)^k = \frac{p_0}{1 - \frac{\lambda}{\mu}}$$

Thus,

$$p_0 = 1 - \frac{\lambda}{\mu}$$

$$p_k = \left(1 - \frac{\lambda}{\mu}\right)\left(\frac{\lambda}{\mu}\right)^k \quad k = 0, 1, 2, \ldots$$

5.3.1 Local Balance Equations

Recall that the steady-state solution of the birth and death process is given by

$$\lambda_0 p_0 = \mu_1 p_1$$
$$(\lambda_i + \mu_i)p_i = \mu_{i+1}p_{i+1} + \lambda_{i-1}p_{i-1} \quad i = 1, 2, \ldots$$
$$\sum_i p_i = 1$$

For $i = 1$, we obtain $(\lambda_1 + \mu_1)p_1 = \mu_2 p_2 + \lambda_0 p_0$. Because we know from the first equation that $\lambda_0 p_0 = \mu_1 p_1$, this equation becomes

$$\lambda_1 p_1 = \mu_2 p_2$$

Similarly, for $i = 2$, we have that $(\lambda_2 + \mu_2)p_2 = \mu_3 p_3 + \lambda_1 p_1$ Applying the last result we obtain

$$\lambda_2 p_2 = \mu_3 p_3$$

Repeated application of this method yields the general result

$$\lambda_i p_i = \mu_{i+1}p_{i+1} \quad i = 0, 1, 2, \ldots$$

This result states that when the process is in the steady state, the rate at which it makes a transition from state i to state $i + 1$, which we refer to the rate of flow from state i to state $i + 1$, is equal to the rate of flow from state $i + 1$ to state i. This property is referred to as *local balance* condition. Recall that it is an application of the reversibility property discussed in Chapter 4. Direct application of the property allows us to solve for the steady-state probabilities of the birth and death process recursively as follows:

$$p_i = \frac{\lambda_{i-1}}{\mu_i} p_{i-1}$$

$$= \frac{\lambda_{i-1}\lambda_{i=2}}{\mu_i\mu_{i-1}} p_{i-2}$$

$$\vdots$$

$$= \frac{\lambda_{i=1}\lambda_{i=2}\cdots\lambda_0}{\mu_i\mu_{i-1}\cdots\mu_1} p_0$$

$$1 = p_0 \left[1 + \sum_{i=1}^{\infty} \frac{\lambda_{i-1}\lambda_{i=2}\cdots\lambda_0}{\mu_i\mu_{i-1}\cdots\mu_1} \right]$$

$$p_0 = \left[1 + \sum_{i=1}^{\infty} \frac{\lambda_{i-1}\lambda_{i=2}\cdots\lambda_0}{\mu_i\mu_{i-1}\cdots\mu_1} \right]^{-1}$$

$$p_i = \frac{\lambda_{i=1}\lambda_{i=2}\cdots\lambda_0}{\mu_i\mu_{i-1}\cdots\mu_1} \left[1 + \sum_{i=1}^{\infty} \frac{\lambda_{i-1}\lambda_{i=2}\cdots\lambda_0}{\mu_i\mu_{i-1}\cdots\mu_1} \right]^{-1} \quad i \geq 1$$

When $\lambda_i = \lambda$ for all i and $\mu_i = \mu$ for all i, we obtain the result

$$p_0 = \left[1 + \sum_{i=1}^{\infty} \left(\frac{\lambda}{\mu} \right)^i \right]^{-1}$$

The sum converges if and only if $\lambda/\mu < 1$, which is equivalent to the condition that $\lambda < \mu$. Under this condition we obtain the solutions

$$p_0 = 1 - \frac{\lambda}{\mu}$$

$$p_i = \left(1 - \frac{\lambda}{\mu} \right) \left(\frac{\lambda}{\mu} \right)^i \quad i \geq 1$$

In Chapter 6, we will refer to this special case of the birth and death process as an M/M/1 queueing system.

5.3.2   Transient Analysis of Birth and Death Processes

Recall from Eq. (5.6) that

$$\frac{dp(t)}{dt} = p(t)Q$$

For the birth and death process we have that

$$Q = \begin{bmatrix} -\lambda_0 & \lambda_0 & 0 & 0 & 0 & \cdots \\ \mu_1 & -(\lambda_1 + \mu_1) & \lambda_1 & 0 & 0 & \cdots \\ 0 & \mu_2 & -(\lambda_2 + \mu_2) & \lambda_2 & 0 & \cdots \\ 0 & 0 & \mu_3 & -(\lambda_3 + \mu_3) & \lambda_3 & 0 \\ \cdots & \cdots & \cdots & \cdots & \cdots & \cdots \end{bmatrix}$$

From this we obtain the following system of differential equations:

$$\frac{dp_0(t)}{dt} = -\lambda_0 p_0(t) + \mu_1 p_1(t)$$

$$\frac{dp_i(t)}{dt} = -(\lambda_i + \mu_i)p_i(t) + \mu_{i+1}p_{i+1}(t) + \lambda_{i-1}p_{i-1}(t) \quad i > 0$$

5.4   First Passage Time

Consider the CTMC $\{X(t), t \geq 0\}$ with state space $1, 2, \ldots,$ The first passage time T_k into state k is defined as follows:

$$T_k = \min \{t \geq 0 | X(t) = k\}$$

Let m_{ik} be defined as follows:

$$m_{ik} = E[T_k | X(0) = i]$$

That is, m_{ik} is the mean first passage time to state k given that the process started in state i. It can be shown that if v_i is the total rate of transition out of state i and v_{ij} is the rate of transition from state i to state j, then in a manner similar to the discrete-time case,

$$v_i m_{ik} = 1 + \sum_{j \neq k} v_{ij} m_{jk}$$

The intuitive meaning of this equation can be understood from recognizing the fact that $1/v_i$ is the mean holding time (or mean waiting time) in state i. Thus, given that the process started in state i, it would spend a mean time of $1/v_i$ in that state and then move into state j with probability $p_{ij} = v_{ij}/v_i$. Then from state j it takes a mean time of m_{jk} to reach state k. Thus, the equation can be rearranged as follows:

$$m_{ik} = \frac{1}{v_i} + \sum_{j \neq k} \frac{v_{ij}}{v_i} m_{jk} = \frac{1}{v_i} + \sum_{j \neq k} p_{ij} m_{jk}$$

This form of the equation is similar to that for the mean first passage time for the discrete-time Markov chain (DTMC) that is discussed in Chapter 4.

Example 5.4

Consider the Markov chain whose state-transition-rate diagram is given in Figure 5.5. Find m_{14}.

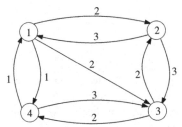

Figure 5.5 State-transition-rate diagram for Example 5.4.

Solution

Because the transition rates v_{ik} are specified in the figure, we first obtain the rates v_i, which are as follows:

$$v_1 = 1 + 2 + 2 = 5$$
$$v_2 = 3 + 3 = 6$$
$$v_3 = 2 + 2 = 4$$
$$v_4 = 1 + 3 = 4$$

Thus,

$$5m_{14} = 1 + m_{14} + 2m_{24} + 2m_{34}$$
$$6m_{24} = 1 + 3m_{14} + 3m_{34}$$
$$4m_{34} = 1 + 2m_{24}$$

The solution to the system of equations is

$$m_{14} = 1.3333$$
$$m_{24} = 1.2778$$
$$m_{34} = 0.8889$$

Thus, it takes 1.3333 units of time to go from state 1 to state 4.

5.5   The Uniformization Method

Uniformization is fundamentally used to transform a CTMC into a discrete-time analog that is more amenable to numerical calculation with respect to transient solutions. Uniformization is sometimes called *randomization,* and the method has probabilistic interpretations that can be used to derive various results.

In our discussion on the CTMC, we have assumed that a transition from any state to itself is not allowed; that is, we assumed that $p_{ii} = 0$ for all i. The uniformization method allows us to remove this restriction. The method works by making the total transition rates v_i the same for all i (that is, it *uniformizes* the transition rates). It chooses a rate v such that

$$v_i \le v \quad \forall i$$

For example, a good choice of v is

$$v = \max_i \{v_i\}$$

Recall that when the process is in state i, it leaves the state at the rate v_i. However, because we have augmented the rate to v, we now suppose that the residual rate $v - v_i$ is the fictitious rate at which it makes a transition from state i back to state i. Thus, we have a Markov chain $\{\hat{X}(t)\}$ that spends an exponentially distributed amount of time with mean $1/v$ in state i and then makes transitions with probabilities that are governed by an imbedded DTMC $\{\hat{X}_n\}$ whose transition probabilities $\hat{p}_{ij}$ are defined as follows:

$$\hat{p}_{ij} = \begin{cases} 1 - \dfrac{v_i}{v} & j = i \\[2mm] \dfrac{v_i}{v} p_{ij} & j \neq i \end{cases}$$

One of the applications of the uniformization method is in computation of the transition probability functions $p_{ij}(t)$. Let $\hat{P}$ denote the transition probability matrix of $\{\hat{X}_n\}$; that is,

$$\hat{P} = \begin{bmatrix} \hat{p}_{11} & \hat{p}_{12} & \hat{p}_{13} & \cdots \\ \hat{p}_{21} & \hat{p}_{22} & \hat{p}_{23} & \cdots \\ \hat{p}_{31} & \hat{p}_{32} & \hat{p}_{33} & \cdots \\ \cdots & \cdots & \cdots & \cdots \end{bmatrix}$$

Then the transition probability functions are given by

$$p_{ij}(t) = \sum_{n=0}^{\infty} e^{-vt} \left\{ \frac{(vt)^n}{n!} \right\} \hat{p}_{ij}(n)$$

where $\hat{p}_{ij}(n)$ is the ijth entry of the n-step transition probability of $\{\hat{X}_n\}$. This provides another way to obtain the transition probability functions $p_{ij}(t)$. In the matrix form this becomes

$$p(t) = \sum_{n=0}^{\infty} e^{-vt} \left\{ \frac{(vt)^n}{n!} \right\} \hat{P}^n$$

5.6 Reversible CTMCs

A CTMC $\{X(t), -\infty < t < \infty\}$ is said to be a reversible Markov chain if for any fixed τ and integer $n \geq 0$ the sequence of states $X(t_1), X(t_2), \ldots, X(t_n)$ has the same probabilistic structure as the sequence $X(\tau - t_1), X(\tau - t_2), \ldots, X(\tau - t_n)$. As discussed in Chapter 4, this means that a sequence of states when looked at backward in time has the same structure as the sequence running forward in time. As in the discrete-time analog discussed in Chapter 4, a CTMC is reversible if

$$v_{ij}p_i = v_{ji}p_j$$

As we discussed earlier, in the steady state the local balance condition for a birth and death process $\{X(t), t \geq 0\}$ states that

$$\lambda_i p_i = \mu_{i+1}p_{i+1} \quad i = 0, 1, 2, \ldots$$

Because for a birth and death process $v_{ij} = \lambda_i$ and $v_{ji} = \mu_{i+1}$, we observe that the reversibility condition has been met. Thus, a birth and death process is a reversible CTMC.

5.7 Problems

5.1 A small company has two identical PCs that are running at the same time. The time until either PC fails is exponentially distributed with a mean of $1/\lambda$. When a PC fails, a technician starts repairing it immediately. The two PCs fail independently of each other. The time to repair a failed PC is exponentially distributed with a mean of $1/\mu$. As soon as the repair is completed the PC is brought back online and is assumed to be as good as new.
a. Give the state-transition-rate diagram of the process.
b. What is the fraction of time that both machines are down?
5.2 Customers arrive at Mike's barber shop according to a Poisson process with rate λ customers per hour. Unfortunately Mike, the barber, has five chairs in his shop for customers to wait when there is already a customer receiving a haircut. Customers who arrive when Mike is busy and all the chairs are occupied leave without waiting for a

haircut. Mike is the only barber in the shop, and the time to complete a haircut is exponentially distributed with a mean of $1/\mu$ hours.

a. Give the state-transition-rate diagram of the process.

b. What is the probability that there are three customers waiting in the shop?

c. What is the probability that an arriving customer leaves without receiving a haircut?

d. What is the probability that an arriving customer does not have to wait?

5.3 A small company has two PCs A and B. The time to failure for PC A is exponentially distributed with a mean of $1/\lambda_A$ hours, and the time to failure for PC B is exponentially distributed with a mean of $1/\lambda_B$ hours. The PCs also have different repair times. The time to repair PC A when it fails is exponentially distributed with a mean of $1/\mu_A$ hours, and the time to repair PC B when it fails is exponentially distributed with a mean of $1/\mu_B$ hours. There is only one repair person available to work on both machines when failure occurs, and each machine is considered to be as good as new after it has been repaired.

a. Give the state-transition-rate diagram of the process.

b. What is the probability that both PCs are down?

c. What is the probability that PC A is the first to fail given that both PCs have failed?

d. What is the probability that both PCs are up?

5.4 Lazy Chris has three identical light bulbs in his living room that he keeps on all the time. Because of his laziness Chris does not replace a light bulb when it fails. (Maybe Chris does not even notice that the bulb has failed!) However, when all three bulbs have failed, Chris replaces them at the same time. The lifetime of each bulb is exponentially distributed with a mean of $1/\lambda$, and the time to replace all three bulbs is exponentially distributed with a mean of $1/\mu$.

a. Give the state-transition-rate diagram of the process.

b. What is the probability that only one light bulb is working?

c. What is the probability that all three light bulbs are working?

5.5 A switchboard has two outgoing lines serving four customers who never call each other. When a customer is not talking on the phone, he or she generates calls according to a Poisson process with rate λ calls per minute. The call durations are exponentially distributed with a mean of $1/\mu$ minutes. If a customer finds the switchboard blocked (i.e., both lines are busy) when attempting to make a call, he or she never tries to make that particular call again; that is, the call is lost.

a. Give the state-transition-rate diagram of the process.

b. What is the fraction of time that the switchboard is blocked?

5.6 A service facility can hold up to six customers who arrive according to a Poisson process with a rate of λ customers per hour. Customers who arrive when the facility is full are lost and never make an attempt to return to the facility. Whenever there are two or fewer customers in the facility, there is only one attendant serving them. The time to service each customer is exponentially distributed with a mean of $1/\mu$ hours. Whenever there are three or more customers, the attendant is joined by a colleague, and the service time is still the same for each customer. When the number of customers goes down to two, the last attendant to complete service will stop serving. Thus, whenever there are two or less customers in the facility, only one attendant can serve.

a. Give the state-transition-rate diagram of the process.

b. What is the probability that both attendants are busy attending to customers?

c. What is the probability that neither attendant is busy?

5.7 A taxicab company has a small fleet of three taxis that operate from the company's station. The time it takes a taxi to take a customer to his or her location and return to the station is exponentially distributed with a mean of $1/\mu$ hours. Customers arrive according to a Poisson process with average rate of λ customers per hour. If a potential customer arrives at the station and finds that no taxi is available, he or she goes to another taxicab company. The taxis always return to the station after dropping off a customer without picking up any new customers on their way back.

   a. Give the state-transition-rate diagram of the process.

   b. What is the probability that an arriving customer sees exactly one taxi at the station?

   c. What is the probability that an arriving customer goes to another taxicab company?

5.8 Consider a collection of particles that act independently in giving rise to succeeding generations of particles. Suppose that each particle, from the time it appears, waits a length of time that is exponentially distributed with a mean of $1/\lambda$ and then either splits into two identical particles with probability p or disappears with probability $1 - p$. Let $X(t), 0 \leq t < \infty$, denote the number of particles that are present at time t.

   a. Find the birth and death rates of the process.

   b. Give the state-transition-rate diagram of the process.

5.9 An assembly line consists of two stations in tandem. Each station can hold only one item at a time. When an item is completed in station 1, it moves into station 2 if the latter is empty; otherwise it remains in station 1 until station 2 is free. Items arrive at station 1 according to a Poisson process with rate λ. However, an arriving item is accepted only if there is no other item in the station; otherwise it is lost from the system. The time that an item spends at station 1 is exponentially distributed with mean $1/\mu_1$, and the time that it spends at station 2 is exponentially distributed with mean $1/\mu_2$. Let the state of the system be defined by (m, n), where m is the number of items in station 1 and n is the number of items in station 2.

   a. Give the state-transition-rate diagram of the process.

   b. Calculate the limiting-state probabilities p_{mn}.

5.10 Trucks bring crates of goods to a warehouse that has a single attendant. It is the responsibility of each truck driver to offload his truck, and the time that it takes to offload a truck is exponentially distributed with mean $1/\mu_1$. When a truck is offloaded, it leaves the warehouse and takes a time that is exponentially distributed with mean $1/\lambda$ to return to the warehouse with another set of crates. When a truck driver is done with offloading his truck, the warehouse attendant takes an additional time that is exponentially distributed with mean $1/\mu_2$ to arrange the crates before the next truck in line can start offloading. Assume that there are N trucks that bring crates to the warehouse. Denote the state of the system by (m, n), where m is the number of trucks in the system and n is the state of the attendant: $n = 1$ if the attendant is busy arranging the crates and $n = 0$ otherwise.

   a. Formulate the problem as a CTMC, specifying the state variables and giving the state-transition-rate diagram of the process.

   b. From the diagram identify the steady-state probability that the attendant is idle with no truck in the warehouse.

5.11 Consider a system consisting of two birth and death processes labeled system 1 and system 2. Customers arrive at system 1 according to a Poisson process with rate λ_1, and customers arrive at system 2 according to a Poisson process with rate λ_2. Each system has two identical attendants. The time it takes an attendant in system 1 to serve a customer is exponentially distributed with mean $1/\mu_1$, and the time it takes an attendant in system 2 to serve a customer is exponentially distributed with mean $1/\mu_2$. Any customer that arrives when the two attendants in its group are busy can receive service from the other group, provided that there is at least one free attendant in that group; otherwise it is lost. Let the state of the system be denoted by (m, n), where m is the number of customers in system 1 and n is the number of customers in system 2. Give the state-transition-rate diagram of the process and specify the probability that a customer receives service from a group that is different from its own group.

5.12 Cars arrive at a parking lot according to a Poisson process with rate λ. There are only four parking spaces, and any car that arrives when all the spaces are occupied is lost. The parking duration of a car is exponentially distributed with mean $1/\mu$. Let $p_k(t)$ denote the probability that k cars are parked in the lot at time t, $k = 0, 1, 2, 3, 4$.

    a. Give the differential equation governing $p_k(t)$.

    b. What are the steady-state values of these probabilities?

    c. What is m_{14}, the mean first passage time to state 4 given that the process started in state 1?

6 Markov Renewal Processes

6.1 Introduction

A Markov renewal process is a stochastic process that is a combination of Markov chains and renewal processes. It can be described as a vector-valued process from which processes, such as the Markov chain, semi-Markov process (SMP), Poisson process, and renewal process, can be derived as special cases of the process. Before we discuss Markov renewal process, we first provide basic introduction to renewal processes and regenerative processes.

6.2 Renewal Processes

Consider an experiment that involves a set of identical light bulbs whose lifetimes are independent. The experiment consists of using one light bulb at a time, and when it fails it is immediately replaced by another light bulb from the set. Each time a failed light bulb is replaced constitutes a *renewal event*. Let X_i denote the lifetime of the ith light bulb, $i = 1, 2, \ldots$, where $X_0 = 0$. Because the light bulbs are assumed to be identical, the X_i are independent and identically distributed with probability density function (PDF) $f_X(x), x \geq 0$, and mean $E[X] < \infty$.

Let $N(t)$ denote the number of renewal events up to and including the time t, where it is assumed that the first light bulb was turned on at time $t = 0$. The time to failure T_n of the first n light bulbs is given by

$$
\begin{aligned}
T_0 &= 0 \\
T_1 &= X_1 \\
T_2 &= X_1 + X_2 \\
&\vdots \\
T_n &= X_1 + X_2 + \cdots + X_n
\end{aligned}
$$

The relationship between the interevent times X_n and the T_n is illustrated in Figure 6.1, where E_k denotes the kth event.

We call T_n the nth *renewal epoch* or the time of the nth renewal, and we have that

$$N(t) = \max\{n \mid T_n \leq t\} \tag{6.1}$$

Markov Processes for Stochastic Modeling. DOI: http://dx.doi.org/10.1016/B978-0-12-407795-9.00006-2

Thus, the process $\{N(t), t \geq 0\}$ is a counting process known as a *renewal process* generated by the interarrival process $\{T_n, n \geq 1\}$ and denotes the number of renewals up to time t. Figure 6.2 shows an example of the sample path of $N(t)$.

Observe that the event that the number of renewals up to and including the time t is less than n is equivalent to the event that the nth renewal occurs at a time that is later than t. Thus, we have that

$$\{N(t) < n\} = \{T_n > t\} \text{ and } \{N(t) \geq n\} = \{T_n \leq t\} \qquad (6.2)$$

Therefore, $P[N(t) < n] = P[T_n > t]$. Let $f_{T_n}(t)$ and $F_{T_n}(t)$ denote the PDF and cumulative distribution function (CDF), respectively, of T_n. Thus, we have that

$$P[N(t) < n] = P[T_n > t] = 1 - F_{T_n}(t) \qquad (6.3)$$

Because $P[N(t) = n] = P[N(t) < n + 1] - P[N(t) < n]$, we obtain the following result for the probability mass function (PMF) of $N(t)$:

$$\begin{aligned} p_{N(t)}(n) = P[N(t) = n] &= P[N(t) < n + 1] - P[N(t) < n] \\ &= P[T_{n+1} > t] - P[T_n > t] = \{1 - F_{T_{n+1}}(t)\} - \{1 - F_{T_n}(t)\} \qquad (6.4) \\ &= F_{T_n}(t) - F_{T_{n+1}}(t) \end{aligned}$$

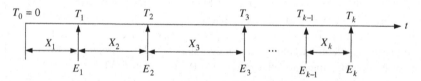

Figure 6.1 Interarrival times of a renewal process.

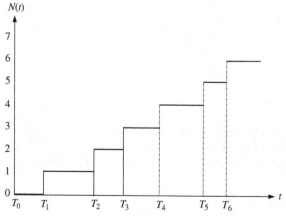

Figure 6.2 Example of sample path of $N(t)$.

This process is called a renewal process because it probabilistically starts over at each arrival epoch, T_n. That is, if the nth arrival occurs at $T_n = \tau$, then, counting from $T_n = \tau$, the jth subsequent arrival epoch is at $T_{n+j} - T_n = X_{n+1} + X_{n+2} + \cdots + X_{n+j}$. Thus, given $T_n = \tau$, the process $\{N(\tau + t) - N(\tau), t \geq 0\}$ is a renewal counting process with independent and identically distributed interarrival intervals of the same distribution as the original renewal process.

A primary reason for studying renewal processes is that many complicated processes have randomly occurring instants at which the system returns to a state probabilistically equivalent to the starting state. These renewal epochs allow us to separate the long-term behavior of the process that can be studied through renewal theory from the behavior within each renewal period.

6.2.1   The Renewal Equation

The expected number of renewals by time t is called the *renewal function*. It is denoted by $H(t)$ and is given by

$$H(t) = E[N(t)] = \sum_{n=0}^{\infty} nP[N(t) = n] = \sum_{n=0}^{\infty} n\{F_{T_n}(t) - F_{T_{n+1}}(t)\}$$

$$= \{F_{T_1}(t) + 2F_{T_2}(t) + 3F_{T_3}(t) + \cdots\} - \{F_{T_2}(t) + 2F_{T_3}(t) + \cdots\}$$

$$= F_{T_1}(t) + F_{T_2}(t) + F_{T_3}(t) + \cdots \tag{6.5}$$

$$= \sum_{n=1}^{\infty} F_{T_n}(t)$$

If we take the derivative of each side, we obtain

$$h(t) = \frac{dH(t)}{dt} = \frac{d}{dt}\sum_{n=1}^{\infty} F_{T_n}(t) = \sum_{n=1}^{\infty} \frac{d}{dt} F_{T_n}(t) = \sum_{n=1}^{\infty} f_{T_n}(t) \tag{6.6}$$

where $h(t)$ is called the *renewal density*. Let $M_h(s)$ denote the Laplace transform of $h(t)$ and $M_{T_n}(s)$ the s-transform of $f_{T_n}(t)$. Because T_n is the sum of n independent and identically distributed random variables, the PDF $f_{T_n}(t)$ is the n-fold convolution of the PDF of X. Thus, we have that

$$M_{T_n}(s) = [M_X(s)]^n$$

From this, we obtain $M_h(s)$ as follows:

$$M_h(s) = \sum_{n=1}^{\infty} M_{T_n}(s) = \sum_{n=1}^{\infty} [M_X(s)]^n = \frac{1}{1 - M_X(s)} - 1$$

$$= \frac{M_X(s)}{1 - M_X(s)} \tag{6.7}$$

This gives

$$M_h(s) = M_X(s) + M_h(s)M_X(s) \tag{6.8}$$

Taking the inverse transform, we obtain

$$h(t) = f_X(t) + \int_{u=0}^{t} h(t-u)f_X(u)\mathrm{d}u \tag{6.9}$$

Finally, integrating both sides of the equation, we obtain

$$H(t) = F_X(t) + \int_{u=0}^{t} H(t-u)f_X(u)\mathrm{d}u \tag{6.10}$$

This equation is called the *fundamental equation of renewal theory*.

Note that we can also obtain the renewal function from the following relationship. Because $N(t)$ is a nonnegative random variable and because $\{N(t) \geq n\} = \{T_n \leq t\}$, we have that

$$E[N(t)] = H(t) = \sum_{n=1}^{\infty} P[N(t) \geq n] = \sum_{n=1}^{\infty} P[T_n \leq t]$$

Since $T_n = T_{n-1} + X_n$ and because T_{n-1} and X_n are independent, we have that

$$P[T_n \leq t] = \int_{x=0}^{t} P[T_{n-1} \leq t - x]f_X(x)\mathrm{d}x \quad n \geq 2$$

Thus, we have that

$$\begin{aligned}
H(t) &= \sum_{n=1}^{\infty} P[T_n \leq t] = P[T_1 \leq t] + \sum_{n=2}^{\infty} P[T_n \leq t] \\
&= F_X(t) + \sum_{n=2}^{\infty} \int_{x=0}^{t} P[T_{n-1} \leq t - x]f_X(x)\mathrm{d}x \\
&= F_X(t) + \int_{x=0}^{t} \sum_{n=2}^{\infty} P[T_{n-1} \leq t - x]f_X(x)\mathrm{d}x \\
&= F_X(t) + \int_{x=0}^{t} \sum_{n=1}^{\infty} P[T_n \leq t - x]f_X(x)\mathrm{d}x \\
&= F_X(t) + \int_{x=0}^{t} H(t-x)f_X(x)\mathrm{d}x \quad t \geq 0
\end{aligned}$$

This is the same as Eq. (6.10). Note that the third equation follows from the fact that $P[T_1 \leq t] = F_X(t)$.

Example 6.1

Assume that X is exponentially distributed with mean $1/\lambda$. Then, we obtain

$$f_X(x) = \lambda e^{-\lambda x}$$

$$M_X(s) = \frac{\lambda}{s+\lambda}$$

$$M_h(s) = \frac{M_X(s)}{1-M_X(s)} = \frac{\lambda}{s}$$

$$h(t) = L^{-1}\{M_h(s)\} = \lambda$$

$$H(t) = \int_{u=0}^{t} h(u)du = \lambda t$$

where $L^{-1}\{M_h(s)\}$ is the inverse Laplace transform of $M_h(s)$.

6.2.2 Alternative Approach

An alternative method of studying the renewal process is as follows. With respect to the light bulbs, the renewal density $h(t)$ represents the conditional failure rate, that is, $h(t)dt$ is the probability that a light bulb will fail in the interval $(t, t + dt)$, given that it has survived up to time t. Thus, we can write

$$h(t)dt = P[X \le t + dt | X > t] = \frac{P[t < X \le t+dt]}{P[X>t]} = \frac{F_X(t+dt) - F_X(t)}{1 - F_X(t)}$$

$h(t)$ is sometimes called the *hazard function* of X and represents the instantaneous rate at which an item will fail, given that it survived up to time t. If we define $\overline{F}_X(t) = 1 - F_X(t)$, then we obtain

$$\{1 - F_X(t)\}h(t)dt = F_X(t+dt) - F_X(t) = F_X(t+dt) - F_X(t) - 1 + 1$$
$$= \{1 - F_X(t)\} - \{1 - F_X(t+dt)\}$$

That is,

$$\overline{F}_X(t) - \overline{F}_X(t+dt) = \overline{F}_X(t)h(t)dt \Rightarrow \frac{\overline{F}_X(t+dt) - \overline{F}_X(t)}{dt} = \frac{d}{dt}\overline{F}_X(t) = -\overline{F}_X(t)h(t)$$

Thus,

$$\frac{(d/dt)\overline{F}_X(t)}{\overline{F}_X(t)} = \frac{d}{dt}\ln\{\overline{F}_X(t)\} = -h(t) \tag{6.11}$$

Integrating on both sides yields

$$[\ln\{\overline{F}_X(u)\}]_0^t = -\int_0^t h(u)\,du$$

Because $\overline{F}_X(0) = 1$, we obtain

$$\ln\{\overline{F}_X(t)\} = -\int_0^t h(u)\,du$$

Thus, we have that

$$\overline{F}_X(t) = \exp\left[-\int_0^t h(u)\,du\right] = e^{-H(t)} \tag{6.12}$$

The PDF of the age $X(t)$ is given by

$$f_X(t) = -\frac{d\overline{F}(t)}{dt} = h(t)\exp\left[-\int_0^t h(u)\,du\right] = h(t)\,e^{-H(t)} \tag{6.13}$$

This implies that $h(t), F_X(t)$, or $f_X(t)$ can be used to characterize the distribution of the lifetimes. Because the renewal function $H(t) = E[N(t)]$, the previous results can be expressed as follows:

$$\overline{F}_X(t) = \exp\left[-\int_0^t h(u)\,du\right] = e^{-H(t)} = e^{-E[N(t)]} \tag{6.14a}$$

$$f_X(t) = \frac{d\overline{F}(t)}{dt} = h(t)\exp\left[-\int_0^t h(u)\,du\right] = h(t)\,e^{-H(t)} = h(t)\,e^{-E[N(t)]} \tag{6.14b}$$

Now, we have that

$$F_{T_1}(t) = P[T_1 \le t] = F_X(t)$$
$$F_{T_2}(t) = P[T_2 \le t] = P[X_1 + X_2 \le t] = F_X^{(2)}(t)$$
$$F_{T_n}(t) = P[T_n \le t] = P[X_1 + X_2 + \cdots + X_n \le t] = F_X^{(n)}(t)$$

where $F_X^{(k)}(t)$ is the k-fold time convolution of $F_X(t)$. From these equations, we obtain

$$P[N(t) \ge k] = P[T_k \le t] = F_X^{(k)}(t)$$
$$F_{N(t)}(k) = P[N(t) \le k] = 1 - P[N(t) \ge k + 1] = 1 - P[T_{k+1} \le t]$$
$$= 1 - F_X^{(k+1)}(t)$$

Thus,

$$
\begin{aligned}
p_{N(t)}(k) = P[N(t) = k] &= P[N(t) \le k] - P[N(t) \le k - 1] \\
&= \{1 - F_X^{(k+1)}(t)\} - \{1 - F_X^{(k)}(t)\} \\
&= F_X^{(k)}(t) - F_X^{(k+1)}(t)
\end{aligned}
$$

$$
p_{N(t)}(0) = 1 - F_X(t)
$$

The renewal function is then given by

$$
\begin{aligned}
H(t) = E[N(t)] &= \sum_{k=0}^{\infty} k p_{N(t)}(k) = \sum_{k=1}^{\infty} k p_{N(t)}(k) \\
&= \{F_X(t) - F_X^{(2)}(t)\} + \{2F_X^{(2)}(t) - 2F_X^{(3)}(t)\} + \{3F_X^{(3)}(t) - 3F_X^{(4)}(t)\} + \cdots \\
&= \sum_{k=1}^{\infty} F_X^{(k)}(t) = \sum_{k=1}^{\infty} F_{T_k}(t)
\end{aligned}
$$

which is the result we obtained earlier. If $F_X(t)$ is absolutely continuous, then $F_X^{(k)}(t)$ is absolutely continuous and we obtain

$$
H(t) = \sum_{k=1}^{\infty} F_X^{(k)}(t) = \sum_{k=1}^{\infty} \int_0^t f_X^{(k)}(u) du \tag{6.15}
$$

Using Fubini's theorem, we obtain

$$
H(t) = \sum_{k=1}^{\infty} F_X^{(k)}(t) = \sum_{k=1}^{\infty} \int_0^t f_X^{(k)}(u) du = \int_0^t \left\{ \sum_{k=1}^{\infty} f_X^{(k)}(u) \right\} du \equiv \int_0^t h(u) du
$$

Thus, we have that

$$
h(t) = \sum_{k=1}^{\infty} f_X^{(k)}(t) \tag{6.16}
$$

Example 6.2

X is a second-order Erlang random variable with parameter λ. Thus, $f_X(t) = \lambda^2 t\, e^{-\lambda t}$ and $M_X(s) = [\lambda/(s+\lambda)]^2$. From this, we obtain

$$
M_h(s) = \frac{M_X(s)}{1 - M_X(s)} = \frac{[\lambda/(s+\lambda)]^2}{1 - [\lambda/(s+\lambda)]^2} = \frac{\lambda^2}{s(s + 2\lambda)} = \frac{\lambda}{2} \left\{ \frac{1}{s} - \frac{1}{s + 2\lambda} \right\}
$$

Thus,

$$
h(t) = \frac{\lambda}{2} \{ 1 - e^{-2\lambda t} \} \quad t > 0
$$

Note that $\lim_{t \to \infty} h(t) = \lambda/2$.

6.2.3 The Elementary Renewal Theorem

We state the following theorem called the elementary renewal theorem without proof:

$$\lim_{t \to \infty} \left\{ \frac{H(t)}{t} \right\} = \frac{1}{E[X]} \qquad (6.17)$$

Another rendition of the theorem is the following. Let $\{N(t), t \ge 0\}$ be a renewal counting process generated by the interarrival process $\{T_n, n \ge 1\}$, where the mean interarrival time is $0 < E[T] < \infty$. Then,

$$\lim_{t \to \infty} \left\{ \frac{N(t)}{t} \right\} = \frac{1}{E[T]} \qquad (6.18)$$

with probability 1. Also, it can be shown that

$$\lim_{t \to \infty} \left\{ \frac{E[N(t)]}{t} \right\} = \frac{1}{E[T]} \qquad (6.19)$$

6.2.4 Random Incidence and Residual Time

Consider a renewal process $N(t)$ in which events (or arrivals) occur at times $0 = T_0, T_1, T_2, \ldots$. As discussed earlier, the interevent times X_k can be defined in terms of the T_k as follows:

$$\begin{aligned}
X_1 &= T_1 - T_0 = T_1 \\
X_2 &= T_2 - T_1 \\
&\vdots \\
X_k &= T_k - T_{k-1}
\end{aligned}$$

Note that the X_k are mutually independent and identically distributed.

Consider the following problem in connection with the X_k. Assume the T_k are the points in time that buses arrive at a bus stop. A passenger arrives at the bus stop at a *random time* and wants to know how long he or she will wait until the next bus arrival. This problem is usually referred to as the *random incidence problem*, because the subject (or passenger in this example) is incident to the process at a random time. Let R be the random variable that denotes the time from the moment the passenger arrived until the next bus arrival. R is referred to as the *residual life* of the renewal process. Also, let W denote the length of the interarrival gap that the passenger entered by random incidence. Figure 6.3 illustrates the random incidence problem.

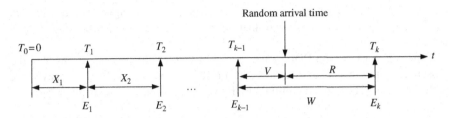

Figure 6.3 Random incidence.

Let $f_X(x)$ denote the PDF of the interarrival times; let $f_W(w)$ denote the PDF of W, the gap entered by random incidence; and let $f_R(r)$ denote the PDF of the residual life, R. The probability that the random arrival occurs in a gap of length between w and $w + dw$ can be assumed to be directly proportional to the length w of the gap and relative occurrence $f_X(w)dw$ of such gaps. That is,

$$f_W(w)dw = \beta w f_X(w)dw$$

where β is a constant of proportionality. Thus, $f_W(w) = \beta f_X(w)$. Because $f_W(w)$ is a PDF, we have that

$$\int_{-\infty}^{\infty} f_W(w)dw = 1 = \beta \int_{-\infty}^{\infty} w f_X(w)dw = \beta E[X]$$

Thus, $\beta = 1/E[X]$, and we obtain

$$f_W(w) = \frac{w f_X(w)}{E[X]} \quad w \geq 0 \tag{6.20}$$

The expected value of W is given by

$$E[W] = \frac{E[X^2]}{E[X]} \tag{6.21}$$

This result applies to all renewal processes.

A Poisson process is an example of a renewal process in which X is exponentially distributed with $E[X] = 1/\lambda$ and $E[X^2] = 2/\lambda^2$. Thus, for a Poisson process, we obtain

$$
\begin{aligned}
f_W(w) &= \lambda w f_X(x) = \lambda^2 w\, e^{-\lambda w} \quad w \geq 0 \\
E[W] &= \frac{2}{\lambda}
\end{aligned}
$$

This means that for a Poisson process the gap entered by random incidence has the second-order Erlang distribution; thus, the expected length of the gap is twice the expected length of an interarrival time. This is often referred to as the *random incidence paradox*. The reason for this fact is that the passenger is more likely to enter a large gap than a small gap; that is, the gap entered by random incidence is not a typical interval.

Next, we consider the PDF of the residual life R of the process. Given that the passenger enters a gap of length w, he or she is equally likely to be anywhere within the gap. Thus, the conditional PDF of R, given that $W = w$, is given by

$$f_{R|W}(r|w) = \frac{1}{w} \quad 0 \le r \le w$$

When we combine this result with the previous one, we get the joint PDF of R and W as follows:

$$f_{RW}(r, w) = f_{R|W}(r|w) f_W(w) = \frac{1}{w} \left\{ \frac{w f_X(w)}{E[X]} \right\} = \frac{f_X(w)}{E[X]} \quad 0 \le r \le w < \infty$$

The marginal PDF of R and its expected value become

$$f_R(r) = \int_{-\infty}^{\infty} f_{RW}(r, w) dw = \int_{r}^{\infty} \frac{f_X(w)}{E[X]} dw$$

$$= \frac{1 - F_X(r)}{E[X]} \quad r \ge 0 \tag{6.22}$$

$$E[R] = \int_0^{\infty} r f_R(r) dr = \frac{1}{E[X]} \int_{r=0}^{\infty} r \int_{w=r}^{\infty} f_X(w) dw \, dr$$

$$= \frac{1}{E[X]} \int_{w=0}^{\infty} \int_{r=0}^{w} r f_X(w) dr \, dw = \frac{1}{E[X]} \int_{w=0}^{\infty} f_X(w) \left[\frac{r^2}{2} \right]_0^w dw \tag{6.23}$$

$$= \frac{1}{2E[X]} \int_{w=0}^{\infty} w^2 f_X(w) dw$$

$$= \frac{E[X^2]}{2E[X]}$$

For the Poisson process, X is exponentially distributed so that $f_X(x) = \lambda e^{-\lambda x}$ and $F_X(x) = 1 - e^{-\lambda x}$. This means that

$$f_R(r) = \frac{1 - F_X(r)}{E[X]} = \frac{e^{-\lambda r}}{(1/\lambda)} = \lambda e^{-\lambda r} \quad r \ge 0$$

Thus, for a Poisson process, the residual life of the process has the same distribution as the interarrival time, which can be expected from the "forgetfulness" property of the exponential distribution.

In Figure 6.3, the random variable V denotes the time between the last bus arrival and the passenger's random arrival. Because $W = V + R$, the expected value of V is $E[V] = E[W] - E[R]$. For a Poisson process, $E[V] = (2/\lambda) - (1/\lambda) = 1/\lambda$.

6.2.5 Delayed Renewal Process

Suppose we allow the epoch at which the first renewal occurs to be arbitrarily distributed. The resulting type of process is a generalization of the class of renewal processes known as *delayed renewal processes*. The word *delayed* simply means that the usual renewal process, with independent and identical interrenewal times, is delayed until after the epoch of the first renewal. Thus, a renewal counting process $\{N_D(t), t \geq 0\}$ generated by the interarrival sequence $\{X_n, n \geq 1\}$ is called a delayed renewal process if the X_n are independent random variables, X_1 has a CDF $F_X(x)$ and $\{X_n, n \geq 2\}$ are identically distributed with CDF $G_X(x)$.

Using the technique used for the ordinary renewal process, the renewal function for the delayed renewal process can be obtained as follows:

$$E[N(t)] = H(t) = \sum_{n=1}^{\infty} P[N(t) \geq n] = \sum_{n=1}^{\infty} P[T_n \leq t]$$

Because $T_n = T_{n-1} + X_n$ and because T_{n-1} and X_n are independent for $n \geq 2$, we have that

$$P[T_n \leq t] = \int_{x=0}^{t} P[T_{n-1} \leq t - x] f_X(x) dx \quad n \geq 2$$

Thus, we have that

$$H_D(t) = \sum_{n=1}^{\infty} P[T_n \leq t] = P[T_1 \leq t] + \sum_{n=2}^{\infty} P[T_n \leq t]$$

$$= F_X(t) + \sum_{n=2}^{\infty} \int_{x=0}^{t} P[T_{n-1} \leq t - x] g_X(x) dx$$

$$= F_X(t) + \int_{x=0}^{t} \sum_{n=2}^{\infty} P[T_{n-1} \leq t - x] g_X(x) dx \qquad (6.24)$$

$$= F_X(t) + \int_{x=0}^{t} \sum_{n=1}^{\infty} P[T_n \leq t - x] g_X(x) dx$$

$$= F_X(t) + \int_{x=0}^{t} H(t - x) g_X(x) dx \quad t \geq 0$$

where $g_X(x) = dG_X(x)/dx$, which we assume to exist; otherwise, we replace $g_X(x)dx$ by $dG_X(x)$. Note that $H_D(t)$ only appears on the left-hand side unlike the renewal function of the ordinary renewal process equation where $H(t)$ appears on both sides.

Let the mean interarrival time of a delayed renewal process be defined by

$$E[X_2] = \int_0^\infty \{1 - F_X(x)\}dx$$

Then, it can be shown that with probability 1 (see, for example, Gallager, 1996),

$$\lim_{t \to \infty} \left\{ \frac{N_D(t)}{t} \right\} = \frac{1}{E[X_2]} \qquad (6.25)$$

This is the elementary renewal theorem for a delayed renewal process.

6.3 Renewal-Reward Process

Consider an environment where a reward is received at the time of each renewal. Specifically, let a reward Y_i be received at the ith renewal. We assume that $Y_1, Y_2, Y_3, \ldots$, are independent and identically distributed, where $E[|Y_i|] < \infty$ and each Y_i may take negative values as well as positive values. The cumulative reward up to time t is given by

$$Y(t) = \sum_{i=1}^{N(t)} Y_i \qquad (6.26)$$

We assume that $Y(t) = 0$ in the event that $N(t) = 0$. The process $\{Y(t), t \ge 0\}$ is called a *reward-renewal process* generated by the sequence $\{(Y_n, X_n)\}$. If Y_i is independent of X_i, $\{Y(t), t \ge 0\}$ is called a *cumulative process*, and if $\{N(t), t \ge 0\}$ is a Poisson process, $\{Y(t), t \ge 0\}$ is a compound Poisson process. An example of a sample path of $Y(t)$ is shown in Figure 6.4.

As an example of a renewal-reward process, consider a company that receives orders for its product. Each time an order is received constitutes a renewal epoch.

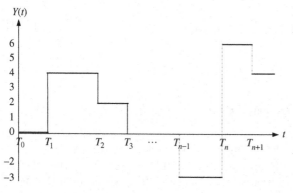

Figure 6.4 Example of a sample path of $Y(t)$.

Let Y_n denote the quantity ordered at the nth renewal epoch. If we assume that the Y_n are independent and identically distributed, the number of items ordered by time t, $Y(t)$, is given by

$$Y(t) = Y_1 + Y_2 + \cdots + Y_{N(t)}$$

and $\{Y(t), t \geq 0\}$ is a renewal-reward process, where $\{N(t), t \geq 0\}$ is a renewal process associated with the order arrival times.

6.3.1 The Reward-Renewal Theorem

The following limit property is called the *renewal-reward theorem*. If $E[|Y_1|] < \infty$ and $E[X_1] < \infty$, then

$$\lim_{t \to \infty} \left\{ \frac{Y(t)}{t} \right\} = \frac{E[Y_1]}{E[X_1]} \tag{6.27}$$

The proof of the theorem is as follows:

$$\frac{Y(t)}{t} = \frac{\sum_{i=1}^{N(t)} Y}{N(t)} \times \frac{N(t)}{t}$$

According to the strong law of large numbers,

$$\lim_{t \to \infty} \left\{ \frac{1}{N(t)} \sum_{i=1}^{N(t)} Y_i \right\} = E[Y_1] \quad \text{and} \quad \lim_{t \to \infty} \left\{ \frac{N(t)}{t} \right\} = \frac{1}{E[X_1]}$$

which proves the theorem. If we define a *cycle* to be completed every time a renewal starts, then Eq. (6.27) can be stated in the following way:

$$\text{Long-run average reward} = \frac{E[\text{rewards earned during one cycle}]}{E[\text{length of one cycle}]}$$

An extension of the theorem that we state without proof is the following:

$$\lim_{t \to \infty} \left\{ \frac{E[Y(t)]}{t} \right\} = \frac{E[Y_1]}{E[X_1]} \tag{6.28}$$

6.4 Regenerative Processes

Informally, a regenerative process is a renewal process in which segments of the process between renewal times are independent and identically distributed. A more formal definition is as follows. A process $\{X(t), t \geq 0\}$ is called a regenerative process if there is a nonnegative random variable R_1 such that

a. $\{X(t + R_1), t \geq 0\}$ is stochastically identical to $\{X(t), t \geq 0\}$.
b. $\{X(t + R_1), t \geq 0\}$ is independent of $\{X(t), 0 \leq t < R_1\}$ and R_1, that is, $\{X(t + R_1), t \geq 0\}$ is independent of R_1 and of the past history of $\{X(t)\}$ prior to R_1.

Thus, the continuation of the process beyond R_1 is a probabilistic replica of the process starting at 0. As in the renewal process, the time intervals of the renewal points are independent and identically distributed. But unlike the renewal process, the process $X(t)$ is not necessarily discrete or nondecreasing; it could be any arbitrary stochastic process. The property of regeneration implies that there are other times $R_2, R_3, \ldots$ that have the same property as R_1.

The foregoing discussion implies that we can visualize a regenerative process as a process that can be partitioned into cycles demarcated by time markers $R_1, R_2, R_3, \ldots$, such that between any two consecutive time markers the process has the same probabilistic behavior. These time markers thus constitute the regeneration points and represent times of occurrence of a particular event of interest. Thus, the process is independent of its history up to the previous regeneration point. For example, assume that $\{X(t), t \geq 0\}$ is a continuous-time Markov chain (CTMC) and we are interested in when the process enters a particular state k. Then each time the process is in state k, the future of the process after that time has the same probabilistic law as the process had starting at time 0.

As stated earlier, the regeneration points $\{R_n\}$ enable us to obtain cycles in the process, and the difference between successive regeneration points is called a *cycle length*. In particular, the time interval $[R_{n-1}, R_n)$ is called the nth cycle of the process. A regenerative process whose cycle lengths have a finite mean is called a *positive recurrent* regenerative process. Since the regeneration points are fundamentally renewal points of the process, we say that every regenerative process has an embedded renewal process.

6.4.1   Inheritance of Regeneration

One of the properties of regenerative processes is that functions of regenerative processes are regenerative. This means that if $\{X(t)\}$ is a regenerative process, the process $Y(t) = f(X(t))$ is a regenerative process with the same regeneration epochs as $\{X(t)\}$. This is one of the differences between a Markov process and a regenerative process. While a function of a Markov process is not guaranteed to be a Markov process, it will be regenerative if the Markov process is a regenerative process.

6.4.2   Delayed Regenerative Process

A stochastic process $\{X(t), t \geq 0\}$ is defined to be a delayed regenerative process if there exists a nonnegative random variable R_1 such that

a. $E[R_1] < \infty$
b. $\{X(t + R_1), t \geq 0\}$ is independent of $\{X(t), 0 \leq t < R_1\}$
c. $\{X(t + R_1), t \geq 0\}$ is a regenerative process.

Thus, in a delayed regenerative process, there is a sequence of regenerative epochs $\{R_n, n \geq 1\}$ such that the behavior of the process over different regenerative cycles is independent, but the behavior of the first cycle may be different from that in later cycles.

6.4.3   Regenerative Simulation

Many systems that we encounter in many applications are too complicated for an exact mathematical analysis. In these situations, we are faced with two options. The first option is to make some simplifying assumptions of the system in order to define a problem that is mathematically tractable. The alternative option is to do a simulation study of the original system.

A simulation is a controlled statistical experimental technique that can be used to study complex stochastic systems that are not amenable to analytic and/or numerical solution. Generally, in order to analytically study a system, it is usually assumed that times between events are exponentially distributed. One of the advantages of a simulation is that we do not have to assume exponentiality. Thus, we can use simulation, for example, to investigate the distribution of the times between events or to validate the exponential distribution assumption.

In order to draw meaningful conclusions from a simulation, it is necessary to make statistically valid statements about the outcomes of the experiment. Suppose, for example, a queuing system is simulated in order to estimate a response variable B, such as the average waiting time in a queueing system. In addition to obtaining a point estimate $\hat{B}$ of B, it is desirable to estimate a confidence interval for B. An estimated $100\alpha\%$*confidence interval* for B is an interval $(\hat{B}_1, \hat{B}_2)$ whose endpoints $\hat{B}_1$ and $\hat{B}_2$ are estimated via simulation and have the property that $P[\hat{B}_1 < B < \hat{B}_2] = \alpha$. Thus, an estimated confidence interval carries with it a statement that the variable of interest is contained in the interval with a given probability.

Let $X_1, X_2, \ldots, X_n$ be independent realizations of a random variable X with unknown mean μ_X and unknown variance σ_X^2. The sample mean and sample variance of X are defined, respectively, by

$$\overline{X} = \frac{1}{n}\sum_{i=1}^{n} X_i$$

$$S^2 = \frac{1}{n-1}\sum_{i=1}^{n}(X_i - \overline{X})^2 \tag{6.29}$$

We can use the sample mean $\overline{X}$ as an estimator for the unknown mean μ_X, but we need to construct a confidence interval for μ_X. To do this, we invoke the central limit theorem, which states that for sufficiently large n,

$$\frac{\sum_{i=1}^{n} X_i - n\mu_X}{\sigma_{\overline{X}}} = \frac{\sum_{i=1}^{n} X_i - n\mu_X}{\sigma_X/\sqrt{n}}$$

is approximately a standard normal random variable. Let z_α be the 100αth percentile of the standard normal distribution; that is,

$$P[|Z| > z_\alpha] = 1 - \alpha$$

where Z is a standard normal random variable. Then,

$$P\left[-z_{\alpha/2} \leq \frac{\sum_{i=1}^{n} X_i - n\mu_X}{\sigma_X/\sqrt{n}} \leq z_{\alpha/2}\right] \approx \alpha$$

This is equivalent to the following condition:

$$P\left[\overline{X} - \frac{z_{\alpha/2}S}{\sqrt{n}} \leq \mu_X \leq \overline{X} + \frac{z_{\alpha/2}S}{\sqrt{n}}\right] \approx \alpha$$

In other words, the interval $(\overline{X} - (z_{\alpha/2}S/\sqrt{n}), \overline{X} + (z_{\alpha/2}S/\sqrt{n}))$ is an approximate $100\alpha\%$ confidence interval for μ_X.

There are two problems associated with simulation experiments. The first problem is that simulation usually estimates steady-state parameters. However, the system is not in a steady-state condition when the simulation run begins, which means that the initial observations are transient observations that cannot be used for the steady-state measure of performance. The traditional approach to dealing with this "initialization effect" is to not start collecting data until it is believed that the simulated system has essentially reached a steady-state condition. Unfortunately, it is difficult to estimate just how long this warm-up period needs to be.

The second problem is that its observations are likely to be highly correlated. The standard statistical procedures for computing the confidence interval for some measure of performance assume that the sample observations are statistically independent random observations from the underlying probability distribution for the measure. One method of dealing with this problem is to execute a series of completely separate and independent simulation runs of equal length and to use the average measure of performance for each run (excluding the initial warm-up period) as an individual observation. The main disadvantage is that each run requires an initial warm-up period for approaching a steady-state condition, as discussed earlier. Another method eliminates this disadvantage by having the runs done consecutively, using the ending condition of one run as the steady-state starting condition for the next run. This means that one continuous overall simulation run (except for the one initial warm-up period) is divided for bookkeeping purposes into a series of equal portions (referred to as *batches*). The average measure of performance for each batch is then treated as an individual observation. However, while this method considerably reduces correlation between observations by making the portions sufficiently long, it does not completely eliminate the problem.

The regenerative simulation method addresses these problems in the following manner. It performs only one simulation run. However, instead of dividing this simulation into N equal-size subruns, it divides the simulation into N unequal-size subruns by identifying the regeneration points of the process at which point the process starts afresh probabilistically. The behavior of the process after the regeneration point is independent of and probabilistically identical to the behavior of the process before the regeneration point. In this way, the method provides a solution to the problems of how to start a simulation and how to deal with correlated output. One disadvantage of the method is that not all systems have regeneration points and that hence the method is not generally applicable. Also, visits to the regeneration points can be very rare, especially for large systems.

As an example of the regenerative simulation, we consider a G/G/1 queue in which we define the regeneration points as those moments that a customer arrives at an empty queue. It must be noted that the indices of customers who arrive to find the n customers present, where $n \geq 1$ and fixed, will not in general be regeneration points because the distribution of remaining service time of the customer in service will be different for successive customers who arrive to find n customers present. However, if the service times are exponentially distributed, the indices are regeneration points due to the memoryless property of the exponential distribution.

Let K_i be the number of customers served in the ith regeneration cycle (or subrun) and let Y_i be the sum of the waiting times of the customers that are served in the ith regeneration cycle; that is,

$$Y_i = \sum_{j=1}^{K_i} W_{ij}$$

where W_{ij} is the waiting time of the jth customer in the ith regeneration cycle. Now, the theory of regenerative processes tells us that the long-term average waiting time of customers can be calculated using information over only one regeneration period. More specifically,

$$E[W] = \frac{E[Y_i]}{E[K_i]}$$

Hence, an estimator of $E[W]$ is given by

$$\overline{W} = \frac{\overline{Y}}{\overline{K}}$$

For the construction of a confidence interval, we define the random variables

$$V_i = Y_i - E[W]K_i \quad i = 1, 2, \ldots, N$$

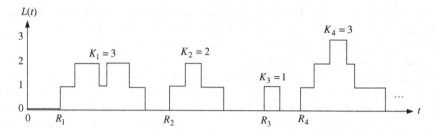

Figure 6.5 A realization of the number of customers in the system.

which are independent and identically distributed random variables with mean 0 and variance

$$\sigma_V^2 = \sigma_Y^2 + \{E[W]\}^2 \sigma_K^2 - 2E[W]\,\text{Cov}(Y, K)$$

Thus, from the central limit theorem, we have that

$$\left[\overline{W} - \frac{z_{\alpha/2}\sigma_V}{\overline{K}\sqrt{N}}, \overline{W} + \frac{z_{\alpha/2}\sigma_V}{\overline{K}\sqrt{N}} \right]$$

is an approximate $100\alpha\%$ confidence interval for $E[W]$. The unknown quantity σ_V^2 can be estimated by the sample variance

$$S_V^2 = S_Y^2 + \overline{W}^2 S_K^2 - 2\overline{W} S_{YK}$$

where S_Y^2 is the sample variance of Y, S_K^2 is the sample variance of K, and S_{YK} is the sample covariance of Y and K. That is,

$$S_Y^2 = \frac{1}{N-1} \sum_{i=1}^{N} (Y_i - \overline{Y})^2$$

$$S_K^2 = \frac{1}{N-1} \sum_{i=1}^{N} (K_i - \overline{K})^2$$

$$S_{YK} = \frac{1}{N-1} \sum_{i=1}^{N} (Y_i - \overline{Y})(K_i - \overline{K})$$

Let $L(t)$ denote the number of customers in the system at time t. Figure 6.5 illustrates a realization of the $\{L(t), t \geq 0\}$ process.

6.5 Markov Renewal Process

We have earlier defined the renewal process $\{N(t), t \geq 0\}$ as a counting process that denotes the number of renewals up to time t. The Markov renewal process is a generalization of the renewal process in which the times between renewals are chosen according to a Markov chain. That is, it is a combination of the theories of Markov chains and renewal processes. It is a vector-valued stochastic process from which the renewal process, Markov chains, and SMPs can be derived.

Consider a random variable X_n that takes values in a countable set Ω, and a random variable T_n that takes values in the interval $[0, \infty)$ such that $0 = T_0 \leq T_1 \leq T_2 \leq \cdots$. The bivariate stochastic process $\{(X_n, T_n), n \in \Omega\}$ is defined to be a Markov renewal process with state space Ω if

$$P[X_{n+1} = j, T_{n+1} - T_n \leq t | X_0, X_1, \ldots, X_n; T_0, T_1, \ldots, T_n]$$
$$= P[X_{n+1} = j, T_{n+1} - T_n \leq t | X_n]$$

for $n = 0, 1, \ldots; j \in \Omega$, and $t \in [0, \infty)$. An alternative definition of the Markov renewal process is as follows. Let $N_k(t)$ denote the number of times the process $\{(X_n, T_n)\}$ visits state $X_n = k$ in the interval $[0, t)$, then $\{N_k(t), k \in \Omega, t \geq 0\}$ is a Markov renewal counting process. In particular, if we assume that the initial state is k, then transitions into state k constitute renewals, which means that successive times between transitions into state k are independent and identically distributed. The interval $H_n = T_{n+1} - T_n, n \geq 0$, is called the *holding time* or *waiting time* in state X_n.

Markov renewal process is a more general class of stochastic processes than the renewal process in the following sense. While the renewal counting process generated by a renewal process keeps track of the number of arrivals (which can be likened to the number of times that the process has visited a particular state) up to time t, the Markov renewal counting process generated by a Markov renewal process records the number of times that the process has visited each of the states up to time t. Thus, a renewal counting process is a scalar-valued stochastic process while a Markov renewal counting process is a vector-valued stochastic process: $N_{\mathrm{MRP}}(t) = \{N_0(t), N_1(t), N_2(t), \ldots\}$.

We define the following function:

$$V_k(n, t) = \begin{cases} 1 & \text{if } X = k, H_0 + H_1 + \cdots + H_{n-1} \leq t; \quad n = 1, 2, \ldots \\ 0 & \text{otherwise} \end{cases}$$

where $k \in \Omega$ and $t \geq 0$. Then

$$N_k(t) = \sum_{n=0}^{\infty} V_k(n, t) \quad k \in \Omega, \quad t \geq 0 \tag{6.30}$$

6.5.1 The Markov Renewal Function

The function

$$M_{ik}(t) = E[N_k(t) | X_0 = i] \quad i, k \in \Omega, \quad t \geq 0 \tag{6.31}$$

is called the *Markov renewal function*. Substituting the value of $N_k(t)$ from Eq. (6.30), we obtain

$$M_{ik}(t) = E\left[\sum_{n=0}^{\infty} V_k(n,t)|X_0 = i\right] = \sum_{n=0}^{\infty} E[V_k(n,t)|X_0 = i]$$

$$= \sum_{n=1}^{\infty} P[X_n = k, J_n \leq t|X_0 = i]$$

where $J_n = H_0 + H_1 + \cdots + H_{n-1}$ is the time from the beginning until the process enters state X_n; alternatively, it is the epoch of the nth transition of the process $\{(X_n, T_n)\}$. We define the *one-step transition probability* $Q_{ij}(t)$ of the SMP by

$$Q_{ij}(t) = P[X_{n+1} = j, H_n \leq t|X_n = i] \quad t \geq 0 \tag{6.32}$$

independent of n. Thus, $Q_{ij}(t)$ is the conditional probability that the process will be in state j next, given that it is currently in state i and the waiting time in the current state i is no more than t. The matrix $Q = [Q_{ij}(t)]_{i,j \in \Omega; t \geq 0}$ is called the *semi-Markov kernel* over Ω. In particular,

$$Q_{ik}(t) = P[X_1 = k, H_0 \leq t|X_0 = i] \quad t \geq 0 \tag{6.33}$$

Thus,

$$M_{ik}(t) = \sum_{n=1}^{\infty} P[X_n = k, J_n \leq t|X_0 = i]$$

$$= P[X_1 = k, J_1 \leq t|X_0 = i] + \sum_{n=2}^{\infty} P[X_n = k, J_n \leq t|X_0 = i]$$

$$= P[X_1 = k, H_0 \leq t|X_0 = i] + \sum_{n=2}^{\infty} P[X_n = k, J_n \leq t|X_0 = i]$$

$$= Q_{ik}(t) + \sum_{n=2}^{\infty} P[X_n = k, J_n \leq t|X_0 = i]$$

If we define

$$Q_{ik}^{(n)}(t) = P[X_n = k, J_n \leq t|X_0 = i]$$

$$Q_{ik}^{(0)}(t) = \begin{cases} 0 & i \neq k \\ 1 & i = k \end{cases}$$

then $Q_{ik}^{(1)}(t) = Q_{ik}(t)$ and

$$Q_{ik}^{(2)}(t) = P[X_2 = k, J_2 \leq t|X_0 = i]$$

$$= \sum_{j \in \Omega} \int_0^t P[X_2 = k, H_1 \leq t - u|X_1 = j] \, dQ_{ij}(u)$$

$$= \sum_{j \in \Omega} \int_0^t Q_{ij}(t - u) \, dQ_{jk}(u)$$

From this, we can easily obtain the following recursive relationship:

$$Q_{ik}^{(n+1)}(t) = \sum_{j \in \Omega} \int_0^t Q_{ij}^{(n)}(t-u) \, dQ_{jk}(u) \tag{6.34}$$

If we define the matrix $Q = [Q_{ik}]$, then the above expression is the convolution of $Q^{(n)}$ and Q. That is,

$$Q^{(n+1)}(t) = Q^{(n)}(t) \cdot Q(t) \tag{6.35}$$

Thus, if we define the matrix $M(t) = [M_{ik}(t)]$, we obtain

$$M(t) = \sum_{n=1}^{\infty} Q^{(n)}(t) \quad t \geq 0 \tag{6.36}$$

We can rewrite this equation as follows:

$$M(t) = Q(t) + \sum_{n=2}^{\infty} Q^{(n)}(t) = Q(t) + \sum_{n=1}^{\infty} Q^{(n)}(t) \cdot Q(t)$$

$$= Q(t) + \left(\sum_{n=1}^{\infty} Q^{(n)}(t) \right) \cdot Q(t)$$

$$= Q(t) + M(t) \cdot Q(t) = Q(t) + Q(t) \cdot M(t)$$

If we take the Laplace transform of both sides, we obtain

$$L_M(s) = L_Q(s) + L_Q(s)L_M(s) \Rightarrow [I - L_Q(s)]L_M(s) = L_Q(s)$$

where $L_M(s)$ is the Laplace transform of $M(t)$ and $L_Q(s)$ is the Laplace transform of $Q(t)$. Thus,

$$L_M(s) = [I - L_Q(s)]^{-1}L_Q(s) = \sum_{n=1}^{\infty} [L_Q(s)]^n = [I - L_Q(s)]^{-1} - I \tag{6.37}$$

Note that from the equation $L_M(s) = L_Q(s) + L_Q(s)L_M(s) = L_Q(s)[I + L_M(s)]$, we also obtain

$$L_Q(s) = L_M(s)[I + L_M(s)]^{-1} \tag{6.38}$$

6.6   Semi-Markov Processes

An SMP is both a Markov renewal process and a generalization of the Markov process. While a Markov renewal process is concerned with the generalized renewal random variables (i.e., it records the number of visits to each state of the process), an SMP is concerned with the random variables that describe the state of the process at some time.

In a discrete-time Markov process, we assume that the amount of time spent in each state before a transition to the next state occurs is a unit time. Similarly, in a CTMC, we assume that the amount of time spent in a state before a transition to the next state occurs is exponentially distributed. An SMP is a process that makes transitions from state to state like a Markov process. However, the amount of time spent in each state before a transition to the next state occurs is an arbitrary random variable that depends on the next state the process will enter. Thus, at transition instants an SMP behaves like a Markov process.

6.6.1 Discrete-Time SMPs

Consider a finite-state discrete-time random process $\{X_n | n = 0, 1, \ldots, N\}$. That is, the state space is $\Omega = \{0, 1, \ldots, N\}$. Assume that when the process enters state i, it chooses its next state as state j with probability p_{ij}, where

$$p_{ij} \geq 0 \quad i, j \in \Omega$$
$$\sum_{j=0}^{N} p_{ij} = 1$$

Let $T_0, T_1, \ldots$ denote the transition epochs on the nonnegative real line such that $0 = T_0 \leq T_1 \leq T_2 \leq \cdots$. Define the interval $T_{i+1} - T_i = H_i$ to be the *holding time* (or *waiting time*) in state $i \in \Omega$. H_i can be explained as follows. After choosing j, the process spends a holding time H_{ij} before making the transition, where H_{ij} is a positive, integer-valued random variable with the PMF

$$p_{H_{ij}}(m) = P[H_{ij} = m] \quad m = 1, 2, \ldots \tag{6.39}$$

It is assumed that $E[H_{ij}] < \infty$ and $p_{H_{ij}}(0) = 0$ for all i and j, which means that the system spends at least one unit of time before making a transition. Note that if we focus only on the transitions and ignore the times between transitions, we will have a Markov process. However, when we include the holding times, the process will no longer satisfy the Chapman–Kolmogorov equation. We call the process that governs transitions between states the *embedded Markov process*. Thus, the PMF of the holding time H_i in state i is given by

$$p_{H_i}(m) = \sum_{j=0}^{N} p_{ij} p_{H_{ij}}(m) \quad m = 1, 2, \ldots \tag{6.40}$$

The mean holding time in state i is given by

$$E[H_i] = \sum_{j=0}^{N} p_{ij} E[H_{ij}] \quad i = 0, 1, 2, \ldots, N \tag{6.41}$$

The two-dimensional stochastic process $\{(X_n, T_n), n = 0, 1, \ldots, N\}$, is called a discrete-time SMP if the following conditions are satisfied:

a. $\{X_n | n = 0, 1, \ldots, N\}$ is a Markov chain, which is called the embedded Markov chain
b. If $H_n = T_{n+1} - T_n$, then

$$P[X_{n+1} = j, T_{n+1} - T_n \leq m | X_0, X_1, \ldots, X_n = i; T_0, T_1, \ldots, T_n]$$
$$= P[X_{n+1} = j, H_n \leq m | X_n = i]$$

We can represent a discrete-time SMP as shown in Figure 6.6, which is the state transition diagram of a discrete-time SMP whose embedded Markov process is a Markov chain with three states. A transition arc contains both the transition probability and the PMF of the conditional holding time.

State Probabilities

Let $\phi_{ij}(n)$ denote the probability that the process is in state j at time n given that it entered state i at time 0. In Howard (1960), $\phi_{ij}(n)$ is referred to as the "interval-transition probabilities of the process." However, a more general name is the *transition probability function* from state i to state j. To get an expression for $\phi_{ij}(n)$, we consider two cases:

a. For $i \neq j$, the process makes a transition to some state k at time m, and then in the remaining time $n - m$ it travels to state j. The probability of this event is

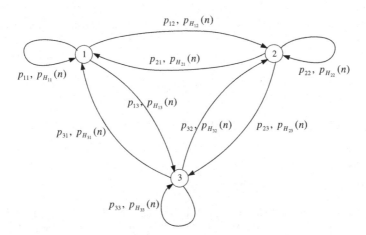

Figure 6.6 State transition diagram of a discrete-time SMP.

$$\phi_{ij}(n) = \sum_{k=0}^{N} p_{ik} \sum_{m=0}^{n} p_{H_{ik}}(m)\phi_{kj}(n-m) \quad i \neq j$$

b. For $i = j$, we have an additional probability that the process never left state i during the interval of interest. The probability of this additional event is

$$P[H_i > n] = 1 - P[H_i \leq n] = 1 - F_{H_i}(n)$$

Let

$$\delta_{ij} = \begin{cases} 1 & i = j \\ 0 & i \neq j \end{cases}$$

Because the event that it never left state i during the interval and the event that it left i for some state k are mutually exclusive, we have that for $i, j \in \Omega$ and $n \geq 0$,

$$\begin{aligned}
\phi_{ij}(n) &= \delta_{ij}\{1 - F_{H_i}(n)\} + \sum_{k=1}^{N} p_{ik} \sum_{m=0}^{n} p_{H_{ik}}(m)\phi_{kj}(n-m) \\
&= \delta_{ij}\{1 - F_{H_i}(n)\} + \sum_{k=1}^{N} p_{ik}p_{H_{ik}}(n) * \phi_{kj}(n)
\end{aligned} \tag{6.42}$$

where $\phi_{ij}(0) = \delta_{ij}$ and $*$ is the convolution operator. The preceding equation is referred to as the *backward Chapman–Kolmogorov equation* for the SMP. Let $P_H(n)$ denote the matrix of the $p_{H_{ik}}(n)$, and let P denote the state transition matrix of the embedded Markov chain. Using the terminology of Howard (1971b) and using a slightly different notation, we define the *discrete-time core matrix* $C(n)$ by

$$C(n) = P \Delta P_H(n) \tag{6.43}$$

where the elements of $C(n)$ are $c_{ij}(n) = p_{ij}p_{H_{ij}}(n)$. Thus, the transition probability function of Eq. (6.42) becomes

$$\phi_{ij}(n) = \delta_{ij}\{1 - F_{H_i}(n)\} + \sum_{k=1}^{N} c_{ik}(n) * \phi_{kj}(n) \tag{6.44}$$

Taking the z-transform of both sides, we obtain

$$\phi_{ij}(z) = \frac{\delta_{ij}[1 - G_{H_i}(z)]}{1 - z} + \sum_{k=1}^{N} c_{ik}(z)\phi_{kj}(z) \tag{6.45}$$

where $G_{H_i}(z)$ is the z-transform of $p_{H_i}(n)$. Let $D(z)$ be the $N \times N$ diagonal matrix whose ith element is $[1 - G_{H_i}(z)]/(1 - z)$. Then, in the matrix form, Eq. (6.45) becomes

$$\Phi(z) = D(z) + C(z)\Phi(z)$$

which gives

$$\Phi(z) = [I - C(z)]^{-1}D(z)$$

If we define

$$\phi_{ij} = \lim_{n \to \infty} \phi_{ij}(n)$$

then from Howard (1971b), it can be shown that

$$\phi_{ij} = \frac{\pi_j E[H_j]}{\sum_{k=1}^{N} \pi_k E[H_k]} \tag{6.46}$$

where π_j is the limiting state probability of being in state j in the embedded Markov chain. Note that the right-hand side of the equation is independent of i. Thus, we define the limiting probability of the SMP by

$$\phi_j = \frac{\pi_j E[H_j]}{\sum_{k=1}^{N} \pi_k E[H_k]} \tag{6.47}$$

The limiting state probability is also called the *occupancy distribution* because ϕ_j gives the long-run fraction of time that the process spends in state j.

First Passage Times

Let T_{ij} denote the first passage time from state i to state j. That is,

$$T_{ij} = \min\{n > 0 | X_n = j, X_0 = i\}$$

Let $m_{ij} = E[T_{ij}]$, that is, m_{ij} is the mean first passage time from state i to state j. We can use the same arguments used in Chapters 4 and 5 to obtain m_{ij} recursively as follows. Because the mean waiting time in state i is $E[H_i]$, then given that the process starts in state i, it will spend a mean time $E[H_i]$ before making a transition to some state k with probability p_{ik}. Then from state k, it takes a mean time m_{kj} to reach state j. Thus, we obtain

$$m_{ij} = E[H_i] + \sum_{k \neq j} p_{ik} m_{kj} \tag{6.48}$$

The mean recurrence time is obtained when $j = i$. Specifically, we note that because the limiting state probabilities of the embedded Markov chain exist, they satisfy the balance equations

$$\pi_i = \sum_{k=0}^{N} \pi_k p_{ki}$$

Thus, multiplying on both sides of Eq. (6.48) by π_i and summing over i, we obtain

$$\sum_{i=0}^{N} \pi_i m_{ij} = \sum_{i=0}^{N} \pi_i E[H_i] + \sum_{i=0}^{N} \pi_i \sum_{k=0, \, k\neq j}^{N} p_{ik} m_{kj}$$

$$= \sum_{i=0}^{N} \pi_i E[H_i] + \sum_{k=0, \, k\neq j}^{N} m_{kj} \sum_{i=0}^{N} \pi_i p_{ik} = \sum_{i=0}^{N} \pi_i E[H_i] + \sum_{k=0, \, k\neq j}^{N} m_{kj} \pi_k$$

$$= \sum_{i=0}^{N} \pi_i E[H_i] + \sum_{k=0, \, k\neq j}^{N} \pi_k m_{kj}$$

$$= \sum_{i=0}^{N} \pi_i E[H_i] + \sum_{k=0}^{N} \pi_k m_{kj} - \pi_j m_{jj}$$

where in the third equality, we have made use of the balance equation. Canceling the like terms on both sides of the equation, we obtain

$$m_{jj} = \frac{\sum_{i=0}^{N} \pi_i E[H_i]}{\pi_i} \quad j = 0, 1, \ldots, N \tag{6.49}$$

From these results, we obtain the following definitions:

a. Two states i and j in an SMP are said to communicate if they communicate in the embedded Markov chain. For two such states, $P[T_{ij} < \infty] \times P[T_{ji} < \infty] > 0$.
b. A state i in an SMP is said to be a recurrent state if it is a recurrent state in the embedded Markov chain. For such a state, $P[T_{ii} < \infty] > 0$.
c. A state i in an SMP is said to be a transient state if it is a transient state in the embedded Markov chain. For such a state, $P[T_{ii} < \infty] = 0$.

6.6.2 Continuous-Time SMPs

Consider a finite-state continuous-time stochastic process $\{X(t), t \geq 0\}$ with state space $\Omega = [0, 1, \ldots, N]$. Assume that the process just entered state i at time $t = 0$, then it chooses the next state j with probability p_{ij}, where

$$p_{ij} \geq 0 \quad i, j \in \Omega$$

$$\sum_{j=0}^{N} p_{ij} = 1$$

Given that the next transition out of state i will be to state j, the time H_{ij} that the process spends in state i until the next transition has the PDF $f_{H_{ij}}(t), t \geq 0$. The random variable H_{ij} is called the *holding time* for a transition from i to j, and it is assumed that $E[H_{ij}] < \infty$. As discussed in the discrete-time case, if we focus only on the transitions and ignore the times between transitions, we will have a Markov process. However, when we include the holding times, the process will no longer satisfy the Chapman−Kolmogorov equation unless the holding times are exponentially distributed. We call the process that governs transitions between states the *embedded Markov process*. The time H_i that the process spends in state i before making a transition is called the *waiting time* in state i, and its PDF is given by

$$f_{H_i}(t) = \sum_{j=0}^{N} p_{ij} f_{H_{ij}}(t) \quad t \geq 0 \tag{6.50}$$

Thus, the mean waiting time in state i is

$$E[H_i] = \sum_{j=0}^{N} p_{ij} E[H_{ij}] \quad i = 0, 1, \ldots, N$$

The two-dimensional stochastic process $\{(X_n, T_n) | n = 0, 1, \ldots, N\}$ is called a continuous-time SMP if the following conditions are satisfied:

a. $\{X_n | n = 0, 1, \ldots, N\}$ is a Markov chain, the embedded Markov chain
b. If we define $H_n = T_{n+1} - T_n$, then

$$P[X_{n+1} = j, T_{n+1} - T_n \leq t | X_0, X_1, \ldots, X_n = i; T_0, \ldots, T_n]$$
$$= P[X_{n+1} = j, H_n \leq t | X_n = i] \quad t \geq 0$$

As in the discrete-time case, we can represent a continuous-time SMP as shown in Figure 6.7, which is the state transition diagram of an SMP whose embedded Markov process is a Markov chain with three states. In this case, a transition arc contains both the transition probability and the PDF of the conditional holding time.

We can now see the difference between a CTMC and a continuous-time SMP. In a CTMC, the holding times (or sojourn times) are exponentially distributed and depend only on the current state. In a continuous-time SMP, the holding times can have an arbitrary distribution and can depend on both the current state and the state to be visited next.

A transition from state i to state j is called a *real transition* if $i \neq j$, and it is called a *virtual transition* if $i = j$. As discussed earlier, the one-step transition probability $Q_{ij}(t)$ of the SMP is defined by

$$Q_{ij}(t) = P[X_{n+1} = j, H_n \leq t | X_n = i] \quad t \geq 0 \tag{6.51}$$

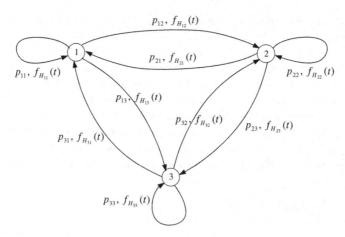

Figure 6.7 State transition diagram of a continuous-time SMP.

independent of n. Thus, $Q_{ij}(t)$ is the conditional probability that the process will be in state j next, given that it is currently in state i and the waiting time in the current state i is no more than t. From this, we obtain the transition probability p_{ij} of the embedded Markov chain and the cumulative distribution function (CDF) of the waiting time at state i, $F_{H_i}(t)$, as follows:

$$p_{ij} = P[X_{n+1} = j | X_n = i] = \lim_{t \to \infty} P[X_{n+1} = j, H_n \le t | X_n = i]$$
$$= Q_{ij}(\infty) \tag{6.52}$$

$$F_{H_i}(t) = P[H_i \le t] = P[H_n \le t | X_n = i] = \sum_{j \in \Omega} P[P[X_{n+1} = j, H_n \le t | X_n = i]$$
$$= \sum_{j \in \Omega} Q_{ij}(t) \tag{6.53}$$

Note that it is normal practice to define the matrix of the $Q_{ij}(t)$ as $Q = [Q_{ij}(t)]$, as defined earlier for the discrete-time SMP. However, this Q matrix is not the same as the intensity matrix or the infinitesimal generator of the CTMC. The convention is historical and the two have historically been defined differently. Note also that

$$Q_{ij}(t) = P[X_{n+1} = j, H_n \le t | X_n = i]$$
$$= P[H_n \le t | X_n = i, X_{n+1} = j] P[X_{n+1} = j | X_n = i]$$

Because $P[X_{n+1} = j | X_n = i] = p_{ij}$, if we define $G_{ij}(t) = P[H_n \le t | X_n = i, X_{n+1} = j]$, then we have that

$$Q_{ij}(t) = p_{ij} G_{ij}(t) \tag{6.54}$$

There are several systems that can be modeled by a continuous-time SMP. These include any system that can be modeled by a CTMC, because from the foregoing discussion we observe that a CTMC is an SMP in which

$$f_{H_{ij}}(t) = \lambda_{ij}\, e^{-\lambda_{ij} t} \quad t \geq 0, \ 0 \leq i, \ j \leq N$$

A major area of application of continuous-time SMPs is in reliability and availability studies, see Osaki (1985) and Limnios and Oprisan (2001). For example, consider a system that is subject to failures and repairs. Assume that when the system is up (or functioning) the time until it fails, which is called the time to failure, has an arbitrary distribution. After it has failed, the time until it is repaired, which is called the time to repair, has another arbitrary distribution. If we denote the upstate by state 1 and the downstate by state 2, then the behavior of the system can be modeled by an SMP with the transition probability matrix of the embedded Markov chain given by

$$P = \begin{bmatrix} 0 & 1 \\ 1 & 0 \end{bmatrix}$$

More sophisticated failure modes can similarly be modeled by the SMP.

State Probabilities

Let $\phi_{ij}(t)$ denote the probability that the process is in state j at time t given that it entered state i at time 0. As mentioned earlier, $\phi_{ij}(t)$ is referred to as the *transition probability function* from state i to state j. Following the technique used for the discrete-time case, we consider two cases:

a. For $i \neq j$, the process makes a transition to some state k at time u, and then in the remaining time $t - u$ it travels to state j. The probability of this event is

$$\phi_{ij}(t) = \sum_{k=0}^{N} p_{ik} \int_0^t f_{H_{ik}}(u)\phi_{kj}(t - u)\mathrm{d}u \quad i \neq j$$

b. For $i = j$, we have an additional probability that the process never left state i during the interval of interest. The probability of this additional event is

$$P[H_i > t] = 1 - P[H_i \leq t] = 1 - F_{H_i}(t)$$

Because the event that it never left state i during the interval and the event that it left i for some state k are mutually exclusive, then for $i, j \in \Omega$ and $t \geq 0$, we have that

$$\phi_{ij}(t) = \delta_{ij}[1 - F_{H_i}(t)] + \sum_{k=0}^{N} p_{ik} \int_0^t f_{H_{ik}}(u)\phi_{kj}(t - u)\mathrm{d}u \tag{6.55}$$

where $\phi_{ij}(0) = \delta_{ij}$. Because the integral term is essentially a convolution integral, we can write

$$
\begin{aligned}
\phi_{ij}(t) &= \delta_{ij}[1 - F_{H_i}(t)] + \sum_{k=0}^{N} p_{ik} f_{H_{ik}}(t) * \phi_{kj}(t) \\
&= \delta_{ij}[1 - F_{H_i}(t)] + \sum_{k=0}^{N} c_{ik}(t) * \phi_{kj}(t)
\end{aligned}
\tag{6.56}
$$

where, as in the discrete-time case, we define the *continuous-time core matrix*

$$
C(t) = [c_{ij}(t)] = P \, \Delta f_H(t) = [p_{ij} f_{H_{ij}}(t)]
$$

We define $\phi_{ij}(s)$ as the s-transform of $\phi_{ij}(t)$, $c_{ij}(s)$ as the s-transform of $c_{ij}(t)$, and note that the s-transform of $F_{H_i}(t)$ is $M_{H_i}(s)/s$, where $M_{H_i}(s)$ is the s-transform of $f_{H_i}(t)$. Thus, taking the s-transform on both sides of Eq. (6.54), we obtain the following:

$$
\begin{aligned}
\phi_{ij}(s) &= \frac{\delta_{ij}}{s}\left[1 - M_{H_i}(s)\right] + \sum_{k=0}^{N} c_{ik}(s)\phi_{kj}(s) \\
&= \frac{\delta_{ij}}{s}\left[1 - \sum_{k=0}^{N} p_{ik}(s) M_{H_{ik}}(s)\right] + \sum_{k=0}^{N} c_{ik}(s)\phi_{kj}(s)
\end{aligned}
\tag{6.57}
$$

Finally, if $D(s)$ is the $N \times N$ diagonal matrix with entries $\{1 - M_{H_i}(s)\}/s$, the matrix form of the equation becomes

$$
\Phi(s) = D(s) + C(s)\Phi(s)
$$

which gives

$$
\Phi(s) = [I - C(s)]^{-1} D(s)
\tag{6.58}
$$

If we define

$$
\phi_{ij} = \lim_{t \to \infty} \phi_{ij}(t)
\tag{6.59}
$$

then from Heyman and Sobel (1982) and Howard (1960, 1971b), it can be shown that if π_j is the limiting state probability of the embedded Markov chain being in state j,

$$
\phi_{ij} = \frac{\pi_j E[H_j]}{\sum_{k=0}^{N} \pi_k E[H_k]}
\tag{6.60}
$$

Because the right-hand side of the equation is independent of i, we obtain the limiting probability of the SMP as

$$\phi_j = \frac{\pi_j E[H_j]}{\sum_{k=0}^{N} \pi_k E[H_k]} \tag{6.61}$$

As stated earlier, the limiting state probability is also called the *occupancy distribution* because ϕ_j gives the long-run fraction of time that the process spends in state j.

First Passage Times

Let T_{ij} denote the first passage time from state i to state j. That is,

$$T_{ij} = \min\{t > 0 | X(t) = j, X(0) = i\}$$

Let $m_{ij} = E[T_{ij}]$, that is, m_{ij} is the mean first passage time from state i to state j. From earlier discussion in the discrete-time case, we have that

$$m_{ij} = E[H_i] + \sum_{k \neq j} p_{ik} m_{kj} \tag{6.62}$$

Finally, using the same method used in the discrete-time case, we obtain the mean recurrence time as

$$m_{jj} = \frac{\sum_{i=0}^{N} \pi_i E[H_i]}{\pi_j} \quad j = 0, 1, \ldots, N \tag{6.63}$$

6.7 Markov Regenerative Process

Markov regenerative processes (MRGPs) constitute a more general class of stochastic processes than traditional Markov processes. Markovian dependency, the first-order dependency, is the simplest and most important dependency in stochastic processes. The past history of a Markov chain is summarized in the current state and the behavior of the system thereafter only depends on the current state. The sojourn time of a homogeneous CTMC is exponentially distributed. However, nonexponentially distributed transitions are common in real-life systems. SMPs have generally distributed sojourn times but lack the ability to capture local behaviors during the intervals between successive regenerative points. MRGPs are discrete-state continuous-time stochastic processes with embedded regenerative time points at which the process enjoys the Markov property. MRGPs provide a natural generalization of SMPs with local behavior accounted for. Thus, the SMP, the discrete-time Markov chain, and the CTMC are special cases of the MRGP.

A stochastic process $\{Z(t), t \geq 0\}$ is called an MRGP, if there exists a Markov renewal sequence $\{(X_n, T_n), n \geq 0\}$ of random variables such that all the conditional finite dimensional distributions of $\{Z(T_n + t), t \geq 0\}$ given $\{Z(u), 0 \leq u \leq T_n, X_n = i\}$ are the same as those of $\{Z(t), t \geq 0\}$ given $X_0 = i$. The above definition implies that

$$P[Z(T_n + t) = j | Z(u), 0 \leq u \leq T_n, X_n = i] = P[Z(t) = j | X_0 = i] \qquad (6.64)$$

The MRGP does not have the Markov property in general, but there is a sequence of time points $\{T_0 = 0, T_1, T_2, \ldots\}$ at which Markov property holds. From the above definition, it is obvious that every SMP is an MRGP. The difference between SMP and MRGP is that in an SMP every state transition is a regeneration point, which is not necessarily true for the MRGP. The requirement of regeneration at every state transition makes the SMP of limited interest for transient analysis of systems that involve deterministic parameters as in a communication system. An example of the sample path of MRGP is shown in Figure 6.8, where it is shown that not every arrival epoch is a regeneration point.

As discussed earlier, MRGPs are a generalization of many stochastic processes including Markov processes. The difference between SMP and MRGP can be seen by comparing the sequence T_n with the sequence obtained from the state transition instants K_n. In an SMP, every state transition is a regeneration point, which means that $K_n = T_n$. For MRGP, every transition point is not a regeneration point and as such T_n is a subsequence of K_n. The fact that every transition is a regeneration point in the SMP limits its use in applications that involve deterministic parameters.

The time instants T_n at which transitions take place are called regeneration points. The behavior of the process can be determined by the latest regeneration point, which can specify the elapsed time and the latest state visited. A regeneration point is independent of the elapsed time up to the instant of jump instant.

For an SMP, no state change occurs between successive regeneration points and the sample paths are piecewise constant and T_n is when the nth jump occurs. But for an MRGP, the stochastic process between T_n and T_{n+1} can be any continuous-time stochastic process, including CTMC, SMP, or another MRGP. This means that the

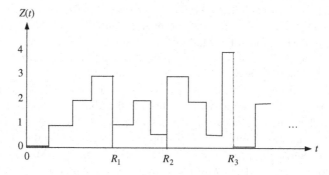

Figure 6.8 An example of the sample path of an MRGP.

sample paths are no longer piecewise constant, because local behaviors exist between consecutive Markov regeneration points. The jumps do not necessarily have to be at the T_n. More information on MRGPs can be found in Kulkarni (2010).

6.8 Markov Jump Processes

Recall from Chapter 3 that a jump process is a stochastic process that makes transitions between discrete states in the following manner: It spends an amount of time called the holding time (or sojourn time) in the current state and then jumps to another state where it spends another holding time, and so on. A Markov jump process is a jump process in which the holding times are exponentially distributed. Thus, if the holding times depend on both the current state and the state to be visited next, then a Markov jump process is an SMP. On the other hand, if the holding times depend only on the current state, then a Markov jump process is a CTMC.

Consider a Markov jump process that starts at state $X(0) = x$ at time $t = 0$ and waits until time $t = T_1$ when it makes a jump of size θ_1 that is not necessarily positive. The process then waits until time $t = T_2$ when it makes another jump of size θ_2, and so on. The jump sizes θ_i are also assumed to be independent and identically distributed. The times $T_1, T_2, \ldots$ are the instants when the process makes jumps, and the intervals $\tau_i = T_i - T_{i-1}, i = 1, 2, \ldots$, called the *waiting times* (or *pausing times*), are assumed to be independent and exponentially distributed with rates λ_i. The time at which the nth jump occurs, T_n, is given by

$$T_n = \sum_{i=1}^{n} \tau_i \quad n = 1, 2, \ldots; \quad \tau_0 = 0$$

The time the process spends in any state x (i.e., the waiting time in state x) and the choice of the next state y are independent. Let $T(t)$ denote the cumulative waiting time from time $t = 0$ to the instant the process changes the current state $X(t)$, and let the function $p_{ij}(s, t)$ denote the probability that the process is in state j at time t, given that it was in state i at time s; that is, for $t \geq s$

$$p_{ij}(s, t) = P[X(t) = j | X(s) = i]$$

Then, we have that for $\tau > t$,

$$
\begin{aligned}
P[X(t) = y, T(t) \leq \tau | X(0) = x] &= P[X(t) = y | X(0) = x, T(t) \leq \tau]P[T(t) \leq \tau] \\
&= P[X(t) = y | X(0) = x]P[T(t) \leq \tau] \\
&= P[X(t) = y | X(0) = x]P[t < w(y) \leq \tau] \\
&= p_{xy}(0, t) \int_{t}^{\tau} \lambda_y \, e^{-\lambda_y u} \, du \\
&= p_{xy}(0, t)\{e^{-\lambda_y t} - e^{-\lambda_y \tau}\}
\end{aligned}
$$

where the second equality follows from the independence of the waiting time and the choice of next state, and $w(y)$ is the waiting time in state y. A realization of the process is illustrated in Figure 6.9.

Let Θ be a random variable that denotes the jump size. The state of the process at time t relative to its initial value is given by

$$\Delta X(t) = \sum_{i=1}^{N(t)} \Theta_i$$

where the upper limit $N(t)$ is a random function of time that denotes the number of jumps up to time t and is given by

$$N(t) = \max\{n | T_n \le t\}$$

We have used the notation $\Delta X(t)$ to denote the change in state by time t relative to the initial value $X(0)$ because $X(0)$ might not be zero; but if it is zero, then $X(t) = X(0) + \Delta X(t)$.

Because $X(t)$ is a Markov process, the Markov property holds. That is, given the times $0 < t_2 < t_2 < \cdots < t_n < s < t$, we have that

$$P[X(t) = y | X(s) = x, X(t_n) = x_n, \ldots, X(t_1) = x_1] = P[X(t) = y | X(s) = x]$$

Let $\pi_k(t)$ denote that probability that the process is in state k at time t. Then, for $s, t > 0$,

$$
\begin{aligned}
P[X(t+s) = j, X(0) = i] &= \sum_k P[X(t+s) = j, X(s) = k, X(0) = i] \\
&= \sum_k P[X(t+s) = j | X(s) = k, X(0) = i] P[X(s) = k, X(0) = i] \\
&= \sum_k P[X(t+s) = j | X(s) = k] P[X(s) = k, X(0) = i] \qquad (6.65) \\
&= \sum_k P[X(t+s) = j | X(s) = k] P[X(s) = k | X(0) = i] P[X(0) = i] \\
&= \sum_k p_{ik}(0, s) p_{kj}(s, t) \pi_i(0)
\end{aligned}
$$

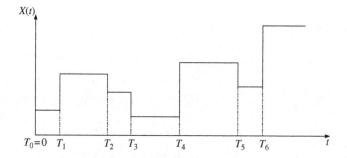

Figure 6.9 Realization of a Markov jump process.

where the third equality is due to the Markov property of $X(t)$. Now, we know that

$$P[X(t+s)=j, X(0)=i] = P[X(t+s)=j|X(0)=i]P[X(0)=i]$$
$$= p_{ij}(0, t+s)\pi_i(0)$$

Combining the two results, we obtain

$$p_{ij}(0, t+s) = \sum_k p_{ik}(0, s)p_{kj}(s, t) \tag{6.66}$$

which is the Chapman–Kolmogorov equation for Markov jump processes.

6.8.1 The Homogeneous Markov Jump Process

We now consider the special case where $X(0) = 0$ and the waiting times are identically distributed with parameter λ. Let T denote the waiting time and Θ the jump size that we assume to be discrete and nonnegative. Similarly, let $p_{X(t)}(x, t)$ denote the PMF of $X(t)$, $p_\Theta(\theta)$ the PMF of Θ, $f_T(t)$ the PDF of T, and $p_{N(t)}(n, t)$ the PMF of $N(t)$. Also, let $G_{X(t)}(z)$ denote the z-transform of $p_{X(t)}(x, t)$, let $G_\Theta(z)$ denote the z-transform of $p_\Theta(\theta)$, and let $G_{N(t)}(z)$ denote the z-transform of $p_{N(t)}(n, t)$, where

$$G_\Theta(z) = E[z^\Theta] = \sum_{\theta=0}^\infty z^\theta p_\Theta(\theta)$$

$$G_{N(t)}(z) = E[z^{N(t)}] = \sum_{n=0}^\infty z^n p_{N(t)}(n, t)$$

Because $X(t) = \sum_{i=1}^{N(t)} \Theta_i$, we know from probability theory (see, for example, Ibe, 2005) that

$$G_{X(t)}(z) = G_{N(t)}(G_\Theta(z)) \tag{6.67}$$

Thus, if $p_{N(t)}(n, t)$ and $p_\Theta(\theta)$ are known, we can determine $G_{X(t)}(z)$ and consequently $p_{X(t)}(x, t)$.

The expected value of $X(t)$ is given by

$$E[X(t)] = E[N(t)]E[\Theta] \tag{6.68}$$

Because the intervals τ_i are independent and exponentially distributed, we note that the process $\{N(t), t \geq 0\}$ is a renewal process and T_n is the time of the nth renewal. Thus, we have that

$$p_{N(t)}(n, t) = P[N(t)=n] = P[N(t)<n+1] - P[N(t)<n] = P[T_{n+1}>t] - P[T_n>t]$$
$$= 1 - F_{T_{n+1}}(t) - \{1 - F_{T_n}(t)\} \tag{6.69}$$
$$= F_{T_n}(t) - F_{T_{n+1}}(t)$$

and the expected value of $N(t)$ is given by

$$
\begin{aligned}
E[N(t)] &= \sum_{n=0}^{\infty} n p_{N(t)}(n, t) = \sum_{n=0}^{\infty} n \left[F_{T_n}(t) - F_{T_{n+1}}(t) \right] \\
&= \left\{ F_{T_1}(t) + 2F_{T_2}(t) + 3F_{T_3}(t) + \cdots \right\} - \left\{ F_{T_2}(t) + 2F_{T_3}(t) + \cdots \right\} \\
&= F_{T_1}(t) + F_{T_2}(t) + F_{T_3}(t) + \cdots \\
&= \sum_{n=1}^{\infty} F_{T_n}(t) = \sum_{n=1}^{\infty} \left\{ 1 - \sum_{k=0}^{n-1} \frac{(\lambda t)^k e^{-\lambda t}}{k!} \right\} \\
&= \sum_{n=1}^{\infty} \sum_{k=n}^{\infty} \frac{(\lambda t)^k e^{-\lambda t}}{k!}
\end{aligned}
\tag{6.70}
$$

Recall that $p_{0x}(0, t)$ is the probability that the state of the process at time t is x, given that it was initially in state 0; that is,

$$
p_{0x}(0, t) = P[X(t) = x | X(0) = 0]
$$

Because T and Θ are assumed to be independent, we can obtain $p_{0x}(0, t)$ as follows:

$$
p_{0x}(0, t) = \delta(x)R(t) + \int_0^t \sum_{k=0}^{\infty} p_{0k}(0, \tau) f_T(t - \tau) p_\Theta(x - k) d\tau
\tag{6.71}
$$

This equation is usually referred to as the *master equation* of the Markov jump process. In the above equation, $\delta(x)$ is the Dirac delta function and $R(t)$ is called the *survival probability* for state $X(t) = 0$; it is the probability that the waiting time when the process is in state 0 is greater than t and is given by

$$
R(t) = P[T_0 > t] = 1 - F_{T_0}(t) = e^{-\lambda t}
\tag{6.72}
$$

Equation (6.71) states that the probability that $X(t) = x$, given that the process was initially at state 0, is equal to the probability that the process was in state $X(t) = 0$ up to time t plus the probability that the process was at some state k at time τ, where $0 < \tau \le t$, and within the waiting time $t - \tau$ a jump of size $x - k$ took place.

We define joint $z - s$ transform of $p_{0x}(0, t)$ as follows:

$$
\tilde{P}(z, s) = \int_0^{\infty} e^{-st} \left\{ \sum_{x=0}^{\infty} z^x p_{0x}(0, t) \right\} dt
\tag{6.73}
$$

Thus, the master equation becomes transformed into the following:

$$\tilde{P}(z,s) = \int_0^\infty e^{-st} \sum_{x=0}^\infty z^x \left\{ \delta(x)R(t) + \int_0^t \sum_{k=0}^\infty p_{0k}(0,\tau)f_T(t-\tau)p_\Theta(x-k)d\tau \right\}dt$$
$$= R(s) + \tilde{P}(z,s)G_\Theta(z)M_T(s)$$

where $R(s)$ is the s-transform of $R(t)$ and $M_T(s)$ is the s-transform of $f_T(t)$. From this, we obtain

$$\tilde{P}(z,s) = \frac{R(s)}{1 - G_\Theta(z)M_T(s)} \tag{6.74}$$

Because

$$R(s) = \frac{1 - M_T(s)}{s}$$

we have that

$$\tilde{P}(z,s) = \frac{R(s)}{1 - G_\Theta(z)M_T(s)} = \frac{1 - M_T(s)}{s[1 - G_\Theta(z)M_T(s)]} \tag{6.75}$$

Similarly, because T is exponentially distributed with a mean of $1/\lambda$, we have that $M_T(s) = \lambda/(s+\lambda)$ and

$$\tilde{P}(z,s) = \frac{1}{s + \lambda - \lambda G_\Theta(z)} \tag{6.76}$$

From this, we obtain the inverse s-transform and the PMF, respectively, as follows:

$$P(z,t) = e^{-\lambda z[1-G_\Theta(z)]} \tag{6.77a}$$

$$p_{0x}(0,t) = \frac{1}{x!}\left[\frac{d^n}{dz^n}P(z,t)\right]_{z=0} \tag{6.77b}$$

This solution assumes that $X(t)$ takes only nonnegative values. When negative values are allowed, the z-transform must be replaced by the characteristic function. For the special case, when Θ is a Poisson random variable, $G_\Theta(z) = e^{-\lambda(1-z)}$, we obtain

$$\tilde{P}(z,s) = \frac{1}{s + \lambda - \lambda G_\Theta(z)} = \frac{1}{s + \lambda - \lambda e^{-\lambda(1-z)}} \tag{6.78}$$

An alternative method of deriving $p_{0x}(0, t)$ is from our earlier observation that $G_{X(t)}(z) = G_{N(t)}(G_\Theta(z))$. We know from earlier discussion that the PMF of $N(t)$ is given by

$$
\begin{aligned}
p_{N(t)}(n, t) &= F_{T_n}(t) - F_{T_{n+1}}(t) \\
&= \left\{ 1 - \sum_{k=0}^{n-1} \frac{(\lambda t)^k e^{-\lambda t}}{k!} \right\} - \left\{ 1 - \sum_{k=0}^{n} \frac{(\lambda t)^k e^{-\lambda t}}{k!} \right\} \\
&= \frac{(\lambda t)^n e^{-\lambda t}}{n!}
\end{aligned}
\tag{6.79}
$$

Thus, the z-transform of $p_{N(t)}(n, t)$ is given by

$$
G_{N(t)}(z) = e^{-\lambda t(1-z)}
$$

This gives

$$
G_{X(t)}(z) \equiv P(z, t) = G_{N(t)}(G_\Theta(z)) = e^{-\lambda(1-G_\Theta(z))}
$$

which is the same result that we obtained earlier, and from this we can obtain $p_{0x}(0, t)$. The conditional expected value $E[X(t)|X(0) = 0]$ is given by

$$
E[X(t)|X(0) = 0] = \left[\frac{dG_{X(t)}(z)}{dz} \right]_{z=1}
$$

Similarly, the conditional variance $\sigma^2_{X(t)|X(0)}$ is given by

$$
\sigma^2_{X(t)|X(0)} = \left[\frac{d^2 G_{X(t)}(z)}{dz^2} + \frac{dG_{X(t)}(z)}{dz} - \left\{ \frac{dG_{X(t)}(z)}{dz} \right\}^2 \right]_{z=1}
$$

We can generalize the result for the case when $X(0) = x$ by defining the probability transition function $p_{xy}(0, t)$ as follows:

$$
\begin{aligned}
p_{xy}(0, t) &= P[X(t) = y | X(0) = x] \\
&= \delta(y - x)R(t) + \int_0^t \sum_{k=0}^{\infty} p_{0k}(0, \tau) f_T(t - \tau) p_\Theta(y - x - k) d\tau
\end{aligned}
\tag{6.80}
$$

Using the same technique we used for the case when $X(0) = 0$, we can obtain the $z - s$ transform of the probability transition function.

6.9 Problems

6.1 Consider a machine that is subject to failure and repair. The time to repair the machine when it breaks down is exponentially distributed with mean $1/\mu$. The time the machine runs before breaking down is also exponentially distributed with mean $1/\lambda$. When repaired, the machine is considered to be as good as new. The repair time and the running time are assumed to be independent. If the machine is in good condition at time 0, what is the expected number of failures up to time t?

6.2 The Merrimack Airlines company runs a commuter air service between Manchester, NH, and Cape Cod, MA. Because the company is a small one, there is no set schedule for their flights, and no reservation is needed for the flights. However, it has been determined that their planes arrive at the Manchester airport according to a Poisson process with an average rate of two planes per hour. Gail arrived at the Manchester airport and had to wait to catch the next flight.

    a. What is the mean time between the instant Gail arrived at the airport until the time the next plane arrived?

    b. What is the mean time between the arrival time of the last plane that took off from the Manchester airport before Gail arrived and the arrival time of the plane that she boarded?

6.3 Victor is a student who is conducting experiments with a series of light bulbs. He started with 10 identical light bulbs, each of which has an exponentially distributed lifetime with a mean of 200 h. Victor wants to know how long it will take until the last bulb burns out (or fails). At noontime, he stepped out to get some lunch with six bulbs still on. Assume that he came back and found that none of the six bulbs has failed.

    a. After Victor came back, what is the expected time until the next bulb failure?

    b. What is the expected length of time between the fourth bulb failure and the fifth bulb failure?

6.4 A machine has three components labeled 1, 2, and 3, whose times between failure are exponentially distributed with mean $1/\lambda_1, 1/\lambda_2$, and $1/\lambda_3$, respectively. The machine needs all three components to work, thus when a component fails the machine is shut down until the component is repaired and the machine is brought up again. When repaired, a component is considered to be as good as new. The time to repair component 1 when it fails is exponentially distributed with mean $1/\mu_1$. The time to repair component 2 when it fails is constant at $1/\mu_2$, and the time to repair component 3 when it fails is a third-order Erlang random variable with parameter μ_3.

    a. What fraction of time is the machine working?

    b. What fraction of time is component 2 being repaired?

    c. What fraction of time is component 3 idle but has not failed?

    d. Given that Bob arrived when component 1 was being repaired, what is the expected time until the machine is operational again?

6.5 A high school student has two favorite brands of bag pack labeled X and Y. She continuously chooses between these brands in the following manner. Given that she currently has brand X, the probability that she will buy brand X again is 0.8, and the probability that she will buy brand Y next is 0.2. Similarly, given that she currently has brand Y, the probability that she will buy brand X next is 0.3, and the probability that she will buy brand Y again is 0.7. With respect to the time between purchases, if she currently has a brand X bag pack, the time until the next purchase is exponentially distributed with a mean of 6 months. Similarly, given that she currently has a brand Y bag pack, the time until the next purchase is exponentially distributed with a mean of 8 months.

a. What is the long-run probability that she has a brand Y bag pack?

b. Given that she has just purchased brand X, what is the probability that t months later her last purchase was brand Y?

6.6 Larry is a student who does not seem to make up his mind whether to live in the city or in the suburb. Every time he lives in the city, he moves to the suburb after one semester. Half of the time he lives in the suburb, he moves back to the city after one semester. The other half of Larry's suburban living results in his moving to a new apartment in the suburb where he lives for a time that is geometrically distributed with a mean of two semesters.

a. Model Larry's living style by a two-state discrete-time SMP giving the state-transition diagram.

b. Assume that Larry lives in the city at the beginning of the current semester. What is the probability that he will be living in the suburb k semesters from now?

c. What is the probability that the total duration of any uninterrupted stay in the suburb is k semesters? What are the mean and variance of the duration of one such stay in the suburb?

6.7 Consider a Markov renewal process with the semi-Markov kernel Q given by

$$Q = \begin{bmatrix} 0.6(1 - e^{-5t}) & 0.4(1 - e^{-2t}) \\ 0.5 - 0.2\,e^{-3t} - 0.3\,e^{-5t} & 0.5 - 0.5\,e^{-2t} - t\,e^{-2t} \end{bmatrix}$$

a. Determine the state-transition probability matrix P for the Markov chain $\{X_n\}$.

b. Determine the conditional distributions $G_{ij}(t)$ of the waiting time in state i given that the next state is j for all i, j.

6.8 A machine can be in one of three states: good, fair, and broken. When it is in a good condition, it will remain in this state for a time that is exponentially distributed with mean $1/\mu_1$ before going to the fair state with probability 4/5 and to the broken state with probability 1/5. When it is in a fair condition, it will remain in this state for a time that is exponentially distributed with mean $1/\mu_2$ before going to the broken state. When it is in the broken state, it will take a time that is exponentially distributed with mean $1/\mu_3$ to be repaired before going to the good state with probability 3/4 and to the fair state with probability 1/4. What is the fraction of time that the machine is in each state?

6.9 Customers arrive at a taxi depot according to a Poisson process with rate λ. The dispatcher sends for a taxi where there are N customers waiting at the station. It takes M units of time for a taxi to arrive at the depot. When it arrives, the taxi picks up all waiting customers. The taxi company incurs a cost at a rate of nk per unit time whenever n customers are waiting. What is the steady-state average cost that the company incurs?

6.10 A component is replaced every T time units and upon its failure, whichever comes first. The lifetimes of successive components are independent and identically distributed random variables with PDF $f_X(x)$. A cost $c_1 > 0$ is incurred for each planned replacement, and a fixed cost $c_2 > c_1$ is incurred for each failure replacement. What is the long-run average cost per unit time?

6.11 In her retirement days, a mother of three grownup children splits her time living with her three children who live in three different states. It has been found that her choice of where to spend her time next can be modeled by a Markov chain. Thus, if the children are labeled by ages as child 1, child 2, and child 3, the transition probabilities are as follows. Given that she is currently staying with child 1, the probability that she will stay with child 1 next is 0.3, the probability that she will stay with child 2 next is 0.2, and the probability that she will stay with child 3 next is 0.5. Similarly, given that she is currently staying with child 2, the probability that she will stay with child 1 next is 0.1, the probability that she will stay with child 2 next is 0.8, and the probability that she will stay with child 3 next is 0.1. Finally, given that she is currently staying with child 3, the probability that she will stay with child 1 next is 0.4, the probability that she will stay with child 2 next is 0.4, and the probability that she will stay with child 3 next is 0.2. The length of time that she spends with child 1 is geometrically distributed with mean 2 months, the length of time she spends with child 2 is geometrically distributed with mean 3 months, and the time she spends with child 3 is geometrically distributed with mean 1 month.

a. Obtain the transition probability functions of the process; that is, obtain the set of probabilities $\{\phi_{ij}(n)\}$, where $\phi_{ij}(n)$ is the probability that the process is in state j at time n given that it entered state i at time 0.

b. What is the occupancy distribution of the process?

6.12 Consider Problem 6.11. Assume that the time she spends with each of her children is exponentially distributed with the same means as specified. Obtain the transition probability functions $\{\phi_{ij}(t)\}$ of the process.

7 Markovian Queueing Systems

7.1 Introduction

A queue is a waiting line. Queues arise in many of our daily activities. For example, we join a queue to buy stamps at the post office, to cash checks or deposit money at the bank, to pay for groceries at the grocery store, to purchase tickets for movies or games, or to get a table at the restaurant. This chapter discusses a class of queueing systems called Markovian queueing systems. They are characterized by the fact that either the service times are exponentially distributed or customers arrive at the system according to a Poisson process, or both. The emphasis in this chapter is on the steady-state analysis.

7.2 Description of a Queueing System

In a queueing system, *customers* from a specified *population* arrive at a *service facility* to receive service. The service facility has one or more *servers* who attend to arriving customers. If a customer arrives at the facility when all the servers are busy attending to earlier customers, the arriving customer joins the queue until a server is free. After a customer has been served, he or she leaves the system and will not join the queue again. That is, service with feedback is not allowed. We consider systems that obey the *work conservation rule*: A server cannot be idle when there are customers to be served. Figure 7.1 illustrates the different components of a queueing system with three servers.

When a customer arrives at a service facility, a server commences service on the customer if the server is currently idle. Otherwise, the customer joins a queue that is attended to in accordance with a specified service policy such as first-come first-served (FCFS), last-come first-served (LCFS), and priority. Thus the time a customer spends waiting for service to begin is dependent on the service policy. Also, because there can be one or more servers in the facility, more than one customer can be receiving service at the same time. The following notation is used to represent the random variables associated with a queueing system:

a. N_q is the number of customers in queue.
b. N_s is the number of customers receiving service.
c. N is the total number of customers in the system: $N = N_q + N_s$.
d. W is the time a customer spends in queue before going to service, i.e., the waiting time.

Markov Processes for Stochastic Modeling. DOI: http://dx.doi.org/10.1016/B978-0-12-407795-9.00007-4

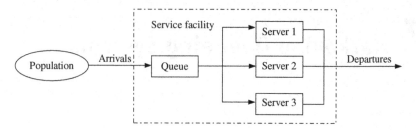

Figure 7.1 Components of a queueing system.

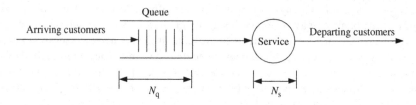

Figure 7.2 The queueing process.

e. X is the time a customer spends in actual service.
f. T is the total time a customer spends in the system (also called the sojourn time): $T = W + X$.

Figure 7.2 is a summary of the queueing process at the service facility.
A queueing system is characterized as follows:

1. *Population*, which is the source of the customers arriving at the service facility. The population can be finite or infinite.
2. *Arriving pattern*, which defines the customer interarrival process.
3. *Service time distribution*, which defines the time taken to serve each customer.
4. *Capacity of the queueing facility*, which can be finite or infinite. If the capacity is finite, customers that arrive when the system is full are lost (or blocked). Thus, a finite-capacity system is a blocking system.
5. *Number of servers*, which can be one or more than one. A queueing system with one server is called a single-server system, otherwise it is called a multiserver system. A single-server system can serve only one customer at a time, while multiserver systems can serve multiple customers simultaneously. In a multiserver system, the servers can be identical, which means that their service rates are identical and it does not matter which server a particular customer receives service from. On the other hand, the servers can be heterogeneous in the sense that some provide faster service than others. In this case, the time a customer spends in service depends on which server provides the service. A special case of a multiserver system is the infinite-server system where each arriving customer is served immediately, that is, there is no waiting in queue.
6. *Queueing discipline*, which is also called the *service discipline*. It defines the rule that governs how the next customer to receive service is selected after a customer who is

currently receiving service leaves the system. Specific disciplines that can be used include the following:

a. FCFS, which means that customers are served in the order they arrived. The discipline is also called first-in first-out (FIFO).

b. LCFS, which means that the last customer to arrive receives service before those who arrived earlier. The discipline is also called last-in first-out (LIFO).

c. Service in random order (SIRO), which means that the next customer to receive service after the current customer has finished receiving service will be selected in a probabilistic manner, such as tossing a coin and rolling a die.

d. Priority, which means that customers are divided into ordered classes such that a customer in a higher class will receive service before a customer in a lower class, even if the higher-class customer arrives later than the lower-class customer. There are two types of priority: *preemptive* and *nonpreemptive*. In preemptive priority, the service of a customer currently receiving service is suspended upon the arrival of a higher-priority customer; the latter goes straight to receive service. The preempted customer goes in to receive service upon the completion of service of the higher-priority customer, if no higher-priority customer arrived while the high-priority customer was being served. How the service of a preempted customer is continued when the customer goes to complete his or her service depends on whether we have a *preemptive repeat* or *preemptive resume* policy. In preemptive repeat, the customer's service is started from the beginning when the customer enters to receive service again, regardless of how many times the customer is preempted. In preemptive resume, the customer's service continues from where he or she stopped before being preempted. Under nonpreemptive priority, an arriving high-priority customer goes to the head of the queue and waits for the current customer's service to be completed before he or she enters to receive service ahead of other waiting lower-priority customers.

Thus, the time a customer spends in the system is a function of the preceding parameters and service policies.

7.3 The Kendall Notation

The Kendall notation is a shorthand notation that is used to describe queueing systems. It is written in the form:

$$A/B/c/D/E/F$$

- "A" describes the arrival process (or the interarrival time distribution), which can be an exponential or nonexponential (i.e., general) distribution.
- "B" describes the service time distribution.
- "c" describes the number of servers at the queueing station.
- "D" describes the system capacity, which is the maximum number of customers allowed in the system including those currently receiving service; the default value is infinity.
- "E" describes the size of the population from where arrivals are drawn; the default value is infinity.
- "F" describes the queueing (or service) discipline. The default is FCFS.

- When default values of D, E, and F are used, we use the notation A/B/c, which means a queueing system with infinite capacity, customers arrive from an infinite population, and service in an FCFS manner. Symbols traditionally used for A and B are:
  - GI, which stands for general independent interarrival time.
  - G, which stands for general service time distribution.
  - M, which stands for memoryless (or exponential) interarrival time or service time distribution. Note that an exponentially distributed interarrival time means that customers arrive according to a Poisson process.
  - D, which stands for deterministic (or constant) interarrival time or service time distribution.

For example, we can have queueing systems of the following form:

- M/M/1 queue, which is a queueing system with exponentially distributed interarrival time, exponentially distributed service time, a single server, infinite capacity, customers are drawn from an infinite population, and service is on an FCFS basis.
- M/D/1 queue, which is a queueing system with exponentially distributed interarrival time, constant service time, a single server, infinite capacity, customers are drawn from an infinite population, and service is on an FCFS basis.
- M/G/3/20 queue, which is a queueing system with exponentially distributed interarrival time, general (or nonexponentially) distributed service time, three servers, a finite capacity of 20 (i.e., a maximum of 20 customers can be in the system, including the three that can be in service at the same time), customers are drawn from an infinite population, and service is on an FCFS basis.

7.4 The Little's Formula

The Little's formula is a statement on the relationship between the mean number of customers in the system, the mean time spent in the system, and the average rate at which customers arrive at the system. Let λ denote the mean arrival rate, $E[N]$ the mean number of customers in the system, $E[T]$ the mean total time spent in the system, $E[N_q]$ the mean number of customers in queue, and $E[W]$ the mean waiting time. Then Little's formula (Little, 1961) states that,

$$E[N] = \lambda E[T] \tag{7.1}$$

which says that the mean number of customers in the system (including those currently being served) is the product of the average arrival rate and the mean time a customer spends in the system. The formula can also be stated in terms of the number of customers in queue, as follows:

$$E[N_q] = \lambda E[W] \tag{7.2}$$

which says that the mean number of customers in queue (or waiting to be served) is equal to the product of the average arrival rate and the mean waiting time.

7.5   The PASTA Property

Markovian queueing systems with Poisson arrivals possess the PASTA (Poisson Arrivals See Time Averages) property. This property, which was proposed by Wolff (1982), asserts that customers with Poisson arrivals see the system as if they arrived at an arbitrary point in time despite the fact that they induce transitions in the system. This phenomenon arises from the lack of memory of exponential interarrival times with the result that the arrival history just before a tagged arrival instant is stochastically identical to that of a random instant as well as that of the arrival instant. PASTA is a powerful tool used in the analysis of many queueing systems.

7.6   The M/M/1 Queueing System

This is the simplest queueing system in which customers arrive according to a Poisson process to a single-server service facility, and the time to serve each customer is exponentially distributed. The model also assumes the various default values: infinite capacity at the facility, customers are drawn from an infinite population, and service is on an FCFS basis.

Because we are dealing with a system that can increase or decrease by at most one customer at a time, it is a birth-and-death process with homogeneous birth rate λ and homogeneous death rate μ. This means that the mean service time is $1/\mu$. Thus, the state transition rate diagram is shown in Figure 7.3.

Let p_n be the limiting-state probability that the process is in state n, $n = 0, 1, 2,\ldots$. Then applying the balance equations, we obtain:

$$\lambda p_0 = \mu p_1 \Rightarrow p_1 = \left(\frac{\lambda}{\mu}\right) p_0 = \rho p_0$$

$$(\lambda + \mu)p_1 = \lambda p_0 + \mu p_2 \Rightarrow p_2 = \rho p_1 = \rho^2 p_0$$

$$(\lambda + \mu)p_2 = \lambda p_2 + \mu p_3 \Rightarrow p_3 = \rho p_2 = \rho^3 p_0$$

where $\rho = \lambda/\mu$. Similarly, it can be shown that

$$p_n = \rho^n p_0 \quad n = 0, 1, 2, \ldots$$

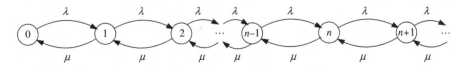

Figure 7.3 State transition rate diagram for M/M/1 queue.

Because

$$\sum_{n=0}^{\infty} p_n = 1 = p_0 \sum_{n=0}^{\infty} \rho^n = \frac{p_0}{1-\rho} \quad \text{where} \quad \rho < 1$$

we obtain

$$
\begin{aligned}
p_0 &= 1 - \rho \\
p_n &= (1-\rho)\rho^n \quad n = 0, 1, 2, \ldots; \rho < 1
\end{aligned}
\tag{7.3}
$$

Because $p_0 = 1 - \rho$, which is the probability that the system is not empty and hence the server is not idle, we call ρ the *server utilization* (or *utilization factor*).

The expected number of customers in the system is given by

$$E[N] = \sum_{n=0}^{\infty} np_n = \sum_{n=0}^{\infty} n(1-\rho)\rho^n = (1-\rho)\sum_{n=0}^{\infty} n\rho^n$$

But

$$\frac{d}{d\rho}\sum_{n=0}^{\infty} \rho^n = \sum_{n=0}^{\infty} \frac{d}{d\rho}\rho^n = \sum_{n=1}^{\infty} n\rho^{n-1} = \frac{1}{\rho}\sum_{n=0}^{\infty} n\rho^n$$

Thus,

$$\sum_{n=0}^{\infty} n\rho^n = \rho\frac{d}{d\rho}\sum_{n=0}^{\infty} \rho^n = \rho\frac{d}{d\rho}\left(\frac{1}{1-\rho}\right) = \frac{\rho}{(1-\rho)^2}$$

Therefore,

$$E[N] = (1-\rho)\sum_{n=0}^{\infty} n\rho^n = (1-\rho)\left\{\frac{\rho}{(1-\rho)^2}\right\} = \frac{\rho}{1-\rho}
\tag{7.4}$$

We can obtain the mean time in the system from Little's formula as follows:

$$E[T] = \frac{E[N]}{\lambda} = \frac{\lambda/\mu}{\lambda(1-\rho)} = \frac{1}{\mu(1-\rho)} = \frac{E[X]}{1-\rho}
\tag{7.5}$$

where the last equality follows from the fact that the mean service time is $E[X] = 1/\mu$. Similarly, the mean waiting time and mean number of customers in queue are given by

$$E[W] = E[T] - E[X] = \frac{E[X]}{1-\rho} - E[X] = \frac{\rho E[X]}{1-\rho} = \frac{\rho}{\mu(1-\rho)}
\tag{7.6}$$

$$E[N_q] = \lambda E[W] = \frac{\lambda\rho}{\mu(1-\rho)} = \frac{\rho^2}{1-\rho}
\tag{7.7}$$

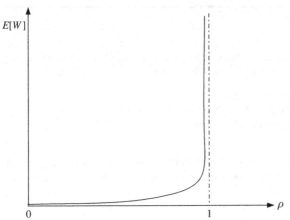

Figure 7.4 Mean waiting time versus server utilization.

Recall that the mean number of customers in service is $E[N_s]$. Using the above results, we obtain

$$E[N_s] = E[N] - E[N_q] = \frac{\rho}{1-\rho} - \frac{\rho^2}{1-\rho} = \rho \tag{7.8}$$

Thus, the mean number of customers in service is ρ, the probability that the server is busy. Note that the mean waiting time, the mean time in the system, the mean number of customers in the system, and the mean number of customers in queue become extremely large as the server utilization ρ approaches 1. Figure 7.4 illustrates this for the case of the expected waiting time.

Example 7.1

Students arrive at the campus post office according to a Poisson process at an average rate of one student every 4 min. The time required to serve each student is exponentially distributed with a mean of 3 min. There is only one postal worker at the counter, and any arriving student who finds the worker busy joins a queue that is served in an FCFS manner.

a. What is the probability that an arriving student has to wait?
b. What is the mean waiting time of an arbitrary student?
c. What is the mean number of waiting students at the post office?

Solution
Example 7.1 is an M/M/1 queue with the following parameters:

$$\lambda = 1/4$$
$$\mu = 1/3 = 1/E[X]$$
$$\rho = \lambda/\mu = 3/4 = 0.75$$

a. P[arriving student waits] = P[server is busy] = $\rho = 0.75$
b. $E[W] = \rho E[X]/(1-\rho) = (0.75)(3)/(0.25) = 9$ min
c. $E[N_q] = \lambda E[W] = (1/4)(9) = 2.25$ students.

Example 7.2

Customers arrive at a checkout counter in a grocery store according to a Poisson process at an average rate of 10 customers per hour. There is only one clerk at the counter, and the time to serve each customer is exponentially distributed with a mean of 4 min.

a. What is the probability that a queue forms at the counter?
b. What is the average time that a customer spends at the counter?
c. What is the average queue length at the counter?

Solution

This is an M/M/1 queueing system. We must first convert the arrival and service rates to the same unit of customers per minute because service time is in minutes. Thus, the parameters of the system are as follows:

$$\lambda = 10/60 = 1/6$$
$$\mu = 1/4 = 1/E[X]$$
$$\rho = \lambda/\mu = 2/3$$

a. P[queue forms] = P[server is busy] = $\rho = 2/3$
b. Average time at the counter = $E[T] = E[X]/(1 - \rho) = 4(1/3) = 12$ min
c. Average queue length = $E[N_q] = \rho^2/(1 - \rho) = (4/9)/(1/3) = 4/3$

7.6.1 Stochastic Balance

A shortcut method of obtaining the steady-state equations in an M/M/1 queueing system is by means of a flow balance procedure called *stochastic balance*. The idea behind stochastic balance is that in any steady-state condition, the rate at which the process moves from left to right across a "probabilistic wall" is equal to the rate at which it moves from right to left across that wall. For example, Figure 7.5 shows three states $n - 1$, n, and $n + 1$ in the state transition rate diagram of an M/M/1 queue.

The rate at which the process crosses wall A from left to right is λp_{n-1}, which is true because this can only happen when the process is in state $n - 1$. Similarly, the rate at which the process crosses wall A from right to left is μp_n. By stochastic

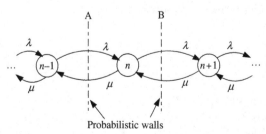

Figure 7.5 Stochastic balance concept.

Probabilistic walls

balance we mean that $\lambda p_{n-1} = \mu p_n$ or $p_n = (\lambda/\mu)p_{n-1} = \rho p_{n-1}$, which is the result we obtained earlier in the analysis of the system.

7.6.2 Total Time and Waiting Time Distributions of the M/M/1 Queueing System

Consider a tagged customer, say customer k, that arrives at the system and finds it in state n. If $n > 0$, then because of the fact that the service time X is exponentially distributed, the service time of the customer receiving service when the tagged customer arrives "starts from scratch." Thus, the total time that the tagged customer spends in the system is given by

$$T_k = \begin{cases} X_k + X_1 + X_2 + \cdots + X_n & n > 0 \\ X_k & n = 0 \end{cases}$$

Because the X_k are identically distributed, the s-transform of the probability density function (PDF) of T_k is given by

$$M_{T|n}(s|n) = [M_X(s)]^{n+1} \quad n = 0, 1, 2, \ldots$$

Thus, the s-transform of the PDF of the unconditional total time in the system is given by

$$M_T(s) = \sum_{n=0}^{\infty} M_{Tn}(s|n)p_n = \sum_{n=0}^{\infty} [M_X(s)]^{n+1}p_n = M_X(s)\sum_{n=0}^{\infty} [M_X(s)]^n(1 - \rho)\rho^n$$

$$= (1 - \rho)M_X(s)\sum_{n=0}^{\infty} [\rho M_X(s)]^n = \frac{(1 - \rho)M_X(s)}{1 - \rho M_X(s)}$$

Because $M_X(s) = \mu/(s + \mu)$, we obtain the following result:

$$M_T(s) = \frac{\mu(1 - \rho)}{1 + \mu(1 - \rho)} \tag{7.9}$$

This shows that T is an exponentially distributed random variable with PDF and cumulative distribution function (CDF) given respectively by

$$f_T(t) = \mu(1 - \rho)e^{-\mu(1-\rho)t} \tag{7.10a}$$

$$F_T(t) = 1 - e^{-\mu(1-\rho)t} \tag{7.10b}$$

and mean $1/\mu(1-\rho)$, as shown earlier. Similarly, the waiting time distribution can be obtained by considering the experience of the tagged customer in the system, as follows:

$$W_k = \begin{cases} X_1 + X_2 + \cdots + X_n & n>0 \\ 0 & n=0 \end{cases}$$

Thus, the s-transform of the PDF of W_k, given that there are n customers when the tagged customer arrives, is given by

$$M_{W|n}(s|n) = \begin{cases} [M_X(s)]^n & n=1,2,\ldots \\ 1 & n=0 \end{cases}$$
$$= [M_X(s)]^n \qquad n=0,1,2,\ldots$$

Therefore,

$$M_W(s) = \sum_{n=0}^{\infty} M_{W|n}(s|n)p_n = \sum_{n=0}^{\infty}[M_X(s)]^n p_n = \sum_{n=0}^{\infty}[M_X(s)]^n(1-\rho)\rho^n$$
$$= (1-\rho)\sum_{n=0}^{\infty}[\rho M_X(s)]^n = \frac{1-\rho}{1-\rho M_X(s)} = \frac{(1-\rho)(s+\mu)}{s+\mu(1-\rho)}$$
$$= \frac{(1-\rho)\{s+\mu(1-\rho)\} + \rho\mu(1-\rho)}{s+\mu(1-\rho)}$$
$$= (1-\rho) + \frac{\rho\mu(1-\rho)}{s+\mu(1-\rho)}$$

(7.11)

Thus, the PDF and CDF of W are given by

$$f_W(w) = (1-\rho)\delta(w) + \rho\mu(1-\rho)e^{-\mu(1-\rho)w}$$
$$= (1-\rho)\delta(w) + \lambda(1-\rho)e^{-\mu(1-\rho)w}$$

(7.12a)

$$F_W(w) = P[W \le w] = (1-\rho) + \rho[1 - e^{-\mu(1-\rho)w}]$$
$$= 1 - \rho e^{-\mu(1-\rho)w}$$

(7.12b)

where $\delta(w)$ is the Kronecker delta. Observe that the mean waiting time is given by

$$E[W] = \int_0^{\infty} \{1 - F_W(w)\}dw = \int_0^{\infty} \rho e^{-\mu(1-\rho)w}\,dw = \frac{\rho}{\mu(1-\rho)}$$

which is the result we obtained earlier. An alternative method of obtaining the preceding results is as follows. Let $r_{n+1}(t)$ denote the probability that $n+1$ service completions are made in a time of less than or equal to t, given that a tagged customer found n customers in the system when it arrived. Then

$$F_T(t) = P[T \le t] = \sum_{n=0}^{\infty} p_n r_{n+1}(t) = \sum_{n=0}^{\infty} p_n \int_0^t \frac{\mu^{n+1} x^n}{n!} e^{-\mu x} \, dx$$

$$= \sum_{n=0}^{\infty} (1-\rho)\rho^n \int_0^t \frac{\mu^{n+1} x^n}{n!} e^{-\mu x} \, dx = (1-\rho) \int_0^t \left\{ \sum_{n=0}^{\infty} \rho^n \frac{\mu^{n+1} x^n}{n!} \right\} e^{-\mu x} \, dx$$

$$= \mu(1-\rho) \int_0^t \left\{ \sum_{n=0}^{\infty} \frac{(\rho \mu x)^n}{n!} \right\} e^{-\mu x} \, dx = \mu(1-\rho) \int_0^t e^{\rho \mu x} e^{-\mu x} \, dx$$

$$= \mu(1-\rho) \int_0^t e^{-\mu(1-\rho)x} \, dx$$

$$= 1 - e^{-\mu(1-\rho)t}$$

$$(7.13)$$

Similarly,

$$F_W(w) = P[W \le w] = \sum_{n=1}^{\infty} p_n r_n(t) + p_0 = \sum_{n=1}^{\infty} p_n \int_0^w \frac{\mu^n x^{n-1}}{(n-1)!} e^{-\mu x} \, dx + p_0$$

$$= \sum_{n=1}^{\infty} (1-\rho)\rho^n \int_0^w \frac{\mu^n x^{n-1}}{(n-1)!} e^{-\mu x} \, dx + (1-\rho)$$

$$= \mu\rho(1-\rho) \int_0^w \left\{ \sum_{n=1}^{\infty} \rho^{n-1} \frac{\mu^{n-1} x^{n-1}}{(n-1)!} \right\} e^{-\mu x} \, dx + (1-\rho)$$

$$= \mu\rho(1-\rho) \int_0^w \left\{ \sum_{n=0}^{\infty} \frac{(\rho \mu x)^n}{n!} \right\} e^{-\mu x} \, dx + (1-\rho)$$

$$= \mu\rho(1-\rho) \int_0^w e^{\rho \mu x} e^{-\mu x} \, dx + (1-\rho)$$

$$= \mu\rho(1-\rho) \int_0^w e^{-\mu(1-\rho)x} \, dx + (1-\rho) = \rho(1 - e^{-\mu(1-\rho)w}) + 1 - \rho$$

$$= 1 - \rho e^{-\mu(1-\rho)w}$$

$$(7.14)$$

7.7 Examples of Other M/M Queueing Systems

The goal of this section is to describe some relatives of the M/M/1 queueing systems without rigorously analyzing them as we did for the M/M/1 queueing system. These systems can be used to model different human behaviors and they include:

a. *Blocking* from entering a queue, which is caused by the fact that the system has a finite capacity and a customer that arrives when the system is full is rejected.

b. *Defections* from a queue, which can be caused by the fact that a customer has spent too much time in queue and leaves out of frustration without receiving service. Defection is also called *reneging* in queueing theory.

c. *Jockeying* for position among many queues, which can arise when in a multiserver system, each server has its own queue and some customer in one queue notices that another queue is being served faster than their own, thus he or she leaves his or her queue and moves into the supposedly faster queue.

d. *Balking* before entering a queue, which can arise if the customer perceives the queue to be too long and chooses not to join it at all.

e. *Bribing* for queue position, which is a form of dynamic priority because a customer pays some "bribe" to improve his or her position in the queue. Usually the more bribe the customer pays, the better position he or she gets.

f. *Cheating* for queue position, which is different from bribing because in cheating, the customer uses trickery rather than his or her personal resources to improve his or her position in the queue.

g. *Bulk service*, which can be used to model table assignment in restaurants. For example, a queue at a restaurant might appear to be too long, but in actual fact when a table is available (i.e., a server is ready for the next customer), it can be assigned to a family of four, which is identical to serve the four people in queue together.

h. *Batch arrival*, which can be used to model how friends arrive in groups at a movie theater, concert show, or a ball game; or how families arrive at a restaurant. Thus, the number of customers in each arriving batch can be modeled by some probabilistic law.

From this list, it can be seen that queueing theory is a very powerful modeling tool that can be applied to all human activities and hence to all walks of life. In the following sections, we describe some of the different queueing models that are based on Poisson arrivals and/or exponentially distributed service times.

7.7.1 The M/M/c Queue: The c-Server System

In this scheme, customers arrive according to a Poisson process with rate λ, and there are c identical servers. When a customer arrives, he or she is randomly assigned to one of the idle servers until all servers are busy when a single queue is formed. Note that if a queue is allowed to form in front of each server, then we have an M/M/1 queue with modified arrival because customers join a server's queue in some probabilistic manner. In the single queue case, we assume that there is an infinite capacity and service is based on an FCFS policy.

The service rate in the system is dependent on the number of busy servers. If only one server is busy, the service rate is μ; if two servers are busy, the service rate is 2μ, and so on until all servers are busy when the service rate becomes $c\mu$. Thus, until all servers are busy, the system behaves like a heterogeneous queueing system in which the service rate in each state is different. When all servers are busy, it behaves like a homogeneous queueing system in which the service rate is the same in each state. The state transition rate diagram of the system is shown in Figure 7.6.

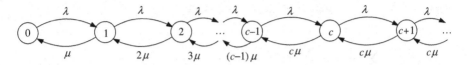

Figure 7.6 State transition rate diagram for the M/M/c queue.

Thus, the service rate is

$$\mu_n = \begin{cases} n\mu & 0 \le n < c \\ c\mu & n \ge c \end{cases} \tag{7.15}$$

Using stochastic balance equations, we obtain

$$\lambda p_{n-1} = \min(n,c)\mu p_n \quad n = 1, 2, \ldots \tag{7.16}$$

Iterating on this equation, we obtain the limiting-state probability of being in state n as

$$p_n = \begin{cases} \left(\dfrac{\lambda}{\mu}\right)^n \left(\dfrac{1}{n!}\right) p_0 & n = 0, 1, \ldots, c \\[2ex] \dfrac{(\lambda/\mu)^n}{c!c^{n-c}} p_0 & n \ge c \end{cases} \tag{7.17a}$$

$$\sum_{n=0}^{\infty} p_n = 1 \Rightarrow p_0 = \frac{1}{\displaystyle\sum_{n=0}^{c-1} \left(\frac{\lambda}{\mu}\right)^n \left(\frac{1}{n!}\right) + \left(\frac{\lambda}{\mu}\right)^c \left(\frac{1}{c!}\right)\left(\frac{c\mu}{c\mu - \lambda}\right)} \quad \lambda < c\mu \tag{7.17b}$$

Note that queues can only form when the process is in state c or any state higher than c. Thus, arriving customers who see the system in any state less than c do not have to wait. The probability that an arriving customer has to wait, which is usually referred to as the *delay probability*, is obtained using PASTA as

$$P_W = p_c + p_{c+1} + p_{c+2} + \cdots = \frac{p_c}{1 - \dfrac{\lambda}{c\mu}} = \frac{p_c}{1 - \rho}$$

$$= \frac{(c\rho)^c}{c!}\left\{(1-\rho)\sum_{n=0}^{c-1}\frac{(c\rho)^n}{n!} + \frac{(c\rho)^c}{c!}\right\}^{-1} \tag{7.18}$$

where $\rho = \lambda/c\mu$. The mean queue length is given by

$$E[N_q] = \sum_{n=c}^{\infty} np_n = p_c \sum_{n=0}^{\infty} n\rho^n = \frac{\rho p_c}{(1-\rho)^2} = \frac{\rho P_W}{1-\rho} \qquad (7.19)$$

From Little's formula, we obtain the mean waiting time as

$$E[W] = \frac{E[N_q]}{\lambda} = \frac{p_c}{c\mu(1-\rho)^2} = \frac{P_W}{c\mu(1-\rho)} \qquad (7.20)$$

We can also obtain the distribution of the waiting time as follows. Let $F_W(w) = P[W \le w]$ denote the CDF of the waiting time. When there are at least c customers in the system, the composite service rate is $c\mu$; thus, the interdeparture times are exponentially distributed with mean $1/c\mu$. Therefore, when there are $n \ge c$ customers in the system, the total service time of the n customers is Erlang of order $n - c + 1$, and the PDF of the total service time, S_n, is given by

$$f_{S_n}(t) = \frac{(c\mu)^{n-c+1} t^{n-c}}{(n-c)!} e^{-c\mu t} \quad t \ge 0; \ n \ge c$$

Now,

$$F_W(t) = F_W(0) + P[0 < W \le t]$$

where

$$F_W(0) = P[N_q = 0] = P[N < c] = p_0 \sum_{n=0}^{c-1} \left(\frac{\lambda}{\mu}\right)^n \left(\frac{1}{n!}\right)$$

Because

$$p_0 \left[\sum_{n=0}^{c-1} \left(\frac{\lambda}{\mu}\right)^n \left(\frac{1}{n!}\right) + \left(\frac{\lambda}{\mu}\right)^c \left(\frac{1}{c!}\right) \left(\frac{c\mu}{c\mu - \lambda}\right)\right] = 1$$

we have that

$$F_W(0) = p_0 \sum_{n=0}^{c-1} \left(\frac{\lambda}{\mu}\right)^n \left(\frac{1}{n!}\right) = 1 - p_0 \left(\frac{\lambda}{\mu}\right)^n \left(\frac{1}{n!}\right) \left(\frac{c\mu}{c\mu - \lambda}\right) = 1 - \frac{(c\mu)^c p_0}{(1-\rho)c!}$$

Similarly,

$$P[0 < W \le t] = \sum_{n=c}^{\infty} p_n \int_0^t f_{S_n}(u)\,du = \sum_{n=c}^{\infty} p_n \int_0^t \frac{(c\mu)^{n-c+1} u^{n-c}}{(n-c)!} e^{-c\mu u}\,du$$

$$= \int_0^t e^{-c\mu u} \left\{ \sum_{n=c}^{\infty} \frac{(\lambda/\mu)^n}{c! c^{n-c}} p_0 \frac{(c\mu)^{n-c+1} u^{n-c}}{(n-c)!} \right\} du$$

$$= \frac{p_0(\lambda/\mu)^c}{c!} \int_0^t c\mu\, e^{-c\mu u} \left\{ \sum_{n=c}^{\infty} \frac{(\lambda u)^{n-c}}{(n-c)!} \right\} du = \frac{p_0(c\rho)^c}{c!} \int_0^t c\mu\, e^{-c\mu u} \left\{ e^{\lambda u} \right\} du$$

$$= \frac{p_0(c\rho)^c}{c!} \int_0^t c\mu\, e^{-c\mu(1-\rho)u}\,du = \frac{p_0(c\rho)^c}{(1-\rho)c!} \left\{ 1 - e^{-c\mu(1-\rho)t} \right\}$$

Thus,

$$F_W(t) \quad = 1 - \frac{(c\mu)^c p_0}{(1-\rho)c!} + \frac{p_0(c\rho)^c}{(1-\rho)c!} \left\{ 1 - e^{-c\mu(1-\rho)t} \right\} = 1 - \frac{p_0(c\rho)^c}{(1-\rho)c!} e^{-c\mu(1-\rho)t}$$

$$= 1 - \frac{p_0(\lambda/\mu)^c}{(1-\rho)c!} e^{-c\mu(1-\rho)t} = 1 - \frac{p_c}{1-\rho} e^{-c\mu(1-\rho)t}$$

$$(7.21)$$

Note that we can obtain the expected waiting time from the preceding equation as follows:

$$E[W] = \int_0^{\infty} \{1 - F_W(t)\}\,dt = \frac{p_c}{1-\rho} \int_0^{\infty} e^{-c\mu(1-\rho)t}\,dt = \frac{p_c}{c\mu(1-\rho)^2}$$

which is the result we obtained earlier.

Example 7.3

Students arrive at a checkout counter in the college cafeteria according to a Poisson process with an average rate of 15 students per hour. There are two cashiers at the counter, and they provide identical service to students. The time to serve a student by either cashier is exponentially distributed with a mean of 3 min. Students that find both cashiers busy on their arrival join a single queue. What is the probability that an arriving student does not have to wait?

Solution

This is an M/M/2 queueing problem with the following parameters with λ and μ in students per minute:

$$\lambda = 15/60 = 1/4$$

$$\mu = 1/3$$

$$p_0 = \cfrac{1}{1 + \cfrac{\lambda}{\mu} + \left(\cfrac{\lambda}{\mu}\right)^2 \left[\cfrac{2\mu}{2(2\mu - \lambda)}\right]} = \frac{5}{11}$$

$$p_1 = \left(\frac{\lambda}{\mu}\right) p_0 = \left(\frac{3}{4}\right)\left(\frac{5}{11}\right) = \frac{15}{44}$$

An arriving student does not have to wait if he or she finds the system either empty or with only one server busy. The probability of this event is $p_0 + p_1 = 35/44$.

7.7.2   The M/M/1/K Queue: The Single-Server Finite-Capacity System

In this system, customers arrive according to a Poisson process with rate λ, there is a single server and there is room for only K customers including the customer receiving service. Thus, arriving customers that see the system in state K are lost. The state transition rate diagram is shown in Figure 7.7.

Using stochastic balance equations, we obtain the steady-state probability that the process is in state n as

$$p_n = \left(\frac{\lambda}{\mu}\right)^n p_0 \quad n = 0, 1, 2, \ldots, K \tag{7.22a}$$

$$\sum_{n=0}^{K} p_n = 1 = \frac{p_0(1 - \rho^{K+1})}{1 - \rho} \Rightarrow p_0 = \frac{1 - \rho}{1 - \rho^{K+1}} \tag{7.22b}$$

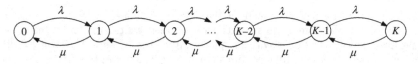

Figure 7.7 State transition rate diagram for the M/M/1/K queue.

where $\rho = \lambda/\mu < 1$. The mean number of customers in the system and the mean number in queue are given respectively by

$$E[N] = \sum_{n=0}^{K} np_n = p_0 \sum_{n=0}^{K} n\rho^n = \rho p_0 \frac{d}{d\rho}\left\{\frac{1 - \rho^{K+1}}{1 - \rho}\right\}$$

$$= \rho p_0 \left\{\frac{1 - \rho^{K+1} - (1 - \rho)(K + 1)\rho^K}{(1-\rho)^2}\right\} = \rho p_0 \left\{\frac{1 - (K + 1)\rho^K + K\rho^{K+1})}{(1-\rho)^2}\right\}$$

$$= \frac{\rho}{1 - \rho^{K+1}}\left\{\frac{1 - (K + 1)\rho^K + K\rho^{K+1})}{1 - \rho}\right\} = \frac{\rho}{1 - \rho} - \frac{(K + 1)\rho^{K+1}}{1 - \rho^{K+1}}$$

(7.23a)

$$E[N_q] = E[N] - (1 - p_0) = E[N] - \frac{\rho(1 - \rho^K)}{1 - \rho^{K+1}} = \frac{\rho}{1 - \rho} - \frac{\rho(1 + K\rho^K)}{1 - \rho^{K+1}} \quad (7.23b)$$

Using L'Hopital's rule, we obtain the values when $\rho = 1$ as follows:

$$\lim_{\rho \to 1} p_n = \frac{1}{K + 1} \quad n = 0, 1, \ldots, K \tag{7.24a}$$

$$\lim_{\rho \to 1} E[N] = \frac{K}{2} \tag{7.24b}$$

$$\lim_{\rho \to 1} E[N_q] = \frac{K(K - 1)}{2(K + 1)} \tag{7.24c}$$

Note that not all the traffic arriving at the system enters the system because customers are not allowed into the system when there are already K customers in the system. That is, customers are turned away with probability

$$p_K = \frac{(1 - \rho)\rho^K}{1 - \rho^{K+1}}$$

Thus, we define the *actual rate* at which customers arrive into the system, λ_A, as

$$\lambda_A = \lambda(1 - p_K) = \frac{\lambda(1 - \rho^K)}{1 - \rho^{K+1}} \tag{7.25}$$

We can then apply Little's formula to obtain

$$E[T] = \frac{E[N]}{\lambda_A} = \frac{1}{\mu}\left\{\frac{1}{1-\rho} - \frac{K\rho^K}{1-\rho^K}\right\} \tag{7.26a}$$

$$E[W] = \frac{E[N_q]}{\lambda_A} = \frac{1}{\mu}\left\{\frac{\rho}{1-\rho} - \frac{K\rho^K}{1-\rho^K}\right\} \tag{7.26b}$$

The CDF of the waiting time can be obtained by noting that when the system is in equilibrium, the probability that an arriving customer joins the queue is $1 - p_K$. Thus, the probability that an arriving customer who finds n customers in the system, where $n < K$, joins the queue is $p_n/(1 - p_K)$. Thus, the CDF of W is given by

$$F_W(t) = P[W \le t] = F_W(0) + P[0 < W \le t]$$

where

$$F_W(0) = \frac{p_0}{1 - p_K} = \left\{\frac{1 - \rho^{K+1}}{1 - \rho^K}\right\}\left\{\frac{1-\rho}{1-\rho^{K+1}}\right\} = \frac{1-\rho}{1-\rho^K}$$

When there are n customers in the system, the time to complete their service is an Erlang random variable of order n. Thus,

$$P[0 < W \le t] = \sum_{n=1}^{K-1} \frac{p_n}{1-p_K} \int_0^t \frac{\mu^n u^{n-1}}{(n-1)!} e^{-\mu u}\, du$$

$$= \sum_{n=1}^{K-1} \frac{p_n}{1-p_K}\left\{1 - \int_t^\infty \frac{\mu^n u^{n-1}}{(n-1)!} e^{-\mu u}\, du\right\}$$

Now,

$$\int_t^\infty \frac{\mu^n u^{n-1}}{(n-1)!} e^{-\mu u}\, du = \sum_{j=0}^{n-1} \frac{(\mu t)^j}{j!} e^{-\mu t}$$

Thus,

$$P[0 < W \le t] = \sum_{n=1}^{K-1} \frac{p_n}{1 - p_K} \left\{ 1 - \sum_{j=0}^{n-1} \frac{(\mu t)^j}{j!} e^{-\mu t} \right\}$$

which gives

$$
\begin{aligned}
F_W(t) &= \frac{1-\rho}{1-\rho^K} + \sum_{n=1}^{K-1} \frac{p_n}{1-p_K} \left\{ 1 - \sum_{j=0}^{n-1} \frac{(\mu t)^j}{j!} e^{-\mu t} \right\} \\
&= \frac{1-\rho}{1-\rho^K} + \sum_{n=1}^{K-1} \frac{p_n}{1-p_K} - \frac{1}{1-p_K} \sum_{n=1}^{K-1} p_n \sum_{j=0}^{n-1} \frac{(\mu t)^j}{j!} e^{-\mu t} \\
&= 1 - \frac{1-\rho}{1-\rho^K} \sum_{n=1}^{K-1} \rho^n \sum_{j=0}^{n-1} \frac{(\mu t)^j}{j!} e^{-\mu t}
\end{aligned}
\tag{7.27}
$$

From this we obtain the mean waiting time as

$$
\begin{aligned}
E[W] &= \int_0^\infty \{1 - F_W(t)\} \, dt = \int_0^\infty \frac{1-\rho}{1-\rho^K} \sum_{n=1}^{K-1} \rho^n \sum_{j=0}^{n-1} \frac{(\mu t)^j}{j!} e^{-\mu t} \, dt \\
&= \frac{1-\rho}{1-\rho^K} \sum_{n=1}^{K-1} \rho^n \sum_{j=0}^{n-1} \frac{1}{j!} \int_0^\infty (\mu t)^j e^{-\mu t} \, dt \\
&= \frac{1-\rho}{\mu(1-\rho^K)} \sum_{n=1}^{K-1} \rho^n \sum_{j=0}^{n-1} \frac{1}{j!} \int_0^\infty (\mu t)^j e^{-\mu t} \, d\mu t \\
&= \frac{1-\rho}{\mu(1-\rho^K)} \sum_{n=1}^{K-1} n\rho^n \\
&= \begin{cases} \dfrac{1}{\mu} \left[\dfrac{\rho}{1-\rho} - \dfrac{K\rho^K}{1-\rho^K} \right] & \rho \ne 1 \\[3mm] \dfrac{K-1}{2\mu} & \rho = 1 \end{cases}
\end{aligned}
\tag{7.28}
$$

where the second to the last equality follows from the fact that

$$\int_0^\infty x^n e^{-x} \, dx = \Gamma(n+1) = n! \quad \text{for } n > 0 \text{ an integer}$$

and $\Gamma(k)$ is the gamma function of k. Similarly, the CDF of the total time in the system can be obtained as follows:

$$
\begin{aligned}
F_T(t) = P[T \le t] &= \sum_{n=0}^{K-1} \frac{p_n}{1-p_K} \int_0^t \frac{\mu^{n+1} u^n}{n!} e^{-\mu u}\, du \\
&= \sum_{n=0}^{K-1} \frac{p_n}{1-p_K} \left\{ 1 - \int_t^\infty \frac{\mu^{n+1} u^n}{n!} e^{-\mu u}\, du \right\} = \sum_{n=0}^{K-1} \frac{p_n}{1-p_K} \left\{ 1 - \sum_{j=0}^n \frac{(\mu t)^j}{j!} e^{-\mu t} \right\} \\
&= \sum_{n=0}^{K-1} \frac{p_n}{1-p_K} - \frac{1}{1-p_K} \sum_{n=0}^{K-1} p_n \sum_{j=0}^n \frac{(\mu t)^j}{j!} e^{-\mu t} = 1 - \frac{1}{1-p_K} \sum_{n=0}^{K-1} p_n \sum_{j=0}^n \frac{(\mu t)^j}{j!} e^{-\mu t} \\
&= 1 - \frac{1-\rho}{1-\rho^K} \sum_{n=0}^{K-1} \rho^n \sum_{j=0}^n \frac{(\mu t)^j}{j!} e^{-\mu t}
\end{aligned}
$$
(7.29)

The mean total time in the system is given by

$$
\begin{aligned}
E[T] &= \int_0^\infty \{1 - F_T(t)\}\, dt = \int_0^\infty \frac{1-\rho}{1-\rho^K} \sum_{n=0}^{K-1} \rho^n \sum_{j=0}^n \frac{(\mu t)^j}{j!} e^{-\mu t}\, dt \\
&= \frac{1-\rho}{1-\rho^K} \sum_{n=0}^{K-1} \rho^n \sum_{j=0}^n \frac{1}{j!} \int_0^\infty (\mu t)^j e^{-\mu t}\, dt \\
&= \frac{1-\rho}{\mu(1-\rho^K)} \sum_{n=1}^{K-1} \rho^n \sum_{j=0}^n \frac{1}{j!} \int_0^\infty (\mu t)^j e^{-\mu t}\, d\mu t \\
&= \frac{1-\rho}{\mu(1-\rho^K)} \sum_{n=0}^{K-1} (n+1)\rho^n \\
&= \begin{cases} \frac{1}{\mu}\left[\frac{1}{1-\rho} - \frac{K\rho^K}{1-\rho^K}\right] & \rho \ne 1 \\ \frac{K+1}{2\mu} & \rho = 1 \end{cases}
\end{aligned}
$$
(7.30)

Example 7.4

Each morning people arrive at Ed's garage to have their cars fixed. Ed's garage can only accommodate four cars. Anyone arriving when there are already four cars in the garage has to go away without leaving his or her car for Ed to fix. Ed's customers arrive according to a Poisson process with a rate of one customer per hour, and the time it takes Ed to service a car is exponentially distributed with a mean of 45 min.

a. What is the probability that an arriving customer finds Ed idle?
b. What is the probability that an arriving customer leaves without getting his or her car fixed?
c. What is the expected waiting time at Ed's garage?

Solution
This is an M/M/1/4 queue with the following parameters:

$$\lambda = 1$$
$$\mu = 60/45 = 4/3$$
$$\rho = \lambda/\mu = 3/4$$
$$p_0 = \frac{1-\rho}{1-\rho^5} = \frac{1-0.75}{1-(0.75)^5} = 0.3278$$

a. The probability that an arriving customer finds Ed idle is $p_0 = 0.3278$.
b. The probability that a customer leaves without fixing the car is the probability that he or she finds the garage full when he or she arrived, which is $p_4 = \rho^4 p_0 = 0.1037$.
c. The expected waiting time at Ed's garage is

$$E[W] = 0.75\left\{\frac{0.75}{1-0.75} - \frac{4(0.75)^4}{1-(0.75)^4}\right\} = 0.8614$$

7.7.3   The M/M/c/c Queue: The c-Server Loss System

This is a very useful model in telephony. It is used to model calls arriving at a telephone switchboard according to a Poisson process, and the switchboard has a finite capacity. It is assumed that the switchboard can support a maximum of c simultaneous calls (i.e., it has a total of c channels available). Any call that arrives when all c channels are busy will be lost. This is usually referred to as the *blocked calls lost* model. The state transition rate diagram is shown in Figure 7.8.

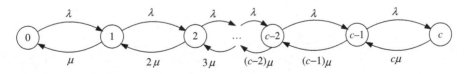

Figure 7.8 State transition rate diagram for the M/M/*c*/*c* queue.

Using the stochastic balance technique we used in earlier models, it can be shown that the steady-state probability that the process is in state n is given by

$$p_n = \frac{1}{n!} \left(\frac{\lambda}{\mu} \right)^n p_0$$

$$1 = \sum_{n=0}^{c} p_n = p_0 \sum_{n=0}^{c} \frac{(\lambda/\mu)^n}{n!}$$

Thus,

$$p_0 = \left[\sum_{n=0}^{c} \frac{(\lambda/\mu)^n}{n!} \right]^{-1}$$

and we obtain

$$p_n = \frac{(\lambda/\mu)^n/n!}{\sum\limits_{n=0}^{c} (\lambda/\mu)^n/n!} \qquad 0 \leq n \leq c \tag{7.31}$$

The probability that the process is in state c, p_c, is called the *Erlang's loss formula*, which is given by

$$p_c = \frac{(\lambda/\mu)^c/c!}{\sum\limits_{n=0}^{c} (\lambda/\mu)^n/n!} = \frac{(c\rho)^c/c!}{\sum\limits_{n=0}^{c} (c\rho)^n/n!} \tag{7.32}$$

where $\rho = \lambda/c\mu$ is the utilization factor of the system.

As in the M/M/1/K queueing system, not all traffic enters the system. The actual average arrival rate into the system is

$$\lambda_A = \lambda(1 - p_c) \tag{7.33}$$

Because no customer is allowed to wait, $E[W]$ and $E[N_q]$ are both zero. However, the mean number of customers in the system is

$$E[N] = \sum_{n=0}^{c} n p_n = p_0 \sum_{n=1}^{c} n \frac{(\lambda/\mu)^n}{n!} = (\lambda/\mu) p_0 \sum_{n=1}^{c} \frac{(\lambda/\mu)^{n-1}}{(n-1)!}$$

$$= (\lambda/\mu) p_0 \sum_{n=0}^{c-1} \frac{(\lambda/\mu)^n}{n!} = (\lambda/\mu)[1 - p_c] \tag{7.34}$$

By Little's formula,

$$E[T] = E[N]/\lambda_A = 1/\mu$$

This confirms that the mean time a customer admitted into the system spends in the system is the mean service time.

Example 7.5

Bob established a dial-up service for Internet access in his cyber cafe. As a small businessman, Bob can only support four lines for his customers. Any of Bob's customers that arrives at the cafe when all four lines are busy is blocked. Bob's studies indicate that customers arrive at the cafe according to a Poisson process with an average rate of eight customers per hour, and the duration of each customer's Internet use is exponentially distributed with a mean of 10 min. If Jay is one of Bob's customers, what is the probability that on one particular trip to the cafe, he could not use the Internet service?

Solution

This is an example of an M/M/4/4 queueing system. The parameters of the model are as follows:

$$\lambda = 8/60 = 2/15$$
$$\mu = 1/E[X] = 1/10$$
$$\rho = \lambda/c\mu = \lambda/4\mu = 1/3$$
$$c\rho = 4/3$$

The probability that Jay was blocked is the probability that he arrived when the process was in state 4. This is given by

$$p_4 = \frac{(c\rho)^4/4!}{\sum\limits_{n=0}^{4}(c\rho)^n/n!} = \frac{(4/3)^4/24}{1 + (4/3) + ((4/3)^2/2) + ((4/3)^3/6) + ((4/3)^4/24)}$$

$$= \frac{0.1317}{1 + 1.3333 + 0.8889 + 0.3951 + 0.1317} = 0.0351$$

7.7.4 The M/M/1//K Queue: The Single-Server Finite Customer Population System

In the previous examples, we assumed that the customers are drawn from an infinite population because the arrival process has a Poisson distribution. Assume that there are K potential customers in the population. An example is where we have a total of K machines that can be either operational or down, needing a serviceman to fix them. If we assume that the customers act independently of each other and that given that a customer has not yet come to the service facility, the time until he

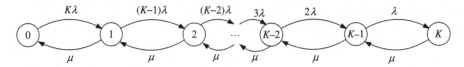

Figure 7.9 State transition rate diagram for the M/M/1//K queue.

or she comes to the facility is exponentially distributed with mean $1/\lambda$, then the number of arrivals when n customers are already in the service facility is Poisson with parameter $\lambda(K - n)$. When $n = K$, there are no more customers left to draw from, which means that the arrival rate becomes zero. Thus, the state transition rate diagram is as shown in Figure 7.9.

The arrival rate when the process is in state n is

$$\lambda_n = \begin{cases} (K - n)\lambda & 0 \le n < K \\ 0 & n \ge K \end{cases} \tag{7.35}$$

It can be shown that the steady-state probabilities are given by

$$p_n = \frac{K!}{(K - n)!} \left(\frac{\lambda}{\mu}\right)^n p_0 \quad n = 0, 1, 2, \ldots, K \tag{7.36a}$$

$$\sum_{n=0}^{K} p_n = 1 \Rightarrow p_0 = \left[K! \sum_{n=0}^{K} \frac{(\lambda/\mu)^n}{(K-n)!} \right]^{-1} \tag{7.36b}$$

Other schemes can easily be derived from the preceding models. For example, we can obtain the state transition rate diagram for the c-server finite population system with population $K > c$ by combining the arriving process on the M/M/1//K queueing system with the service process of the M/M/c queueing system.

Example 7.6

A small organization has three old PCs, each of which can be working (or operational) or down. When any PC is working, the time until it fails is exponentially distributed with a mean of 10 h. When a PC fails, the repairman immediately commences servicing it to bring it back to the operational state. The time to service each failed PC is exponentially distributed with a mean of 2 h. If there is only one repairman in the facility and the PCs fail independently, what is the probability that the organization has only two PCs working?

Solution

This is an M/M/1//3 queueing problem in which the arrivals are PCs that have failed and the single server is the repairman. Thus, when the process is in state 0, all PCs are working; when it is in state 1, two PCs are working; when it is in state 2, only one PC is working; and when it is in state 3, all PCs are down awaiting repair. The parameters of the problem are:

$$\lambda$$

$$\mu = 1/2$$

$$\lambda/\mu = 0.2$$

$$p_0 = \frac{1}{3!\displaystyle\sum_{n=0}^{3}((\lambda/\mu)^n/(K-n)!)} = \frac{1}{6[(1/6) + (0.2/2) + ((0.2)^2/1) + ((0.2)^3/1)]}$$

$$= \frac{1}{1 + 0.6 + 0.24 + 0.048} = 0.5297$$

As stated earlier, the probability that two computers are working is the probability that the process is in state 1, which is given by

$$p_1 = \left(\frac{\lambda}{\mu}\right)p_0 = (0.2)(0.5297) = 0.1059$$

7.8   M/G/1 Queue

In this system, customers arrive according to a Poisson process with rate λ and are served by a single server with a general service time X whose PDF is $f_X(x), x \geq 0$, finite mean is $E[X]$, second moment is $E[X^2]$, and finite variance is σ_X^2. The capacity of the system is infinite, and the customers are served on an FCFS basis. Thus, the service time distribution does not have the memoryless property of the exponential distribution, and the number of customers in the system time t, $N(t)$, is not a Poisson process. Therefore, a more appropriate description of the state at time t includes both $N(t)$ and the residual life of the service time of the current customer. That is, if R denotes the residual life of the current service, then the set of pairs $\{(N(t), R)\}$ provides the description of the state space. Thus, we have a two-dimensional state space, which is a somewhat complex way to proceed with the analysis. However, the analysis is simplified if we can identify those points in time where the state is easier to describe. Such points are usually chosen to be those time instants at which customers leave the system, which means that $R = 0$.

To obtain the steady-state analysis of the system, we proceed as follows. Consider the instant the kth customer arrives at the system. Assume that the ith customer was receiving service when the kth customer arrived. Let R_i denote the residual life of the service time of the ith customer at the instant the kth customer arrived, as shown in Figure 7.10.

Assume that N_{qk} customers were waiting when the kth customer arrived. Because the service times are identically distributed, the waiting time W_k of the kth customer is given by

$$W_k = R_i u(k) + N_{qk} X$$

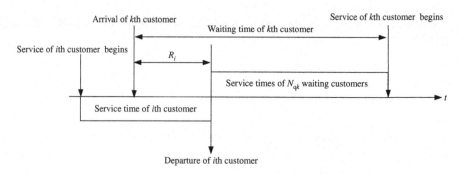

Figure 7.10 Service experience of the ith customer in M/G/1 queue.

where $u(k)$ is an indicator function that has a value of 1 if the server was busy when the kth customer arrived and zero otherwise. That is, if N_k defines the total number of customers in the system when the kth customer arrived, then

$$u(k) = \begin{cases} 1 & N_k > 0 \\ 0 & \text{otherwise} \end{cases}$$

Thus, taking expectations on both sides and noting that N_{qk} and X are independent random variables, and also that $u(k)$ and R_i are independent random variables, we obtain

$$E[W_k] = E[R_i]E[u(k)] + E[N_{qk}]E[X]$$

In Chapter 6, it was shown from the principle of random incidence that

$$E[R_i] = \frac{E[X^2]}{2E[X]}$$

Also,

$$E[u(k)] = 0P[N_q = 0] + 1P[N_k > 0] = P[N_k > 0] = 1 - p_0 = \rho$$

Finally, from Little's formula, $E[N_{qk}] = \lambda E[W_k]$. Thus, the mean waiting time of the kth customer is given by

$$E[W_k] = \frac{\rho E[X^2]}{2E[X]} + \lambda E[W_k]E[X] = \frac{\rho E[X^2]}{2E[X]} + \rho E[W_k]$$

From this we obtain

$$E[W_k] = \frac{\rho E[X^2]}{2(1 - \rho)E[X]} = \frac{\lambda E[X^2]}{2(1 - \rho)]}$$

Because the experience of the kth customer is a typical experience, we conclude that the mean waiting time in an M/G/1 queue is given by

$$E[W] = \frac{\rho E[X^2]}{2(1 - \rho)E[X]} = \frac{\lambda E[X^2]}{2(1 - \rho)} \tag{7.37}$$

Thus, the expected number of customers in the system is given by

$$E[N] = \rho + E[N_q] = \rho + \lambda E[W] = \rho + \frac{\lambda^2 E[X^2]}{2(1 - \rho)} \tag{7.38}$$

This expression is called the *Pollaczek–Khinchin formula*. It is sometimes written in terms of the coefficient of variation C_X of the service time. The square of C_X is defined as follows:

$$C_X^2 = \frac{\sigma_X^2}{(E[X])^2} = \frac{E[X^2] - (E[X])^2}{(E[X])^2} = \frac{E[X^2]}{(E[X])^2} - 1$$

Thus, the second moment of the service time becomes

$$E[X^2] = (1 + C_X^2)(E[X])^2$$

and the Pollaczek–Khinchin formula becomes

$$E[N] = \rho + \frac{\lambda^2 E[X^2]}{2(1 - \rho)} = \rho + \frac{\lambda^2 (1 + C_X^2)(E[X])^2}{2(1 - \rho)} = \rho + \frac{\rho^2 (1 + C_X^2)}{2(1 - \rho)} \tag{7.39}$$

Similarly, the mean waiting time becomes

$$E[W] = \frac{\lambda E[X^2]}{2(1 - \rho)} = \frac{\lambda (1 + C_X^2)(E[X])^2}{2(1 - \rho)} = \frac{\rho (1 + C_X^2)E[X]}{2(1 - \rho)} \tag{7.40}$$

7.8.1 Waiting Time Distribution of the M/G/1 Queue

We can obtain the distribution of the waiting time as follows. Let N_k denote the number of customers left behind by the kth departing customer, and let A_k denote the number of customers that arrive during the service time of the kth customer. Then we obtain the following relationship:

$$N_{k+1} = \begin{cases} N_k - 1 + A_{k+1} & N_k > 0 \\ A_{k+1} & N_k = 0 \end{cases}$$

Thus, we see that $\{N_k, k = 0, 1, \ldots\}$ forms a Markov chain called the *imbedded* M/G/1 Markov chain. Let the transition probabilities of the imbedded Markov chain be defined as follows:

$$p_{ij} = P[N_{k+1} = j | N_k = i]$$

Because N_k cannot be greater than $N_{k+1} + 1$, we have that $p_{ij} = 0$ for all $j < i - 1$. For $j \geq i - 1$, p_{ij} is the probability that exactly $j - i + 1$ customers arrived during the service time of the $(k + 1)$th customer, $i > 0$. Also, because the kth customer left the system empty in state 0, p_{0j} represents the probability that exactly j customers arrived while the $(k + 1)$th customer was being served. Similarly, because the kth customer left one customer behind in state 1, which is the $(k + 1)$th customer, p_{1j} is also the probability that exactly j customers arrived while the $(k + 1)$th customer was being served. Thus, $p_{0j} = p_{1j}$ for all j. Let the random variable A_S denote the number of customers that arrive during a service time. Then the PMF of A_S is given by

$$p_{A_S}(n) = P[A_S = n] = \int_{x=0}^{\infty} \frac{(\lambda x)^n}{n!} e^{-\lambda x} f_X(x) dx \quad n = 0, 1, \ldots$$

If we define $\alpha_n = P[A_S = n]$, then the state transition matrix of the imbedded Markov chain is given as follows:

$$P = \begin{bmatrix} \alpha_0 & \alpha_1 & \alpha_2 & \alpha_3 & \cdots & \cdots \\ \alpha_0 & \alpha_1 & \alpha_2 & \alpha_3 & \cdots & \cdots \\ 0 & \alpha_0 & \alpha_1 & \alpha_2 & \cdots & \cdots \\ 0 & 0 & \alpha_0 & \alpha_1 & \cdots & \cdots \\ 0 & 0 & 0 & \alpha_0 & \cdots & \cdots \\ \cdots & \cdots & \cdots & \cdots & \cdots & \cdots \\ \cdots & \cdots & \cdots & \cdots & \cdots & \cdots \end{bmatrix}$$

The state transition rate diagram is shown in Figure 7.11. Observe that the z-transform of the PMF of A_S is given by

$$G_{A_S}(z) = \sum_{n=0}^{\infty} z^n p_{A_S}(n) = \sum_{n=0}^{\infty} z^n \int_{x=0}^{\infty} \frac{(\lambda x)^n}{n!} e^{-\lambda x} f_X(x) dx$$

$$= \int_{x=0}^{\infty} \left\{ \sum_{n=0}^{\infty} \frac{(\lambda x z)^n}{n!} \right\} e^{-\lambda x} f_X(x) dx = \int_{x=0}^{\infty} e^{\lambda x z} e^{-\lambda x} f_X(x) dx \qquad (7.41)$$

$$= \int_{x=0}^{\infty} e^{-(\lambda - \lambda z)x} f_X(x) dx$$

$$= M_X(\lambda - \lambda z)$$

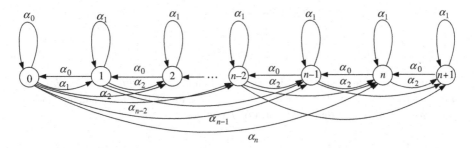

Figure 7.11 Partial state transition diagram for M/G/1 imbedded Markov chain.

where $M_X(s)$ is the s-transform of the PDF of X, the service time. That is, the z-transform of the PMF of A_S is equal to the s-transform of the PDF of X evaluated at the point $s = \lambda - \lambda z$. Let $f_T(t)$ denote the PDF of T, the total time in the system. Let K be a random variable that denotes the number of customers that a tagged customer leaves behind. This is the number of customers that arrived during the total time that the tagged customer was in the system. Thus, the PMF of K is given by

$$p_K(n) = P[K = n] = \int_{x=0}^{\infty} \frac{(\lambda t)^n}{n!} e^{-\lambda t} f_T(t) \mathrm{d}t \quad n = 0, 1, 2, \ldots$$

As in the case of A_S, it is easy to show that the z-transform of the PMF of K is given by

$$G_K(z) = M_T(\lambda - \lambda z)$$

Recall that N_k, the number of customers left behind by the kth departing customer, satisfies the relationship

$$N_{k+1} = \begin{cases} N_k - 1 + A_{k+1} & N_k > 0 \\ A_{k+1} & N_k = 0 \end{cases}$$

where A_k denotes the number of customers that arrive during the service time of the kth customer. Thus, K is essentially the value of N_k for our tagged customer. In Kleinrock (1975), it is shown that the z-transform of the PMF of K is given by

$$G_K(z) = \frac{(1 - \rho)M_X(\lambda - \lambda z)(1 - z)}{M_X(\lambda - \lambda z) - z}$$

Thus, we have that

$$M_T(\lambda - \lambda z) = \frac{(1 - \rho)M_X(\lambda - \lambda z)(1 - z)}{M_X(\lambda - \lambda z) - z}$$

If we set $s = \lambda - \lambda z$, we obtain the following:

$$M_T(s) = \frac{\frac{s}{\lambda}(1 - \rho)M_X(s)}{M_X(s) - \left\{1 - \frac{s}{\lambda}\right\}} = \frac{s(1 - \rho)M_X(s)}{s - \lambda + \lambda M_X(s)} \tag{7.42}$$

This is one of the equations that is usually called the *Pollaczek–Khinchin formula*. Finally, because $T = W + X$, which is the sum of two independent random variables, we have that the s-transform of T is given by

$$M_T(s) = M_W(s)M_X(s)$$

From this we obtain the s-transform of the PDF of W as

$$M_W(s) = \frac{M_T(s)}{M_X(s)} = \frac{s(1 - \rho)}{s - \lambda + \lambda M_X(s)} \tag{7.43}$$

This is also called the Pollaczek–Khinchin formula.

7.8.2 The M/E$_k$/1 Queue

The M/E$_k$/1 queue is an M/G/1 queue in which the service time has the Erlang-k distribution. It is usually modeled by a process in which service consists of a customer passing, stage by stage, through a series of k independent and identically distributed subservice centers, each of which has an exponentially distributed service time with mean $1/k\mu$. Thus, the total mean service time is $k \times (1/k\mu) = 1/\mu$. The state transition rate diagram for the system is shown in Figure 7.12. Note that the states represent service stages. Thus, when the system is in state 0, an arrival causes it to enter state k; when the system is in state 1, an arrival causes it to enter state $k + 1$, and so on. A completion of service at state j leads to a transition to state $j - 1, j \geq 1$.

While we can analyze the system from scratch, we can also apply the results obtained for the M/G/1 queue. We know that for an Erlang-k random variable X, the following results can be obtained:

$$f_X(x) = \frac{(k\mu)^k x^{k-1} e^{-k\mu x}}{(k - 1)!}$$

$$M_X(s) = \left(\frac{k\mu}{s + k\mu}\right)^k$$

$$E[X] = \frac{1}{\mu}$$

$$\sigma_X^2 = \frac{k}{(k\mu)^2} = \frac{1}{k\mu^2}$$

$$C_X^2 = \frac{\sigma_X^2}{(E[X])^2} = \frac{1}{k}$$

$$\rho = \lambda E[X] = \lambda/\mu < 1$$

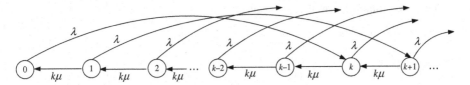

Figure 7.12 State transition rate diagram for the M/E_k/1 queue.

Thus, we obtain the following results:

a. The mean waiting time is

$$E[W] = \frac{\rho(1 + C_X^2)E[X]}{2(1 - \rho)} = \frac{\rho(k + 1)}{2k\mu(1 - \rho)}$$

b. The s-transform of the waiting time is

$$M_W(s) = \frac{s(1 - \rho)}{s - \lambda + \lambda M_X(s)} = \frac{s(1 - \rho)}{s - \lambda + \lambda(k\mu/(s + k\mu))^k}$$

c. The mean total number of customers in the system is

$$E[N] = \rho + \frac{\rho^2(1 + C_X^2)}{2(1 - \rho)} = \rho + \frac{\rho^2(k + 1)}{2k(1 - \rho)}$$

d. The s-transform of the total time in the system is

$$M_T(s) = \frac{s(1 - \rho)M_X(s)}{s - \lambda + \lambda M_X(s)} \bigg|_{M_X(s) = \left(\frac{k\mu}{s + k\mu}\right)^k}$$

7.8.3 The M/D/1 Queue

The M/D/1 queue is an M/G/1 queue with a deterministic (or fixed) service time. We can analyze the queueing system by applying the results for M/G/1 queueing system as follows:

$$f_X(x) = \delta(x - 1/\mu)$$
$$M_X(s) = e^{-s/\mu}$$
$$E[X] = \frac{1}{\mu}$$
$$\sigma_X^2 = 0$$
$$C_X^2 = \frac{\sigma_X^2}{(E[X])^2} = 0$$
$$\rho = \lambda E[X] = \lambda/\mu < 1$$

Thus, we obtain the following results:

a. The mean waiting time is

$$E[W] = \frac{\rho(1 + C_X^2)E[X]}{2(1 - \rho)} = \frac{\rho}{2\mu(1 - \rho)}$$

b. The s-transform of the waiting time is

$$M_W(s) = \frac{s(1 - \rho)}{s - \lambda + \lambda M_X(s)} = \frac{s(1 - \rho)}{s - \lambda + \lambda e^{-s/\mu}}$$

c. The mean total number of customers in the system is

$$E[N] = \rho + \frac{\rho^2(1 + C_X^2)}{2(1 - \rho)} = \rho + \frac{\rho^2}{2(1 - \rho)}$$

d. The s-transform of the total time in the system is

$$M_T(s) = \frac{s(1 - \rho)e^{-s/\mu}}{s - \lambda + \lambda e^{-s/\mu}}$$

7.8.4 The M/M/1 Queue Revisited

The M/M/1 queue is also an example of the M/G/1 queue with the following parameters.

$$f_X(x) = \mu e^{-\mu x}$$

$$M_X(s) = \frac{\mu}{s + \mu}$$

$$E[X] = \frac{1}{\mu}$$

$$\sigma_X^2 = \frac{1}{\mu^2}$$

$$C_X^2 = \frac{\sigma_X^2}{(E[X])^2} = 1$$

$$\rho = \lambda E[X] = \lambda/\mu < 1$$

When we substitute for these parameters in the equations for M/G/1, we obtain the results previously obtained for M/M/1 queueing system.

7.8.5 The M/H$_k$/1 Queue

This is a single-server, infinite-capacity queueing system with Poisson arrivals with hyperexponentially distributed service time of order k. That is, with probability θ_j,

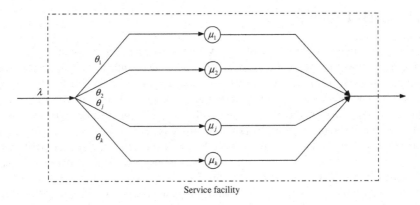

Service facility

Figure 7.13 A k-stage hyperexponential server.

an arriving customer will choose to receive service from server j whose service time is exponentially distributed with a mean of $1/\mu_j$, $1 \le j \le k$, where

$$\sum_{j=1}^{k} \theta_j = 1$$

The system is illustrated in Figure 7.13.
For this system, we have that

$$f_X(x) = \sum_{j=1}^{k} \theta_j \mu_j \, e^{-\mu_j x}$$

$$M_X(s) = \sum_{j=1}^{k} \theta_j \left\{ \frac{\mu_j}{s + \mu_j} \right\}$$

$$E[X] = \sum_{j=1}^{k} \left\{ \frac{\theta_j}{\mu_j} \right\}$$

$$E[X^2] = 2 \sum_{j=1}^{k} \left\{ \frac{\theta_j}{\mu_j^2} \right\}$$

$$C_X^2 = \frac{E[X^2]}{(E[X])^2} - 1 = \frac{2 \sum_{j=1}^{k} \left\{ \dfrac{\theta_j}{\mu_j^2} \right\}}{\left(\sum_{j=1}^{k} \{\theta_j/\mu_j\} \right)^2}$$

$$\rho = \lambda E[X] = \lambda/\mu < 1$$

7.9   G/M/1 Queue

The G/M/1 queue is the dual of the M/G/1 queue. In this system, customers arrive according to a general arrival process with independent and identically distributed interarrival times A with PDF $f_A(t)$ and mean $1/\lambda$. The facility has a single server, and the time X to serve arriving customers is exponentially distributed with mean $1/\mu$. As in the case of the M/G/1 queue, the number $N(t)$ of customers in the system is not Markovian. In this case, the reason is that to completely define a state, we need both $N(t)$ and the time that has elapsed since the last arrival. Thus, if Y is the time that has elapsed since the last arrival, then the state of the G/M/1 queue can be defined by the set of pairs $\{(N, Y)\}$, which means that we need a complicated two-dimensional state description. As in the M/G/1 queue, we look for those special points in time where a much easier state description can be formulated.

Because of the memoryless nature of the service process, such points that provide an easier state description are those time instants at which customers arrive. At these points $Y = 0$, which means the state description is captured by N only. Let N_k denote the number of customers that the kth arriving customer sees upon joining the queue, where $N_k = 0, 1, 2, \ldots$. Let T_k denote the time between the kth arrival and the $(k+1)$th arrival. Let S_k denote the number of service completions during T_k, as illustrated in Figure 7.14.

Thus, we obtain the following equation:

$$N_{k+1} = \max\{N_k + 1 - S_k, 0\} \quad k = 1, 2, \ldots$$

The initial condition is $N_1 = 0$, and we see that the sequence $\{N_k\}_{k=1}^{\infty}$ forms a Markov chain called the G/M/1 *imbedded Markov chain*. Let the transition probabilities be defined by

$$p_{ij} = P[N_{k+1} = j | N_k = i]$$

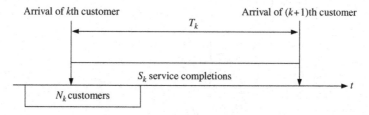

Figure 7.14 Service experience of the kth customer in G/M/1 queue.

It is clear that $p_{ij} = 0$ for all $j > i + 1$. For $j \le i + 1$, p_{ij} represents the probability that exactly $i + 1 - j$ customers are served during the interval between the kth arrival and the $(k + 1)$th arrival, given that the server is busy during this interval. Let r_n denote the probability that n customers are served during an interarrival time. Then r_n is given by

$$r_n = \int_{t=0}^{\infty} \frac{(\mu t)^n}{n!} e^{-\mu t} f_A(t) dt$$

Thus, the transition probability matrix is given by

$$P = [p_{ij}] = \begin{bmatrix} 1 - r_0 & r_0 & 0 & 0 & 0 & 0 & \cdots \\ 1 - \sum_{k=0}^{1} r_k & r_1 & r_0 & 0 & 0 & 0 & \cdots \\ 1 - \sum_{k=0}^{2} r_k & r_2 & r_1 & r_0 & 0 & 0 & \cdots \\ 1 - \sum_{k=0}^{3} r_k & r_3 & r_2 & r_1 & r_0 & 0 & \cdots \\ \cdots & \cdots & \cdots & \cdots & \cdots & \cdots & \cdots \end{bmatrix}$$

The partial state transition rate diagram is illustrated in Figure 7.15. In the figure,

$$p_{m0} = 1 - \sum_{k=0}^{m} r_k$$

$$p_{00} = 1 - r_0$$

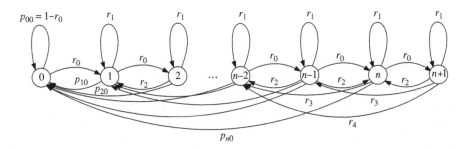

Figure 7.15 Partial state transition diagram for G/M/1 imbedded Markov chain.

Let π_n denote the limiting-state probability that the queue is in state n. Then these probabilities must satisfy the following balance equation:

$$\pi_0 = \pi_0 p_{00} + \pi_1 p_{10} + \pi_2 p_{20} + \cdots = \sum_{i=0}^{\infty} \pi_i p_{i0}$$

$$\pi_n = \pi_{n-1} r_0 + \pi_n r_1 + \pi_{n+1} r_2 + \cdots = \sum_{i=0}^{\infty} \pi_{n-1+i} r_i \quad n = 1, 2, \ldots$$

The solution to this system of equations is of the form

$$\pi_n = c\beta^n \quad n = 0, 1, 2, \ldots \tag{7.44}$$

where c is some constant. Thus, substituting this in the previous equation, we obtain

$$\beta^n = \sum_{i=0}^{\infty} \beta^{n-1+i} r_i \Rightarrow \beta = \sum_{i=0}^{\infty} \beta^i r_i$$

Because

$$r_n = \int_{t=0}^{\infty} \frac{(\mu t)^n}{n!} e^{-\mu t} f_A(t) dt$$

we obtain

$$\beta = \sum_{i=0}^{\infty} \beta^i \int_{t=0}^{\infty} \frac{(\mu t)^i}{i!} e^{-\mu t} f_A(t) dt = \int_{t=0}^{\infty} \left\{ \sum_{i=0}^{\infty} \frac{(\beta \mu t)^i}{i!} \right\} e^{-\mu t} f_A(t) dt$$

$$= \int_{t=0}^{\infty} e^{\beta \mu t} e^{-\mu t} f_A(t) dt = \int_{t=0}^{\infty} e^{-(\mu - \mu \beta)t} f_A(t) dt \tag{7.45}$$

$$= M_A(\mu - \mu \beta)$$

where $M_A(s)$ is the s-transform of the PDF $f_A(t)$. Because we know that $M_A(0) = 1$, $\beta = 1$ is a solution to the functional equation $\beta = M_A(\mu - \mu \beta)$. It can be shown that as long as $\lambda/\mu = \rho < 1$, there is a unique real solution for β in the range $0 < \beta < 1$, which is the solution we are interested in. Now, we know that

$$\sum_{n=0}^{\infty} r_n = c \sum_{n=0}^{\infty} \beta^n = 1 = \frac{c}{1 - \beta}$$

Thus, $c = 1 - \beta$, and we obtain

$$\pi_n = (1 - \beta)\beta^n \quad n \geq 0 \tag{7.46}$$

To find the mean total time in the system (or the sojourn time), let n denote the number of customers in the system when some tagged customer arrived. Because the service time is exponentially distributed, the total time that the tagged customer spends in the system is the time to serve $n + 1$ customers, including the tagged customer, which is given by the following random sum of random variables:

$$T = \sum_{k=1}^{n+1} X_k$$

where the X_k are independent and identically distributed with the PDF $f_X(x) = \mu\,e^{-\mu x}, x \geq 0$. Thus, the s-transform of the PDF of T, $M_T(s)$, can be obtained as follows:

$$T|_{N=n} = X_1 + X_2 + \cdots + X_{n+1}$$
$$M_{T|_{N=n}}(s|n) = \{M_X(s)\}^{n+1}$$

$$
\begin{aligned}
M_T(s) &= \sum_{n=0}^{\infty} M_{T|_{N=n}}(s|n)\pi_n = \sum_{n=0}^{\infty} \{M_X(s)\}^{n+1}\pi_n = \sum_{n=0}^{\infty} \left\{\frac{\mu}{s+\mu}\right\}^{n+1} (1-\beta)\beta^n \\
&= \frac{\mu(1-\beta)}{s+\mu} \sum_{n=0}^{\infty} \left\{\frac{\beta\mu}{s+\mu}\right\}^n = \frac{\mu(1-\beta)}{s+\mu} \left\{\frac{1}{1-(\beta\mu/(s+\mu))}\right\} \\
&= \frac{\mu(1-\beta)}{s+\mu} \left\{\frac{s+\mu}{s+\mu-\beta\mu}\right\} \\
&= \frac{\mu(1-\beta)}{s+\mu(1-\beta)}
\end{aligned}
\tag{7.47}
$$

This means that the sojourn time is exponentially distributed with mean $\frac{1}{\mu(1-\beta)}$. From this we obtain the following results:

$$E[W] = \frac{1}{\mu(1-\beta)} - \frac{1}{\mu} = \frac{\beta}{\mu(1-\beta)} \tag{7.48a}$$

$$E[N_q] = \lambda E[W] = \frac{\rho\beta}{1-\beta} \tag{7.48b}$$

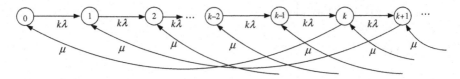

Figure 7.16 State transition rate diagram for the $E_k/M/1$ queue.

7.9.1 The $E_k/M/1$ Queue

The $E_k/M/1$ queue is a G/M/1 queue in which the interarrival time has the Erlang-k distribution. It is usually modeled by a process in which an arrival consists of a customer passing, stage by stage, through a series of k independent and identically distributed substations, each of which has an exponentially distributed service time with mean $1/k\mu$. Thus, the total mean interarrival time is $k \times (1/k\lambda) = 1/\lambda$. The state transition rate diagram for the system is shown in Figure 7.16. Note that the states represent arrival stages. Thus, when the system is in state 0, an arrival is complete only when the system enters state k and then service can commence. A completion of service causes the system to jump k states to the left.

As an example of a G/M/1 queue, we analyze the system by applying the results of the G/M/1 analysis as follows:

$$f_A(t) = \frac{(k\lambda)^k t^{k-1} e^{-k\lambda t}}{(k-1)!}$$

$$M_A(s) = \left(\frac{k\lambda}{s+k\lambda}\right)^k$$

$$\beta = M_A(\mu - \mu\beta) = \left(\frac{k\lambda}{\mu - \mu\beta + k\lambda}\right)^k = \left(\frac{k\rho}{1 - \beta + k\rho}\right)^k$$

For the special case of $k = 1$, the equation $\beta = M_A(\mu - \mu\beta)$ has the solutions $\beta = 1$ and $\beta = \rho$. Because we seek a solution $0 < \beta < 1$, we accept the solution $\beta = \rho$. Similarly, when $k = 2$, the equation becomes

$$(\beta - 1)\{\beta^2 - \beta(1 + 4\rho) + 4\rho^2\} = 0$$

whose solutions are

$$\beta_1 = 1$$

$$\beta_2 = \frac{1 + 4\rho + \sqrt{1 + 8\rho}}{2} = \left\{2\rho + \frac{1}{2} + \sqrt{2\rho + \frac{1}{4}}\right\} \in (1, 4)$$

$$\beta_3 = \frac{1 + 4\rho - \sqrt{1 + 8\rho}}{2} = \left\{2\rho + \frac{1}{2} - \sqrt{2\rho + \frac{1}{4}}\right\} \in (0, 1)$$

Because there exist values of $0 \le \rho < 1$ for which β_2 has values greater than 1 and we seek a solution $0 < \beta < 1$, we accept β_3 as the only valid solution. Thus, with this value of β, we can obtain all the relevant performance parameters.

7.9.2 The D/M/1 Queue

In this case, the interarrival times are deterministic with a constant value of $1/\lambda$, and service times are exponentially distributed with mean $1/\mu$. There is one server and infinite capacity. This can be used to model a system with a periodic arrival stream of customers, as is the case with time-division multiplexing voice communication systems. Thus, we have that

$$f_A(t) = \delta(t - 1/\lambda)$$
$$M_A(s) = e^{-s/\lambda}$$
$$\beta = M_A(\mu - \mu\beta) = e^{-(\mu - \mu\beta)/\lambda} = e^{-(1-\beta)/\rho} \Rightarrow \rho \ln(\beta) = \beta - 1$$

The last equation can be solved iteratively for a fixed value of ρ to obtain a solution for β in the range $0 < \beta < 1$ that can be used to obtain the performance parameters of the system. Table 7.1 gives the solutions to the preceding equation for different values of ρ.

That is, for a given value of ρ, we can obtain the mean total time in the system, the mean waiting time, and the mean number of customers in queue, respectively, as

$$E[T] = \frac{1}{\mu(1 - \beta)}$$

$$E[W] = \frac{1}{\mu(1 - \beta)} - \frac{1}{\mu} = \frac{\beta}{\mu(1 - \beta)}$$

$$E[N_q] = \lambda E[W] = \frac{\rho\beta}{(1 - \beta)}$$

Table 7.1 Values of β for Different Values of ρ

ρ	β
0.1	0.00004
0.2	0.00698
0.3	0.04088
0.4	0.10735
0.5	0.20319
0.6	0.32424
0.7	0.46700
0.8	0.62863
0.9	0.80690

7.10 M/G/1 Queues with Priority

Usually, all customers do not have the same urgency. Some customers require immediate attention while others can afford to wait. Thus in many situations, arriving customers are grouped into different priority classes numbered 1 to P, such that priority 1 is the highest priority, followed by priority 2, and so on with priority P being the lowest priority.

As discussed earlier in the chapter, there are two main classes of priority queues. These are *preemptive priority* and *nonpreemptive priority*. In a nonpreemptive priority queue, if a higher-priority customer arrives while a lower-priority customer is being served, the arriving higher-priority customer will wait until the lower-priority customer's service is completed. Thus, any customer that enters for service will complete the service without interruption. In a preemptive priority queue, if a higher-priority customer arrives when a lower-priority customer is being served, the arriving customer will preempt the customer being served and begin service immediately. When the preempted customer returns for service, the service can be completed in one of two ways. Under the *preemptive resume* policy, the customer's service will resume from the point at which it was suspended due to the interruption. Under the *preemptive repeat* policy, the customer's service will start from the beginning, which means that all the work that the server did prior to the preemption is lost. There are two types of preemptive repeat priority queueing: *preemptive repeat without resampling* (also called *preemptive repeat identical*) and *preemptive repeat with resampling* (also called *preemptive repeat different*). In preemptive repeat identical, the interrupted service must be repeated with an identical requirement, which means that its associated random variable must not be resampled. In preemptive repeat different, a new service time is drawn from the underlying distribution function.

We assume that customers from priority class p, where $p = 1, 2, \ldots, P$, arrive according to a Poisson process with rate λ_p, and the time to serve a class p customer has a general distribution with mean $E[X_p]$. Define

$$\rho_p = \lambda_p E[X_p]$$

$$\lambda = \sum_{p=1}^{P} \lambda_p$$

$$E[X] = \sum_{p=1}^{P} \frac{\lambda_p E[X_p]}{\lambda}$$

$$\rho = \lambda E[X] = \sum_{p=1}^{P} \rho_p < 1$$

$$\gamma_k = \sum_{p=1}^{k} \rho_p$$

Thus, ρ_p is the utilization of the server by priority class p customers, λ is the aggregate arrival rate of all customers, and γ_k is the utilization of the server by priority classes 1 to k customers. Let W_p be the waiting time of priority class p customers, and let T_p denote the total time that a priority class p customer spends in the system.

7.10.1 Nonpreemptive Priority

Consider a tagged priority class 1 customer that arrives for service. Assume that the number of priority class 1 customers waiting when the tagged priority class 1 customer arrived is L_1. The arriving tagged priority class 1 customer has to wait for the customer receiving service when it arrived to complete its service. Assume that the customer receiving service is a priority class p customer whose residual service time is R_p. Thus, the expected waiting time of the tagged customer is given by

$$E[W_1] = E[L_1]E[X_1] + \sum_{p=1}^{P} \rho_p E[R_p]$$

From Little's formula, we have that $E[L_1] = \lambda_1 E[W_1]$. If we define

$$E[R] = \sum_{p=1}^{P} \rho_p E[R_p]$$

then we have that

$$E[W_1] = \lambda_1 E[W_1]E[X_1] + E[R] = \rho_1 E[W_1] + E[R]$$

This gives

$$E[W_1] = \frac{E[R]}{1 - \rho_1}$$

The term $E[R]$ is the expected residual service time. From renewal theory, as discussed in Chapter 6, we know that

$$E[R_p] = \frac{E[X_p^2]}{2E[X_p]} \Rightarrow E[R] = \frac{1}{2}\sum_{p=1}^{P} \frac{\rho_p E[X_p^2]}{E[X_p]} = \frac{1}{2}\sum_{p=1}^{P} \lambda_p E[X_p^2]$$

Thus,

$$E[W_1] = \frac{1}{2(1 - \rho_1)}\sum_{p=1}^{P} \lambda_p E[X_p^2] \tag{7.49}$$

Following the same approach used for a tagged priority class 1 customer, for a priority class 2 customer, we have that

$$E[W_2] = E[L_1]E[X_1] + E[L_2]E[X_2] + \lambda_1 E[W_2]E[X_1] + \sum_{p=1}^{P} \rho_p E[R_p]$$

where L_p is the number of class p customers waiting when the tagged class 2 customer arrived, $p = 1, 2$. The first term on the right is the mean time to serve the priority class 1 customers that were in queue when the tagged customer arrived, the second term is the mean time to serve the priority class 2 customers that were in queue when the tagged customer arrived, the third term is the mean time to serve those priority class 1 customers that arrived while the tagged priority class 2 customer is waiting to be served, and the fourth term is the mean residual service time of the customer receiving service when the tagged customer arrived. By Little's formula, $E[L_1] = \lambda_1 E[W_1]$ and $E[L_2] = \lambda_2 E[W_2]$. Thus, we have that

$$E[W_2] = \rho_1 E[W_1] + \rho_2 E[W_2] + \rho_1 E[W_2] + E[R]$$

This gives

$$E[W_2] = \frac{\rho_1 E[W_1] + E[R]}{1 - \rho_1 - \rho_2} = \frac{E[R]}{(1 - \rho_1)(1 - \rho_1 - \rho_2)} \tag{7.50}$$

By continuing in the same way, we can obtain the mean waiting time of a class p customer as

$$E[W_p] = \frac{E[R]}{(1 - \rho_1 - \cdots - \rho_{p-1})(1 - \rho_1 - \cdots - \rho_p)} = \frac{E[R]}{(1 - \gamma_{p-1})(1 - \gamma_p)} \tag{7.51}$$

7.10.2 Preemptive Resume Priority

In preemptive priority in general, the lower-priority class customers are "invisible" to the higher-priority class customers. This means that the lower-priority class customers do not affect the queues of higher-priority class customers. Thus, the mean residual time experienced by class p customers is given by

$$E[R^p] = \frac{1}{2} \sum_{k=1}^{p} \lambda_k E[X_k^2] \tag{7.52}$$

Under the preemptive resume policy, when a customer's service is interrupted by the arrival of a higher-priority class customer, the service will be resumed from the point of interruption when all the customers from higher-priority classes have been served. Thus, the mean service time of a customer is not affected by the interruptions since service always resumes from where it was interrupted. The mean waiting time of a tagged priority class p customer is made up of two parts:

a. The mean time to serve the priority classes $1, 2, \ldots, p$ customers that are ahead of the tagged customer. This component of the mean waiting time is denoted by $E[W_p^a]$. To a priority class p customer, $E[W_p^a]$ is the mean waiting time in an M/G/1 queue without priority when lower-priority classes are neglected. Thus,

$$E\left[W_p^a\right] = \frac{E[R^p]}{1 - \rho_1 - \cdots - \rho_p}$$

where $E[R^p]$ is the mean residual time experienced by customers of priority class p, as defined earlier.

b. The mean time to serve those customers in priority classes $1, 2, \ldots, p-1$ who arrive while the tagged customer is in queue, which is

$$E[W_p^b] = \sum_{k=1}^{p-1} \lambda_k E[T_p] E[X_k] = \sum_{k=1}^{p-1} \rho_k E[T_p]$$

where $E[T_p] = E[W_p] + E[X_p]$ is the mean time a priority class p customer spends in the system waiting for and receiving service. We refer to T_p as the *sojourn time* of a priority class p customer. Thus, the mean sojourn time of a priority class p customer is given by

$$E[T_p] = E[X_p] + E[W_p^a] + E[W_p^b]$$

$$= E\left[X_p\right] + \frac{E[R^p]}{1 - \rho_1 - \cdots - \rho_p} + E\left[T_p\right] \sum_{k=1}^{p-1} \rho_k$$

From this we obtain

$$E\left[T_p\right] = \frac{E[X_p]}{1 - \rho_1 - \cdots - \rho_{p-1}} + \frac{E[R^p]}{(1 - \rho_1 - \cdots - \rho_p)(1 - \rho_1 - \cdots - \rho_{p-1})} \tag{7.52}$$

and the mean waiting time is given by

$$E\left[W_p\right] = E\left[T_p\right] - E\left[X_p\right] = \frac{\gamma_{p-1} E[X_p]}{1 - \gamma_{p-1}} + \frac{E[R^p]}{(1 - \gamma_{p-1})(1 - \gamma_p)} \tag{7.53}$$

7.10.3 Preemptive Repeat Priority

Under the preemptive repeat policy, when a customer's service is interrupted by the arrival of a higher-priority class customer, the service will be restarted from the beginning when all the customers from higher-priority classes have been served. Let K_p be a random variable that denotes the number of times that a priority class p customer is preempted until its service is completed. Let q_p denote the probability that a priority class p customer enters for service and completes the service without being preempted. Then, K_p is a geometrically distributed random variable with PMF and mean given by

$$p_{K_p}(k) = q_p(1-q_p)^{k-1} \quad k \geq 1$$
$$E[K_p] = 1/q_p$$

The mean waiting time of a tagged priority class p customer is made up of two parts:

a. The mean time to serve the priority classes $1, 2, \ldots, p$ customers that are ahead of the tagged customer. As in the preemptive resume priority case, this component of the waiting time is given by

$$E\left[W_p^a\right] = \frac{E[R^p]}{1 - \rho_1 - \cdots - \rho_p} = \frac{E[R^p]}{1 - \gamma_p}$$

b. The mean time to serve those customers in priority classes $1, 2, \ldots, p-1$ who arrive while the tagged customer is in queue, which is

$$E[W_p^b] = \sum_{k=1}^{p-1} \lambda_k E[T_p]E[X_k] = \sum_{k=1}^{p-1} \rho_k E[T_p]$$

where $E[T_p] = E[W_p] + E[Y_p]$ is the mean sojourn time of a priority class p customer and Y_p is the effective service time, which consists of the actual uninterrupted service time and duration of the $K_p - 1$ times that elapsed before the service was interrupted. Let V_p denote the time that elapses from the instant a priority class p customer commences service until it is preempted. From renewal theory, we know that

$$E\left[V_p\right] = E\left[R_p\right] = \frac{E[X_p^2]}{2E[X_p]}$$

Thus,

$$E[Y_p] = E\left[X_p\right] + \{E\left[K_p\right] - 1\} E\left[V_p\right] = E\left[X_p\right] + \frac{\{E[K_p] - 1\}E[X_p^2]}{2E[X_p]}$$

$$= E\left[X_p\right] + \frac{\{(1/q_p) - 1\}E[X_p^2]}{2E[X_p]}$$

Therefore, the mean sojourn time of a priority class p customer is given by

$$E[T_p] = E[Y_p] + E[W_p] = E[Y_p] + E[W_p^a] + E[W_p^b]$$

$$= E[X_p] + \frac{\{(1/q_p) - 1\}E[X_p^2]}{2E[X_p]} + \frac{E[R^p]}{1 - \gamma_p} + \sum_{k=1}^{p-1} \rho_k E[T_p]$$

This gives

$$E[T_p] = \frac{E[X_p]}{1 - \rho_1 - \cdots - \rho_{p-1}} + \frac{\{1 - q_p\}E[X_p^2]}{2q_p(1 - \rho_1 - \cdots - \rho_{p-1})E[X_p]}$$

$$+ \frac{E[R^p]}{(1 - \lambda_p)(1 - \rho_1 - \cdots - \rho_{p-1})}$$

$$= \frac{E[X_p]}{(1 - \gamma_{p-1})} + \frac{(1 - q_p)E[X_p^2]}{2q_p(1 - \gamma_{p-1})E[X_p]} + \frac{E[R^p]}{(1 - \gamma_p)(1 - \gamma_{p-1})}$$

We next compute q_p by noting that it is the probability that no higher-priority class customer arrives over the time required to serve a priority class p customer. Since arrivals occur according to a Poisson process, the aggregate arrival rate of priority classes $1, 2, \ldots, p - 1$ is given by

$$\Lambda_{p-1} = \lambda_1 + \cdots + \lambda_{p-1}$$

Thus,

$$q_p = \int_0^\infty e^{-\Lambda_{p-1}x} f_{X_p}(x)\mathrm{d}x = M_{X_p}(\Lambda_{p-1})$$

where $M_{X_p}(s)$ is the s-transform of the PDF of X_p. Finally, the mean waiting time is given by

$$E[W_p] = E[T_p] - E[X_p]$$

$$= \frac{\gamma_{p-1}E[X_p]}{(1 - \gamma_{p-1})} + \frac{(1 - q_p)E[X_p^2]}{2q_p(1 - \gamma_{p-1})E[X_p]} + \frac{E[R^p]}{(1 - \gamma_p)(1 - \lambda_{p-1})} \qquad (7.54)$$

7.11 Markovian Networks of Queues

Many complex systems can be modeled by networks of queues in which customers sequentially receive service from one or more servers, where each server has its own queue. Thus, a network of queues is a collection of individual queueing systems that have common customers. In such a system, a departure from one queueing system is often an arrival to another queueing system in the network. Such networks are commonly encountered in communication and manufacturing systems. An interesting example of networks of queues is the activities of patients in an outpatient hospital facility where no prior appointments are made before a patient can see the doctor.

In such a facility, patients first arrive at a registration booth where they are first processed on an FCFS basis. Then they proceed to a waiting room to see a doctor on an FCFS basis. After seeing the doctor, a patient may be told to go for a laboratory test, which requires another waiting, or he or she may be given a prescription to get the medication. If he or she chooses to fill the prescription from the hospital pharmacy, he or she joins another queue, otherwise, leaves the facility and gets the medication elsewhere. Each of these activities is a queueing system that is fed by the same patients. Interestingly, after a laboratory test is completed, a patient may be required to take the result back to the doctor, which means that he or she rejoins a previous queue. After seeing the doctor a second time, the patient may either leave the facility without visiting the pharmacy or will visit the pharmacy and then leave the facility.

A queueing network is completely defined when we specify the external arrival process, the routing of customers among the different queues in the network, the number of servers at each queue, the service time distribution at each queue, and the service discipline at each queue. The network is modeled by a connected directed graph whose nodes represent the queueing systems and the arcs between the nodes have weights that are the *routing probabilities*, where the routing probability from node A to node B is the probability that a customer who leaves the queue represented by node A will next go to the queue represented by node B with the probability labeled on the arc (A, B). If no arc exits between two nodes, then when a customer leaves one node it does not go directly to the other node; alternatively, we say that the weight of such an arc is zero.

A network of queues is classified as either *open* or *closed*. In an open network, there is at least one node through which customers enter the network and at least one node through which customers leave the network. Thus, in an open network, a customer cannot be prevented from leaving the network. Figure 7.17 is an illustration of an open network of queues.

By contrast, in a closed network, there are no external arrivals or departures allowed. In this case, there is a fixed number of customers who are circulating forever among the different nodes. Figure 7.18 is an illustration of a closed network of queues. Alternatively, it can be used to model a finite-capacity system in which a new customer enters the system as soon as one customer leaves the system. One example of a closed network is a computer system that at any time has a fixed number of programs that use the CPU and I/O resources.

A network of queues is called a *Markovian network of queues* (or Markovian queueing network) if it can be characterized as follows. First, the service times at the different queues are exponentially distributed with possibly different mean values. Second, if they are open networks of queues, then external arrivals at the

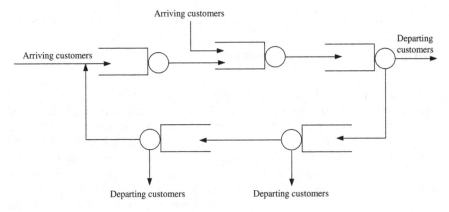

Figure 7.17 Example of open network of queues.

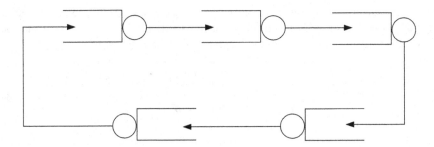

Figure 7.18 Example of closed network of queues.

network are according to Poisson processes. Third, the transitions between successive nodes (or queues) are independent. Finally, the system reaches a steady state.

To analyze the Markovian network of queues, we first consider a very important result called the *Burke's output theorem* by Burke (1956).

7.11.1 Burke's Output Theorem and Tandem Queues

Burke proved that the departure process of an M/M/c queue is Poisson. Specifically, Burke's theorem states that for an M/M/c queue in the steady state with arrival rate λ, the following results hold:

a. The departure process is a Poisson process with rate λ. Thus, the departure process is statistically identical to the arrival process.

b. At each time t, the number of customers in the system is independent of the sequence of departure times prior to t.

Figure 7.19 A network of N queues in tandem.

An implication of this theorem is as follows. Consider a queueing network with N queues in tandem, as shown in Figure 7.19. The first queue is an M/M/1 queue through which customers arrive from outside with rate λ. The service rate at queue i is μ_i, $1 \le i \le N$, such that $\lambda/\mu_i = \rho_i < 1$. That is, the system is stable. Thus, the other queues are x/M/1 queues, and for now we assume that we do not know precisely what "x" is. According to Burke's theorem, the arrival process at the second queue, which is the output process of the first queue, is Poisson with rate λ. Similarly, the arrival process at the third queue, which is the output process of the second queue, is Poisson with rate λ, and so on. Thus, x = M and each queue is essentially an M/M/1 queue.

Assume that customers are served on an FCFS basis at each queue, and let K_i denote the steady-state number of customers at queue i, $1 \le i \le N$. Since a departure at queue i results in an arrival at queue $i+1$, $i = 1, 2, \ldots, N-1$, it can be shown that the joint probability of queue lengths in the network is given by

$$
\begin{aligned}
P[K_1 = k_1, K_2 = k_2, \ldots, K_N = k_N] &= (1 - \rho_1)\rho_1^{k_1}(1 - \rho_2)\rho_2^{k_2} \cdots (1 - \rho_N)\rho_N^{k_N} \\
&= \prod_{i=1}^{N}(1 - \rho_i)\rho_i^{k_i}
\end{aligned}
\tag{7.55}
$$

Since the quantity $(1 - \rho_i)\rho_i^{k_i}$ denotes the steady-state probability that the number of customers in an M/M/1 queue with utilization factor ρ_i is k_i, the above result shows that the joint probability distribution of the number of customers in the network of N queues in tandem is the product of the N individual probability distributions. That is, the numbers of customers at distinct queues at a given time are independent. For this reason, the solution is said to be a *product-form solution*, and the network is called a *product-form queueing network*.

A further generalization of this system of tandem queues is a *feed-forward queueing network*, where customers can enter the network at any queue but a customer cannot visit a previous queue, as shown in Figure 7.20.

Let η_i denote the external Poisson arrival rate at queue i, and let p_{ij} denote the probability that a customer that has finished receiving service at queue i will next proceed to queue j, where p_{i0} denotes the probability that the customer leaves the network after receiving service at queue i. Then the rate λ_j at which customers arrive at queue j (from both outside and from other queues in the network) is given by

$$
\begin{aligned}
\lambda_1 &= \eta_1 \\
\lambda_j &= \eta_j + \sum_{i<j}\lambda_i p_{ij} \quad j = 2, \ldots, M
\end{aligned}
\tag{7.56}
$$

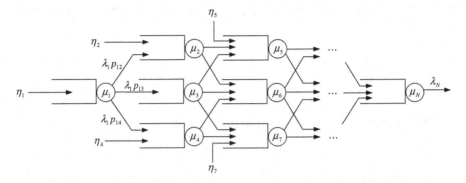

Figure 7.20 A feed-forward queueing network.

It can be shown that if the network is stable (i.e., $\lambda_j/\mu_j = \rho_j < 1$ for all j), then each queue is essentially an M/M/1 queue and a product-form solution exists. That is,

$$P[N_1 = n_1, N_2 = n_2, \ldots, N_M = n_M] = (1 - \rho_1)\rho_1^{n_1}(1 - \rho_2)\rho_2^{n_2} \cdots (1 - \rho_M)\rho_M^{n_N}$$

$$= \prod_{i=1}^{M}(1 - \rho_i)\rho_i^{k_i}$$

When a customer is allowed to visit a previous queue (i.e., a feedback loop exists), the output process of an M/M/1 queue is no longer Poisson because the arrival process and the departure process are no longer independent. For example, if with probability 0.9 a customer returns to the queue after receiving service, then after each service completion there is a very high probability of an arrival (from the same customer that was just served). Markovian queueing networks with feedback are called *Jackson networks*.

7.11.2 Jackson or Open Queueing Networks

As stated earlier, Jackson networks allow feedback loops, which means that a customer can receive service from the same server multiple times before exiting the network. The service time at each queue is exponentially distributed. As previously discussed, because of the feedback loop, the arrival process at each such queue is no longer Poisson. The question becomes this: Is a Jackson network a product-form queueing network?

Consider an open network of N queues in which customers arrive from outside the network to queue i according to a Poisson process with rate η_i, and queue i has c_i identical servers. The time to serve a customer at queue i is exponentially distributed with mean $1/\mu_i$. After receiving service at queue i, a customer proceeds to queue j with probability p_{ij} or leaves the network with probability p_{i0}, where

$$p_{i0} = 1 - \sum_{j=1}^{N} p_{ij} \quad i = 1, 2, \ldots, N$$

Thus, λ_i, the aggregate arrival rate at queue i, is given by

$$\lambda_i = \eta_j + \sum_{j=1}^{N} \lambda_j p_{ji} \quad j = 2, \ldots, N \tag{7.57}$$

Let $\lambda = [\lambda_1, \lambda_2, \ldots, \lambda_N]$, $\eta = [\eta_1, \eta_2, \ldots, \eta_N]$, and let P be the $N \times N$ routing matrix $P = [p_{ij}]$. Thus, we may write

$$\lambda = \eta + \lambda P \Rightarrow \lambda[I - P] = \eta$$

Because the network is open and any customer in any queue eventually leaves the network, each entry of the matrix P^n converges to zero as $n \to \infty$. This means that the matrix $I - P$ is invertible and the solution of the preceding equation is

$$\lambda = \eta[I - P]^{-1}$$

We assume that the system is stable, which means that $\rho_i = \lambda_i / c_i \mu_i < 1$ for all i. Let K_i be a random variable that defines the number of customers at queue i in the steady state. We are interested in joint probability mass function $p_{K_1 K_2 \cdots K_N}(k_1, k_2, \ldots, k_N)$. Jackson's theorem (Jackson, 1963) states that this joint probability mass function is given by the following product-form solution:

$$\begin{aligned} p_{K_1 K_2 \cdots K_N}(k_1, k_2, \ldots, k_N) &= P[K_1 = k_1, K_2 = k_2, \ldots, K_N = k_N] \\ &= P[K_1 = k_1]P[K_2 = k_2] \cdots P[K_N = k_N] \\ &= \prod_{i=1}^{N} P[K_i = k_i] \end{aligned}$$

where $P[K_i = k_i] = p_{K_i}(k_i)$ is the steady-state probability that there are k_i customers in an M/M/c_i queueing system with arrival rate λ_i and service rate $c_i \mu_i$. That is, the theorem states that the network acts as if each queue i is an independent M/M/c_i queueing system. Equivalently, we may write as follows for the case of M/M/1 queueing system:

$$p_K(k) = p_{K_1 K_2 \cdots K_N}(k_1, k_2, \ldots, k_N) = \prod_{i=1}^{N}(1 - \rho_i)\rho_i^{k_i} = \frac{1}{G(N)} \prod_{i=1}^{N} \rho_i^{k_i} \tag{7.58a}$$

$$G(N) = \prod_{i=1}^{N}(1 - \rho_i)^{-1} \tag{7.58b}$$

Example 7.7

Consider an open Jackson network with $N = 2$, as shown in Figure 7.21. This can be used to model a computer system in which new programs arrive at a CPU according to a Poisson process with rate η. After receiving service with an exponentially distributed time with a mean of $1/\mu_1$ at the CPU, a program proceeds to the I/O with probability p or leaves the system with probability $1 - p$. At the I/O, the program receives an exponentially distributed service with a mean of $1/\mu_2$ and immediately returns to the CPU for further processing. Assume that programs are processed in a FIFO manner.

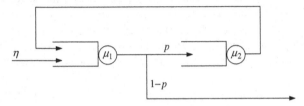

Figure 7.21 An example of open Jackson network.

Solution

The aggregate arrival rates of programs to the two queues are

$$\lambda_1 = \eta + \lambda_2$$
$$\lambda_2 = p\lambda_1$$

From these two equations, we obtain the solutions $\lambda_1 = \eta/(1-p)$ and $\lambda_2 = p\eta/(1-p)$. Thus, since $c_1 = c_2 = 1$, we have that

$$P[K_1 = k_1, K_2 = k_2] = P[K_1 = k_1]P[K_2 = k_2] = (1 - \rho_1)\rho_1^{k_1}(1 - \rho_2)\rho_2^{k_2}$$

where $\rho_1 = \lambda_1/\mu_1$ and $\rho_2 = \lambda_2/\mu_2$.

7.11.3 Closed Queueing Networks

A *closed network*, which is also called a *Gordon–Newell network* or a *closed Jackson queueing network*, is obtained by setting $\eta = 0$ and $p_{i0} = 0$ for all i. It is assumed that K customers continuously travel through the network and each node i has c_i server. Gordon and Newell (1967) proved that this type of network also has a product-form solution of the form

$$P[K_1 = k_1, K_2 = k_2, \ldots, K_N = k_N] = \frac{1}{G_N(K)} \prod_{i=1}^{N} \frac{\rho_i^{k_i}}{d_i(k_i)} \qquad (7.59)$$

where

$$\rho_i = \alpha_i/\mu_i$$

$$d_i(k_i) \quad \begin{cases} k_i! & k_i \leq c_i \\ (c_i)!c_i^{k_i-c_i} & k_i > c_i \end{cases}$$

$$G_N(K) = \sum_{k_1 + k_2 + \cdots + k_N = K} \prod_{i=1}^{N} \frac{\rho_i^{k_i}}{d_i(k_i)}$$

and $\alpha = (\alpha_1, \alpha_2, \ldots, \alpha_N)$ is any nonzero solution of the equation $\alpha = \alpha P$, where P is the routing matrix

$$P = \begin{bmatrix} p_{11} & p_{12} & \cdots & p_{1N} \\ p_{21} & p_{22} & \cdots & p_{2N} \\ \vdots & \vdots & \ddots & \vdots \\ p_{N1} & p_{N2} & \cdots & p_{NN} \end{bmatrix}$$

Note that the flow balance equation can be expressed in the form $[I - P]\alpha = 0$. Thus, the vector α is the eigenvector of the matrix $I - P$ corresponding to its zero eigenvalue. This means that the traffic intensities α_i can only be determined to within a multiplicative constant. However, while the choice of α influences the computation of the normalizing constant $G_N(K)$, it does not affect the occupancy probabilities.

$G_N(K)$ depends on the values of N and K. The possible number of states in a closed queueing network with N queues and K customers is

$$\binom{K + N - 1}{K} = \binom{K + N - 1}{N - 1}$$

Thus, it is computationally expensive to evaluate the parameter even for a moderate network. An efficient recursive algorithm called the *convolution algorithm* is used to evaluate $G_N(K)$ for the case where the μ_i is constant. Once $G_N(K)$ is evaluated, the relevant performance measures can be obtained. The convolution algorithm and other algorithms for evaluating $G_N(K)$ are discussed in many texts such as Ibe (2011).

Example 7.8

Consider the two-queue network shown in Figure 7.22. Obtain the joint probability $P[K_1 = k_1, K_2 = k_2]$ when $K = 3$ and $c_1 = c_2 = 1$.

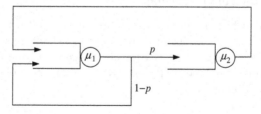

Figure 7.22 An example of closed Jackson network.

Solution
Here $N = 2$ and

$$P = \begin{bmatrix} 1 - p & p \\ 1 & 0 \end{bmatrix}$$

Thus, if $\alpha = (\alpha_1, \alpha_2)$, then

$$\alpha = \alpha P = [\alpha_1, \alpha_2] \begin{bmatrix} 1-p & p \\ 1 & 0 \end{bmatrix} = [\alpha_1(1-p) + \alpha_2, \alpha_1 p]$$

One nonzero solution to the equation $[\alpha_1, \alpha_2] = [\alpha_1(1-p) + \alpha_2, \alpha_1 p]$ is $\alpha_1 = 1$, $\alpha_2 = p$. Thus,

$$\rho_1 = 1/\mu_1$$
$$\rho_2 = p/\mu_2$$
$$d_1(k_i) = 1$$
$$G_2(3) = \sum_{k_1 + k_2 = 3} \rho_1^{k_1} \rho_2^{k_2} = \rho_1^3 + \rho_1^2 \rho_2 + \rho_1 \rho_2^2 + \rho_2^3$$

From this we obtain

$$P[K_1 = k_1, K_2 = k_2] = P[K_1 = k_1, K_2 = 3 - k_1]$$
$$= \frac{\rho_1^{k_1} \rho_2^{3-k_1}}{\rho_1^3 + \rho_1^2 \rho_2 + \rho_1 \rho_2^2 + \rho_2^3} \quad k_1 = 0, 1, 2, 3$$

7.12   Applications of Markovian Queues

The application areas of queueing theory have changed over the years. In the early days of the development of the theory, many of the application areas were service networks in operations research. These include inventory systems, dams, and reliability studies. Today, the theory is widely used in computer systems, telecommunication networks, and manufacturing systems. Much of the current research on queueing theory is motivated by the need to understand and control the behavior of these systems in order to improve their design and performance. Many queueing problems that arise in practice involve making one or more of three basic decisions:

a. Number of servers c required in a service facility to provide an acceptable quality of service.
b. Efficiency of the servers, which is reflected by ρ, the system utilization.
c. How to minimize the mean waiting time of customers for a fixed number of servers.

M/M/1 queue and its variants are used in many telecommunication systems to model the arrival of messages at a transmission channel with input buffers. Here, the service time is the message length divided by the channel capacity. Multiserver queues are used for capacity planning of telephone networks, checkout counters, and banking systems. The M/M/c queue is used to model the so-called *blocked calls delayed* systems, and the M/M/c/c queue is used to model the so-called *blocked calls lost* systems.

Networks of queues are used in many manufacturing, communications, and storage systems. A communication network consists of an interconnection of routers each of which is essentially a queueing system. Different routing policies can be used to handle the incoming messages called packets including FCFS and priority scheme. Similarly, manufacturing systems consist of different processing stages of the product line, and each of these stages is a queueing system. Thus, queueing networks are used in many applications. Other non-Markovian networks of queues are discussed by Ibe (2011).

7.13 Problems

7.1 People arrive to buy tickets at a movie theater according to a Poisson process with an average rate of 12 customers per hour. The time it takes to complete the sale of a ticket to each person is exponentially distributed with a mean of 3 min. There is only one cashier at the ticket window, and any arriving customer that finds the cashier busy joins a queue that is served in an FCFS manner.
  a. What is the probability that an arriving customer does not have to wait?
  b. What is the mean number of waiting customers at the window?
  c. What is the mean waiting time of an arbitrary customer?

7.2 Cars arrive at a car wash according to a Poisson process with a mean rate of eight cars per hour. The policy at the car wash is that the next car cannot pass through the wash procedure until the car in front of it is completely finished. The car wash has a capacity to hold 10 cars, including the car being washed, and the time it takes a car to go through the wash process is exponentially distributed with a mean of 6 min. What is the average number of cars lost to the car wash company every 10-h day as a result of the capacity limitation?

7.3 A shop has five identical machines that break down independently of each other. The time until a machine breaks down is exponentially distributed with a mean of 10 h. There are two repairmen who fix the machines when they fail. The time to fix a machine when it fails is exponentially distributed with a mean of 3 h, and a failed machine can be repaired by either of the two repairmen. What is the probability that exactly one machine is operational at any one time?

7.4 People arrive at a phone booth according to a Poisson process with a mean rate of five people per hour. The duration of calls made at the phone booth is exponentially distributed with a mean of 4 min.
  a. What is the probability that a person arriving at the phone booth will have to wait?
  b. The phone company plans to install a second phone at the booth when it is convinced that an arriving customer would expect to wait at least 3 min before using the phone. At what arrival rate will this occur?

7.5 People arrive at a library to borrow books according to a Poisson process with a mean rate of 15 people per hour. There are two attendants at the library, and the time to serve each person by either attendant is exponentially distributed with a mean of 3 min.
  a. What is the probability that an arriving person will have to wait?
  b. What is the probability that one or both attendants are idle?

7.6 A clerk provides exponentially distributed service to customers who arrive according to a Poisson process with an average rate of 15 per hour. If the service facility has an infinite capacity, what is the mean service time that the clerk must provide in order that the mean waiting time shall be no more than 10 min?

7.7 A company is considering how much capacity K to provide in its new service facility. When the facility is completed, customers are expected to arrive at the facility according to a Poisson process with a mean rate of 10 customers per hour, and customers that arrive when the facility is full are lost. The company hopes to hire an attendant to serve at the facility. Because the attendant is to be paid by the hour, the company hopes to get its money's worth by making sure that the attendant is not idle for more than 20% of the time he or she should be working. The service time is expected to be exponentially distributed with a mean of 5.5 min.

a. How much capacity should the facility have to achieve this goal?

b. With the capacity obtained in part (a), what is the probability that an arriving customer is lost?

7.8 A small PBX serving a start-up company can only support five lines for communication with the outside world. Thus, any employee who wants to place an outside call when all five lines are busy is blocked and will have to hang up. A blocked call is considered to be lost because the employee will not make that call again. Calls to the outside world arrive at the PBX according to a Poisson process with an average rate of six calls per hour, and the duration of each call is exponentially distributed with a mean of 4 min.

a. What is the probability that an arriving call is blocked?

b. What is the actual arrival rate of calls to the PBX?

7.9 A machine has four identical components that fail independently. When a component is operational, the time until it fails is exponentially distributed with a mean of 10 h. There is one resident repairman at the site so that when a component fails, the repairman immediately swaps it out and commences servicing it to bring it back to the operational state. When the repairman has finished repairing a failed component, the repairman immediately swaps it back into the machine. The time to service each failed component is exponentially distributed with a mean of 2 h. Assume the machine needs at least two operational components to work.

a. What is the probability that only one component is working?

b. What is the probability that the machine is down?

7.10 A cyber cafe has six PCs that customers can use for Internet access. These customers arrive according to a Poisson process with an average rate of six per hour. Customers who arrive when all six PCs are being used are blocked and have to go elsewhere for their Internet access. The time that a customer spends using a PC is exponentially distributed with a mean of 8 min.

a. What is the fraction of arriving customers that are blocked?

b. What is the actual arrival rate of customers at the cafe?

c. What fraction of arriving customers would be blocked if one of the PCs is out of service for a very long time?

7.11 Consider a birth-and-death process representing a multiserver finite population system with the following birth and death rates:

$$\lambda_k = (4 - k)\lambda \quad k = 0, 1, \ldots, 4$$
$$\mu_k = k\mu \quad k = 1, \ldots, 4$$

a. Find, in terms of λ and μ, p_k, the probability that there are k customers in the system, $k = 0, 1, \ldots, 4$.

b. Find the average number of customers in the system.

7.12 Customers arrive at a checkout counter in a grocery store according to a Poisson process with an average rate of 10 customers per hour. There are two clerks at the counter, and the time either clerk takes to serve each customer is exponentially distributed with an unspecified mean. If it is desired that the probability that both cashiers are idle is to be no more than 0.4, what will be the mean service time?

7.13 Consider an M/M/1/5 queueing system with mean arrival rate λ and mean service time $1/\mu$ that operates in the following manner. When any customer is in queue, the time until he or she defects (i.e., leaves the queue without receiving service) is exponentially distributed with a mean of $1/\beta$. It is assumed that when a customer begins receiving service, he or she does not defect.
a. Give the state transition rate diagram of the process.
b. What is the probability that the server is idle?
c. Find the average number of customers in the system.

7.14 Consider an M/M/1 queueing system with mean arrival rate λ and mean service time $1/\mu$ that operates in the following manner. When the number of customers in the system is greater than three, a newly arriving customer joins the queue with probability p and balks (i.e., leaves without joining the queue) with probability $1 - p$.
a. Give the state transition rate diagram of the process.
b. What is the probability that the server is idle?
c. Find the actual arrival rate of customers in the system.

7.15 Consider an M/M/1 queueing system with mean arrival rate λ and mean service time $1/\mu$. The system provides bulk service in the following manner. When the server completes any service, the system returns to the empty state if there are no waiting customers. Customers who arrive while the server is busy join a single queue. At the end of a service completion, all waiting customers enter the service area to begin receiving a common service.
a. Give the state transition rate diagram of the process.
b. What is the probability that the server is idle?

7.16 Consider an M/M/2 queueing system with hysteresis. Specifically, the system operates as follows. Customers arrive according to a Poisson process with rate λ customers per second. There are two identical servers, each of which serves at the rate of μ customers per second, but as long as the number of customers in the system is less than eight, only one server is busy serving them. When the number of customers exceeds eight, the second server is brought in, and the two will continue to serve until the number of customers in the system decreases to four when the server that has just completed a service is retired and only one server is allowed in the system.
a. Give the state transition rate diagram of the process.
b. What is the probability that both servers are idle?
c. What is the probability that exactly one server is busy?
d. What is the expected waiting time in the system?

7.17 Consider an M/G/1 queueing system where service is rendered in the following manner. Before a customer is served, a biased coin whose probability of heads is p is flipped. If it comes up heads, the service time is exponentially distributed with mean $1/\mu$. If it comes up tails, the service time is constant at d. Calculate the following:
a. The mean service time, $E[X]$
b. The coefficient of variation, C_X, of the service time
c. The expected waiting time, $E[W]$
d. The s-transform of the PDF of W

7.18 Consider a finite-capacity G/M/1 queueing that allows at most three customers in the system including the customer receiving service. The time to serve a customer is exponentially distributed with mean $1/\mu$. As usual, let r_n denote the probability that n customers are served during an interarrival time, $n = 0, 1, 2, 3$, and let $f_A(t)$ be the PDF of the interarrival times. Find r_n in terms of μ and $M_A(s)$, where $M_A(s)$ is the s-transform of $f_A(t)$.

7.19 Consider a queueing system in which the interarrival times of customers are the third-order Erlang random variable with parameter λ. The time to serve a customer is exponentially distributed with parameter μ. What is the expected waiting time of a customer?

7.20 Consider a queueing system in which customers arrive according to a Poisson process with rate λ. The time to serve a customer is a third-order Erlang random variable with parameter μ. What is the expected waiting time of a customer?

7.21 Consider a two-priority queueing system in which priority class 1 (i.e., high-priority) customers arrive according to a Poisson process with rate two customers per hour and priority class 2 (i.e., low-priority) customers arrive according to a Poisson process with rate five customers per hour. Assume that the time to serve class 1 customers is exponentially distributed with a mean of 9 min, and the time to serve class 2 customers is constant at 6 min. Determine the following:

    a. The mean waiting times of class 1 and class 2 customers when nonpreemptive priority is used.

    b. The mean waiting times of class 1 and class 2 customers when preemptive resume priority is used.

    c. The mean waiting times of class 1 and class 2 customers when preemptive repeat priority is used.

7.22 Consider the network shown in Figure 7.23, which has three exponential service stations with rates $\mu_1, \mu_2,$ and μ_3, respectively. External customers arrive at the station labeled Queue 1 according to a Poisson process with rate η. Let N_1 denote the steady-state number of customers at Queue 1, N_2 the steady-state number of customers at Queue 2, and N_3 the steady-state number of customers at Queue 3. Find the joint PMF $p_{N_1 N_2 N_3}(n_1, n_2, n_3)$.

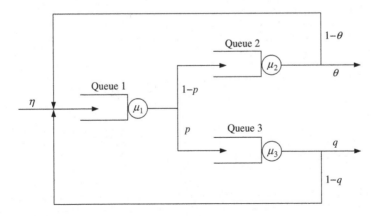

Figure 7.23 Figure for Problem 7.22.

7.23 Consider the acyclic Jackson network of queues shown in Figure 7.24, which has the property that a customer cannot visit a node more than once. Specifically, assume that there are four exponential single nodes with service rates μ_1, μ_2, μ_3, and μ_4, respectively, such that external arrivals occur at nodes 1 and 2 according to Poisson processes with rates η_1 and η_2, respectively. Upon completion of service at node 1, a customer proceeds to node 2 with probability p, to node 3 with probability r, and to node 4 with probability $1 - p - r$. Similarly, upon completion of service at node 2, a customer proceeds to node 3 with probability q and to node 4 with probability $1 - q$. After receiving service at node 3 or node 4, a customer leaves the system. Find the expected total time a customer spends in the system, and the total service time that a customer receives in the system.

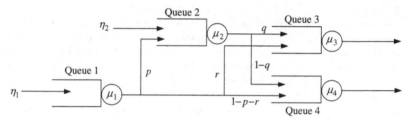

Figure 7.24 Figure for Problem 7.23.

7.24 Consider the closed network of queues shown in Figure 7.25. Assume that the number of customers inside the network is $K = 3$. Find the joint PMF $p_{N_1 N_2 N_3}(n_1, n_2, n_3)$, if there is a single exponential server at each queue with the service rate indicated.

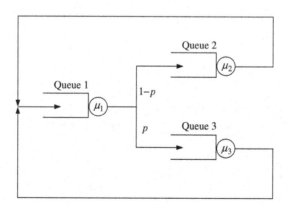

Figure 7.25 Figure for Problem 7.24.

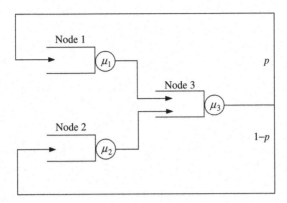

Figure 7.26 Figure for Problem 7.25.

7.25 Consider a closed network with $K = 3$ circulating customers, as shown in Figure 7.26. There is a single exponential server at nodes 1, 2, and 3 with service rates μ_1, μ_2, and μ_3, respectively. After receiving service at node 1 or node 2, a customer proceeds to node 3. Similarly, after receiving service at node 3, a customer goes to node 1 with probability p, and to node 2 with probability $1 - p$. Find the joint PMF $p_{N_1 N_2 N_3}(n_1, n_2, n_3)$.

8 Random Walk

8.1 Introduction

A random walk is derived from a sequence of Bernoulli trials. It is used in many fields including thermodynamics, biology, physics, chemistry, and economics where it is used to model fluctuations in the stock market.

Consider a Bernoulli trial in which the probability of success is p and the probability of failure is $1 - p$. Assume that the experiment is performed every T time units, and let the random variable X_k denote the outcome of the kth trial. Furthermore, assume that the probability mass function (PMF) of X_k is as follows:

$$p_{X_k}(x) = \begin{cases} p & x = 1 \\ 1 - p & x = -1 \end{cases}$$

That is, X_k are independent and identically distributed Bernoulli random variables. Finally, let the random variable Y_n be defined as follows:

$$\begin{aligned} Y_0 &= 0 \\ Y_n &= \sum_{k=1}^{n} X_k = Y_{n-1} + X_n \quad n = 1, 2, \ldots \end{aligned} \tag{8.1}$$

If X_k models a process in which we take a step to the right when the outcome of the kth trial is a success and a step to the left when the outcome is a failure, then the random variable Y_n represents the location of the process relative to the starting point (or origin) at the end of the nth trial. The resulting trajectory of the process $\{Y_n\}$ as it moves through the x-y plane, where the x-coordinate represents the time and the y-coordinate represents the location at a given time, is called a one-dimensional *random walk* generated by $\{X_k\}$. If we define the stochastic process $Y(t) = \{Y_n | n \le t < n + 1\}$, then Figure 8.1 shows an example of the sample path of $Y(t)$, where the length of each step is s. It is a staircase with discontinuities at $t = kT, k = 1, 2, \ldots$.

Suppose that at the end of the nth trial there are exactly k successes. Then there are k steps to the right and $n - k$ steps to the left. Thus,

$$Y(nT) = ks - (n - k)s = (2k - n)s = rs$$

Markov Processes for Stochastic Modeling. DOI: http://dx.doi.org/10.1016/B978-0-12-407795-9.00008-6

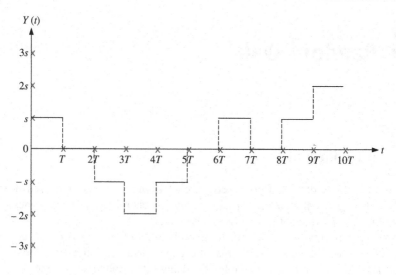

Figure 8.1 A sample path of the random walk.

where $r = 2k - n$. This implies that $Y(nT)$ is a random variable that assumes values rs, where $r = n, n - 2, n - 4, \ldots, -n$. Because the event $\{Y(nT) = rs\}$ is the event $\{k \text{ successes in } n \text{ trials}\}$, where $k = (n + r)/2$, we have that

$$P[Y(nT) = rs] = P\left[\frac{n + r}{2} \text{ successes}\right] = \binom{n}{\frac{n+r}{2}} p^{\frac{n+r}{2}}(1-p)^{\frac{n-r}{2}} \tag{8.2}$$

Note that $(n + r)$ must be an even number. Also, because $Y(nT)$ is the sum of n independent Bernoulli random variables, its mean and variance are given as follows:

$$E[Y(nT)] = nE[X_k] = n[ps - (1 - p)s] = (2p - 1)ns$$

$$E[X_k^2] = ps^2 + (1 - p)s^2 = s^2 \tag{8.3}$$

$$\text{Var}[Y(nT)] = n\sigma_{X_k}^2 = n[s^2 - s^2(2p-1)^2] = 4p(1 - p)ns^2$$

In the special case where $p = 1/2$, $E[Y(nT)] = 0$, and $\text{Var}[Y(nT)] = ns^2$, the random walk is called a *symmetric random walk*, which is also known as a *Bernoulli random walk*.

The one-dimensional random walk possesses the so-called *skip-free property*, which means that to go from state a to b, the process must pass through all intermediate states because its value can change by at most 1 at each step. This is because a random walk of length n is essentially a system of integers $\{X_0, X_1, \ldots, X_n\}$ such that $X_0 = 0$ and $|X_i - X_{i-1}| = 1$.

8.2 Occupancy Probability

Let the random variable $Y_n = X_1 + X_2 + \cdots + X_n$, where X_k are independent and identically distributed Bernoulli random variables. We would like to find the value of $P[Y_n = m]$, the probability of being at location m at the end of the nth step.

We assume that $X_k = 1$ corresponds to a step to the right in the kth trial and $X_k = -1$ corresponds to a step to the left. Suppose that out of the n steps the process takes r steps to the right and, therefore, $n - r$ steps to the left. Thus, we have that $Y_n = m = r - (n - r) = 2r - n$, which gives $r = (n + m)/2$. Using the binomial distribution we obtain

$$P[Y_n = m | Y_0 = 0] = \begin{cases} \binom{n}{\frac{n+m}{2}} p^{\frac{n+m}{2}} (1-p)^{\frac{n-m}{2}} & n + m \text{ even} \\ 0 & n + m \text{ odd} \end{cases} \tag{8.4}$$

For the special case of $m = 0$ we have that

$$u_n \equiv P[Y_n = 0 | Y_0 = 0] = \begin{cases} \binom{n}{\frac{n}{2}} [p(1-p)]^{n/2} & n \text{ even} \\ 0 & n \text{ odd} \end{cases}$$

which makes sense because this means that the process took as many steps to the right as it did to the left for it to return to the origin. Thus, we may write

$$u_{2n} = \binom{2n}{n} [p(1-p)]^n \quad n = 0, 1, 2, \ldots \tag{8.5}$$

The value of $P[Y_n = m | Y_0 = 0]$ when n is large can be obtained from the Stirling's approximation as follows:

$$n! \sim (2\pi)^{\frac{1}{2}} n^{n+\frac{1}{2}} e^{-n} \Rightarrow \log(n!) \sim (n + \tfrac{1}{2}) \log n - n + \tfrac{1}{2} \log 2\pi$$

Thus,

$$\lim_{n\to\infty} P[Y_n = m|Y_0 = 0]$$

$$= \binom{n}{\frac{n+m}{2}} p^{(n+m)/2}(1-p)^{(n-m)/2} = \frac{n!}{(n+m/2)!(n-m/2)!} p^{(n+m)/2}(1-p)^{(n-m)/2}$$

$$\approx \frac{\left\{(2\pi)^{1/2} n^{(n+1)/2} e^{-n}\right\} p^{(n+m)/2}(1-p)^{(n-m)/2}}{\left\{(2\pi)^{1/2}(n+m/2)^{((n+m+1)/2)} e^{-(n+m/2)}\right\}\left\{(2\pi)^{1/2}(n-m/2)^{((n-m+1)/2)} e^{-(n-m/2)}\right\}}$$

$$= \frac{2^{n+1}\{n^{n+(1/2)}\} p^{((n+m)/2)}(1-p)^{((n-m)/2)}}{(2\pi)^{1/2}(n+m)^{(n+m+1)/2}(n-m)^{(n-m+1)/2}}$$

(8.6)

Taking logarithm on both sides we obtain

$$\lim_{n\to\infty} \log\{P[Y_n = m|Y_0 = 0]\}$$

$$= \left(n+\tfrac{1}{2}\right)\log 2 - \tfrac{1}{2}\log \pi + \left(n+\tfrac{1}{2}\right)\log n - \left(\frac{n+m+1}{2}\right)\log\left\{n\left(1+\frac{m}{n}\right)\right\}$$

$$- \left(\frac{n-m+1}{2}\right)\log\left\{n\left(1-\frac{m}{n}\right)\right\} + \left(\frac{n+m}{2}\right)\log p + \left(\frac{n-m}{2}\right)\log q$$

$$= \left(n+\tfrac{1}{2}\right)\log 2 - \tfrac{1}{2}\log \pi - \tfrac{1}{2}\log n - \left(\frac{n+m+1}{2}\right)\log\left\{1+\frac{m}{n}\right\}$$

$$- \left(\frac{n-m+1}{2}\right)\log\left\{1-\frac{m}{n}\right\} + \left(\frac{n+m}{2}\right)\log p + \left(\frac{n-m}{2}\right)\log q$$

where $q = 1 - p$. Now, from the Taylor series expansion we know that

$$\log\left(1-\frac{m}{n}\right) = -\frac{m}{n} - \frac{m^2}{2n^2} + O(n^3) \quad -1 \le \frac{m}{n} < 1$$

$$\log\left(1+\frac{m}{n}\right) = \frac{m}{n} - \frac{m^2}{2n^2} + O(n^3) \quad -1 < \frac{m}{n} \le 1$$

Thus,

$$\lim_{n\to\infty} \log\{P[Y_n = m|Y_0 = 0]\}$$

$$= \left(n+\tfrac{1}{2}\right)\log 2 - \tfrac{1}{2}\log\pi - \tfrac{1}{2}\log n + \left(\frac{n+m}{2}\right)\log p + \left(\frac{n-m}{2}\right)\log q$$

$$- \left(\frac{n+m+1}{2}\right)\left\{\frac{m}{n} - \frac{m^2}{2n^2}\right\} + \left(\frac{n-m+1}{2}\right)\left\{\frac{m}{n} + \frac{m^2}{2n^2}\right\}$$

$$= (n+1)\log 2 - \tfrac{1}{2}\log 2\pi n - \frac{m^2(n-1)}{2n^2} + \left(\frac{n+m}{2}\right)\log p + \left(\frac{n-m}{2}\right)\log q$$

$$\approx n\log 2 - \tfrac{1}{2}\log 2\pi n - \frac{m^2}{2n} + \left(\frac{n+m}{2}\right)\log p + \left(\frac{n-m}{2}\right)\log q$$

which gives

$$\lim_{n\to\infty} P[Y_n = m|Y_0 = 0] = \left\{\frac{2^n}{\sqrt{2n\pi}}\right\}p^{(n+m)/2}q^{(n-m)/2}e^{-m^2/2n}$$

$$\approx \left\{\frac{2^n}{\sqrt{n\pi}}\right\}p^{(n+m)/2}q^{(n-m)/2}e^{-m^2/2n} \qquad (8.7)$$

$$\lim_{n\to\infty} u_{2n} \approx \left\{\frac{2^n}{\sqrt{n\pi}}\right\}p^{n/2}q^{n/2}$$

In the special case where $p = q = 1/2$, we obtain

$$\lim_{n\to\infty} P[Y_n = m|Y_0 = 0] = \left\{\frac{1}{\sqrt{n\pi}}\right\}e^{-m^2/2n}$$

$$\lim_{n\to\infty} u_{2n} = \frac{1}{\sqrt{n\pi}}$$

8.3   Random Walk as a Markov Chain

Let Y_n be as defined earlier. Then

$$
\begin{aligned}
P[Y_{n+1} = k | Y_1, Y_2, \ldots, Y_{n-1}, Y_n = m] \\
= P[X_{n+1} = k - m | Y_1, Y_2, \ldots, Y_{n-1}, Y_n = m] \\
= P\left[X_{n+1} = k - m | X_1, X_1 + X_2, \ldots, \sum_{i=1}^{n-1} X_i, \sum_{i=1}^{n} X_i = m \right] \\
= P[X_{n+1} = k - m] \quad \text{by the independence of } X_{n+1} \\
= P\left[X_{n+1} = k - m \Big| \sum_{i=1}^{n} X_i = m \right] \\
= P[X_{n+1} = k - m | Y_n = m] \\
= P[Y_{n+1} = k | Y_n = m]
\end{aligned}
$$

where the third equality is due to the independence of X_{n+1} and the other n random variables. Thus, the future of a random walk depends only on the most recent past outcome. Figure 8.2 shows the state-transition diagram of the Markov chain for a one-dimensional random walk.

If with probability 1 a random walker revisits its starting point, the walk is defined to be a *recurrent random walk*; otherwise, it is defined to be nonrecurrent. Later in this chapter we shall show that

$$
P[Y_n = 0 | Y_0 = 0, n = 1, 2, \ldots] = 1 - |p - q|
$$

Thus, a random walk is recurrent only if it is a symmetric random walk; that is, only if $p = q = 1/2$.

8.4   Symmetric Random Walk as a Martingale

Consider the random walk $\{Y_n | n = 0, 1, 2, \ldots\}$ where $Y_0 = 0$ and

$$
Y_n = \sum_{k=1}^{n} X_k = Y_{n-1} + X_n \quad n = 1, 2, \ldots
$$

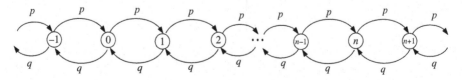

Figure 8.2 State-transition diagram of a random walk.

Assume that $P[X_k = 1] = p$ and $P[X_k = -1] = 1 - p$. Then

$$
\begin{aligned}
E[Y_n|Y_0, Y_1, \ldots, Y_k] &= E[Y_n - Y_k + Y_k|Y_0, Y_1, \ldots, Y_k] \\
&= E[Y_n - Y_k|Y_0, Y_1, \ldots, Y_k] + E[Y_k|Y_0, Y_1, \ldots, Y_k] \\
&= E[Y_n - Y_k|Y_0, Y_1, \ldots, Y_k] + Y_k \\
&= E\left[\sum_{j=k+1}^{n} (X_j|Y_0, Y_1, \ldots, Y_k)\right] + Y_k \\
&= \sum_{j=k+1}^{n} E[X_j|X_1, \ldots, X_k] + Y_k = \sum_{j=k+1}^{n} E[X_j] + Y_k \\
&= (n - k)(2p - 1) + Y_k
\end{aligned}
$$

The last equality follows from the fact that $E[X_k] = (1)p + (-1)(1 - p) = 2p - 1$. Now, for a symmetric random walk, $p = 1/2$ and the result becomes

$$
E[Y_n|Y_0, Y_1, \ldots, Y_k] = Y_k
$$

Thus, a symmetric random walk is a martingale, but a nonsymmetric random walk is not a martingale.

8.5   Random Walk with Barriers

The random walk previously described assumes that the process can continue forever; in other words, it is unbounded. Sometimes the walker cannot go outside some defined boundaries, in which case the walk is said to be a restricted random walk and the boundaries are called *barriers*. These barriers can impose different characteristics on the walk process. For example, they can be *reflecting barriers*, which means that on hitting them the walk turns around and continues. They can also be *absorbing barriers*, which means that the walk ends when a barrier is hit. Figure 8.3 shows the different types of barriers where it is assumed that the number of states is finite.

When absorbing barriers are present we obtain an absorbing Markov chain. One example of the random walk with barriers is the gambler's ruin problem that is discussed in the next section.

8.6   Gambler's Ruin

Consider the following random walk with two absorbing barriers, which is generally referred to as the gambler's ruin. Suppose a gambler plays a sequence of independent games against an opponent. He starts out with k, and in each game he wins $1 with probability p and loses $1 with probability $q = 1 - p$. When $p > q$,

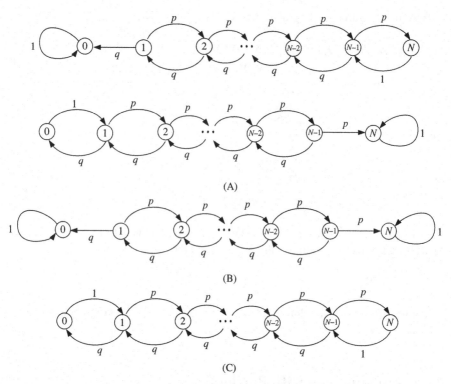

Figure 8.3 Random walks with barriers. (A) Random walk with one absorbing barrier and one reflecting barrier. (B) Random walk with two absorbing barriers. (C) Random walk with two reflecting barriers.

the game is advantageous to the gambler either because he is more skilled than his opponent or the rules of the game favor him. If $p = q$, the game is fair; if $p < q$, the game is disadvantageous to the gambler.

8.6.1 Ruin Probability

Assume that the gambler stops when he has a total of $0 or N. In the latter case he has additional $(N - k)$ over his initial k. (Another way to express this is that he plays against an opponent who starts out with $(N - k)$, and the game stops when either player has lost all his money.) We are interested in computing the probability r_k that the player will be ruined (or he has lost all his money) after starting with k. The state-transition diagram of the process is illustrated in Figure 8.3(B) where the states represent the total amount the gambler currently has.

To solve the problem, we note that at the end of the first game, the player will have the sum of $(k + 1)$ if he wins the game (with probability p) and the sum of $(k - 1)$ if he loses the game (with probability q). Thus, if he wins the first game, the probability that he will eventually be ruined is r_{k+1}; and if he loses his first game, the probability that he will be ruined is r_{k-1}. There are two boundary conditions in this problem. First, $r_0 = 1$ because he cannot gamble when he has no money. Second, $r_N = 0$ because he cannot be ruined since his opponent is already ruined. Thus, we obtain the following difference equation:

$$r_k = qr_{k-1} + pr_{k+1} \quad 0 < k < N \tag{8.8}$$

Because $p + q = 1$, Eq. (8.8) becomes

$$(p + q)r_k = qr_{k-1} + pr_{k+1} \quad 0 < k < N$$

which we can write as

$$p(r_{k+1} - r_k) = q(r_k - r_{k-1})$$

From this we obtain the following:

$$r_{k+1} - r_k = (q/p)(r_k - r_{k-1}) \quad 0 < k < N \tag{8.9}$$

We observe that

$$r_2 - r_1 = (q/p)(r_1 - r_0) = (q/p)(r_1 - 1)$$
$$r_3 - r_2 = (q/p)(r_2 - r_1) = (q/p)^2(r_1 - 1)$$
$$r_4 - r_3 = (q/p)(r_3 - r_2) = (q/p)^3(r_1 - 1)$$

and so on; thus, we obtain the following:

$$r_{k+1} - r_k = (q/p)^k(r_1 - 1) \quad 0 < k < N \tag{8.10}$$

Now,

$$r_k - 1 = r_k - r_0 = (r_k - r_{k-1}) + (r_{k-1} - r_{k-2}) + \cdots + (r_2 - r_1) + (r_1 - 1)$$
$$= [(q/p)^{k-1} + (q/p)^{k-2} + \cdots + (q/p) + 1](r_1 - 1)$$
$$= \begin{cases} \dfrac{1 - (q/p)^k}{1 - (q/p)}(r_1 - 1) & p \neq q \\ k(r_1 - 1) & p = q \end{cases}$$

The boundary condition that $r_N = 0$ implies that

$$r_1 = \begin{cases} 1 - \dfrac{1 - (q/p)}{1 - (q/p)^N} & p \neq q \\ 1 - \dfrac{1}{N} & p = q \end{cases}$$

Thus,

$$r_k = \begin{cases} 1 - \dfrac{1 - (q/p)^k}{1 - (q/p)^N} = \dfrac{(q/p)^k - (q/p)^N}{1 - (q/p)^N} & p \neq q \\[4mm] 1 - \dfrac{k}{N} & p = q \end{cases} \tag{8.11}$$

8.6.2 Alternative Derivation of Ruin Probability

An alternative method of deriving r_k is to try a solution of the form $r_k = z^k$ for the equation

$$r_k = qr_{k-1} + pr_{k+1} \quad 0 < k < N$$

This gives

$$pz^{k+1} + qz^{k-1} = z^k \Rightarrow pz^2 + q = z \Rightarrow pz^2 - z + q = 0$$

Thus,

$$z = \frac{1 \pm \sqrt{1 - 4pq}}{2p} = \frac{1 \pm \sqrt{1 - 4p(1-p)}}{2p} = \frac{1 \pm \sqrt{1 - 4p + 4p^2}}{2p}$$

$$= \frac{1 \pm \sqrt{(1-2p)^2}}{2p} = \frac{1 \pm (1 - 2p)}{2p}$$

The solutions are the two roots $z = 1$ and $z = (1 - p)/p = q/p$. Thus, we have that

$$r_k = a(1)^k + b(q/p)^k = a + b(q/p)^k$$

where a and b are arbitrary constants. If we assume that $p \neq q$, we have that $r_0 = 1 = a + b \Rightarrow a = 1 - b$. Similarly, the condition $r_N = 0 = a + b(q/p)^N$ gives $a = -b(q/p)^N$. Solving the two equations $a = 1 - b = -b(q/p)^N$ gives

$$b = \frac{1}{1 - (q/p)^N}, a = 1 - b = \frac{-(q/p)^N}{1 - (q/p)^N}$$

which gives the result in Eq. (8.11) for the case of $p \neq q$. Similarly, when $p = q$, we obtain the double root $z = 1$, which gives

$$r_k = a(1)^k + bk(1)^k = a + bk$$

The condition $r_0 = 1 \Rightarrow a = 1$, and the condition $r_N = 0 \Rightarrow b = -1/N$, which gives the result in Eq. (8.11) for the case of $p = q$.

8.6.3 Duration of a Game

Let d_k denote the expected time until a gambler that starts with \$$k$ is ruined. Clearly, $d_0 = d_N = 0$. If the gambler wins the first game, the expected duration of the game is $d_{k+1} + 1$; and if he loses the first game, the expected duration of the game is $d_{k-1} + 1$. Thus, d_k satisfies the following difference equation:

$$d_k = q(d_{k-1} + 1) + p(d_{k+1} + 1) = 1 + qd_{k-1} + pd_{k+1} \quad 0 < k < N$$

Because $p + q = 1$, we can rewrite the above difference equation as follows:

$$pd_{k+1} - pd_k - qd_k + qd_{k-1} + 1 = 0 = p(d_{k+1} - d_k) - q(d_k - d_{k-1}) + 1 \quad 0 < k < N$$

That is,

$$p(d_{k+1} - d_k) = q(d_k - d_{k-1}) - 1 \quad 0 < k < N$$

Let $m_k = d_k - d_{k-1}$. Then we have that

$$pm_{k+1} = qm_k - 1$$

Solving the preceding equation iteratively we obtain

$$pm_2 = qm_1 - 1 \Rightarrow m_2 = \frac{q}{p}m_1 - \frac{1}{p}$$

$$pm_3 = qm_2 - 1 \Rightarrow m_3 = \frac{q}{p}m_2 - \frac{1}{p} = \left(\frac{q}{p}\right)^2 m_1 - \frac{1}{p}\left\{1 + \frac{q}{p}\right\}$$

$$pm_4 = qm_3 - 1 \Rightarrow m_4 = \frac{q}{p}m_3 - \frac{1}{p} = \left(\frac{q}{p}\right)^3 m_1 - \frac{1}{p}\left\{1 + \frac{q}{p} + \left(\frac{q}{p}\right)^2\right\}$$

$$pm_5 = qm_4 - 1 \Rightarrow m_5 = \frac{q}{p}m_4 - \frac{1}{p} = \left(\frac{q}{p}\right)^4 m_1 - \frac{1}{p}\left\{1 + \frac{q}{p} + \left(\frac{q}{p}\right)^2 + \left(\frac{q}{p}\right)^3\right\}$$

Thus, in general we have that

$$
m_k = \left(\frac{q}{p}\right)^{k-1} m_1 - \frac{1}{p}\left\{1 + \frac{q}{p} + \left(\frac{q}{p}\right)^2 + \left(\frac{q}{p}\right)^3 + \cdots + \left(\frac{q}{p}\right)^{k-2}\right\}
$$

$$
= \left(\frac{q}{p}\right)^{k-1} m_1 - \frac{1}{p}\sum_{j=0}^{k-2}\left(\frac{q}{p}\right)^j
$$

Now, $m_1 = d_1 - d_0 = d_1$. Thus,

$$
d_k = \sum_{j=1}^{k} m_j = \sum_{j=1}^{k}\left\{\left(\frac{q}{p}\right)^{j-1} m_1 - \frac{1}{p}\sum_{j=0}^{j-2}\left(\frac{q}{p}\right)^j\right\}
$$

$$
= \begin{cases} \dfrac{1-(q/p)^k}{1-(q/p)}\left\{d_1 + \dfrac{1}{p-q}\right\} - \dfrac{k}{p-q} & p \neq q \\ k\{d_1 - (k-1) & p = q \end{cases}
$$

Because $d_N = 0$, we have that

$$
d_1 = \begin{cases} \dfrac{1-(q/p)}{1-(q/p)^N}\left\{\dfrac{N}{p-q}\right\} - \dfrac{1}{p-q} & p \neq q \\ N-1 & p = q \end{cases}
\tag{8.12}
$$

Thus, we finally obtain

$$
d_k = \begin{cases} \dfrac{1-(q/p)^k}{1-(q/p)}\left\{d_1 + \dfrac{1}{p-q}\right\} - \dfrac{k}{p-q} & p \neq q \\ k(N-k) & p = q \end{cases}
\tag{8.13}
$$

8.7 Random Walk with Stay

Consider a random walk in which the walker moves to the right with probability p, to the left with probability q or remains in the current position with probability $v = 1 - p - q$. This is called a *random walk with stay* or *random walk with hesitation*. It can be used to model a gambling problem in which a round of the game can result in a tie with probability v if each round is independent of other rounds. The state-transition diagram for the process is shown in Figure 8.4.

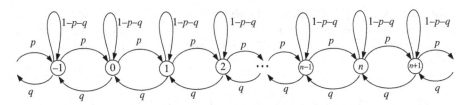

Figure 8.4 State-transition diagram for random walk with stay.

We consider the gambler's ruin problem with stay. Let r_k denote the probability that the player will be ruined after starting in state k. As in the classical gambler's ruin problem without ties, $r_0 = 1$ and $r_N = 0$. Thus, we obtain the following difference equation:

$$r_k = qr_{k-1} + (1 - p - q)r_k + pr_{k+1} \quad 0 < k < N$$

If we assume a solution of the form $r_k = z^k$, the equation becomes

$$z = pz^2 + (1 - p - q)z + q \Rightarrow pz^2 - (p + q)z + q = 0$$

Thus,

$$z = \frac{(p + q) \pm \sqrt{(p+q)^2 - 4pq}}{2p} = \frac{(p + q) \pm \sqrt{(p-q)^2}}{2p} = \frac{(p + q) \pm (p - q)}{2p}$$

That is, $z = 1$ and $z = q/p$, which are the solutions for the classical ruin problem, except that in this case, $p + q \neq 1$.

8.8 First Return to the Origin

Let T_0 denote the time until a random walk returns to the origin for the first time. That is, given that $Y_n = Y_{n-1} + X_n, Y_0 = 0, n = 1, 2, \ldots,$

$$T_0 = \min\{n \geq 1 : Y_n = 0\} \tag{8.14}$$

is the first return to zero. Its PMF is defined by

$$p_{T_0}(n) = P[T_0 = n] = P[Y_1 \neq 0, Y_2 \neq 0, \ldots, Y_{n-1} \neq 0, Y_n = 0]$$

The z-transform of the PMF $p_{T_0}(n)$ is given by the following theorem:

Theorem 8.1 The z-transform of the PMF of the first return to zero is given by

$$G_{T_0}(z) = 1 - (1 - 4pqz^2)^{1/2}$$

Proof Let $p_0(n) = P[Y_n = 0]$ be the probability that the process is at the origin after n steps. Let A be the event that $Y_n = 0$, and let B_k be the event that the first return to the origin occurs at the kth step. Because the B_k are mutually exclusive events, we have that

$$P[A] = \sum_{k=1}^{n} P[A|B_k]P[B_k] = \sum_{k=1}^{n} P[Y_n = 0|T_0 = k]P[T_0 = k] \quad n \geq 1$$

Now, $P[A|B_k] = P[Y_n = 0|T_0 = k] = p_0(n - k)$, which means that

$$p_0(n) = \sum_{k=1}^{n} p_0(n - k)p_{T_0}(k) \quad n \geq 1$$

Let the z-transform of $p_0(n)$ be $G_{Y_n}(z)$. Because the z-transform of $p_{T_0}(k)$ is given by $G_{T_0}(z)$, the preceding equation becomes

$$G_{Y_n}(z) = \sum_{n=0}^{\infty} p_0(n)z^n = p_0(0) + \sum_{n=1}^{\infty}\sum_{k=1}^{n} p_0(n - k)p_{T_0}(k)z^n$$
$$= 1 + G_{Y_n}(z)G_{T_0}(z)$$

where the last equality follows from the fact that $p_0(0) = 1$. Thus, we obtain

$$G_{T_0}(z) = 1 - \frac{1}{G_{Y_n}(z)}$$

To obtain $P[Y_n = 0]$ we must move an equal number of steps to the right and to the left. That is,

$$p_0(n) = \begin{cases} \binom{n}{\frac{n}{2}}(pq)^{n/2} & n \text{ even} \\ 0 & \text{otherwise} \end{cases}$$

Thus, we have that

$$G_{Y_n}(z) = \sum_{n=0}^{\infty} \binom{2n}{n}(pq)^n z^{2n} = \sum_{n=0}^{\infty} \binom{2n}{n}(pqz^2)^n = (1 - 4pqz^2)^{-1/2}$$

where the last equality follows from the identity (Wilf, 1990)

$$\sum_{n=0}^{\infty} \binom{2n}{n}x^n = \frac{1}{\sqrt{1 - 4x}}$$

From this we obtain

$$G_{T_0}(z) = 1 - (1 - 4pqz^2)^{1/2}$$

which completes the proof.

By expanding $G_{T_0}(z)$, as a power series we obtain the distribution of T_0 as follows. We know that according to the binomial theorem

$$(1+a)^n = \sum_{k=0}^{n} \binom{n}{k} a^k$$

If we define $\binom{n}{k}$ to be equal to zero when $k > n$, then we can omit the upper limit in the sum and write the quantity $(1-a)^n$ as follows:

$$(1-a)^n = \sum_{k \geq 0} \binom{n}{k} (-a)^k$$

Thus,

$$G_{T_0}(z) = 1 - (1 - 4pqz^2)^{1/2} = 1 - \sum_{k \geq 0} \binom{1/2}{k} (-4pqz^2)^k$$

$$= \sum_{k \geq 1} \binom{1/2}{k} (-1)^{k-1} (4pq)^k z^{2k}$$

which implies that

$$p_{T_0}(2k) = (-1)^{k-1} \binom{1/2}{k} (4pq)^k \quad k = 1, 2, \ldots \tag{8.15}$$

If we define v_0 as the probability that the process ever returns to the origin, then we have that

$$v_0 = \sum_{n=1}^{\infty} p_{T_0}(n) = G_{T_0}(1) = P[T_0 < \infty] = 1 - (1 - 4pq)^{1/2}$$

Because $p + q = 1$, we have that

$$(1 - 4pq)^{1/2} = (1 - 4p\{1-p\})^{1/2} = (1 - 4p + 4p^2)^{1/2} = \{(1 - 2p)^2\}^{1/2} = |1 - 2p|$$

which means that

$$v_0 = 1 - |1 - 2p| = 1 - |p - q| \tag{8.16}$$

For the symmetric random walk (i.e., $p = q = 1/2$), we have that $\upsilon_0 = 1$ and

$$G_{T_0}(z) = 1 - (1 - z^2)^{1/2}$$

Because $dG_{T_0}(z)/dz|_{z=1} = \infty$, we have that $E[T_0] = \infty$ for a symmetric random walk.

8.9 First Passage Times for Symmetric Random Walk

Let T_{ij} denote the time a symmetric random walk takes to reach the state j for the first time, given that is started in state i. Thus, T_{ij} is referred to as the first passage time of the random walk from i to j; that is,

$$T_{ij} = \min\{n \ge 0 : Y_0 = i, Y_n = j\}$$

Let the PMF of T_{ij} be $p_{T_{ij}}(k) = P[T_{ij} = k]$. We consider two methods of obtaining $p_{T_{ij}}(k)$. The first method is via the generating function, and the second method is via the reflection principle.

8.9.1 First Passage Time via the Generating Function

Consider T_{01}. In the first step, the walker can reach state 1 (i.e., $X_1 = 1$) with probability 1/2 or state -1 (i.e., $X_1 = -1$) with probability 1/2. If $X_1 = 1$, then $T_{01} = 1$. On the other hand, if $X_1 = -1$, then the walker will first reach 0 and then reach 1. That is, $T_{01} = 1 + T_{-1,1}$. But $T_{-1,1} = T_{-1,0} + T_{01}$ since it has to pass through 0 to get to 1. Because of Markov property, $T_{-1,0}$ and T_{01} have the same distribution and they are independent. Thus,

$$\begin{aligned}
G_{T_{01}}(z) &= E[z^{T_{01}}] = E[z^{T_{01}}|X_1 = 1]P[X_1 = 1] + E[z^{T_{01}}|X_1 = -1]P[X_1 = -1]\\
&= E[z^1|X_1 = 1]P[X_1 = 1] + E[z^{1+T_{-1,0}+T_{01}}]P[X_1 = -1]\\
&= zP[X_1 = 1] + zE[z^{T_{-1,0}+T_{01}}]P[X_1 = -1]\\
&= zP[X_1 = 1] + zE[z^{T_{-1,0}}]E[z^{T_{01}}]P[X_1 = -1]\\
&= \frac{z}{2} + \frac{z}{2}\{G_{T_{01}}(z)\}^2
\end{aligned}$$

Solving the quadratic equation $z/2\{G_{T_{01}}(z)\}^2 - G_{T_{01}}(z) + z/2 = 0$ we obtain

$$G_{T_{01}}(z) = \frac{1 \pm \sqrt{1 - 4(1/2)(z)(1/2)(z)}}{z} = \frac{1 \pm \sqrt{1 - z^2}}{z}$$

Because $G_{T_{01}}(z)$ must take values between 0 and 1 when $0 < z < 1$, we must take the negative part of the $\pm$, which gives

$$G_{T_{01}}(z) = \frac{1 - \sqrt{1 - z^2}}{z}$$

From Newton's binomial formula,

$$\sqrt{1 - z^2} = \sum_{k=0}^{\infty} \binom{1/2}{k}(-z^2)^k = \sum_{k=0}^{\infty}(-1)^k \binom{1/2}{k}(z^2)^k \qquad (8.17)$$

From this it can be shown that

$$P[T_{01} = 2k - 1] = (-1)^{k-1} \binom{1/2}{k}$$

Now, T_{0j} is the sum of $T_{01}, T_{12}, T_{23}, \ldots, T_{j-1,j}$ which are independent and identically distributed; that is,

$$T_{0j} = T_{01} + T_{12} + \cdots + T_{j-1,j}$$

Thus, the z-transform of T_{0j} is

$$G_{T_{0j}}(z) = E[z^{T_{0j}}] = E[z^{T_{01} + T_{12} + \cdots + T_{j-1,j}}] = \{G_{T_{01}}(z)\}^j \qquad (8.18)$$

Similarly,

$$G_{T_{ij}}(z) = \{G_{T_{01}}(z)\}^{|j-i|} \qquad (8.19)$$

8.9.2   First Passage Time via the Reflection Principle

The computation of the probability of an event associated with a random walk is essentially the counting of the number of paths that define that event. These probabilities can often be derived from the *reflection principle*, which states as follows:

Reflection principle: Let $k, v > 0$. Any path from $A = (a, k)$ to $N = (n, v)$ that touches or crosses the x-axis in between corresponds to a path from $A' = (a, -k)$ to $N = (n, v)$.

Thus, the x-axis can be thought of as a mirror that casts a "shadow path" of the original path by reflecting it on this mirror until it hits the x-axis for the first time. After the first time the original path hits the x-axis at $B = (b, 0)$, the shadow path is exactly the same as the original path. In Figure 8.5, the segment $A'B$ is the shadow path of the original segment AB. After B the two segments converge and continue

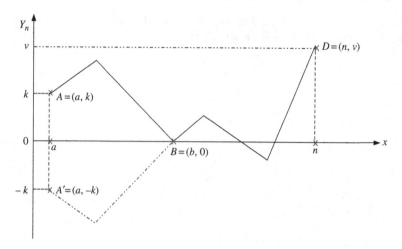

Figure 8.5 Illustration of the reflection principle.

as one path to N. Thus, the number of paths from A to N that touch or cross the x-axis is the same as the number of paths from A' to N.

Let $N_{a,n}(k, v)$ denote the number of possible paths between the points (a, k) and (n, v). (A guide to understanding this parameter is that the subscript indicates the starting and ending times while the argument indicates the initial and final positions.) Then $N_{a,n}(k, v)$ can be computed as follows. Let a path consist of m steps to the right and l steps to the left. Thus, the total number of steps is $m + l = n - a$, and the difference between the number of rightward steps and the leftward steps is $m - l = v - k$. From this we obtain

$$m = \tfrac{1}{2}\{n - a + v - k\} \tag{8.20}$$

Because $N_{a,n}(k, v)$ can be defined as the number of "successes," m, in $m + l = n - a$ binomial trials, we have that

$$N_{a,n}(k, v) = \binom{n - a}{m} \tag{8.21}$$

where m is as defined in Eq. (8.20).

Consider the event $\{Y_n = x | Y_0 = 0\}$; that is, the position of the walker after n steps is x, given that he started at the origin. In this case we have that $a = 0, k = 0, v = x$. Thus, $m = (n - x)/2$ and

$$N_{0,n}(0, x) = \binom{n - a}{m} = \binom{n}{\dfrac{n + x}{2}} = \binom{n}{\dfrac{n - x}{2}}$$

where $(n + x)/2$ is an integer. Thus, if p is the probability of a step of 1 and q is the probability of a step of -1,

$$P[Y_n = x] = N_{0,n}(0, x)p^{(n+x)/2}q^{(n-x)/2} = \binom{n}{\dfrac{n + x}{2}} p^{(n+x)/2}q^{(n-x)/2}$$

as we derived earlier in the chapter.

According to the reflection principle, the number of paths from A to N that touch or cross the time axis is equal to the number of paths from A' to N; that is, if we denote the number of paths that touch or cross the time axis in Figure 8.5 by $N_{a,n}^1(k, v)$, then

$$N_{a,n}^1(k, v) = N_{a,n}(-k, v)$$

If we assume that k and v are positive numbers as shown in Figure 8.5, then the number of paths from (a, k) to (n, v) that do not touch or intersect the time as, denoted by $N_{a,n}^0(k, v)$, is the complement of the number of paths, $N_{a,n}^1(k, v)$, that touch or intersect the time axis. Thus,

$$N_{a,n}^0(k, v) = N_{a,n}(k, v) - N_{a,n}^1(k, v) = N_{a,n}(k, v) - N_{a,n}(-k, v)$$

To apply this principle to the first passage time problem we proceed as follows. Consider a random walk that starts at $A = (0, 0)$, crosses or touches a line $y > v$, and then ends at $D = (n, v)$. This is illustrated in Figure 8.6. From our earlier discussion, the total number of paths between $(0, 0)$ and (n, x) is $N_{0,n}(0, v)$.

Consider a reflection on the line $Y = y$, as shown in Figure 8.6. Assume that the last point at which the path from $(0, 0)$ to (n, v) intersects this line is the point $B = (k, y)$. Then the reflection of the path from $(0, 0)$ to (n, v) on this line from the

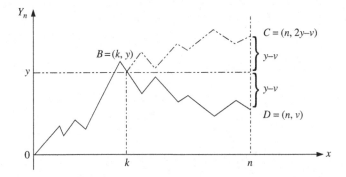

Figure 8.6 Reflection principle illustrated for first passage time.

point B is shown in dotted line. The terminal point of this line is $C = (n, 2y - v)$. According to the reflection principle, the number of paths from A to D that intersect or touch the line $Y = y$ is

$$N_{0,n}^1(0, v) = N_{0,n}(0, 2y - v)$$

To compute the first passage time, we note that for the walker to be at the point v at time n, he must be either at the point $v - 1$ at time $n - 1$ or at the point $v + 1$ at time $v - 1$. Since his first time of reaching the point v is n, we conclude that he must be at $v - 1$ at time $n - 1$. Thus, the number of paths from A to D that do not touch or cross $Y = v$ before time $n - 1$, $N_{0,n}^0(0, v)$, is

$$N_{0,n}^0(0, v) = N_{0,n}(0, v) - N_{0,n}^1(0, v)$$

$$= N_{0,n-1}(0, v - 1) - N_{0,n-1}(0, 2v - (v - 1))$$

$$= N_{0,n-1}(0, v - 1) - N_{0,n-1}(0, v + 1)$$

$$= \binom{n-1}{\frac{n+v-2}{2}} - \binom{n-1}{\frac{n+v}{2}} = \frac{(n-1)!}{\left(\frac{n+v-2}{2}\right)! \left(\frac{n-v}{2}\right)!} - \frac{(n-1)!}{\left(\frac{n+v}{2}\right)! \left(\frac{n-v-2}{2}\right)!}$$

$$= \frac{(n-1)!}{\left(\frac{n+v}{2} - 1\right)! \left(\frac{n-v}{2}\right)!} - \frac{(n-1)!}{\left(\frac{n+v}{2}\right)! \left(\frac{n-v}{2} - 1\right)!} = \frac{(n-1)!}{\left(\frac{n+v}{2}\right)! \left(\frac{n-v}{2}\right)!} \left\{ \frac{n+v}{2} - \frac{n-v}{2} \right\}$$

$$= \left\{ \frac{v}{n} \right\} \frac{n!}{\left(\frac{n+v}{2}\right)! \left(\frac{n-v}{2}\right)!} = \left\{ \frac{v}{n} \right\} \binom{n}{\frac{n+v}{2}}$$

$$= \frac{v}{n} N_{0,n}(0, v)$$

where $N_{0,n}(0, v)$ is the total number of paths from A to D. Thus, the probability that the first passage time from A to D occurs in n steps is

$$p_{T_v}(n) = P[T_v = n] = \frac{v}{n} \binom{n}{\frac{n+v}{2}} p^{\frac{n+v}{2}} (1-p)^{\frac{n-v}{2}}$$

Because a similar result can be obtained when $v < 0$, we have that

$$p_{T_v}(n) = P[T_v = n] = \frac{|v|}{n} \binom{n}{\frac{n+v}{2}} p^{\frac{n+v}{2}} (1-p)^{\frac{n-v}{2}}$$

Note that $n + v$ must be an even number. Also, recall that the probability that the walker is at location v after n steps is given by

$$P[Y_n = v] = \binom{n}{\frac{n+v}{2}} p^{\frac{n+v}{2}} (1-p)^{\frac{n-v}{2}}$$

Thus, the PMF for the first passage time can be written as follows:

$$p_{T_v}(n) = P[T_v = n] = \frac{|v|}{n} P[Y_n = v]$$

8.9.3 Hitting Time and the Reflection Principle

One way to state the reflection principle is that if point m is hit at time $T_m < n$, then it is equally likely, by symmetry, that Y_n will be either above or below m. Alternatively, for every sample path starting at T_m and ending above m at time n, there exists a symmetrical path starting at T_m and ending below m at time n. That is,

$$P[Y_n \geq m | T_m \leq n] = P[Y_n \leq m | T_m \leq n] = 0.5$$

From the law of total probability we have that

$$P[Y_n \geq m] = P[Y_n \geq m | T_m \leq n] P[T_m \leq n] + P[Y_n \geq m | T_m > n] P[T_m > n]$$

But $P[Y_n \geq m | T_m > n] = 0$. Thus, we have that

$$P[Y_n \geq m] = 0.5 P[T_m \leq n] \Rightarrow P[T_m \leq n] = 2P[Y_n \geq m] \qquad (8.22)$$

8.10 The Ballot Problem and the Reflection Principle

Consider Figure 8.5 and assume that $n > a \geq 0$ and $k > 0, v > 0$. According to the reflection principle, the number of paths from A to D that touch or cross the time axis is equal to the number of paths from A' to D. To obtain a path $(Y_0 = 0, Y_1, \ldots, Y_k = x)$ from $(0,0)$ to (k, x) such that $Y_1 > 0, \ldots, Y_k > 0$, we must have $Y_1 = 1$. Thus, the probability that a path from $(0,0)$ to (n, x) does not touch or cross the time axis is the ratio of the number of paths from $(1, 1)$ to (n, x) that do not touch or cross the time axis, $N_{1,n}^0(1, x)$, to the total number of paths from $(1, 1)$ to (n, x). But from the reflection principle, the number of paths that touch or cross the time axis from $(1, 1)$ and end at (n, x), $N_{1,n}^1(1, x)$, is the number of paths from $(1, -1)$ to (n, x). That is,

$$N_{1,n}^1(1, x) = N_{1,n}(-1, x)$$

Thus,

$$N_{1,n}^0(1,x) = N_{1,n}(1,x) - N_{1,n}^1(1,x) = N_{1,n}(1,x) - N_{1,n}(-1,x)$$

$$= \binom{n-1}{\frac{n+x-2}{2}} - \binom{n-1}{\frac{n+x}{2}} = \frac{(n-1)!}{\left(\frac{n-x}{2}\right)!\left(\frac{n+x}{2}-1\right)!} - \frac{(n-1)!}{\left(\frac{n-x}{2}-1\right)!\left(\frac{n+x}{2}\right)!}$$

$$= \frac{(n-1)!}{\left(\frac{n-x}{2}\right)!\left(\frac{n+x}{2}\right)!}\left\{\frac{n+x}{2} - \frac{n-x}{2}\right\} \tag{8.23}$$

$$= \frac{n!}{n\left(\frac{n-x}{2}\right)!\left(\frac{n+x}{2}\right)!}\{x\} = \left\{\frac{x}{n}\right\}\frac{n!}{\left(\frac{n-x}{2}\right)!\left(\frac{n+x}{2}\right)!} = \frac{x}{n}\binom{n}{\frac{n+x}{2}}$$

$$= \frac{x}{n}N_{0,n}(0,x)$$

Thus, the probability of a path from $(0,0)$ to (k,x) not to touch or cross the time axis is

$$\frac{N_{1,n}^0(1,x)}{N_{0,n}(0,x)} = \frac{x}{n}$$

This result can be used to solve the ballot problem, which can be defined as follows. Consider an election involving only two candidates: A and B. Assume that candidate A receives m votes and candidate B receives k votes, where $m > k$. What is the probability that candidate A is always ahead of candidate B in vote counts?

Let the probability that A always leads B in vote counts be $P_{m,k}$. Since the total number of votes cast is $n = m + k$ and the difference in votes is $m - k = x$, then from the above result we have that

$$P_{m,k} = \frac{x}{n} = \frac{m-k}{m+k} \tag{8.24}$$

8.10.1   The Conditional Probability Method

Let the probability that A is always ahead, given that A received the last vote, be denoted by $P[A|AL]$; let the probability that A is always ahead, given that B received the last vote, be $P[A|BL]$. If $P[AL]$ denotes the probability that

A received the last vote, which is $m/(m + k)$, and $P[BL]$ denotes the probability that B received the last vote, which is $k/(m + k)$, then we have that

$$P_{m,k} = P[A|AL]P[AL] + P[A|BL]P[BL] = P[A|AL]\left\{\frac{m}{m + k}\right\} + P[A|BL]\left\{\frac{k}{m + k}\right\}$$

$$= P_{m-1,k}\left\{\frac{m}{m + k}\right\} + P_{m,k-1}\left\{\frac{k}{m + k}\right\}$$

With the initial condition as $P_{1,0} = 1$, we derive the answer by induction. Define $m + k = L$. Then the equation is true for $L = 1$ because $P_{1,0} = (1 - 0)/(1 + 0) = 1$. Similarly, it is true for $L = 2$ because

$$P_{2,0} = P_{1,0}\frac{2}{2 + 0} = 1 = \frac{2 - 0}{2 + 0}$$

Also, when $L = 3$ the valid configurations are $(m, k) = \{(2, 1), (3, 0)\}$. Now, because we want $P_{1,1} = 0$, we obtain

$$P_{2,1} = P_{2,0}\left(\frac{1}{3}\right) = \frac{1}{3} = \frac{2 - 1}{2 + 1}$$

$$P_{3,0} = P_{2,0}\left(\frac{3}{3}\right) = 1 = \frac{3 - 0}{3 + 0}$$

Assume that the solution holds for L. Let m and k be such that $m \geq k$ and $m + k = L + 1$. Then by the induction we have that

$$P_{m,k} = \left\{\frac{m - 1 - k}{m - 1 + k}\right\}\left\{\frac{m}{m + k}\right\} + \left\{\frac{m - k + 1}{m - 1 + k}\right\}\left\{\frac{k}{m + k}\right\} = \frac{m - k}{m + k}$$

Thus, the proposition is also true for $m + k = L + 1$, which implies that it is valid for all m and k.

8.11   Returns to the Origin and the Arc-Sine Law

Consider a symmetric random walk (with $p = 1/2$). Recall that $u_{2n} = P[Y_{2n} = 0|Y_0 = 0]$ is the probability that the walker returns to the origin. Let f_{2n} denote the probability of the first return to zero at time $2n$; that is,

$$f_{2n} = P[Y_{2n} = 0|Y_0 = 0, Y_k \neq 0, k = 1, \ldots, 2n - 1]$$

As stated earlier, in order to return to the origin at time $2n$ the walker must take n steps to the right and n steps to the left. Thus,

$$u_{2n} = \binom{2n}{n}\left(\frac{1}{2}\right)^{2n}$$

To compute f_{2n} we observe that a visit to the origin at time $2n$ may be the first return or the first return occurs at time $2k < 2n$ followed by a return to the origin at time $2n$, which is $2n - 2k$ time units later. Thus, we have the following relationship:

$$u_{2n} = f_{2n}u_0 + f_{2n-2}u_2 + \cdots + f_4u_{2n+4} + f_2u_{2n-2}$$
$$= \sum_{k=1}^{n} f_{2k}u_{2n-2k} \tag{8.25}$$

We define the z-transforms

$$F(z) = \sum_{n=1}^{\infty} f_{2n}z^n$$
$$U(z) = \sum_{n=0}^{\infty} u_{2n}z^n$$

Since $u_0 = 1$ and $F(z)$ is defined for $n \geq 1$, then from Eq. (8.25) we obtain

$$U(z) = 1 + U(z)F(z) \Rightarrow F(z) = \frac{U(z) - 1}{U(z)} \tag{8.26}$$

Now, we know that

$$u_{2n} = \binom{2n}{n}\left(\frac{1}{2}\right)^{2n} \Rightarrow U(z) = \sum_{n=0}^{\infty}\binom{2n}{n}\left(\frac{1}{2}\right)^{2n}z^n = \sum_{n=0}^{\infty}\binom{2n}{n}(2^{-2}z)^n$$

As discussed earlier, it is known that

$$\sum_{n=0}^{\infty}\binom{2n}{n}x^n = \frac{1}{\sqrt{1-4x}}$$

Thus,

$$U(z) = \frac{1}{\sqrt{1-4(z/4)}} = \frac{1}{\sqrt{1-z}}$$

From Eq. (8.26) we obtain

$$F(z) = \frac{U(z) - 1}{U(z)} = \frac{(1-z)^{-1/2} - 1}{(1-z)^{-1/2}} = 1 - (1-z)^{1/2} = 1 - \sqrt{1-z}$$

We recall from the binomial theorem that

$$\sqrt{1-z} = \sum_{n=0}^{\infty} \binom{\frac{1}{2}}{n} (-1)^n z^n = 1 + \sum_{n=1}^{\infty} \binom{\frac{1}{2}}{n} (-1)^n z^n$$

From this we obtain

$$F(z) = -\sum_{n=1}^{\infty} \binom{\frac{1}{2}}{n} (-1)^n z^n$$

An aside: We transform the fractional binomial component $\binom{\frac{1}{2}}{n}$ into nonfractional component as follows:

$$\binom{\frac{1}{2}}{n} = \frac{\frac{1}{2}(\frac{1}{2}-1)(\frac{1}{2}-2)...(\frac{1}{2}-(n-1))(\frac{1}{2}-n)!}{n!(\frac{1}{2}-n)!} = \frac{\frac{1}{2}(\frac{1}{2}-1)(\frac{1}{2}-2)...(\frac{1}{2}-(n-1))}{n!}$$

$$= \frac{\frac{1}{2}(-\frac{1}{2})(-\frac{3}{2})...(-\frac{2n-3}{2})}{n!} = \frac{\frac{1}{2}(-\frac{1}{2})^{n-1}(1)(3)(5)...(2n-5)(2n-3)}{n!}$$

$$= \frac{1}{2}\left(-\frac{1}{2}\right)^{n-1} \frac{(1)(3)(5)...(2n-5)(2n-3)}{n!}$$

$$= \frac{1}{2}\left(-\frac{1}{2}\right)^{n-1} \frac{(1)(2)(3)(4)(5)...(2n-5)(2n-4)(2n-3)(2n-2)}{n!(2)(4)(6)...(2n-4)(2n-2)}$$

$$= \frac{1}{2}\left(-\frac{1}{2}\right)^{n-1} \frac{(1)(2)(3)(4)(5)...(2n-5)(2n-4)(2n-3)(2n-2)}{n!2^{n-1}(1)(2)(3)...(n-2)(n-1)}$$

$$= \frac{1}{2}\left(-\frac{1}{2}\right)^{n-1} \frac{(2n-2)!}{n!2^{n-1}(n-1)!} = (-1)^{n-1} \frac{(2n-2)!}{n!2^{2n-1}(n-1)!} \quad - - \ - - \ - - \ - - \ - - \ -(*)$$

$$= (-1)^{n-1} \frac{2n(2n-1)(2n-2)!n}{n!2^{2n-1}n(n-1)!2n(2n-1)} = (-1)^{n-1} \frac{(2n)!}{n!n!2^{2n}(2n-1)}$$

$$= (-1)^{n-1} \binom{2n}{n} \frac{1}{2^{2n}(2n-1)}$$

Thus,

$$F(z) = -\sum_{n=1}^{\infty} \binom{\frac{1}{2}}{n} (-1)^n z^n = \sum_{n=1}^{\infty} \binom{2n}{n} \frac{1}{2^{2n}(2n-1)} z^n \tag{8.27}$$

$$f_{2n} = \binom{2n}{n} \frac{1}{2^{2n}(2n-1)} \quad n = 1, 2, \ldots$$

$$= \frac{u_{2m}}{2n-1} \tag{8.28}$$

Observe from (*) that

$$\binom{\frac{1}{2}}{n} = (-1)^{n-1} \frac{(2n-2)!}{n!2^{2n-1}(n-1)!} = (-1)^{n-1} \frac{(2n-2)!}{n(n-1)!2^{2n-1}(n-1)!}$$

$$= (-1)^{n-1} \binom{2n-2}{n-1} \left(\frac{1}{2}\right)^{2n-2} \left(\frac{1}{2n}\right) = (-1)^{n-1} \frac{u_{2n-2}}{2n}$$

Thus,

$$F(z) = -\sum_{n=1}^{\infty} \binom{\frac{1}{2}}{n} (-1)^n z^n = \sum_{n=1}^{\infty} \frac{u_{2n-2}}{2n} z^n$$

This means that

$$f_{2n} = \frac{u_{2n-2}}{2n} \tag{8.29}$$

Proposition 8.1 The probability that from time 1 to $2n$ the walker does not return to the origin is u_{2n}; that is,

$$P[Y_1 \neq 0, Y_2 \neq 0, \ldots, Y_n \neq 0] = u_{2n} \tag{8.30}$$

Let $P[Y_{2n} = 2k] = p_{2n,2k}$ denote the probability that the latest return to the origin of a walk of length $2n$ occurs at time $2k$.

Proposition 8.2

$$p_{2n,2k} = u_{2k}u_{2n-2k} = \binom{2k}{k} \binom{2n-2k}{n-k} \left(\frac{1}{2}\right)^{2n} \quad k = 0, 1, \ldots, n$$

Proof

$$P[Y_{2n} = 2k] = P[Y_{2k} = 0, Y_{2k+1} \neq 0, Y_{2k+2} \neq 0, \ldots, Y_{2n} \neq 0]$$
$$= P[Y_{2k} = 0]P[Y_{2k+1} \neq 0, Y_{2k+2} \neq 0, \ldots, Y_{2n} \neq 0 | Y_{2k} = 0]$$
$$= P[Y_{2k} = 0]P[Y_1 \neq 0, Y_2 \neq 0, \ldots, Y_{2n-2k} \neq 0]$$
$$= P[Y_{2k} = 0]P[Y_{2n-2k} = 0]$$
$$= u_{2k}u_{2n-2k}$$

where the second to the last equation follows from Eq. (8.30). This proposition simply states that the walk consists of a loop of length $2k$ followed by a path of length $2n - 2k$ with no return. Observe that $p_{2n,2k}$ is symmetrical in the sense that

$$p_{2n,2k} = p_{2n-2k,2n}, \quad k = 0, 1, \ldots, n$$

That is, the value of $p_{2n,2k}$ is the same for k and $n - k$. This implies that if the walker is on the positive side at half time, then with probability one half he will return to the positive side for the remainder of the walk.

We can use the Stirling's formula to obtain the limiting value of $p_{2n,2k}$ as $n \to \infty$ as follows. Recall that $n! \sim (n/e)^n \sqrt{2\pi n}$ for $n \to \infty$. Thus,

$$p_{2n,2k} = \binom{2k}{k}\binom{2n-2k}{n-k}\left(\frac{1}{2}\right)^{2n} = \frac{(2k)!(2n-2k)!}{(k!)^2((n-k)!)^2}\left(\frac{1}{2}\right)^{2n}$$

$$\sim \frac{(2k/e)^{2k}\{\sqrt{4\pi k}\}\{(2n-2k)/e\}^{2n-2k}\{\sqrt{4\pi(n-k)}\}}{(k/e)^{2k}\{\sqrt{2\pi k}\}^2\{(n-k)/e\}^{2n-2k}\{\sqrt{2\pi(n-k)}\}^2}\left(\frac{1}{2}\right)^{2n} \tag{8.31}$$

$$= \frac{1}{\pi\sqrt{k(n-k)}} = \frac{1}{n\pi\sqrt{\frac{k}{n}\left(1-\frac{k}{n}\right)}}$$

This is the so-called arc-sine law. If we define $l = k/n$, the last function becomes

$$f(l) = \frac{1}{n\pi\sqrt{l(1-l)}}$$

which is a U-shaped function that is flat in the middle of $[0, 1]$ and goes to infinity at the endpoints, as shown in Figure 8.7.

8.12 Maximum of a Random Walk

Let M_n denote the maximum of a symmetric random walk up to time n; that is,

$$M_n = \max\{Y_k : k = 1, 2, \ldots, n\}$$

Theorem 8.2 For a symmetric random walk and any $r \geq 1$,

$$P[M_n \geq r, Y_n = b] = \begin{cases} P[Y_n = b] & b \geq r \\ P[Y_n = 2r - b] & b < r \end{cases}$$

Proof Let T_r be the time at which the path touches the line $Y = r$ for the first time. For every path with $Y_n \geq r$, there exists, according to the reflection principle, another path obtained by reflection through the line $Y_n = r$ from T_r onward up to n, as shown illustrated in Figure 8.8.

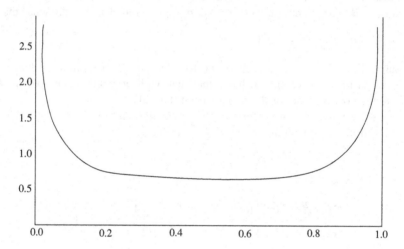

Figure 8.7 The arc-sine distribution.

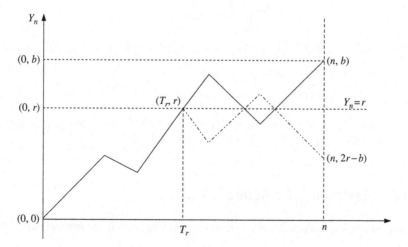

Figure 8.8 Illustration of reflection principle for maximum of random walk.

From the figure we observe that the case when $b \geq r$ is obvious. For the case when $b < r$, we proceed as follows. Let $N_{0,n}^r(0,b)$ denote the number of paths from $(0,0)$ to (n,b) that includes some points on the line $Y_n = r$; that is, some point $(i,r), 0 < i < n$. For such a path we reflect the segment with $T_r \leq x \leq n$ on the line $Y = r$ to obtain a path from $(0,0)$ to $(n, 2r - b)$. Let $N_{0,n}(0, 2r - b)$ be the number of paths from $(0,0)$ to $(n, 2r - b)$. Then according to the reflection principle, $N_{0,n}^r(0,b) = N_{0,n}(0, 2r - b)$. Thus,

$$P[M_n \geq r, Y_n = b] = N_{0,n}^r(0,b)\left(\frac{1}{2}\right)^n$$

$$= N_{0,n}(0, 2r - b)\left(\frac{1}{2}\right)^n$$

$$= P[Y_n = 2r - b]$$

This completes the proof. From this we obtain

$$P[M_k \geq r] = \sum_{b=-\infty}^{\infty} P[M_k \geq r, Y_k = b]$$

$$= \sum_{b=r}^{\infty} P[M_k \geq r, Y_k = b] + \sum_{b=-\infty}^{r-1} P[M_k \geq r, Y_k = b]$$

$$= \sum_{b=r}^{\infty} P[Y_k = b] + \sum_{b=-\infty}^{r-1} P[Y_k = 2r - b]$$

$$= P[Y_k \geq r] + \sum_{m=r+1}^{\infty} P[Y_k = m]$$

$$= P[Y_k = r] + \sum_{m=r+1}^{\infty} P[Y_k = m] + \sum_{m=r+1}^{\infty} P[Y_k = m]$$

$$= P[Y_k = r] + 2\sum_{m=r+1}^{\infty} P[Y_k = m]$$

$$= P[Y_k = r] + 2P[Y_k > r]$$

8.13   Random Walk on a Graph

A graph $G = (V, E)$ is a pair of sets V (or $V(G)$) and E (or $E(G)$) called vertices (or nodes) and edges (or arcs), respectively, where the edges join different pairs of vertices. The vertices are represented by points, and the edges are represented by lines joining the vertices. A graph is a mathematical concept that captures the notion of connection. A graph is defined to be a *simple graph* if there is at most one edge connecting any pair of vertices and an edge does not loop to connect a vertex to itself. When multiple edges are allowed between any pair of vertices, the graph is called a *multigraph*. Examples of a simple graph, a multigraph and a graph with loop are shown in Figure 8.9.

Two vertices are said to be *adjacent* if they are joined by an edge. For example, in Figure 8.9, vertices 1 and 2 are adjacent. An edge e that connects vertices a and b is denoted by (a, b). Such an edge is said to be *incident* with vertices a and b; the vertices a and b are called the *ends* or *endpoints* of e. If the edge $e = (a, b)$ exists, we sometimes call vertex b a *neighbor* of vertex a. The set of neighbors of vertex a is usually denoted by $\Gamma(a)$.

The *degree* (or *valency*) of a vertex x, which is denoted by $d(x)$, is the number of edges that are incident with x. For example, in Figure 8.9(a), $d(3) = 4$ and $d(4) = 2$. If a node x has $d(x) = 0$, then x is said to be *isolated*. It can be shown that

$$\sum_{x \in V(G)} d(x) = 2|E(G)|$$

where $|E(G)|$ is the number of edges in the graph.

A *subgraph* of G is a graph H such that $V(H) \subseteq V(G)$ and $E(H) \subseteq E(G)$, and the endpoints of an edge $e \in E(H)$ are the same as its endpoints in G. A *complete graph* K_n on n vertices is the simple graph that has all $\binom{n}{2}$ possible edges.

A *walk* in a graph is an alternating sequence $x_0, e_1, x_1, e_2, \ldots, x_{k-1}, e_k, x_k$ of vertices x_i, which are not necessarily distinct, and edges e_i such that the endpoints of e_i are x_{i-1} and $x_i, i = 1, \ldots, k$. A *path* is a walk in which the vertices are distinct. For example, in Figure 8.9(a), the path $\{1, 3, 5\}$ connects vertices 1 and 5. When a path can be found between every pair of distinct vertices, we say that the graph is a *connected graph*. A graph that is not connected can be decomposed into two or more connected subgraphs, each pair of which has no node in common. That is, a disconnected graph is the union of two or more disjoint subgraphs.

Another way to describe a graph is in terms of the *adjacency matrix* $A(x, y)$, which has a value 1 in its cell if x and y are neighbors and zero otherwise, for all $x, y \in V$. Then the degree of vertex x is given by

$$d(x) = \sum_{y} A(x, y)$$

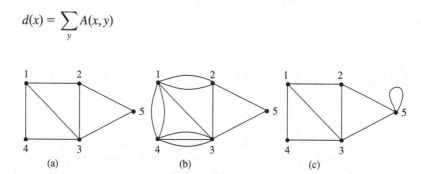

(a)   (b)   (c)

Figure 8.9 Examples of (a) simple graph, (b) multigraph, and (c) graph with loop.

The proximity measures for connected graphs include the following:

- The *hitting time* from node v_i to node v_j is denoted by $H(v_i, v_j)$ and defined as the expected number of steps required to reach v_j for the first time from v_i. The hitting time is not symmetric because generally $H(v_i, v_j) \neq H(v_j, v_i)$.
- The *cover time* $C(v_i)$ from node v_i is the expected number of steps required to visit all the nodes starting from v_i. The cover time for a graph is the maximum $C(v_i)$ over all nodes v_i and denoted by $C(G)$.
- The *commute time* $C(v_i, v_j)$ between node v_i and node v_j is the expected number of steps that it takes to go from v_i to v_j and back to v_i. That is,

$$C(v_i, v_j) = H(v_i, v_j) + H(v_j, v_i)$$

The commute time is symmetric in the sense that $C(v_i, v_j) = C(v_j, v_i)$.

A bound for $C(G)$ was obtained by Kahn et al. (1989) as $C(G) \leq 4n^2 d_{ave}/d_{min}$, where n is the number of nodes in the graph, d_{ave} is the average degree of the graph, and d_{min} is the minimum degree of the graph. In Bollobas (1998) it is shown that in a connected graph with m edges the mean *return time* to a vertex v, which is denoted by $H(v, v)$, is given by

$$H(v, v) = \frac{2m}{d(v)}$$

Graphs are often used to model relationships. When there is a special association in these relationships, the *undirected graphs* we have described so far do not convey this information; a *directed graph* is required. A directed graph (or *digraph*) is a graph in which an edge consists of an ordered vertex pair, giving it a direction from one vertex to the other. Generally in a digraph the edge (a, b) has a direction from vertex a to vertex b, which is indicated by an arrow in the direction from a to b. Figure 8.10 illustrates a simple digraph. When the directions are ignored, we obtain the underlying undirected graph shown in Figure 8.9(a).

Let $G = (V, E)$ be a connected undirected graph with n vertices and m edges. A random walk on G can be described as follows. We start at vertex v_0 and arrive at vertex v_i in the kth step. We move to vertex v_j, which is one of the neighbors of vertex v_i, with probability $1/d(v_i)$. The sequence of random vertices $\{v_t, t = 0, 1, \ldots\}$ is a Markov chain with transition probabilities p_{ij} given by

$$p_{ij} = \begin{cases} 1/d(i) & i, j \in V \\ 0 & \text{otherwise} \end{cases}$$

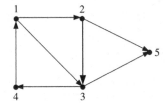

Figure 8.10 Example of a digraph.

Let $P = [p_{ij}]_{i,j \in V}$ be the state-transition probability matrix. Then for the simple graph in Figure 8.9(a), we have that

$$P = \begin{bmatrix} 0 & 1/3 & 1/3 & 1/3 & 0 \\ 1/3 & 0 & 1/3 & 0 & 1/3 \\ 1/4 & 1/4 & 0 & 1/4 & 1/4 \\ 1/2 & 0 & 1/2 & 0 & 0 \\ 0 & 1/2 & 1/2 & 0 & 0 \end{bmatrix}$$

which corresponds to the state-transition diagram shown in Figure 8.11.

Random walk on a graph is used as a search technique in which a search proceeds from a start node by randomly selecting one of its neighbors, say k. At k the search randomly selects one of its neighbors, making an effort not to reselect the node from where it reached k, and so on. If the goal is to reach a particular destination node, the search terminates when this destination is reached.

We can construct the Markov chain of the multigraph in a similar manner. In this case,

$$p_{ij} = \begin{cases} n_{ij}/d(i) & i, j \in V \\ 0 & \text{otherwise} \end{cases}$$

where n_{ij} is the number of edges between nodes i and j. For example, in the multigraph of Figure 8.9(a), we have that

$$P = \begin{bmatrix} 0 & 2/5 & 1/5 & 2/5 & 0 \\ 1/2 & 0 & 1/4 & 0 & 1/4 \\ 1/6 & 1/6 & 0 & 1/2 & 1/6 \\ 2/5 & 0 & 3/5 & 0 & 0 \\ 0 & 1/2 & 1/2 & 0 & 0 \end{bmatrix}$$

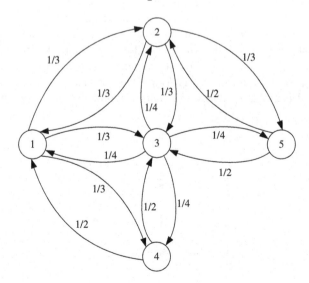

Figure 8.11 State-transition diagram of graph in Figure 8.9(a).

The Markov chain of the multigraph is shown in Figure 8.12.

The Markov chain associated with a random walk on a graph is irreducible if and only if the graph is connected. Let m denote the number of edges in an undirected connected graph $G = (V, E)$, and let $\{\pi_k, k \in V\}$ be the stationary distribution of the Markov chain associated with the graph. The stationary distribution of the Markov chain associated with $G = (V, E)$ is given by the following theorem:

Theorem 8.3 The stationary distribution of the Markov chain associated with the connected graph $G = (V, E)$ is given by $\pi_i = d(i)/2m, i = 1, \ldots, n$; where m is the number of edges in the graph, as defined earlier.

Proof The proof consists in our showing that the distribution $\pi = (\pi_1, \ldots, \pi_n)$ satisfies the equation $\pi P = \pi$. We prove the theorem with a multigraph, which is more general than the simple graph. Thus, we have that with respect to node j,

$$(\pi P)_j = \sum_i \pi_i p_{ij} = \sum_i \left\{ \frac{d(i)}{2m} \times \frac{n_{ij}}{d(i)} \right\} = \frac{1}{2m} \sum_i n_{ij} = \frac{d(j)}{2m} = \pi_j$$

This implies that by definition π is the stationary distribution of the unique Markov chain defined by P. This completes the proof.

Note that for the simple graph we have that $n_{ij} = 1$, and the same result holds.

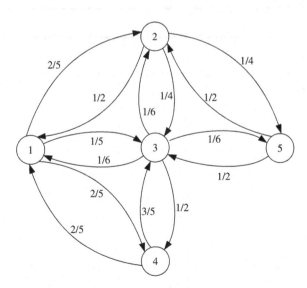

Figure 8.12 State-transition diagram of multigraph in Figure 8.9(b).

Example 8.1

Consider the simple graph of Figure 8.9(a). We have that $m = 7$, which means that the stationary distribution is given by

$$\pi = \left(\frac{3}{14}, \frac{3}{14}, \frac{4}{14}, \frac{2}{14}, \frac{2}{14}\right)$$

Similarly, for the multigraph of Figure 8.9(b), the number of edges is $m = 11$. Thus, the stationary distribution of the Markov chain in Figure 8.11 is given by

$$\pi = \left(\frac{5}{22}, \frac{4}{22}, \frac{6}{22}, \frac{5}{22}, \frac{2}{22}\right)$$

Note that a loop is considered to contribute twice to the degree of a node. Thus, in the case of the graph with loop shown in Figure 8.9(c), $m = 8$, and because $d(5) = 4$, we obtain the stationary distribution as follows:

$$\pi = \left(\frac{3}{16}, \frac{3}{16}, \frac{4}{16}, \frac{2}{16}, \frac{4}{16}\right)$$

Recall that the mean return time to a node v in a connected graph with m edges is given by $H(v, v) = 2m/d(v)$. From the results on the stationary distributions we may then write

$$H(v, v) = 1/\pi_v$$

Example 8.2

Consider a random walk on a two-dimensional lattice consisting of the 4×4 checkerboard shown in Figure 8.13.

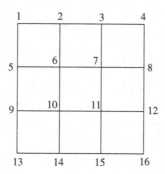

Figure 8.13 Checkerboard.

The number of edges is $m = 24$, and the degrees of the nodes are as follows:

$$d(1) = d(4) = d(13) = d(16) = 2$$
$$d(2) = d(3) = d(5) = d(8) = d(9) = d(12) = d(14) = d(15) = 3$$
$$d(6) = d(7) = d(10) = d(11) = 4$$

Thus, the stationary probabilities are

$$\pi_1 = \pi_2 = \pi_{13} = \pi_{16} = 1/24$$
$$\pi_2 = \pi_3 = \pi_5 = \pi_8 = \pi_9 = \pi_{12} = \pi_{14} = \pi_{15} = 1/16$$
$$\pi_6 = \pi_7 = \pi_{10} = \pi_{11} = 1/12$$

8.13.1 Random Walk on a Weighted Graph

A more general random walk on a graph is that performed on a weighted graph. Specifically, we consider a connected graph $G = (V, E)$ with positive weight w_e assigned to edge $e \in E$. The weighted random walk is a random walk where the transition probabilities are proportional to the weights of the edges; that is,

$$p_{ij} = \frac{w_{(i,j)}}{\sum_{k \sim i} w_{(i,k)}} \equiv \frac{w_{ij}}{\sum_{k \sim i} w_{ik}}$$

If all the weights are 1, we obtain a simple random walk. Let the total weight of the edges emanating from node i be w_i, which is given by

$$w_i = \sum_{k \sim i} w_{ik}$$

Then the sum of the weights of all edges is

$$w = \sum_{i,k|k>i} w_{ik}$$

where the inequality in the summation is used to avoid double counting. It is easy to show that the stationary distribution is given by

$$\pi_i = \frac{w_i}{2w}$$

The proof consists as usual in verifying that the preceding distribution satisfies the relationship $\pi P = \pi$, which can be seen as follows:

$$\sum_i \pi_i p_{ij} = \sum_i \left\{ \frac{w_i}{2w} \times \frac{w_{ij}}{w_i} \right\} = \sum_i \frac{w_{ij}}{2w} = \frac{1}{2w} \sum_i w_{ij}$$
$$= \frac{w_j}{2w} = \pi_j$$

Thus, the stationary probability of state i is proportional to the weight of the edges emanating from node i.

8.14 Correlated Random Walk

In the classical random walk described earlier, there is a constant probability p of incrementing the current state by 1 and, therefore, a constant probability $q = 1 - p$ of decrementing the current state by 1. This is due to the fact that the outcomes of the experiments are independent. A more general case is when there is a correlation between results of successive trials; that is, the outcome of the current trial depends on the outcome of the previous trial. Such random walks are called *correlated random walks* (CRWs) or *persistent random walks* and have been studied by many authors including Goldstein (1951), Gillis (1955), Mohan (1955), Seth (1963), Renshaw and Henderson (1981), Kehr and Argyrakis (1986), Lal and Bhat (1989), Argyrakis and Kehr (1992), Hanneken and Franceschetti (1998), and Bohm (2000).

To illustrate this process, we consider an experiment with two possible outcomes: a win or a loss. The probability of the first outcome is assumed to be fixed, and thereafter each subsequent outcome is governed by the following rule. Given that the current trial results in a win, the probability that the next trial will result in a win is p_1, and the probability that it will result in a loss is $q_1 = 1 - p_1$. Similarly, given the current trial results in a loss, the probability that the next trial will result in a loss is p_0, and the probability that it will result in a win is $q_0 = 1 - p_0$. Thus, the conditional probabilities are as follows:

$$P[\text{win}|\text{win}] = p_1$$
$$P[\text{loss}|\text{win}] = q_1 = 1 - p_1$$
$$P[\text{loss}|\text{loss}] = p_0$$
$$P[\text{win}|\text{loss}] = q_0 = 1 - p_0$$

The CRW can be modeled as a bivariate Markov chain with the location of the walker and the result of the previous walk as the two variables. Specifically, we represent a state by (k, l), where k is the location of the walker, where a move to the right indicates a success and a move to the left indicates a failure; and l is the result index that is defined by

$$l = \begin{cases} 1 & \text{if the previous trial resulted in a successs} \\ 0 & \text{if the previous trial resulted in a failure} \end{cases}$$

The range of k depends on the type of CRW. When the walk is unrestricted, then $k = \ldots, -2, -1, 0, 1, 2, \ldots$. Similarly, when barriers exist, we have that $k = 0, 1, 2, \ldots, N$, where N can be infinite. Thus, the state-transition diagrams in Figure 8.14 show the cases of unrestricted CRW, CRW with reflecting barriers, and CRW with absorbing barriers.

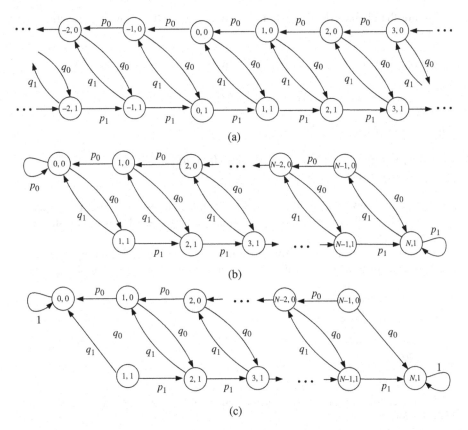

Figure 8.14 State-transition diagrams for CRWs. (a) Unrestricted CRW. (b) CRW with reflecting barriers. (c) CRW with absorbing barriers.

As can be seen from Figure 8.14, the CRW is a quasi-birth-and-death process. Let $\pi_{k,0}(n)$ denote the probability of being in state $(k,0)$ after n trials, and let $\pi_{k,1}(n)$ denote the probability of being in state $(k,1)$ after n trials. Then we have the following difference equations:

$$\pi_{k,0}(n+1) = p_0\pi_{k+1,0}(n) + q_1\pi_{k+1,1}(n) \tag{8.32}$$

$$\pi_{k,1}(n+1) = q_0\pi_{k-1,0}(n) + p_1\pi_{k-1,1}(n) \tag{8.33}$$

Let the characteristic function of $\pi_{k,l}(n)$ be defined by

$$\Pi_l(w,n) = \sum_{k=-\infty}^{\infty} w^k\pi_{k,l}(n) \quad l=0,1; n=1,2,\ldots$$

Taking the characteristic functions of both sides of the difference equations we obtain

$$\Pi_0(w, n+1) = \frac{p_0}{w} \Pi_0(w, n) + \frac{q_1}{w} \Pi_1(w, n)$$

$$\Pi_1(w, n+1) = wq_0 \Pi_0(w, n) + wp_1 \Pi_1(w, n)$$

which can be arranged in the following form:

$$\begin{bmatrix} \Pi_0(w, n+1) \\ \Pi_1(w, n+1) \end{bmatrix} = \begin{bmatrix} w^{-1}p_0 & w^{-1}q_1 \\ wq_0 & wp_1 \end{bmatrix} \begin{bmatrix} \Pi_0(w, n) \\ \Pi_1(w, n) \end{bmatrix}$$

The initial states $\pi_{k,0}(0)$ and $\pi_{k,1}(0)$ can be obtained by noting that the sequence of result indices $\{l\}$ of the state (k, l) constitutes a two-state Markov chain shown in Figure 8.15.

Thus, if we let $\pi_0(0)$ denote the probability that the process initially started with the first trial being a failure and $\pi_1(0)$ the probability that the first trial was a success, then we have that

$$\pi_0(0) = \frac{q_1}{q_0 + q_1}$$

$$\pi_1(0) = \frac{q_0}{q_0 + q_1}$$

The exact solution of the transform equation of the correlated walk after n trials is of the form

$$\Pi(w, n) = \begin{bmatrix} \Pi_0(w, n) \\ \Pi_1(w, n) \end{bmatrix} = \begin{bmatrix} w^{-1}p_0 & w^{-1}q_1 \\ wq_0 & wp_1 \end{bmatrix}^n \begin{bmatrix} \pi_0(0) \\ \pi_1(0) \end{bmatrix}$$

Let A be the coefficient matrix; that is,

$$A = \begin{bmatrix} w^{-1}p_0 & w^{-1}q_1 \\ wq_0 & wp_1 \end{bmatrix}$$

Then we have that

$$\Pi(w, n) = A^n \pi(0) \tag{8.34}$$

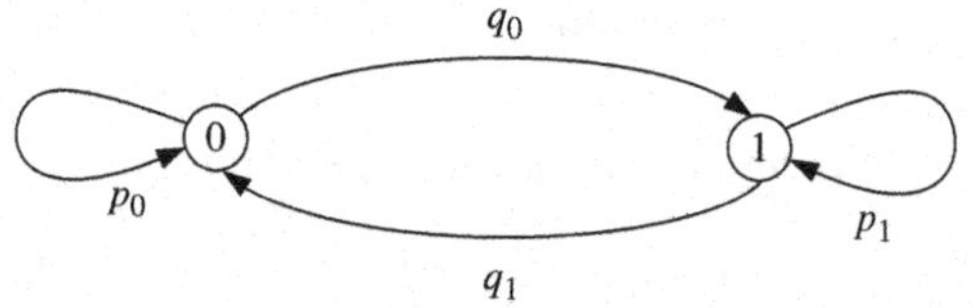

Figure 8.15 State-transition diagrams for result sequence.

We solve the equation via the diagonalization method, which requires our obtaining the eigenvalues of A. The eigenvalues λ of A satisfy the equation

$$|\lambda I - A| = 0$$

which gives

$$\lambda^2 - (wp_1 + w^{-1}p_0)\lambda + p_0p_1 - q_0q_1 = 0$$

from which we obtain

$$\lambda = \frac{(wp_1 + w^{-1}p_0) \pm \sqrt{(wp_1 - w^{-1}p_0)^2 + 4q_0q_1}}{2}$$

Thus, the eigenvalues are

$$\lambda_1 = \frac{(wp_1 + w^{-1}p_0) + \sqrt{(wp_1 - w^{-1}p_0)^2 + 4q_0q_1}}{2}$$

$$\lambda_2 = \frac{(wp_1 + w^{-1}p_0) - \sqrt{(wp_1 - w^{-1}p_0)^2 + 4q_0q_1}}{2}$$

Let the matrix X be defined by $X = (X_1, X_2)$, where $X_i = (x_{1i}, x_{2i})^T$, $i = 1, 2$. The eigenvectors belonging to λ_1 are obtained from

$$(\lambda_1 I - A)X_1 = 0$$

That is,

$$(\lambda_1 - w^{-1}p_0)x_{11} - w^{-1}q_1x_{21} = 0$$
$$-wq_0x_{11} + (\lambda_1 - wp_1)x_{211} = 0$$

If we set $x_{11} = 1$ in the first equation, we obtain $x_{21} = (\lambda_1 - w^{-1}p_0)/w^{-1}q_1 = (\lambda_1 w - p_0)/q_1$. Note that $(\lambda_1 - w^{-1}p_0)/w^{-1}q_1 = q_0/(\lambda_1 - wp_1)$ because this relationship yields the equation

$$\lambda_1^2 - (wp_1 + w^{-1}p_0)\lambda_1 + p_0p_1 - q_0q_1 = 0$$

that we know to be true. Thus,

$$X_1 = (x_{11}, x_{21})^T = [1, (\lambda_1 w - p_0)/q_1]^T$$

is an eigenvector. Similarly, the eigenvectors belonging to λ_2 are obtained from

$$(\lambda_2 I - A)X_2 = 0$$

which is similar to the equation for the eigenvectors belonging to λ_1. Thus, another eigenvector is

$$X_2 = (x_{12}, x_{22})^T = [1, (\lambda_2 w - p_0)/q_1]^T$$

Then using the diagonalization method we obtain

$$\Pi(w, n) = A^n \pi(0) = S \Lambda^n S^{-1} \pi(0) \tag{8.35}$$

where $A = S \Lambda S^{-1}$, S is the matrix whose columns are the eigenvectors and Λ is the diagonal matrix of the corresponding eigenvalues. That is,

$$S = \begin{bmatrix} 1 & 1 \\ (\lambda_1 w - p_0)/q_1 & (\lambda_2 w - p_0)/q_1 \end{bmatrix}$$

$$S^{-1} = \frac{q_1}{w(\lambda_2 - \lambda_1)} \begin{bmatrix} (\lambda_2 w - p_0)/q_1 & -1 \\ -(\lambda_1 w - p_0)/q_1 & 1 \end{bmatrix}$$

$$\Lambda = \begin{bmatrix} \lambda_1 & 0 \\ 0 & \lambda_2 \end{bmatrix} \Rightarrow \Lambda^n = \begin{bmatrix} \lambda_1^n & 0 \\ 0 & \lambda_2^n \end{bmatrix}$$

From this we can obtain the $\pi_{k,1}(n)$, which are the inverse transforms of the components of $\Pi(w, n)$.

CRW is popularly used in ecology to model animal and insect movement. For example, it is used in Kareiva and Shigesada (1983) to analyze insect movement. Similarly, it has been used in Byers (2001) to model animal dispersal and in Jonsen et al. (2005) to model animal movement. CRW has also been used to model mobility in mobile *ad hoc* networks (MANETs) in Bandyopadhyay et al. (2006).

8.15 Continuous-Time Random Walk

In the random walk models described earlier, a walker takes steps in a periodic manner, such as every second or minute or hour, or any other equal time interval. More importantly, a classical random walk is a Bernoulli process that allows only two possible events that have values of ± 1. A more general case is when the time between steps is a random variable and the step size is a random variable. In this case we obtain a continuous-time random walk (CTRW), which was introduced by Montroll and Weiss (1965). CTRW has been applied in many physical phenomena. It is used in physics to model diffusion with instantaneous jumps from one position to the next. For example, it is used in Scher and Montroll (1975) to model the transport of amorphous materials, in Montroll and Shlensinger (1984) to model the transport in disordered media, and in Weiss et al. (1998) to model the transport in turbid media. In Helmstetter and Sornette (2002) it is used to model earthquakes. Recently it has been used in econophysics to model the financial market, particularly to describe the movement of log-prices. Examples of the application of CTRW in econophysics are discussed in Masoliver and Montero (2003), Masoliver et al. (2006), and Scalas (2006a,b).

As discussed earlier, CTRW is a random walk that permits intervals between successive walks to be independent and identically distributed. Thus, the walker starts at the point zero at time $T_0 = 0$ and waits until time T_1 when he makes a jump of size θ_1, which is not necessarily positive. The walker then waits until time T_2 when he makes another jump of size θ_2, and so on. The jump sizes θ_i are also assumed to be independent and identically distributed. Thus, we assume that the times $T_1, T_2, \ldots$ are the instants when the walker makes jumps. The intervals $\tau_i = T_i - T_{i-1}$, $i = 1, 2, \ldots$, are called the *waiting times* (or *pausing times*) and are assumed to be independent and identically distributed. The time at which the nth walk occurs, T_n, is given by

$$T_n = T_0 + \sum_{i=1}^{n} \tau_i \quad n = 1, 2, \ldots; T_0 = 0 \tag{8.36}$$

Thus, one major difference between the classical or discrete-time random walk (DTRW) and the CTRW is that in CTRW the waiting time between jumps is not constant as in the case of DTRW, and the step size is a random variable unlike the DTRW where the step size is ± 1.

The position of the walker at time t is given by

$$X(t) = \sum_{i=1}^{N(t)} \theta_i \tag{8.37}$$

where the upper limit $N(t)$ is a random function of time that denotes the number of jumps up to time t and is given by

$$N(t) = \max\{n : T_n \le t\}$$

Let T denote the waiting time and let Θ denote the jump size. Similarly, let $f_{X(t)}(x, t)$ denote the probability density function (PDF) of $X(t)$, $f_\Theta(\theta)$ the PDF of Θ, $f_T(t)$ the PDF of T, and $p_{N(t)}(n, t)$ the PMF of $N(t)$.

Let $P(x, t)$ denote the probability that the position of the walker at time t is x, given that it was in position 0 at time $t = 0$; that is,

$$P(x, t) = P[X(t) = x | X(0) = 0] \tag{8.38}$$

For an *uncoupled* CTRW, it is assumed that Θ and T are independent so that the joint PDF of the jump size and waiting time $f_{\Theta T}(\theta, t) = f_\Theta(\theta)f_T(t)$. In a *coupled* CTRW, Θ and T are not independent so that

$$f_\Theta(\theta) = \int_0^\infty f_{\Theta T}(\theta, t)dt \tag{8.39}$$

$$f_T(t) = \int_{-\infty}^\infty f_{\Theta T}(\theta, t)d\theta \tag{8.40}$$

Let $\Phi_{X(t)}(w)$ denote the characteristic function of $X(t)$, let $\Phi_\Theta(w)$ denote the characteristic function of Θ, and let $G_{N(t)}(z)$ denote the z-transform of the PMF of $N(t)$, where

$$\Phi_{X(t)}(w) = E[e^{jwX(t)}] = \int_{-\infty}^\infty e^{jwx}f_{X(t)}(x, t)dx$$

$$\Phi_\Theta(w) = E[e^{jw\Theta}] = \int_{-\infty}^\infty e^{jw\theta}f_\Theta(\theta)d\theta$$

$$G_{N(t)}(z) = E[z^{N(t)}] = \sum_{n=0}^\infty z^n p_{N(t)}(n, t)$$

Because $X(t)$ is a random sum of random variables, we know from probability theory (see Ibe (2005), for example) that

$$\Phi_{X(t)}(w) = G_{N(t)}(\Theta(w)) \tag{8.41}$$

Thus, if $p_{N(t)}(n, t)$ and $f_\Theta(\theta)$ are known, we can determine $\Phi_{X(t)}(w)$ and consequently $f_{X(t)}(x, t)$. The expected value of $X(t)$ is given by

$$E[X(t)] = E[N(t)]E[\Theta] \tag{8.42}$$

The expected value of $N(t)$ can be obtained as follows. Because the intervals τ_i are independent and identically distributed, we note that the process $\{N(t):t \geq 0\}$ is a renewal process and T_n is the time of the nth renewal. Now

$$p_{N(t)}(n, t) = P[N(t) = n] = P[N(t) < n + 1] - P[N(t), n] = P[T_{n+1} < t] - P[T_n < t]$$

$$= 1 - F_{T_{n+1}}(t) - \{1 - F_{T_n}(t)\}$$

$$= F_{T_n}(t) - F_{T_{n+1}}(t)$$

and the expected value of $N(t)$ is given by

$$E[N(t)] = \sum_{n=0}^{\infty} n p_{N(t)}(n,t) = \sum_{n=0}^{\infty} n[F_{T_n}(t) - F_{T_{n+1}}(t)]$$

$$= \{F_{T_1}(t) + 2F_{T_2}(t) + 3F_{T_3}(t) + \ldots\} - \{F_{T_2}(t) + 2F_{T_3}(t) + 3F_{T_5}(t) + \ldots\}$$

$$= F_{T_1}(t) + F_{T_2}(t) + F_{T_3}(t) + \ldots$$

$$= \sum_{n=1}^{\infty} F_{T_n}(t)$$

(8.43)

8.15.1 The Master Equation

The relationship between $P(x,t)$ and $f_{\Theta T}(\theta,t)$ is given by the following equation that is generally called the *master equation* of the CTRW:

$$P(x,\ t) = \delta(x)R(t) + \int_0^t \int_{-\infty}^{\infty} P(u,\tau)f_{\Theta T}(x-u,t-\tau)du\ d\tau$$

(8.44)

where $\delta(x)$ is the Dirac delta function and $R(t) = P[T > t] = 1 - F_T(t)$ is called the *survival probability*, which is the probability that the waiting time when the process is in a given state is greater than t. The equation states that the probability that $X(t) = x$ is equal to the probability that the process was in state 0 up to time t, plus the probability that the process was at some state u at time τ, where $0 < \tau \le t$, and within the waiting time $t - \tau$, a jump of size $x - u$ took place. Note that

$$R(t) = \int_t^{\infty} f_T(v)dv = 1 - \int_0^t f_T(v)dv$$

$$f_T(t) = - \frac{dR(t)}{dt}$$

For the uncoupled CTRW, the master equation becomes

$$P(x,t) = \delta(x)R(t) + \int_0^t \int_{-\infty}^{\infty} P(u,\tau)f_{\Theta}(x-u)f_T(t-\tau)du\ d\tau$$

(8.45)

Let the joint Fourier–Laplace transform of $P(x,t)$ be defined as follows:

$$P^*(w,s) = \int_0^{\infty} e^{-st}dt \int_{-\infty}^{\infty} e^{jwx}P(x,t)dx$$

(8.46)

Then the master equation is transformed into

$$
\begin{aligned}
P^*(w,s) &= \int_0^\infty e^{-st} \int_{-\infty}^\infty e^{jwx}[\delta(x)R(t) + \int_0^t \int_{-\infty}^\infty P(u,\tau)f_{\Theta T}(x-u, t-\tau)\mathrm{d}u\,\mathrm{d}\tau]\mathrm{d}x\,\mathrm{d}t \\
&= \tilde{R}(s) + P^*(w,s)\Phi_{\Theta T}(w,s)
\end{aligned}
$$

where $\tilde{R}(s)$ is the Laplace transform of $R(t)$. This gives

$$
P^*(w,s) = \frac{\tilde{R}(s)}{1 - \Phi_{\Theta T}(w,s)} \tag{8.47}
$$

where $\Phi_{\Theta T}(w,s)$ is the joint Fourier–Laplace transform of $f_{\Theta T}(\theta,\tau)$. For an uncoupled CTRW, $\Phi_{\Theta T}(w,s) = \Phi_\Theta(w)\Phi_T(s)$. Thus,

$$
P^*(w,s) = \frac{\tilde{R}(s)}{1 - \Phi_\Theta(w)\Phi_T(s)} \tag{8.48}
$$

Also,

$$
\tilde{R}(s) = \frac{1 - \Phi_T(s)}{s}
$$

Thus, for the uncoupled CTRW we have that

$$
P^*(w,s) = \frac{\tilde{R}(s)}{1 - \Phi_\Theta(w)\Phi_T(s)} = \frac{1 - \Phi(s)T}{s[1 - \Phi_\Theta(w)\Phi_T(s)]} \tag{8.49}
$$

Because $|\Phi_\Theta(w)| < 1$ if $w \neq 0$ and $|\Phi_T(s)| < 1$ if $s \neq 0$, we can rewrite Eq. (8.48) as follows:

$$
P^*(w,s) = \tilde{R}(s) \sum_{n=0}^\infty [\Phi_\Theta(w)\Phi_T(s)]^n \tag{8.50}
$$

Taking the inverse Fourier and Laplace transforms we obtain:

$$
P(x,t) = \sum_{n=0}^\infty P(n,t)f_\Theta^{(n)}(x) \tag{8.51}
$$

where $P(n,t)$ is the probability that n jumps occur up to time t, and $f_\Theta^{(n)}(x)$ is the n-fold convolution of the number of jumps. $P(n,t)$ is given by

$$
P(n,t) = \int_0^t f_T^{(n)}(t-u)R(u)\mathrm{d}u
$$

where $f_T^{(n)}(t)$ is the n-fold convolution of the PDF of the waiting time.

Consider the special case where T is exponentially distributed with a mean of $1/\lambda$, that is,

$$f_T(t) = \lambda\, e^{-\lambda t} \quad t \geq 0$$

This means that

$$R_T(t) = e^{-\lambda t} \quad t \geq 0$$

In this case, we have that $\Phi_T(s) = \lambda/(s + \lambda)$. Thus, for the uncoupled CTRW we obtain

$$P^*(w,s) = \frac{1 - \lambda/(s+\lambda)}{s[1 - \lambda\Phi_\Theta(w)/(s+\lambda)]} = \frac{1}{s + \lambda - \lambda\Phi_\Theta(w)} \tag{8.52}$$

Taking the inverse Laplace transform for this special case we obtain

$$P(w,t) = \exp\{-\lambda t[1 - \Phi_\Theta(w)]\} \tag{8.53}$$

Note that for the special case when $N(t)$ is a Poisson process, the CTRW becomes a compound Poisson process and the analysis is simplified as the characteristic function of $X(t)$ becomes

$$\Phi_{X(t)}(w) = \exp\{-\lambda t[1 - \Phi_\Theta(w)]\} = P(w,t) \tag{8.54}$$

which is not surprising because the fact that $N(t)$ is a Poisson process implies that T is exponentially distributed. The inverse Laplace transform of $P^*(w,s)$ for the uncoupled system is given by

$$P(w,t) = \Phi_{X(t)}(w) = G_{N(t)}(\Phi_\Theta(w)) \tag{8.55}$$

8.16 Self-Avoiding Random Walk

A self-avoiding walk (SAW) is a form of CRW in which the walker is not allowed to visit a previously visited site. This type of walk is typically rendered on a lattice and serves as a model for linear polymer molecules. Polymers are the fundamental building blocks in biological systems. A polymer is a long chain of monomers (i.e., groups of atoms) joined to one another by chemical bonds. These polymer molecules form together randomly with the restriction that no overlaps can occur.

One of the problems with SAW is that although it is a model of random walk its trajectory cannot be described in terms of transition probabilities. Thus, it is a non-Markovian process and consequently it is mathematically difficult to analyze. Currently, its analysis is carried out either through simulation or a direct enumeration of a small number of steps.

It is possible for the random walker to get into a *trap*; that is, he reaches a site whose neighbors have already been visited at which point the walk ends. The SAW in Figure 8.16 illustrates such a SAW called *self-trapping walk* where any movement from the last site the walker visits will result in revisiting a previous site. At this point the walk will terminate.

A related type of SAW is the *self-avoiding polygon* (SAP), which is a closed SAW on a lattice. This means that SAP is a SAW whose last site is adjacent to the first site, as illustrated in Figure 8.17.

There are several reasons for the interest in SAW. First, it is an interesting mathematical concept that is regarded as a classical combinatorial problem. Also, its origin lies in polymer science where it represents a realistic model of long-chain polymers in dilute solution. Polymers are long-chain molecules consisting of a large number of monomers held together by chemical bonds. The self-avoiding property reflects the fact that two monomers cannot occupy the same point in space. One measure of a polymer is number of monomers (i.e., its length); another measure is the average distance from one end to the other. Thus, while SAW is an interesting mathematical concept, it is an important tool in polymer science.

Two issues that are usually addressed in the analysis of self-avoiding random walk are the following:

1. How many possible SAWs of length n can be found?
2. When does the walker get trapped?

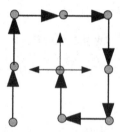

Figure 8.16 Example of self-trapping walk.

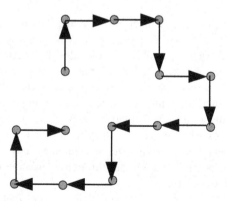

Figure 8.17 Example of SAP.

With respect to the first question, let C_n denote the number of SAWs of n steps that begin at the origin. Many attempts have been made by several authors to obtain C_n. Table 8.1 shows some values of C_n for different values of n on two-dimensional square lattice as given in Slade (1994). Values of C_n for n as high as 39 have been obtained.

Because of the intractability of the problem the preferred method of solution is by simulation. Bounds on C_n can be obtained as follows. The number of simple random walks in a two-dimensional lattice of length n is 4^n since at each site the walker can choose from four possible directions. But for a SAW, he can choose from four directions in the first step, and thereafter he choose from no more than three other locations. Thus, an upper bound on C_n is $C_n \leq 4(3^{n-1})$. A lower bound is obtained by noting that one way to avoid a previously visited location is by moving only in the positive direction that involves only two choices at each location. Thus, we have that

$$2^n \leq C_n \leq 4(3^{n-1})$$

These bounds seem to suggest that we can express C_n by

$$C_n = \alpha(n)\mu^n$$

for some positive number $\mu > 0$ and

$$\lim_{n \to \infty} [\alpha(n)]^{1/n} = 1.$$

This is the so-called Hammersely—Morton theorem, see Hughes (1995). Any SAW of $m + n$ steps can be decomposed into a SAW of m steps followed by a SAW of n steps. Thus C_n satisfies the submultiplicative inequality

$$C_{m+n} \leq C_m C_n$$

Table 8.1 Values of C_n on Two-Dimensional Square Lattice, $n \leq 24$

n	C_n	n	C_n
1	4	13	881,500
2	12	14	2,374,444
3	36	15	6,416,596
4	100	16	17,245,332
5	284	17	46,466,676
6	780	18	124,658,732
7	2172	19	335,116,620
8	5916	20	897,697,164
9	16,258	21	2,408,806,028
10	44,100	22	6,444,560,484
11	120,292	23	17,266,613,812
12	324,932	24	46,146,397,316

Note that the reverse inequality is not true as concatenating two SAWs does not always yield a SAW. If we define $\beta_n = \log C_n$, then β_n satisfies the "subadditive inequality"

$$\beta_{m+n} \leq \beta_m + \beta_n$$

Just as we do not know exactly how many n-step SAWs that start at the origin, we do not know exactly the answer to the second question, which deals with when a self-avoiding walker gets trapped. In the absence of exact analytical methods, asymptotic estimation from exact enumeration of shorter walks has been used to study this property of self-avoiding random walks. The most important statistic is the mean square length of the walk. Let R_n denote the endpoint of a SAW. The mean square length is $E[R_n^2]$, which has been shown by many studies to be of the form

$$E[R_n^2] = An^\nu \quad n \to \infty$$

where $\nu = 1.5$ for a two-dimensional SAW and A is a constant, see Barber and Ninham (1970) and Hughes (1995).

With respect to when a walker becomes trapped, let $P_T(n)$ denote the probability that a self-avoiding walker is trapped after exactly n steps. In Hemmer and Hemmer (1984) it is shown that

$$P_T(n) \propto (n-6)^{0.6} e^{-n/40} \quad n > 6$$
$$P_T(n) = 0 \quad\quad\quad\quad\quad n \leq 6$$

and the average length of SAWs before the walker is trapped is about 71. Thus, when $n \leq 6$, the probability of being trapped is negligible. Traps begin to appear from the seventh step onward. In the two-dimensional SAW, there are eight different walks that form traps at the seventh step. Let K_n denote the number of SAWs that become trapped on the nth step. Table 8.2 shows the values of K_n for different values of n.

Figure 8.18 shows all the 8 different walks that form traps at the 7th step and four of the 16 different walks that form traps at the 8th step.

Table 8.2 Values of K_n on Two-Dimensional Square Lattice, $7 \leq n \leq 14$

n	K_n
7	8
8	16
9	88
10	200
11	760
12	1824
13	6016
14	14,880

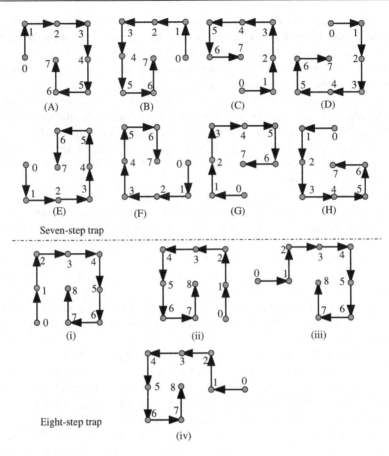

Figure 8.18 Walks trapped at the seventh and eighth steps.

8.17   Nonreversing Random Walk

The nonreversing random walk (NRW) is an intermediate stage between a purely random walk and a SAW. It is a CRW that allows the walker to go left, right or forward at each intersection but not to make a U-turn and go back the way he just came. Thus on a square lattice a nonreversing walker has four choices for the first step, but only three choices for each step thereafter. Another way to view NRW is as follows. A random walk is a walk with no memory while SAW is a walk with unlimited memory because the walker has to remember all the sites he has visited. NRW lies between a walk with no memory and a walk with unlimited memory; there are bounds on the amount of memory possessed by a walker because he only avoids the last site visited.

One reason for the study of NRW is the following. The symmetric two-dimensional random walk assumes that each step is taken independently of the previous steps. This is probably the case for nonhuman walkers but not for humans. A human walker is not likely to go back immediately to the location from where he came, which suggests that his walk has some memory. Therefore, a more reasonable model is one in which the walker will not immediately go back to the location from where he came to the current location. This means that, if the previous step was a west-to-east walk, the next step cannot be an east-to-west walk and vice versa. Similarly, if the previous step was a north-to-south walk, the next step cannot be a south-to-north walk, and vice versa. We assume that if the walker came to the current location by performing a west-to-east movement, he will choose to go east with probability $1/3$, north with probability $1/3$, or south with probability $1/3$, and so for other directions.

Note that in a self-avoidance walk, the walker cannot visit a previously visited location. However, in the NRW the walker can visit a previously visited location, except that he cannot do so immediately after leaving the location since immediate U-turns are forbidden. He can revisit a previously visited location only through visits to at least one other location before doing so. In other words, a visit to a previously visited location is possible via a sequence of locations that loops back to the location.

The one-dimensional nonreversing walk is a trivial walk in which movement is along one direction only. If we assume that the walker has left the origin (or equivalently, the process is in the steady state), the two-dimensional case can be modeled by a Markov chain, as discussed in Narayanan (2011). Let U be the state that represents the event that the last walk was an upward step, D the state that represents the event that the last walk was a downward step, R the state that represents the event that the last walk was a step to the right, and L the state that represents the event that the last walk was a step to the left. Then Figure 8.19 is the Markov chain for the walk. For example, when it is in state U, it cannot make a transition to state D because this will cause it to return the immediate past state, but it can with equal probability go to U, L, or R, and so on for the other states.

If we denote state U by 1, state R by 2, state D by 3, and state L by 4, then the probability transition matrix for the process is given by

$$P = \begin{bmatrix} \frac{1}{3} & \frac{1}{3} & 0 & \frac{1}{3} \\ \frac{1}{3} & \frac{1}{3} & \frac{1}{3} & 0 \\ 0 & \frac{1}{3} & \frac{1}{3} & \frac{1}{3} \\ \frac{1}{3} & 0 & \frac{1}{3} & \frac{1}{3} \end{bmatrix}$$

which is a doubly stochastic matrix. With this matrix we can compute the different parameters of interest.

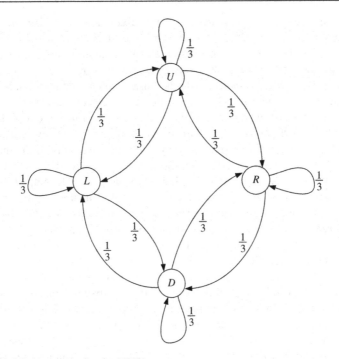

Figure 8.19 Markov chain for the NRW.

8.18 Applications of Random Walk

In this section, we discuss some of the applications of random walk. Earlier, we discussed how it can be used to solve the ballot problem. In this section, we consider its application in the following systems:

1. Web search
2. Insurance risk
3. Content of a dam
4. Cash management
5. Mobility models in mobile networks.

8.18.1 Web Search

Information retrieval seeks to find all documents that are relevant to a user's query. Prior to the advent of the World Wide Web, information retrieval used only word-based techniques; but with the advent of the web, new techniques have been developed. One example is the *hyperlink*, which is a reference of a web page B that is contained in a web page A. When the hyperlink for page B is clicked on page A, page B is displayed. Because almost every web user can publish his own web page, search engines face many difficult tasks. One such task is how to find and index the documents in the web. Also, because the pages contain heterogeneous information with different formats and styles, a great deal of effort must be expended to create reliable and efficient indices. Another task is to provide high quality and relevant results to users who merely use simple keyword-based interfaces. This is an important task because when several documents match a user's query, he can be flooded with so much information, some of which might not be very useful to him.

The analysis of hyperlink provides a way to derive the quality of the information on the web. A great deal of effort has been made by several authors to provide structure to the hyperlink on the web. Algorithms have been developed to compute reliable measures of authority from the topological structure of interconnections among the web pages. One such algorithm is the PageRank, which is used by the Google search engine. PageRank uses a topologically based ranking criterion that computes the authority of a page recursively as a function of the authorities of the pages that link to a target page.

We can view the web as a graph whose nodes correspond to the web pages and whose arcs are defined by hyperlinks between the web pages. Consider a web surfer who jumps from web page to web page in a random manner. The behavior of the surfer on each page depends on the contents of that page. Thus, from our earlier discussion, such an action is essentially a random walk on a graph. Different models of the surfer's behavior have been analyzed in the literature. Examples of these models can be found in Page et al. (1998), Henzinger (2001), Greco et al. (2001), Diligenti et al. (2004), and Bianchini et al. (2005). Most of these models assign a weight to each hyperlink according to the relevance of the link to the user's query, thereby converting the problem to that of a random walk on a weighted graph.

8.18.2 Insurance Risk

Consider an insurance company that starts at time 0 with a capital X_0. Assume that the company receives insurance premiums $Y_1, Y_2, \ldots$ from its customers and pays out compensations $V_1, V_2, \ldots$ at times $1, 2, \ldots$. Thus, at time n the actual capital available to the company is

$$X_n = X_0 + (Y_1 - V_2) + \cdots + (Y_n - V_n)$$

The company is bankrupt if $X_n < 0$. Let the random variable W_k be defined by $W_k = Y_k - V_k$. If we assume that the sequences $\{Y_k\}$ and $\{V_k\}$ are independent, then the company's capital behaves like a process starting at X_0 and having jumps W_k. Thus, the dynamics of the process is given by

$$X_n = \begin{cases} X_{n-1} + W_n & X_{n-1} > 0, X_{n-1} + W_n > 0 \\ 0 & \text{otherwise} \end{cases}$$

If we know the probability distribution of W_k we will be able to solve the problem as a random walk. One question that one might want to ask is the probability that the company becomes bankrupt.

8.18.3   Content of a Dam

Consider a basin behind a dam that has X_n amount of water at the end of the nth time period, which we assume to be a day. During day k a total of Y_k units of water flow into the basin, where Y_k has a well-defined probability distribution. The basin has a finite capacity b so that overflow occurs when the total inflow exceeds b. Thus, the volume of water in the basin at the end of day n is

$$X_n = \min\{X_{n-1} + Y_n, b\}$$

Assume that the demand for water on day k is V_k. Then the effective additional volume of water in the basin is $U_k = Y_k - V_k$, where we assume that the sequences $\{Y_n\}$ and $\{V_n\}$ are independent. Because the content cannot be negative, we have that

$$X_n = \min\{\max\{X_{n-1} + U_n, 0\}, b\}$$

Thus, we have that

$$X_n = \begin{cases} X_{n-1} + U_n & 0 < X_{n-1} + U_n < b \\ 0 & X_{n-1} + U_n \leq 0 \\ b & X_{n-1} + U_n \geq b \end{cases}$$

This means that $\{X_n\}$ is a random walk with absorbing barriers at 0 and b. An overflow might occur when the process reaches level b. The probability distribution of the process can be determined if the probability distribution of U_n is known.

8.18.4   Cash Management

Consider a company that attempts to maintain just enough cash balance to meet its operational needs. It periodically intervenes to ensure that it neither has too much cash on hand nor insufficient cash. To do this, the company has set a goal to periodically manage its cash as follows. If the cash available in a period exceeds the value K, the company buys treasury bills to reduce its cash level to x, where $0 < x < K$. Similarly, if the available cash is 0, the company sells sufficient treasury bills to boost the cash level to x. Thus, the cash level fluctuates as the process goes through a series of cycles, where the duration of each cycle is a random variable, as illustrated in Figure 8.20.

We might want to determine the mean length of a cycle. Also, given that an intervention is made, we might want to know that it is from a given level, say, level K.

This is the probability that the process hits level K before level 0. Thus, if we assume that the cash demand is an asymmetric random walk, we can use the same techniques used to obtain the first passage time and the probability of a maximum to solve the problems associated with this model.

8.18.5 Mobility Models in Mobile Networks

Several random walk models have been proposed for mobility in mobile networks. In this section we consider the CRW model of mobility in *ad hoc* networks (MANETs). As discussed earlier, this model has been used in Bandyopadhyay et al. (2006). A MANET is a collection of mobile users that communicate over wireless links that are not part of a preexisting network infrastructure. Because the nodes are mobile, the network topology usually changes unpredictably over time. An example of a node in a MANET is a wireless-equipped vehicle plying the streets of a city and communicating with similar vehicles.

Consider a MANET whose nodes move in a two-dimensional grid. We assume that a node's motion is subject to the following rules:

- There is a reflecting barrier at 0 such that on reaching the barrier a node is reflected with probability 1.
- A node takes a step in the same direction as its previous step with probability p_0, if the direction is away from the barrier, and it takes a step in the opposite direction with probability q_0. Similarly, if the direction is toward the reflecting barrier, it takes a step in the same direction as its previous step with probability p_1, and takes a step in the opposite direction with probability q_1.
- It takes a step in each of the two orthogonal directions with probability r, where $p_i + q_i + 2r = 1, i = 0, 1$.

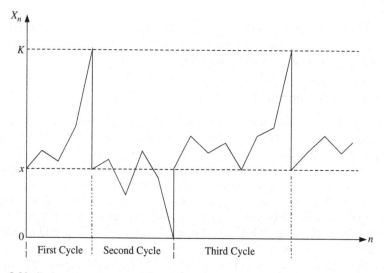

Figure 8.20 Cash management model.

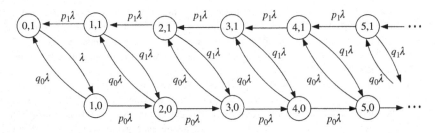

Figure 8.21 State-transition-rate diagram for the CRW.

- The times between epochs at which steps are taken are exponentially distributed with mean $1/\lambda$. This condition amounts to the fact that the node moves from one intersection to a neighboring intersection with a random velocity.

Thus, we can model the system by a continuous-time CRW. We first consider the one-dimensional walk in which $r = 0$. As discussed earlier, we represent a state by (k, l), where k is the location of the walker, where a move to the previous direction indicates a success and a move to the opposite direction indicates a failure, and l is the result index, which is defined by

$$l = \begin{cases} 1 & \text{if the previous trial resulted in a success} \\ 0 & \text{if the previous trial resulted in a failure} \end{cases}$$

Figure 8.21 is the state-transition-rate diagram for the process.

Thus, we can analyze the process using the CRW method discussed earlier. The case where r is not zero can be analyzed by decomposing it into two independent one-dimensional CRWs: one for the east–west direction and the other for the north–south direction. The probability of being in location (x, y) is the product of the probability of being in location x and the probability of being in location y in the east–west and north–south directions, respectively.

8.19 Summary

In this chapter, we have discussed different types of random walk, from the simple gambler's ruin to random walk on graphs, CRWs and CTRW. We have also discussed different applications of random walk. These include its use in solving the ballot problem, its application in web search, insurance risk, content of a dam, cash management and mobility management in mobile networks.

8.20 Problems

8.1 A bag contains four red balls, three blue balls, and three green balls. Jim plays a game in which he bets $1 to draw a ball from the bag. If he draws a red ball, he wins $1;

otherwise he loses \$1. Assume that the balls are drawn with replacement and that Jim starts with \$50 with the hope of reaching \$100 at which point he stops playing. However, if he loses all his money before this, the game also ends and he becomes bankrupt. What is the probability that the game ends with Jim being \$50 richer than he had at the beginning?

8.2 Mark and Kevin play a series of games of cards. During each game each player bets \$1, and whoever wins the game gets \$2. Sometimes a game ends in a tie in which case neither player loses his money. Mark is a better player than Kevin and has a probability 0.5 of wining each game, a probability 0.3 of losing each game, and a probability 0.2 of tying with Kevin. Initially Mark had \$9 and Kevin had \$6, and the game is over when either player is bankrupt.
  a. Give the state-transition diagram of the process.
  b. If r_k denotes the probability that Mark is ruined, given that the game is currently in state k (that is, Mark has \$$k$), obtain an expression for r_k.

8.3 Chris has \$20 and Dana has \$30. They decide to play a game in which each pledges \$1 and flips a fair coin. If both coins come up on the same side, Chris wins the \$2, and if they come up on different sides, Dana wins the \$2. The game ends when either of them has all the money. What is the probability that Chris wins the game?

8.4 Consider a single-server discrete-time queueing system that operates in the following manner. Let X_n denote the number of customers in the system at time $n \in \{0, 1, 2, \ldots\}$. If a customer is receiving service in time n, then the probability that he finishes receiving service before time $n + 1$ is q, where $0 \le q \le 1$. Let the random variable Y_n denote the number of customers that arrive between time n and $n + 1$, where the PMF of Y_n is given by

$$p_{Y_n}(k) = P[Y_n = k] = e^{-\lambda} \frac{\lambda^k}{k!} \quad k = 0, 1, \ldots$$

  a. Give an expression for the relationship between X_{n+1}, X_n and Y_n.
  b. Find the expression for the transition probabilities $P[X_{n+1} = j | X_n = i]$.

8.5 Consider the random walk $S_n = X_1 + X_2 + \cdots + X_n$, where the X_i are independent and identically distributed Bernoulli random variables that take on the value 1 with probability $p = 0.6$ and the value -1 with probability $q = 1 - p = 0.4$.
  a. Find the probability $P[S_8 = 0]$.
  b. What value of p maximizes $P[S_8 = 0]$?

8.6 Let N denote the number of times that an asymmetric random walk that takes a step to the right with probability p and a step to the left with probability $q = 1 - p$ revisits its starting point. Show that the PMF of N is given by

$$p_N(n) = P[N = n] = \beta(1 - \beta)^n \quad n = 0, 1, \ldots$$

where $\beta = |p - q|$.

8.7 Consider an asymmetric random walk that takes a step to the right with probability p and a step to the left with probability $q = 1 - p$. Assume that there are two absorbing barriers, a and b, and that the walk starts at the point k, where $b < k < a$.
  a. What is the probability that the walk stops at b?
  b. What is the mean number of steps until the walk stops?

8.8 Consider a cash management scheme in which a company needs to maintain the available cash to be no more than \$$K$. Whenever the cash level reaches K, the company buys treasury bills and reduces the cash level to x. Whenever the cash level reaches 0, the company sells enough treasury bills to bring the cash level up to x, where $0 < x < K$. Assume that in any given period the probability that the cash level increases by \$1 is p, and the probability that it decreases by \$1 is $q = 1 - p$. We define an intervention cycle as the period from the point when the cash level is x until the point when it is either 0 or K. Let T denote the time at which the available cash first reaches 0 or K, given that it starts at level x.

a. What is the expected value of T?

b. What is the mean number of visits to level m up to time T, where $0 < m < K$?

8.9 Consider a correlated random walk with stay. That is, a walker can move to the right, to the left, or not move at all. Given that the move in the current step is to the right, then in the next step it will move to the right again with probability a, to the left with probability b and remain in the current position with probability $1 - a - b$. Given that the walker did not move in the current step, then in the next step it will move to the right with probability c, to the left with probability d, and not move again with probability $1 - c - d$. Finally, given that the move in the current step is to the left, then in the next step it will move to the right with probability g, to the left again with probability h, and remain in the current position with probability $1 - g - h$. Let the process be represented by the bivariate process $\{X_n, Y_n\}$, where X_n is the location of the walker after n steps and Y_n is the nature of the nth step (i.e., right, left, or no move). Let π_1 be the limiting probability that the process is in "right" state, π_0 the limiting probability that it is in the "no move" state, and π_{-1} the limiting probability that it is in the "left" state, where $\pi_1 + \pi_0 + \pi_{-1} = 1$. Let $\Pi = \{\pi_1, \pi_0, \pi_{-1}\}$.

a. Find the values of π_1, π_0, and π_{-1}.

b. Obtain the transition probability matrix of the process.

8.10 Consider a CTRW $\{X(t)|t \geq 0\}$ in which the jump size, Θ, is normally distributed with mean μ and variance σ^2, and the waiting time, T, is exponentially distributed with mean $1/\lambda$, where Θ and T are independent. Obtain the master equation, $P(x, t)$, which is the probability that the position of the walker at time t is $X(t) = x$, given that it was in position 0 at time $t = 0$.

9 Brownian Motion

9.1 Introduction

Brownian motion is a stochastic process that has applications in fields as vast and different as economics, biology, and management science. Mathematically, it can be thought of as a continuous-time process in which over every infinitely small time interval Δt the entity under consideration moves one "step" in a certain direction. This suggests that Brownian motion can be viewed as a "random walk" process, which we will demonstrate shortly.

The physical manifestation of Brownian motion was observed by the Scottish botanist Robert Brown in 1827. His interpretation of this process was based on the movement of small pollen particles suspended in a drop of water. In his experiments, the pollen particles appeared to move in a completely random fashion, stumping Brown and his colleagues. Upon further investigation, Brown and others verified that this phenomenon was not unique to pollen particles but rather was exhibited by many different types of microscopic particles suspended in a fluid.

Brown's initial observations of Brownian motion went almost 80 years without any significant mathematical or even physical explanation until 1905 when Einstein published a paper that claimed that the motion of these microscopic particles stemmed from the constant forces exerted on the particles from the surrounding fluid, thanks to individual fluid molecules bumping into the particles, thus sending them in motion. Since the pollen grains were completely surrounded by these fluid molecules, they experience these forces in every conceivable direction, which explains why the particles do not move in a set pattern or direction.

The actual development of Brownian motion as a stochastic process did not surface until 1923 when Norbert Weiner, an MIT mathematician, established the modern mathematical framework of what is known today as the Brownian motion random process. This is why Brownian motion is sometimes referred to as the *Wiener process*. In fact, the study of Brownian motion today mostly involves the stochastic process pioneered by Weiner rather than the physical process studied by Brown.

9.2 Mathematical Description

We can gain some intuition to the behavior of Brownian motion by comparing it with a simple random walk. Specifically, assume that a random walk $X(t)$ increases by an infinitesimal step size Δx over each infinitesimal time increment Δt with probability p and decreases by Δx with probability $1 - p$. This allows the process

Markov Processes for Stochastic Modeling. DOI: http://dx.doi.org/10.1016/B978-0-12-407795-9.00009-8

$X(t)$ to be treated as a continuous function of t in which the change in x is without discontinuities. Then, for each step, we have that

$$E[X(t + \Delta t) - X(t)] = p\Delta x + (1 - p)(-\Delta x) = (2p - 1)\Delta x$$
$$E[\{X(t+\Delta t)-X(t)\}^2] = p(\Delta x)^2 + (1 - p)(-\Delta x)^2 = (\Delta x)^2$$

The number of time units of duration Δt in the time t is $[t/\Delta t]$, where $[t/\Delta t]$ is the largest integer that is less than or equal to $t/\Delta t$. Thus, the location of $X(t)$ at time t is given by the sum of $[t/\Delta t]$ random variables, each of which has a value Δx with probability p and a value $-\Delta x$ with probability $1 - p$. The expected value of the sum of these random variables is the sum of their expected values, which is $[t/\Delta t](2p - 1)\Delta x$. Because these random variables are independent and identically distributed, the variance of their sum is the sum of their variances, which is $[t/\Delta t](\Delta x)^2$.

If this process is to approximate the Brownian motion $W(t) \sim N(\mu t, \sigma^2 t)$, we must have that in the limit that

$$\left[\frac{t}{\Delta t}\right](2p - 1)\Delta x = \mu t$$

$$\left[\frac{t}{\Delta t}\right](\Delta x)^2 = \sigma^2 t$$

If for large t, we approximate $[t/\Delta t]$ by $t/\Delta t$, then from these two equations we obtain

$$\Delta x = \sigma\sqrt{\Delta t}$$

$$p = \frac{1}{2}\left(1 + \frac{\mu\Delta t}{\Delta x}\right) = \frac{1}{2}\left(1 + \frac{\mu}{\sigma}\sqrt{\Delta t}\right)$$

Let $T > 0$ be some fixed interval and let $\Delta t = T/n$, where $n = 1, 2, \ldots$. Assume that the random variables $Y_k, k = 1, \ldots, n$, independently take values $\sigma\sqrt{\Delta t}$ with probability $p = \{1 + (\mu\sqrt{\Delta t})/\sigma\}/2$ and $-\sigma\sqrt{\Delta t}$ with probability $1 - \{1 + (\mu\sqrt{\Delta t})/\sigma\}/2$. Define S_n by

$$S_n = \sum_{k=1}^{n} Y_k$$

Then from the central limit theorem, $S_n \to W(t)$ as $n \to \infty$. Thus, the Brownian motion can be regarded as a random walk defined over an infinitesimally small step size Δx and infinitesimally small time intervals Δt between walks such that both Δx and Δt go to zero in such a way that $(\Delta x)^2/\Delta t$ remains constant.

More formally, the Brownian motion $\{W(t), t \geq 0\}$ is a stochastic process that models random continuous motion. It is considered to be the continuous-time analog of the random walk and can also be considered as a continuous-time Gaussian process with independent increments. In particular, the Brownian motion has the following properties:

1. $W(0) = 0$, that is, it starts at zero.
2. $W(t)$ is continuous at $t \geq 0$, that is, it has continuous sample paths with no jumps.
3. It has both stationary and independent increments.
4. For $0 \leq s < t$, the random variable $W = W(t) - W(s)$ has a normal distribution with mean 0 and variance $\sigma_W^2 = \sigma^2(t - s)$. That is, $W \sim N(0, \sigma^2(t - s))$.

Brownian motion is an important building block for modeling continuous-time stochastic processes. In particular, it has become an important framework for modeling financial markets. The path of a Brownian motion is always continuous, but it is nowhere smooth; consequently, it is nowhere differentiable. The fact that the path is continuous means that a particle in Brownian motion cannot jump instantaneously from one point to another.

Because $W(0) = 0$, according to property 4,

$$W(t) = W(t) - W(0) \sim N(0, \sigma^2(t - 0)) = N(0, \sigma^2 t) \qquad (9.1)$$

Thus, $W(t - s) \sim N(0, \sigma^2(t - s))$; that is, $W(t) - W(s)$ has the same distribution as $W(t - s)$. This also means that another way to define a Brownian motion $\{W(t), t \geq 0\}$ is that it is a process that satisfies conditions 1, 2, and 3 along with the condition $W(t) \sim N(0, \sigma^2 t)$.

There are many reasons for studying the Brownian motion. As stated earlier, it is an important building block for modeling continuous-time stochastic processes because many classes of stochastic processes contain Brownian motion. It is a Markov process, a Gaussian process, a martingale, a diffusion process, as well as a Levy process. Over the years, it has become a rich mathematical object. For example, it is the central theme of stochastic calculus.

A Brownian motion is sometimes called a *Wiener process*. A sample function of the Weiner process is shown in Figure 9.1.

Let $B(t) = W(t)/\sigma$. Then $E[B(t)] = 0$ and $\sigma_{B(t)}^2 = 1$. The stochastic process $\{B(t), t \geq 0\}$ is called the *standard Brownian motion*, which has the property that when sampled at regular intervals it produces a symmetric random walk. Note that $B(t) \sim N(0, t)$. In the remainder of this chapter we refer to the Weiner process $W(t) \sim N(0, \sigma^2 t)$ as the *classical Brownian motion* and use the two terms interchangeably.

9.3   Brownian Motion with Drift

Brownian motion is used to model stock prices. Because stock prices do not generally have a zero mean, it is customary to include a *drift* measure that makes the

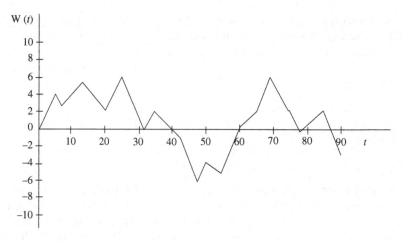

Figure 9.1 Sample function of a Wiener process.

following model with a *drift rate* $\mu > 0$ a better model than the classical Brownian motion:

$$Y(t) = \mu t + W(t) \quad t \geq 0 \tag{9.2}$$

where $W(t)$ is the classical Brownian motion. Note that $E[Y(t)] = \mu t$ and $\sigma^2_{Y(t)} = \sigma^2 t$, which means that $Y(t) \sim N(\mu t, \sigma^2 t)$. Note also that we can express $Y(t)$ in terms of the standard Brownian motion as follows:

$$Y(t) = \mu t + \sigma B(t) \quad t \geq 0 \tag{9.3}$$

9.4 Brownian Motion as a Markov Process

Let $W(t)$ be a classical Brownian motion, and let $\Im_s$ denote the information that is being revealed by watching the process up through the time $s < t$. Then the conditional expected value of $W(t)$ given $\Im_s$ can be obtained as follows:

$$E[W(t)|\Im_s] = E[W(s)|\Im_s] + E[\{W(t) - W(s)\}|\Im_s]$$

The first term on the right-hand side is equal to $W(s)$ because it is already revealed through $\Im_s$. Also, the increment $W(t) - W(s)$ is independent of $\Im_s$; thus $E[\{W(t) - W(s)\}|\Im_s] = E[W(t) + W(s)]$, which is zero. Therefore, we obtain

$$E[W(t)|\Im_s] = W(s) + E[W(t)|W(s)]$$

This means that to predict $W(t)$ given all the information up through time s, we only need to consider the value of the process at time s. Thus, a Brownian motion is a Markov process. This is not surprising because it is an independent increment process. As stated earlier, all independent increment processes have the Markov property.

Because for the classical Brownian motion, the increment over an interval of length X has the Gaussian distribution with the probability density function (PDF)

$$f_X(x) = \frac{1}{\sqrt{2\pi\sigma^2}}e^{-x^2/2\sigma^2} \quad -\infty < x < \infty \qquad (9.4)$$

the classical Brownian motion is a Markov process with transition PDF given by

$$f_{Y|X}(y|x) = \frac{1}{\sqrt{2\pi\sigma^2}}e^{-(y-x)^2/2\sigma^2} \quad -\infty < y < \infty \qquad (9.5)$$

9.5 Brownian Motion as a Martingale

Let $0 \le s \le t$ and let $v \ge 0$. We show that $E[W(t+v)|W(s)] = W(t)$. We recall the Markov property that for $0 \le s \le t$, $E[W(t+v)|W(s)] = E[W(t+v)|W(t)]$. Therefore,

$$E[W(t+v)|W(s)] = E[W(t+v)|W(t)] = E[W(t) + \{W(t+v) - W(t)\}|W(t)]$$
$$= W(t) + E[\{W(t+v) - W(t)|W(t)]$$
$$= W(t) + E[W(t+v) - W(t)] = W(t) + E[W(v)] = W(t) + 0$$
$$= W(t)$$

where the fifth equality follows from the independent increments property.

9.6 First Passage Time of a Brownian Motion

Let T_k denote the time it takes for a classical Brownian motion to go from $W(0) = 0$ to $W(t) = k \neq 0$, that is,

$$T_k = \min\{t > 0 : W(t) = k\} \qquad (9.6)$$

Suppose $k > 0$. Then we obtain the PDF of T_k as follows:

$$P[W(t) \ge k] = P[W(t) \ge k|T_k \le t]P[T_k \le t] + P[W(t) \ge k|T_k > t]P[T_k > t]$$

From the definition of the first passage time, $P[W(t) \ge k|T_k > t] = 0$ for all t. Also, if $T_k \le t$, then we assume that there exists a $t_0 \in (0,t)$ with the property that

$W(t_0) = k$. We know that the process $\{W(t)|W(t_0) = k\}$ has a normal distribution with mean k and variance $\sigma^2(t - t_0)$. That is, the random variable

$$\{W(t)|W(t_0) = k\} \sim N(k, \sigma^2(t - t_0)) \quad \text{for all } t \geq t_0$$

Thus, from the symmetry about k, we have that

$$P[W(t) \geq k|T_k \leq t] = \frac{1}{2}$$

which means that the CDF of T_k is given by

$$F_{T_k}(t) = P[T_k \leq t] = 2P[W(t) \geq k] = \frac{2}{\sigma\sqrt{2\pi t}} \int_k^\infty \exp\{-v^2/2\sigma^2 t\}dv$$

By symmetry, T_k and T_{-k} are identically distributed random variables, which means that

$$F_{T_k}(t) = \frac{2}{\sigma\sqrt{2\pi t}} \int_{|k|}^\infty \exp\{-v^2/2\sigma^2 t\}dv = 2\left\{1 - \Phi\left(\frac{|k|}{\sigma\sqrt{t}}\right)\right\} \quad t > 0$$

where $\Phi(\cdot)$ is the CDF of the standard normal random variable. Let $y^2 = v^2/\sigma^2 t$, which means that $dv = 2\,dy\sqrt{t}$. Then the CDF becomes

$$F_{T_k}(t) = \frac{2}{\sqrt{2\pi}} \int_{|k|/\sigma\sqrt{t}}^\infty \exp\{-y^2/2\}dy \quad t > 0 \tag{9.7}$$

Differentiating with respect to t, we obtain the PDF of T_k as follows:

$$f_{T_k}(t) = \frac{|k|}{\sqrt{2\pi\sigma^2 t^3}} \exp\left\{-\frac{k^2}{2\sigma^2 t}\right\} \quad t > 0 \tag{9.8}$$

Because for the standard Brownian motion $\sigma = 1$, the PDF of T_k for the standard Brownian motion is given by

$$f_{T_k}(t) = \frac{|k|}{\sqrt{2\pi t^3}} \exp\left\{-\frac{k^2}{2t}\right\} \quad t > 0 \tag{9.9}$$

9.7 Maximum of a Brownian Motion

Let $M(t)$ denote the maximum value of the classical Brownian motion $\{W(t)\}$ in the interval $[0, t]$, that is,

$$M(t) = \max\{W(u) \quad 0 \leq u \leq t\} \tag{9.10}$$

The PDF of $M(t)$ is obtained by noting that

$$P[M(t) > x] = P[T_x < t]$$

Thus, we have that

$$F_{M(t)}(x) = P[M(t) \le x] = 1 - P[M(t) > x] = 1 - P[T_x < t] = 1 - P[T_x \le t]$$

$$= 1 - F_{T_x}(t) = 1 - 2\left\{ 1 - \Phi\left(\frac{|x|}{\sigma\sqrt{t}}\right) \right\}$$

$$= 1 - \frac{2}{\sqrt{2\pi}} \int_{|x|/\sigma\sqrt{t}}^{\infty} \exp\{-y^2/2\} dy$$

where the fourth equality follows from the fact that T_x is a continuous random variable. Upon differentiation, we obtain the PDF as

$$f_{M(t)}(x) = \sqrt{\frac{2}{\pi\sigma^2 t}} \exp\left\{ -\frac{x^2}{2\sigma^2 t} \right\} \quad x \ge 0 \tag{9.11}$$

9.8   First Passage Time in an Interval

Let T_{ab} denote the time at which the classical Brownian motion $\{W(t), t \ge 0\}$ for the first time hits either the value a or the value b, where $b < 0 < a$, as shown in Figure 9.2.

Thus, we can write

$$T_{ab} = \min\{t : W(t) = a \quad \text{or} \quad W(t) = b\} \quad b < 0 < a < \infty$$

Let p_{ab} be the probability that $\{W(t)\}$ assumes the value a first, that is,

$$p_{ab} = P[W(T_{ab}) = a]$$

Now, T_{ab} is a stopping time whose mean $E[T_{ab}]$ is finite. Thus, according to the stopping time theorem, $E[W(T_{ab})] = E[W(0)] = 0$. This means that

$$E[W(T_{ab})] = ap_{ab} + b(1 - p_{ab}) = 0$$

From this, we obtain the probability that the process hits a before b as

$$p_{ab} = \frac{|b|}{a + |b|} \tag{9.12}$$

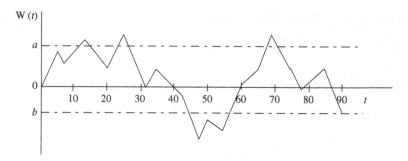

Figure 9.2 First passage time in an interval.

An alternative method of solving the problem is by seeing the Brownian motion as a limit of the symmetric random walk. The probability that the process hits a before b can then be likened to the gambler's ruin problem in which the process is equally likely to go up or down by a distance of Δh. Thus, with respect to the gambler's ruin problem of Chapter 8, $N = (a + |b|)/\Delta h$ and $i = |b|/\Delta h$. Because p_{ab} is the ruin probability, and for the symmetric walk, the ruin probability is i/N, we have that

$$p_{ab} = \frac{i}{N} = \frac{|b|/\Delta h}{(a + |b|)/\Delta h} = \frac{|b|}{a + |b|}$$

9.9 The Brownian Bridge

The Brownian bridge is a classical Brownian motion defined on the interval $[0, 1]$ and conditioned on the event $W(1) = 0$. Thus, the Brownian bridge is the process $\{W(t), t \in [0, 1] \mid W(1) = 0\}$. One way to realize the process is by defining $X(t)$, the Brownian bridge, as follows:

$$X(t) = W(t) - tW(1) \quad 0 \leq t \leq 1 \tag{9.13}$$

The Brownian bridge is sometimes called the *tied-down Brownian motion* (or *tied-down Wiener process*). It is useful for modeling a system that starts at some given level and is expected to return to that level at some specified future time. We note that

$$X(0) = W(0) - 0W(1) = 0$$
$$X(1) = W(1) - 1W(1) = 0$$

Thus, $E[X(t)] = 0$. For $0 \leq s < t \leq 1$, the covariance of $X(t)$ and $X(s)$ is given by

$$
\begin{aligned}
\text{Cov}\{X(s)X(t)\} &= E[\{X(s) - E[X(s)]\}\{X(t) - E[X(t)]\}] = E[X(s)X(t)] \\
&= E[\{W(s) - sW(1)\}\{W(t) - tW(1)\}] \\
&= E[W(s)W(t) - tW(s)W(1) - sW(t)W(1) + stW^2(1)] \\
&= \sigma^2(s \wedge t) - \sigma^2 t(s \wedge 1) - \sigma^2 s(t \wedge 1) + \sigma^2 st \\
&= \sigma^2\{s - st - st + st\} = \sigma^2(s - st) \\
&= \sigma^2 s(1 - t)
\end{aligned}
\tag{9.14}
$$

where we have used the fact that $E[W(s)W(t)] = \sigma^2\min(s, t) = \sigma^2(s \wedge t)$. Thus, the Brownian bridge is not a wide-sense stationary process because the covariance $\text{Cov}\{X(s)X(t)\}$ is not a function of only the difference between s and t.

9.10 Geometric Brownian Motion

Let $\{X(t), t \geq 0\}$ be a Brownian motion with drift, that is, $X(t) = \mu t + \sigma B(t)$. The process $\{Y(t), t \geq 0\}$, which is defined by

$$
Y(t) = e^{X(t)}
\tag{9.15}
$$

is called the geometric Brownian motion (GBM). We may also write

$$
\log(Y(t)) = X(t) = \mu t + \sigma B(t)
\tag{9.16}
$$

Thus, we may define a GBM as a continuous-time stochastic process in which the logarithm of a randomly varying parameter follows a Brownian motion. GBM is used to analyze stock prices in the Black–Scholes model. We will discuss the process later after our discussion on stochastic differential equations (SDEs).

9.11 Introduction to Stochastic Calculus

Let $X(t)$ represent the size of a population at time t. We assume that during each unit of time, say Δt, a constant fraction of the population will have an offspring and a constant fraction of the population will die. Let k_b denote the fraction of the population having children during the interval and let k_d denote the fraction of the population that dies during the interval. We will assume that at time $t = 0$, the population level is specified, that is, $X(0) = x_0$ is some given constant. Thus, the change in the population during the interval Δt is

$$
\Delta x = k_b X(t)\Delta t - k_d X(t)\Delta t = (k_b - k_d)X(t)\Delta t = kX(t)\Delta t
$$

where $k = k_b - k_d$. The rate of change of the population is $\Delta x / \Delta t$. In the limit that Δt becomes infinitesimally small, we have that

$$\lim_{\Delta t \to 0} \frac{\Delta X}{\Delta t} \equiv \frac{dX}{dt} = (k_b - k_d)X(t) = kX(t)$$

Thus, the derivative defines the rate of change of the population, which is proportional to the size of the population at time t. We can distinguish two possible cases: $k_b > k_d$ means that there are more births than deaths, so we expect the population to grow. $k_b < k_d$ means that there are more deaths than births, so the population will eventually go extinct. There is also a marginal case that $k_b = k_d$, for which $k = 0$, where the population does not change at all. To summarize, this simple model of unlimited growth leads to the following ordinary differential equation (ODE) and initial condition:

$$\frac{dX}{dt} = kX \quad X(0) = x_0$$

In general, if the system state at time t is $X(t)$, we can describe the dynamics of the system by the ODE

$$\frac{dX}{dt} = a(X, t) \tag{9.17}$$

The solution to this equation is the trajectory of the system, which is

$$X(t) = x_0 + \int_0^t a(X, u)du$$

The parameter $a(\cdot)$ is a function that ensures that a unique solution $X(t) = X(t; x_0, t_0)$ exists, which satisfies the initial condition $X(t_0) = x_0$. The differential equation enables us to determine how the system behaves at all times t, even though we may not be able to find an analytical solution. An ODE together with an initial condition is called an *initial value problem*.

9.11.1 Stochastic Differential Equation and the Ito Process

In many applications, the experimentally measured trajectories of systems modeled by the preceding ODE do not behave as predicted due to some random effects that disturb the system. This implies that the ODE needs to be modified to include these random effects. A stochastic differential equation (SDE) enables us to extend the deterministic ODE to systems where the system parameters are modeled as random processes. In this case, it is assumed that some degree of noise is present in the dynamics of the process. The common model for such noise is the Weiner process, and the associated differential equation consists of two parts: a deterministic part and a stochastic part. The rationale for modeling the noise by a Weiner process is based on the central limit theorem. Measurement noise comes from different independent sources. According to the central limit theorem, the sum of these noise components has a normal or Gaussian distribution, if we assume that each of them has a finite variance. Thus, we may rewrite the system equation as follows:

$$\frac{dX(t)}{dt} = a(X(t), t) + b(X(t), t)\xi(t) \quad t \geq 0, X(0) = x_0 \tag{9.18}$$

where $a(\cdot)$ and $b(\cdot)$ are given functions, and $\xi(t)$ represents a white noise that is commonly represented by the time derivative of the Wiener process (or Brownian motion). That is,

$$\xi(t) = \frac{dW(t)}{dt}$$

Thus, Eq. (9.18) becomes

$$\frac{dX(t)}{dt} = a(X(t), t) + b(X(t), t)\frac{dW(t)}{dt} \quad X(0) = x_0 \tag{9.19}$$

When we multiply the equation by "dt," we obtain

$$dX(t) = a(X(t), t)dt + b(X(t), t)dW(t) \quad X(0) = x_0 \tag{9.20}$$

It is a common practice to rewrite Eq. (9.20) in the following form:

$$dX_t = \mu(X_t, t)dt + \sigma(X_t, t)dW_t \tag{9.21}$$

where $X_t = X(t)$ is a stochastic process and $W_t = W(t)$ is a Wiener process that models the noise. Any process whose dynamics can be modeled by Eq. (9.21) is called an *Ito process*. The function $\mu(\cdot)$ is called the *drift parameter* (or deterministic component), while the function $\sigma(\cdot)$ is called the *diffusion* or *volatility parameter* (or the stochastic component). In time-homogeneous SDE, $\mu(\cdot)$ and $\sigma(\cdot)$ do not depend on t.

We know that the path of Brownian motion takes sharp turns everywhere and thus Brownian motion is nondifferentiable. Stochastic calculus deals with integration of a stochastic process with respect to another stochastic process. Because Brownian motion is nowhere differentiable, any stochastic process that is driven by Brownian motion is nowhere differentiable. The driving force behind stochastic calculus was the attempt to understand the motion driven by a series of small random impulses, which are modeled by Brownian motion. Because the net distance traveled in time Δt by a particle in Brownian motion is proportional to $\sqrt{\Delta t}$, in the SDE that represents the dynamic behavior of the process, future changes are expressed as differentials, not as derivatives. For a process $\{X(t), t \geq 0\}$, the differential $dX(t)$ is defined by

$$dX_t = dX(t) = X(t + dt) - X(t)$$

The integral form is the forward sum of uncountable and random increments over time and is given by

$$X(t) = \int_0^t \mathrm{d}X(u)$$

Stochastic calculus is the mathematics used for modeling financial options. It is used to model investor behavior and asset pricing. It has also found applications in fields such as control theory and mathematical biology. Observe that $X(t)$ is a random variable, and we would like to obtain such statistics as its mean and variance.

9.11.2   The Ito Integral

The Ito integral deals with the integration of the expression

$$\int_0^t X(u)\mathrm{d}B(u)$$

where $X(t)$ is a stochastic process and $B(t)$ is the Brownian motion. Thus, the integral is referred to as the Ito integral of $X(t)$ with respect to the Brownian motion. If $B(t)$ were the function $b(t)$ that is differentiable, we would write $\mathrm{d}b(u) = b_1(u)\mathrm{d}u$, where $b_1(u) = \mathrm{d}b(u)/\mathrm{d}u$, and thus obtain the expression

$$\int_0^t X(u)\mathrm{d}b(u) = \int_0^t X(u)b_1(u)\mathrm{d}u$$

that is known to be a standard integral. When the process $B(t)$ is not differentiable, as is the case when it is a Brownian motion, the integral becomes an unfamiliar integral. To evaluate the integral when $B(t)$ is not differentiable, we first divide the interval $[0, t]$ into n disjoint intervals at points $0 = t_0 < t_1 < \cdots < t_n = t$ and obtain

$$\int_0^t X(u)\mathrm{d}B(u) = \lim_{n \to \infty} \sum_{k=0}^{n-1} X(u_k)\{B(u_{k+1}) - B(u_k)\} \tag{9.22}$$

which is the *Ito integral*. However, for this to work, $X(t)$ and $B(t)$ are required to satisfy certain conditions, which include the fact that $X(t)$ must be smooth enough for $X(t_k)$ to represent $X(t)$ in the interval (t_k, t_{k+1}). Also, the $X(t_k)$ are required to be independent of the increments $B(t_{k+1}) - B(t_k)$. In addition, $X(t)$ is required to be an *adapted process*, which is sometimes called a *nonanticipating process* because it cannot "see into the future."

To fully understand an adapted process, we first define the concept of *filtration*. A filtration is a family of σ-algebras $\{F_t, 0 \leq t < \infty\}$ that is increasing, that is, if

$s \le t$, then $F_s \subset F_t$. The property that a filtration is increasing implies that information is not forgotten. The space $(\Omega, F, \{F_t\}, P)$ is called a *filtered probability space*. A stochastic process $\{X(t), 0 \le t \le T\}$ on a filtered probability space is called an adapted process if for any $t \in [0, T]$, F_t contains all the information about the random variable $X(t)$; alternatively, it is an adapted process if the random variable $X(t)$ is F_t-adapted.

Thus, a stochastic process $\{X(t)\}$ is defined to be *Ito integrable* on the interval $[0, T]$ if the following conditions are satisfied:

1. The random variable $X(t)$ is adapted for $t \in [0, T]$.
2. $\int_0^t E[X^2(u)] du < \infty$.

We state the following propositions without proof; the proofs can be found in many books on stochastic processes such as Capasso and Bakstein (2005), Steele (2001), Oksendal (2005), Benth (2004), and Klebaner (2005).

Proposition 9.1 Let $f(t)$ and $g(t)$ be continuous functions in the interval $[a, b]$, and let $B(t)$ be a standard Brownian motion. Then

1. $E\left[\int_a^b f(t) dB(t) \right] = 0$

2. $E\left[\int_a^b f(t) dB(t) \int_a^b g(t) dB(t) \right] = \int_a^b E[f(t)g(t)] dt$

3. $E\left[\left\{ \int_a^b f(t) dB(t) \right\}^2 \right] = \mathrm{Var}\left(\int_a^b f(t) dB(t) \right) = \int_a^b E[\{f(t)\}^2] dt$; this is called the *Ito isometry* property.

Proposition 9.2

$$\int_a^b B(t) dB(t) = \frac{1}{2} \left[B^2(b) - B^2(a) \right] - \frac{b-a}{2} = \frac{B^2(b) - B^2(a) - (b-a)}{2}$$

Proposition 9.3

$$\int_a^b t \, dB(t) = bB(b) - aB(a) - \int_a^b B(t) dt$$

We also state the following stochastic differential identities:

$$(dt)^2 = 0$$
$$(dt)(dB(t)) = 0 \tag{9.23}$$
$$(dW(t))^2 = dt \Rightarrow (dB(t))^2 = \sigma^2 \, dt$$

Example 9.1

We are given $dX(t) = \mu\,dt + \sigma\,dB(t)$ with $X(0) = x_0$. The solution is

$$X(t) = X(0) + \int_0^t \mu\,du + \int_0^t \sigma\,dB(u) = X(0) + \mu t + \sigma B(t)$$

since $B(0) = 0$.

Example 9.2

We are given $dX(t) = \mu\,dt + \sigma t\,dB(t)$ with $X(0) = x_0$. The solution is

$$X(t) = X(0) + \int_0^t \mu\,du + \int_0^t \sigma u\,dB(u)$$

$$= X(0) + \mu t + \sigma t B(t) - \sigma 0 B(0) - \int_0^t B(u)du$$

$$= X(0) + \mu t + \sigma t B(t) - \int_0^t B(u)du$$

9.11.3 The Ito's Formula

The Ito's formula serves as a bridge between classical theory and stochastic theory. It is the stochastic equivalent of Taylor's theorem about the expansion of functions. Consider the stochastic process $X(t)$ with the stochastic differential

$$dX(t) = \mu(t)dt + \sigma(t)dB(t)$$

in the interval $[0, T]$. Let $f(t, y)$ be a continuous function in the interval $[0, T]$ that is twice differentiable in y. Then the Ito's formula gives the stochastic differential of the function $f(t, X(t))$ as follows:

$$df(t,X(t)) = \left\{ \frac{\partial f(t,X(t))}{\partial t} + \mu(t,X(t))\frac{\partial f(t,X(t))}{\partial X} + \frac{1}{2}\sigma^2(t,X(t))\frac{\partial^2 f(t,X(t))}{\partial X^2} \right\} dt$$

$$+ \sigma(t,X(t))\frac{\partial f(t,X(t))}{\partial X}dB(t)$$

$$(9.24)$$

Note that the first part of the formula is the consequence of the Taylor formula. The last term is essentially a new term in the "stochastic Taylor series." Also, the

Ito's formula can be viewed as the stochastic analog of the chain rule of classical calculus.

9.12   Solution of Stochastic Differential Equations

As discussed earlier, a differential equation is a rule that allows us to calculate the value of some quantity at a later time given the value at some earlier time. Thus, an SDE can be viewed as the representation of the dynamic behavior of a stochastic process. Almost all important continuous-time continuous-state stochastic processes are Ito processes, which means that their dynamic behavior can be expressed by an equation of the form

$$dX(t) = \mu(X(t), t)dt + \sigma(X(t), t)dB(t) \quad X(0) = x_0 \tag{9.25}$$

where the functions $\mu(X(t), t)$ and $\sigma(X(t), t)$ are given, $X(t)$ is an unknown process, and $B(t)$ is the standard Brownian motion. Such an SDE is said to be *driven by Brownian motion*. As stated earlier, Brownian motion-driven systems have become an important framework for modeling financial markets. The coefficients $\mu(X(t), t)$ and $\sigma(X(t), t)$ can be interpreted as measures of short-term growth and short-term variability, respectively. Thus, adjusting them permits a modeler to construct stochastic processes that reflect real-life behavior. A solution to the preceding SDE is

$$X(t) = X(0) + \int_0^t \mu(X(u), u)du + \int_0^t \sigma(X(u), u)dB(u) \tag{9.26}$$

If the two integrals exist for all $t > 0$, then $X(t)$ is called a *strong solution* of the SDE. In some processes, the coefficients $\mu(X(t), t)$ and $\sigma(X(t), t)$ are defined as follows: $\mu(X(t), t) = \mu X(t)$, where $-\infty < \mu < \infty$, and $\sigma(X(t), t) = \sigma X(t)$, where $\sigma > 0$. Thus, the SDE for such processes is given by

$$dX(t) = \mu X(t)dt + \sigma X(t)dB(t) \quad X(0) = x_0 > 0 \tag{9.27}$$

The Ito's formula is the key to the solution of many SDEs. We illustrate the solution with the GBM.

9.13   Solution of the Geometric Brownian Motion

As discussed earlier, we can define the GBM $\{Y(t), t \geq 0\}$ by the following SDE:

$$dY(t) = \mu Y(t)dt + \sigma Y(t)dB(t) \tag{9.28}$$

where

$$Y(t) = e^{X(t)}$$
$$X(t) = \mu t + \sigma B(t)$$

The process $\{X(t), t \geq 0\}$ is a Brownian motion with drift. The solution to the SDE of the process is obtained as follows. We define $f(Y(t)) = \ln\{Y(t)\} = X(t)$, so that

$$\frac{\partial f}{\partial t} = 0, \quad \frac{\partial f}{\partial Y} = \frac{1}{Y}, \quad \frac{\partial^2 f}{\partial Y^2} = -\frac{1}{Y^2}$$

Since $\mu(Y(t), t) = \mu Y(t)$ and $\sigma(Y(t), t) = \sigma Y(t)$, from the Ito's formula we have that

$$df(Y(t), t) = d\left[\ln\left\{Y(t)\right\}\right] = \left[(0) + \frac{\mu Y(t)}{Y(t)} + \frac{1}{2}\left\{-\frac{1}{Y^2(t)}\right\}\sigma^2 Y^2(t)\right]dt + \frac{\sigma Y(t)}{Y(t)}dB(t)$$

$$= \left\{\mu - \frac{1}{2}\sigma^2\right\}dt + \sigma\, dB(t)$$

Now we know that

$$d\left[\ln\{Y(t)\}\right] = \frac{dY(t)}{Y(t)} = \left\{\mu - \frac{1}{2}\sigma^2\right\}dt + \sigma\, dB(t)$$

From this we obtain

$$\int_0^t d\ln\{Y(u)\} = \int_0^t \left\{\mu - \frac{1}{2}\sigma^2\right\}du + \int_0^t \sigma\, dB(u)$$

Thus, since $B(0) = 0$,

$$\ln\left\{\frac{Y(t)}{Y(0)}\right\} = \left\{\mu - \frac{1}{2}\sigma^2\right\}t + \sigma B(t) \tag{9.29}$$

Finally, we get

$$Y(t) = Y(0)\exp\left\{\left[\mu - \frac{1}{2}\sigma^2\right]t + \sigma B(t)\right\}$$

$$= Y(0)\exp\left\{\left[\mu - \frac{1}{2}\sigma^2\right]t + W(t)\right\} \tag{9.30}$$

where $W(t)$ is the classical Brownian motion. Thus, we may express the solution to the equation as $Y(t) = Y_0 \, e^{X(t)}$, where

$$X(t) = \left[\mu - \frac{1}{2}\sigma^2\right]t + \sigma B(t) \Rightarrow dX(t) = \left[\mu - \frac{1}{2}\sigma^2\right]dt + \sigma \, dB(t)$$

This means that $X(t)$ is a Brownian motion with drift.

Example 9.3

Stock X follows a GBM where the drift factor is 0.96 and the variance factor is 0.55. At some particular time t, it is known that $dt = 0.04$ and $dB(t) = 0.45$. At time t, the stock trades for $200 per share. What is the instantaneous rate of change in the price of stock X?

Solution
The instantaneous rate of change in the price of the stock is given by

$$dX(t) = \mu X(t)dt + \sigma X(t)dB(t)$$

where $X(t) = 200$, $\mu = 0.96$, and $\sigma = 0.55$. Thus, we have that

$$dX(t) = (0.96)(200)(0.04) + (0.55)(200)(0.45) = 57.18$$

Example 9.4

A stock X follows a GBM with a drift factor of 0.35 and a volatility of 0.43. Given that $X(4) = 2$, what is the probability that $X(13) > 9$?

Solution
We are required to find $P[X(13) > 9 | X(4) = 2]$. First, we use the formula

$$X(t) = X(u)\exp\left\{\left[\mu - \frac{1}{2}\sigma^2\right](t - u) + \sigma\left(\sqrt{t - u}\right)B(t)\right\}$$

Thus,

$$X(13) = X(4)\exp\left\{[0.35 - (0.5)(0.43)^2](13 - 4) + 0.43\left(\sqrt{13 - 4}\right)B(13)\right\}$$

$$= 2 \exp\{2.31795 + 1.29B(13)\}$$

From this we obtain

$$
\begin{aligned}
P[X(13) > 9 \,|\, X(4) = 2] &= P[2 \exp\{2.31795 + 1.29B(13)\} > 9] \\
&= P[\exp\{2.31795 + 1.29B(13)\} > 4.5] \\
&= P[2.31795 + 1.29B(13) > \ln(4.5)] \\
&= P[2.31795 + 1.29B(13) > 1.504077397] \\
&= P[1.29B(13) > -0.8138726] = P[B(13) > -0.63091] \\
&= 1 - P[B(13) \leq -0.63091] = 1 - \Phi(-0.63091) \\
&= 1 - \{1 - \Phi(0.63091)\} = \Phi(0.63091) \approx \Phi(0.63) \\
&= 0.7357
\end{aligned}
$$

where we have used the fact that $B(t) \sim N(0, 1)$.

9.14 The Ornstein–Uhlenbeck Process

The Brownian motion is used to construct the Ornstein–Uhlenbeck (OU) process, which has become a popular tool for modeling interest rates. Recall that the derivative of the Brownian motion $X(t)$ does not exist at any point in time. Thus, if $X(t)$ represents the position of a particle, we might be interested in obtaining its velocity, which is the derivative of the motion. The OU process is an alternative model to the Brownian motion that overcomes the preceding problem. It does this by considering the velocity $V(t)$ of a Brownian motion at time t. Over a small time interval, two factors affect the change in velocity: the frictional resistance of the surrounding medium whose effect is proportional to $V(t)$ and the random impact of neighboring particles whose effect can be represented by a standard Wiener process. Thus, because mass times velocity equals force, we have that

$$
m \, dV(t) = -\gamma V(t)dt + dB(t)
$$

where $\gamma > 0$ is called the *friction coefficient* and $m > 0$ is the mass. If we define $\alpha = \gamma/m$ and $\beta = 1/m$, we obtain the OU process with the following differential equation:

$$
dV(t) = -\alpha V(t)dt + \beta \, dB(t) \tag{9.31}
$$

The OU process is used to describe the velocity of a particle in a fluid and is encountered in statistical mechanics. It is the model of choice for random movement toward a concentration point. It is sometimes called a *continuous-time Gauss–Markov process*, where a Gauss–Markov process is a stochastic process that satisfies the requirements for both a Gaussian process and a Markov process. Because a Wiener process is both a Gaussian process and a Markov process, in addition to being a stationary independent increment process, it can be considered a Gauss−Markov process with independent increments.

The OU process can be obtained from the standard Brownian process $B(t)$ by scaling and time change, as follows:

$$V(t) = e^{-\alpha t} B\left(\frac{\beta^2}{2\alpha} e^{2\alpha t}\right) \qquad (9.32)$$

Thus, $E[V(t)] = 0$, and because $B(t)$ is Gaussian, its covariance is

$$\begin{aligned}
\mathrm{Cov}\{V(t+\tau)V(t)\} &= E\left[e^{-\alpha(t+\tau)} B\left(\frac{\beta^2}{2\alpha} e^{2\alpha(t+\tau)}\right) e^{-\alpha t} B\left(\frac{\beta^2}{2\alpha} e^{2\alpha t}\right)\right] \\
&= e^{-\alpha(2t+\tau)} E\left[B\left(\frac{\beta^2}{2\alpha} e^{2\alpha(t+\tau)}\right) B\left(\frac{\beta^2}{2\alpha} e^{2\alpha t}\right)\right] \\
&= e^{-\alpha(2t+\tau)} \min\left\{\frac{\beta^2}{2\alpha} e^{2\alpha(t+\tau)}, \frac{\beta^2}{2\alpha} e^{2\alpha t}\right\} \\
&= e^{-\alpha(2t+\tau)} \frac{\beta^2}{2\alpha} e^{2\alpha t} = \frac{\beta^2}{2\alpha} e^{-\alpha\tau}
\end{aligned} \qquad (9.33)$$

Thus, $V(t)$ is a stationary process. However, unlike the Wiener process, it does not have independent increments. For a *standard OU process*, $\alpha = \beta = 1$ so that

$$V(t) = e^{-t} B\left(\frac{1}{2} e^{2t}\right) \qquad (9.34)$$

This OU process is referred to as the *Brownian motion-driven OU process*. Other types of the process are the Poisson-driven and gamma-driven OU processes.

9.14.1 Solution of the OU SDE

To solve the OU SDE $dV(t) = -\alpha V(t)dt + \beta\, dB(t)$, we define $f(t, V(t)) = e^{\alpha t} V(t)$. This gives

$$\frac{\partial f}{\partial t} = \alpha\, e^{\alpha t} V(t), \qquad \frac{\partial f}{\partial V} = e^{\alpha t}, \qquad \frac{\partial^2 f}{\partial V^2} = 0$$

Thus, applying the Ito's formula we obtain

$$df = d(e^{\alpha t} V(t)) = \left\{ \frac{\partial f}{\partial t} + (-\alpha V) \frac{\partial f}{\partial V} + \frac{1}{2} \beta^2 \frac{\partial^2 f}{\partial V^2} \right\} dt + \beta \frac{\partial f}{\partial V} dB$$

$$= \left\{ \alpha\, e^{\alpha t} V(t) - \alpha\, e^{\alpha t} V(t) + \frac{1}{2} \beta^2(0) \right\} dt + \beta\, e^{\alpha t}\, dB$$

$$= \beta\, e^{\alpha t}\, dB(t)$$

Given that $V(0) = v_0$, we obtain the solution as

$$e^{\alpha t} V(t) = v_0 + \int_0^t e^{\alpha u} \beta\, dB(u)$$

which gives

$$V(t) = v_0\, e^{-\alpha t} + e^{-\alpha t} \int_0^t e^{\alpha u} \beta\, dB(u)$$

Because $B(t)$ is a Brownian process, we have that the mean of $V(t)$ as

$$E[V(t)] = v_0\, e^{-\alpha t} + E\left[\int_0^t e^{-\alpha(t-u)} \beta\, dB(u) \right] = v_0\, e^{-\alpha t}$$

Using the Ito isometry, we obtain the variance of $V(t)$ as

$$\sigma^2_{V(t)} = E\left[\left\{ \int_0^t e^{-\alpha(t-u)} \beta\, dB(u) \right\}^2 \right] = \int_0^t [\beta\, e^{-\alpha(t-u)}]^2 du = \beta^2 \int_0^t e^{-2\alpha(t-u)}\, du$$

$$= \frac{\beta^2}{2\alpha} (1 - e^{-2\alpha t})$$

Thus, we have that

$$V(t) \sim N\left(v_0\, e^{-\alpha t}, \frac{\beta^2}{2\alpha} \left[1 - e^{-2\alpha t} \right] \right) \tag{9.35}$$

From this result we observe that as $t \to \infty$, the influence of the initial value decays exponentially and $\sigma^2_{V(t)} \to \beta^2/2\alpha$. Thus, as $t \to \infty$, $V(t)$ converges exponentially to a Gaussian distribution with mean zero and variance $\beta^2/2\alpha$, that is,

$$\lim_{t \to \infty} V(t) \sim N(0, \beta^2/2\alpha) \tag{9.36}$$

9.14.2 First Alternative Solution Method

Another method of solution of the OU SDE $dV(t) = -\alpha V(t)dt + \beta\, dB(t)$ is as follows. As in the previous method, we consider the function $f(t, V(t)) = e^{\alpha t}V(t)$ whose differential can be obtained directly as

$$d(e^{\alpha t}V(t)) = \alpha\, e^{\alpha t}V(t)dt + e^{\alpha t}dV(t)$$
$$= \alpha\, e^{\alpha t}V(t)dt + e^{\alpha t}\{-\alpha V(t)dt + \beta\, dB(t)\}$$
$$= e^{\alpha t}\beta\, dB(t)$$

which is the result we obtained earlier.

9.14.3 Second Alternative Solution Method

Because the OU process is a Gaussian process that is completely characterized by the mean and the variance, an alternative method of analyzing the process is due to Gillespie (1996), which is obtained by rewriting the SDE for the process as follows:

$$dV(t) = V(t + dt) - V(t) = -\alpha V(t)dt + \beta\, dB(t)$$

Thus, taking expectations and remembering that $B(t)$ is a zero-mean process, we obtain

$$E[V(t+dt)] - E[V(t)] = -\alpha E[V(t)]dt$$
$$\Rightarrow \lim_{dt \to 0} \frac{E[V(t+dt)] - E[V(t)]}{dt} = \frac{dE[V(t)]}{dt} = -\alpha E[V(t)]$$

The solution to the second equation is

$$E[V(t)] = v_0\, e^{-\alpha t} \quad t \geq 0$$

Also, $V(t + dt) = V(t) - \alpha V(t)dt + \beta\, dB(t)$; squaring on both sides gives

$$V^2(t + dt) = V^2(t) + \alpha^2 V^2(t)\{dt\}^2 + \beta^2\{dB(t)\}^2 - 2\alpha V^2(t)dt + 2\beta V(t)dB(t)$$
$$- 2\alpha\beta V(t)dB(t)dt$$

Taking expectations on both sides and recalling that $E[\{dB(t)\}^2] = \sigma^2_{dB(t)} = dt$ and $E[dB(t)] = 0$, we obtain

$$E[V^2(t + dt)] = E[V^2(t)] + \alpha^2 E[V^2(t)](dt)^2 + \beta^2 dt - 2\alpha E[V^2(t)]dt$$

where we have assumed that $V(t)$ is statistically independent of $B(u)$ for all $t \leq u$ so that $E[V(t)dB(t)] = E[V(t)]E[dB(t)] = 0$. Thus, we obtain

$$E[V^2(t + dt)] - E[V^2(t)] = E[V^2(t)]\{\alpha^2 \, dt - 2\alpha\}dt + \beta^2 \, dt$$

$$\frac{E[V^2(t + dt)] - E[V^2(t)]}{dt} = E[V^2(t)]\{\alpha^2 \, dt - 2\alpha\} + \beta^2$$

Taking the limits as $dt \to 0$, we obtain

$$\lim_{dt \to 0} \frac{E[V^2(t + dt)] - E[V^2(t)]}{dt} = \frac{dE[V^2(t)]}{dt} = \beta^2 - 2\alpha E[V^2(t)]$$

If we assume that $V^2(0) = v_0^2$, then the solution to the second equation is given by

$$E[V^2(t)] = v_0^2 + \frac{\beta^2}{2\alpha}\{1 - e^{-2\alpha t}\}$$

Thus, using the previous result for $E[V(t)]$, the variance of $V(t)$ is given by

$$\text{Var}\left\{V(t)\right\} = E[V^2(t)] - \{E[V(t)]\}^2 = \frac{\beta^2}{2\alpha}\{1 - e^{-2\alpha t}\}$$

This means that

$$V(t) \sim N\left(v_0 \, e^{-\alpha t}, \frac{\beta^2}{2\alpha}\{1 - e^{-2\alpha t}\}\right)$$

as we obtained earlier.

9.15 Mean-Reverting OU Process

A mean-reverting process is a process that, over time, tends to drift toward its long-term mean. The differential formula for a mean-reverting process $\{Y_t\}$ is as follows:

$$dY_t = \theta(K - Y_t)dt + \sigma \, dW_t$$

where $\theta > 0$ is the *rate of mean reversion* and K is the value around which Y_t tends to oscillate. The coefficient of dt is called the *drift term*. Specifically, when $Y_t > K$, the drift term is negative, which results in the process pulling back toward the equilibrium level. Similarly, when $Y_t < K$, the drift term is positive, which results in Y_t pulling up to the higher equilibrium level. Thus, another way to define a mean-reverting process is one whose changes in its values are negatively correlated. In this way, the process does not wander off to infinity. Instead, it always tries to come back to a well-defined asymptotic mean value. For this reason, it is some-times used to model processes such as prices, interest rates, and volatilities that tend to return to a mean or average value after reaching extremes.

A mean-reverting OU process $X(t)$ is the solution to the following SDE

$$dX(t) = \alpha\{\mu - X(t)\}dt + \beta \, dB(t) \tag{9.37}$$

where $B(t)$ is the standard Brownian process, μ is the long-run mean of $X(t)$, and α is the rate of mean reversion.

As in the previous sections, we solve this equation by considering $e^{\alpha t}X(t)$ and taking the differential

$$d(e^{\alpha t}X(t)) = \alpha \, e^{\alpha t}X(t)dt + e^{\alpha t}dX(t) = \alpha \, e^{\alpha t}X(t)dt + e^{\alpha t}[\alpha\{\mu - X(t)\}dt + \beta \, dB(t)]$$
$$= \alpha\mu \, e^{\alpha t} \, dt + \beta \, e^{\alpha t} \, dB(t)$$

Given that $X(0) = x_0$, we obtain the solution as

$$e^{\alpha t}X(t) = x_0 + \int_0^t \alpha\mu \, e^{\alpha u} \, du + \int_0^t \beta \, e^{\alpha u} \, dB(u)$$

which gives

$$X(t) = x_0 \, e^{-\alpha t} + \int_0^t \alpha\mu \, e^{-\alpha(t-u)} \, du + \int_0^t e^{-\alpha(t-u)}\beta \, dB(u)$$
$$= x_0 e^{-\alpha t} + \mu(1 - e^{-\alpha t}) + \int_0^t e^{-\alpha(t-u)}\beta \, dB(u)$$
$$= \mu + e^{-\alpha t}(x_0 - \mu) + \int_0^t e^{-\alpha(t-u)}\beta \, dB(u)$$

Because $B(t)$ is a Brownian motion, we have that the mean of $X(t)$ is

$$E[X(t)] = \mu + e^{-\alpha t}(x_0 - \mu) + E\left[\int_0^t e^{-\alpha(t-u)}\beta \, dB(u)\right]$$
$$= \mu + e^{-\alpha t}(x_0 - \mu)$$

Using the Ito isometry, we obtain the variance of $X(t)$ as

$$\sigma^2_{X(t)} = E\left[\left\{\int_0^t \beta \, e^{-\alpha(t-u)} \, dB(u)\right\}^2\right] = \int_0^t \{\beta \, e^{-\alpha(t-u)}\}^2 du = \beta^2 \int_0^t e^{-2\alpha(t-u)} \, du$$
$$= \frac{\beta^2}{2\alpha}(1 - e^{-2\alpha t})$$

Thus, we have that

$$X(t) \sim N\left(\mu + e^{-\alpha t}(x_0 - \mu), \frac{\beta^2}{2\alpha}(1 - e^{-2\alpha t})\right) \tag{9.38}$$

and

$$\lim_{t \to \infty} X(t) \sim N\left(\mu, \frac{\beta^2}{2\alpha}\right) \tag{9.39}$$

The difference between this process and the standard OU process is that as $t \to \infty$, the mean of the mean-reverting scheme is nonzero; in fact, it is μ. In the case of the traditional scheme, the mean is zero.

9.16   Fractional Brownian Motion

The fractional Brownian motion (FBM) $\{B_H(t), t \geq 0\}$ is a generalization of the Brownian motion. Although the main principles of FBM were introduced earlier by Kolmogorov, the name was introduced by Mandelbrot and van Ness (1968) who defined the process by the stochastic integral

$$B_H(t) = \frac{1}{\Gamma(H + 1/2)}\left\{\int_{-\infty}^{0}[(t-u)^{H-1/2} - (-u)^{H-1/2}]dB(u) + \int_{0}^{t}(t-u)^{H-1/2}dB(u)\right\} \tag{9.40}$$

for $t \geq 0, 0 < H < 1$, and $B(t)$ is a Brownian motion. FBM is a centered Gaussian process with stationary but not independent increments; it has independent increments only when it is a standard Brownian motion. Before we discuss the details of the process, we first examine the following processes that are used to describe FBM: self-similar processes and long-range dependence (LRD) processes.

9.16.1   Self-Similar Processes

From our knowledge of geometry, when two objects have the same shape, we say that they are similar. For example, we talk of similar triangles, which are triangles that have the same shape but can be of different sizes. A geometric shape is defined to be self-similar if the same geometric structures are observed independently of the distance from which the shape is looked at. That is, self-similarity is a term used to describe an object that looks roughly the same when viewed at different degrees of magnification or different scales on a dimension (space or time).

Self-similar processes were introduced by Kolmogorov in the early 1940s, but they were brought to the attention of statisticians and others in related fields in the late 1960s and early 1970s by Mandelbrot and van Ness. They are used to model the presence of long-term correlation in a data set.

More formally, a self-similar process is a stochastic process that is invariant in distribution under suitable scaling of time and space; that is, it is a process that is invariant against changes in scale or size. Specifically, the stochastic process $\{X(t), t \geq 0\}$ is said to be self-similar if for any $a > 0$, there exists some $b > 0$ such that $\{X(at), t \geq 0\}$ has the same distribution as $\{bX(t), t \geq 0\}$. Generally, a and b are related by $b = a^H$, where H is a parameter called the *Hurst index*. Thus, a self-similar stochastic process is one in which $\{X(at), t \geq 0\}$ has the same distribution as $\{a^H X(t), t \geq 0\}$ for all $a > 0$. That is,

$$X(at) \sim a^H X(t) \tag{9.41}$$

Self-similarity is an attribute of many laws of nature and is a unifying concept that underlies power laws and fractals.

9.16.2 Long-Range Dependence

Long-range dependence (LRD) describes the memory of a process. In an LRD process (also called a *long-memory process*), the current state has significant influence on its subsequent states far into the future with the result that there is a high variability in the behavior of the process over multiple timescales. In other words, when we aggregate the data to produce the average behavior of the process over large time intervals, its observed behavior does not become smooth. Another way to describe LRD is that there is a significant correlation between observations of samples of the process separated by large time spans. This means that what happens at one instance has an effect on what happens much later. LRD processes and their statistics have applications in areas such as finance, econometrics, hydrology, meteorology, turbulence, geophysics, statistical physics, communication networks, neuroscience, and analysis of DNA sequences.

Quantitatively, we say that a stationary process $X(t)$ is *long-range dependent* if its autocorrelations decay to zero so slowly that their sum does not converge, that is,

$$\sum_{k=1}^{\infty} \rho_X(k) = \infty$$

where $\rho_X(k)$ is the autocorrelation function (ACF) of $X(t)$, which is given by

$$\rho_X(k) = \frac{E[\{X(t) - \mu_X\}\{X(t+k) - \mu_X\}]}{\sigma_X^2}$$

where μ_X and σ_X are the respective mean and standard deviation of $X(t)$. The reason why the sum of the ACFs does not converge is because in LRD processes,

$\rho(n) \approx n^{-\alpha}$ for some $0 < \alpha < 1$ when $n \to \infty$. In short-range dependent processes, we have that $\rho(n) = \beta^n$ as n tends to infinity, where $\beta < 1$, which means that $\sum_{n=1}^{\infty} \rho(n) < \infty$. Thus, intuitively, the dependence between widely separated values in an LRD process does not decay to zero, even across infinitely large time shifts.

In summary, we state that short-range dependent processes are characterized by an ACF that decays exponentially fast. On the other hand, processes with LRD exhibit a much slower decay of the correlations; their ACFs typically obey some power law.

9.16.3   Self-Similarity and Long-Range Dependence

A stationary process $X(t)$ is defined to be a *second-order self-similar process* if its ACF satisfies the condition

$$\rho_X(k) = \frac{1}{2}\left[(k+1)^{2H} - 2k^{2H} + (k-1)^{2H}\right] \quad 0.5 < H < 1 \tag{9.42}$$

$X(t)$ is defined to be asymptotically exactly self-similar if

$$\lim_{k \to \infty} \rho_X(k) = \frac{1}{2}\left[(k+1)^{2H} - 2k^{2H} + (k-1)^{2H}\right] \quad 0.5 < H < 1 \tag{9.43}$$

Second-order self-similar processes are characterized by a hyperbolically decaying ACF and are extensively used to model long-range dependent processes. Some self-similar processes may exhibit LRD, but not all processes that have LRD are self-similar. The Hurst parameter H is a measure of the extent of LRD in a time series. H takes on values from 0 to 1. A value of 0.5 or less indicates the absence of LRD. The closer H is to 1, the greater is the degree of persistence or LRD. Thus, a self-similar process $X(t)$ exhibits LRD when $0.5 < H < 1$. As $H \to 1$, the dependence becomes stronger. While self-similarity does not generally imply LRD when $0.5 < H < 1$, a continuous-time self-similar process is exactly (and asymptotically) second-order self-similar and hence possesses LRD.

9.16.4   FBM Revisited

More formally, a random process $\{B_H(t), t \geq 0\}$ is an FBM (also called a fractal Brownian motion) if, for an index $0 < H < 1$, the following conditions hold:

1. $B_H(t)$ is continuous with probability 1 and $B_H(0) = 0$.
2. For any $t \geq 0$ and $H > 0$, the increment $B_H(t + \theta) - B_H(t)$ is normally distributed with mean zero and variance θ^{2H}, which means that the CDF of $B_H(t + \theta) - B_H(t)$ is given by

$$P\left[\{B_H(t+\theta) - B_H(t)\} \leq x\right] = \frac{1}{\sqrt{2\pi\theta^{2H}}} \int_{-\infty}^{x} \exp\left(-\frac{u^2}{2\theta^{2H}}\right) du \tag{9.44}$$

As discussed earlier, the index H is called the *Hurst index* (or *Hurst parameter*), which is used to model the dependence property of the increments. If $H = 1/2$, then FBM becomes the normal Brownian motion. Thus, the main difference between the regular Brownian motion and the FBM is that while the increments in the regular Brownian motion are independent, they are dependent in the FBM. This dependence means that if there is an increasing pattern in the previous "step," then the current step is likely to be increasing too. While FBM permits correlation between events, it retains the short tails of the normal distribution.

FBM satisfies the following condition:

$$B_H(0) = E[B_H(t)] = 0 \quad t \geq 0 \tag{9.45}$$

The covariance function of FBM, $R_H(s,t)$, is given by

$$R_H(s,t) = E[B_H(s)B_H(t)] = \frac{1}{2}(s^{2H} + t^{2H} - |t-s|^{2H}) \quad s,t \geq 0 \tag{9.46}$$

This result, which implies that the FBM is nonstationary, can be established as follows.

$$E[\{B_H(s) - B_H(t)\}^2] = E[\{B_H(s-t) - B_H(0)\}^2] = E[\{B_H(s-t)\}^2] = \sigma^2(s-t)^{2H}$$

Also,

$$\begin{aligned} E[\{B_H(s) - B_H(t)\}^2] &= E[\{B_H(s)\}^2] + E[\{B_H(t)\}^2] - 2E[B_H(s)B_H(t)] \\ &= \sigma^2(s)^{2H} + \sigma^2(t)^{2H} - 2E[B_H(s)B_H(t)] \\ &= \sigma^2(s-t)^{2H} \end{aligned}$$

Thus,

$$E[B_H(s)B_H(t)] = \frac{\sigma^2}{2}\{(s)^{2H} + (t)^{2H} - (s-t)^{2H}\}$$

where $\sigma = 1$ for the standard Brownian motion. The value of H determines the kind of process of the FBM. Generally, three types of processes can be captured by the model:

1. If $H > 1/2$, the increments of the process are positively correlated, and the process is said to be *persistent* or has a *long memory* (or *strong after-effects*). In this case, consecutive increments tend to have the same sign; that is, a random step in one direction is preferentially followed by another step in the same direction.
2. If $0 < H < 1/2$, the increments are negatively correlated. In this case, the process is said to be *antipersistent*, which means that consecutive increments are more likely to have opposite signs; that is, a random step in one direction is preferentially followed by a reversal of direction.

3. If $H = 1/2$, the process is a regular Brownian motion in which the steps are statistically independent of one another. Thus, successive Brownian motion increments are as likely to have the same sign as the opposite, and thus there is no correlation.

Thus, a process has independent increments if and only if $H = 0.5$; that is, it is a Brownian motion. If $H < 0.5$, the increments will be negatively correlated, and if $H > 0.5$, the increments will be positively correlated. Also, the Hurst index is a measure of the roughness of the sample path of the FBM. As H decreases, the FBM appears rougher, which is why H is sometimes called the "roughness exponent."

Another property of FBM is self-similarity. As discussed earlier, a self-similar process is a stochastic process that is invariant in distribution under suitable scaling of time and space. Specifically, the stochastic process $\{X(t), t \geq 0\}$ is said to be self-similar if for any $a > 0$, there exists some $b > 0$ such that $\{X(at), t \geq 0\}$ has the same distribution as $\{bX(t), t \geq 0\}$. In the case of the FBM with a Hurst index H, a and b are related by $b = a^H$. Thus, FBM is a self-similar stochastic process in the sense that $\{B_H(at), t \geq 0\}$ has the same distribution as $\{a^H B_H(t), t \geq 0\}$ for all $a > 0$.

Also, FBM exhibits LRD when $H > 1/2$, which means that if we define the ACF $\rho_H(n) = E[B_H(t_0)B_H(t_n)]$, then

$$\sum_{n=1}^{\infty} \rho_H(n) = \infty$$

When $H < 1/2$, $\rho_H(n)$ tends to decay exponentially; that is, $\rho_H(n) = \beta^n$ as n tends to infinity, where $\beta < 1$, which means that the process exhibits short-range dependence.

FBM has been used by Norros (1995) and Mikosch et al. (2002) to model aggregated connectionless traffic. It has also been used by Rogers (1997), Dasgupta (1998), and Sottinen (2001) to model financial markets. The motivation for using FBM to model financial markets is that earlier empirical studies indicated that the so-called logarithmic returns $r_n = \log\{S(t_n)/S(t_{n-1})\}$, where $S(t)$ is the observed price of a given stock at time t, is normal, which means that $H = 0.5$. However, new empirical studies indicate that r_n tends to have a strong after-effect with $H > 0.5$. Thus, advocates of FBM claim that it is a better model than traditional Brownian motion by using the Hurst index to capture dependency.

9.17 Fractional Gaussian Noise

Fractional Gaussian noise describes a time series that is somewhat random but less random than white Gaussian noise. In an FGN process, there is LRD or long-time memory; that is, there is a significant correlation between observations of signals separated by large time spans. This means that what happens at one instance has an effect on what happens later—even much later. Whereas the random walk is the discrete-time analog of the Brownian motion, the FGN is the discrete-time analog of the FBM.

FGN is the process $\{G_H(t), t > 0\}$ that is obtained from the FBM increments as follows:

$$G_H(t) = B_H(t + 1) - B_H(t) \tag{9.47}$$

That is, FGN is the increment of the FBM. It is a stationary Gaussian process with zero mean and covariance given by

$$\rho(k) = E[G_H(t)G_H(t+k)] = \frac{1}{2}\left[|k+1|^{2H} - 2|k|^{2H} + |k-1|^{2H}\right] \quad k>0 \quad (9.48)$$

When $H = 1/2$, all covariances are zero, except for $k = 0$, and $\{G_H(t), t>0\}$ represents white noise. We can rewrite Eq. (9.48) as follows:

$$\rho(k) = \frac{k^{2H-2}}{2}\left\{k^2\left[\left(1+\left|\frac{1}{k}\right|\right)^{2H} - 2 + \left(1-\left|\frac{1}{k}\right|\right)^{2H}\right]\right\}$$

Using the binomial series expansion,

$$(1+x)^n = 1 + nx + \frac{n(n-1)}{2!}x^2 + \frac{n(n-1)(n-2)}{3!}x^2 + \cdots$$

$$(1-x)^n = 1 - nx + \frac{n(n-1)}{2!}x^2 - \frac{n(n-1)(n-2)}{3!}x^2 + \cdots$$

we see that with $n = 2H$ and $x = 1/k$,

$$\lim_{k\to\infty} \rho(k) \approx H(2H-1)|k|^{2H-2} \quad\quad (9.49)$$

For $1/2 < H < 1$, we have that $2H - 2 = 2(H - 1) = p < 0$ and $|p| < 1$. Thus, for large k,

$$\sum_{k=-\infty}^{\infty} \rho(k) \approx H(2H-1)\sum_{k=-\infty}^{\infty}\frac{1}{|k|^p} = H(2H-1)\left\{\rho(0) + 2\sum_{k=1}^{\infty}\frac{1}{|k|^p}\right\} = \infty$$

where the last equality follows from the fact that the Riemann zeta function

$$\zeta(v) = \sum_{n=1}^{\infty}\frac{1}{n^v}$$

is divergent for $v \leq 1$ and convergent for $v > 1$. Thus, when $1/2 < H < 1$, the process has LRD (or long memory).

When $0 < H < 1/2$, $2H - 2 = 2(H - 1) = p < 0$ and $|p| > 1$. This means that $\sum_{k=-\infty}^{\infty} \rho(k) < \infty$ and the process has short-range dependence (or short memory).

9.18 Multifractional Brownian Motion

One of the disadvantages of FBM is that because the Hurst index, which governs the pointwise regularity of a system, is constant, it can only be used to model phenomena that have the same irregularities globally; that is, phenomena that have monofractal structure. Thus, when a system has intricate structures with variations in irregularities, FBM becomes inadequate, and a model that permits the Hurst index to vary as a function of time or position becomes necessary. The multifractional Brownian motion (MBM) is used in this case. In MBM, the Hurst index is replaced by the *Holder function* $H(t)$ that describes the local variations in the system, where $0 < H(t) < 1$. Thus, MBM generalizes FBM by replacing the Hurst index with the Holder function. This generalization was introduced independently by Peltier and Vehel (1995) and Benassi (1997). It can be defined as follows:

$$B_{H(t)}(t) = \frac{1}{\Gamma\left(H(t) + \frac{1}{2}\right)} \left\{ \int_{-\infty}^{0} [(t-u)^{H(t)-1/2} - (-u)^{H(t)-1/2}] dB(u) + \int_{0}^{t} (t-u)^{H(t)-1/2} dB(u) \right\}$$

where $B(t)$ is a Brownian motion.

The fact that the local regularity of MBM may be tuned via a functional parameter has made it a useful model in various areas such as finance, biomedicine, geophysics, and image analysis. For example, when innovations of the time series of the rates of return in the stock market are independent, the time series can be modeled by a Brownian motion. If the series has long memory, then it is better modeled by an FBM. However, because stock prices experience some "quiet" periods that correspond to a higher Holder exponent and some "erratic" periods that correspond to a low Holder exponent, such financial time series are better modeled by an MBM.

9.19 Problems

9.1 Assume that X and Y are independent random variables such that $X \sim N(0, \sigma^2)$ and $Y \sim N(0, \sigma^2)$. Consider the random variables $U = (X + Y)/2$ and $V = (X - Y)/2$. Show that U and V are independent with $U \sim N(0, \sigma^2/2)$ and $V \sim N(0, \sigma^2/2)$.

9.2 Suppose $X(t)$ is a standard Brownian motion and $Y(t) = tX(1/t)$. Show that $Y(t)$ is a standard Brownian motion.

9.3 Let $\{X(t), t \geq 0\}$ be a Brownian motion with drift rate μ and variance parameter σ^2. What is the conditional distribution of $X(t)$ given that $X(u) = b, u < t$?

9.4 Let $T = \min\{t|B(t) = 5 - 3t\}$. Use the martingale stopping theorem to find $E[T]$.

9.5 Let $Y(t) = \int_0^t B(u)du$, where $\{B(t), t \geq 0\}$ is the standard Brownian motion. Find
   a. $E[Y(t)]$
   b. $E[Y^2(t)]$
   c. The conditional distribution of $Y(t)$, given that $B(t) = x$.

9.6 Consider the Brownian motion with drift

$$Y(t) = \mu t + \sigma B(t) + x$$

where $Y(0) = x$ and $b < x < a$. Let $p_a(x)$ denote the probability that hits a before b.

a. Show that

$$\frac{1}{2}\frac{d^2 p_a(x)}{dx^2} + \mu \frac{dp_a(x)}{dx} = 0$$

b. Deduce that

$$p_a(x) = \frac{e^{-2\mu b} - e^{-2\mu x}}{e^{-2\mu b} - e^{-2\mu a}}$$

c. What is $p_a(x)$ when $\mu = 0$?

9.7 What is the mean value of the first passage time of the reflected Brownian motion $\{|B(t)|, t \ge 0\}$ with respect to a positive level x, where $B(t)$ is the standard Brownian motion? Determine the CDF of $|B(t)|$.

9.8 Let the process $\{X(t), t \ge 0\}$ be defined by $X(t) = B^2(t) - t$, where $\{B(t), t \ge 0\}$ is a standard Brownian motion.

a. What is $E[X(t)]$?

b. Show that $\{X(t), t \ge 0\}$ is a martingale.

Hint: Start by computing $E[X(t)|B(v), 0 \le v \le t]$.

9.9 Let $\{B(t), t \ge 0\}$ be a standard Brownian motion and define the process $Y(t) = e^{-t}B(e^{2t}), t \ge 0$; that is, $\{Y(t), t \ge 0\}$ is the OU process.

a. Show that $Y(t)$ is a Gaussian process.

b. Find the covariance function of the process.

9.10 The price of a certain stock follows Brownian motion. The price at time $t = 3$ is 52. Determine the probability that the price is more than 60 at time $t = 10$.

9.11 Let T_a be the time until a standard Brownian motion process hits the point a. Calculate $P[T_2 \le 8]$.

10 Diffusion Processes

10.1 Introduction

Diffusion processes are continuous-time, continuous-state processes whose sample paths are everywhere continuous but nowhere differentiable. They arise in many natural phenomena and in some sense are generalized versions of the Brownian motion. The study of diffusion processes originated from the field of statistical physics, but diffusion processes have been used to model many physical, biological, engineering, economic, and social phenomena because diffusion is one of the fundamental mechanisms for transport of materials in physical, chemical, and biological systems. Also, many functions can be calculated explicitly for one-dimensional diffusion process. In the fields where diffusion has been applied, it has been used to model phenomena evolving randomly and continuously in time under certain conditions, for example, security price fluctuations in a perfect market, variations of population growth in ideal conditions, and communication systems with noise.

As stated earlier, a well-known example of a diffusion process is the Brownian motion. In the Brownian view, diffusion is a process that causes particle mixing because of random collisions among themselves or with other particles. We can model diffusion the same way we did the movement of a single particle in Brownian motion, but with diffusion we just have more particles to deal with. If a number of particles subject to Brownian motion are present in a given medium and there is no preferred direction for the random oscillations, then over a period of time the particles will tend to be spread evenly throughout the medium. Thus the end result of diffusion would be a constant particle concentration throughout the environment.

This means that at a macroscopic level, diffusion results in particles moving from areas of high concentrations to areas of low concentration. Essentially, if there is no *concentration gradient*, then there will be no diffusion. Once particles have diffused to even out the concentration everywhere, the concentration gradient is now zero, and they do not move in any particular direction anymore. Thus, we can define diffusion as the spontaneous "spreading" of particles from a region of high concentration to one of lower concentration. Alternatively, diffusion can also be viewed as occurring when a system is not in equilibrium and random motion tends to bring everything toward uniformity. For example, diffusion enables heat to flow from a hot part of a conductor to a cold part of the conductor. Similarly, when we put a drop of dye in a jar of water, the dye spreads throughout the water, and we

Markov Processes for Stochastic Modeling. DOI: http://dx.doi.org/10.1016/B978-0-12-407795-9.00010-4

can say that the dye *diffused* through the water. Another example is when you are cooking food and the smell of the food spreads from the oven throughout the kitchen, and eventually throughout your home. In this case, the odor *diffused* through the air.

Diffusion processes are used to model the price movements of financial instruments. The Black–Scholes model for pricing options assumes that the underlying instrument follows a traditional diffusion process with small, continuous, random movements.

10.2   Mathematical Preliminaries

Consider a continuous-time continuous-state Markov process $\{X(t), t \geq 0\}$ whose transition probability distribution is given by

$$F(y, t|x, s) = P[X(t) \leq y | X(s) = x] \quad s < t \tag{10.1}$$

If the derivative

$$f(y, t|x, s) = \frac{\partial}{\partial y} F(y, t|x, s) \tag{10.2}$$

exists, then it is called the *transition density function* of the diffusion process. Since $\{X(t)\}$ is a Markov process, $f(y, t|x, s)$ satisfies the Chapman–Kolmogorov equation:

$$f(y, t|x, s) = \int_{-\infty}^{\infty} f(y, t|z, u) f(z, u|x, s) dz \tag{10.3}$$

We assume that the process $\{X(t), t \geq 0\}$ satisfies the following conditions:

1. $P[|X(t + \Delta t) - X(t)| > \varepsilon | X(t)] = o(\Delta t)$, for $\varepsilon > 0$, which states that the sample path is continuous; alternatively, we say that the process is continuous in probability.
2. $E[X(t + \Delta t) - X(t)|X(t) = x] = a(x, t)\Delta t + o(\Delta t)$ so that

$$\lim_{\Delta t \to 0} \frac{E[X(t + \Delta t) - X(t)|X(t) = x]}{\Delta t} = \lim_{\Delta t \to 0} \frac{1}{\Delta t} \int_{-\infty}^{\infty} (y - x) f(y, t + \Delta t)|x, t) dy$$

$$= a(x, t)$$

3. $E[\{X(t + \Delta t) - X(t)\}^2 | X(t) = x] = b(x, t)\Delta t + o(\Delta t)$ is finite so that

$$\lim_{\Delta t \to 0} \frac{E[\{X(t + \Delta t) - X(t)\}^2 | X(t) = x]}{\Delta t} = \lim_{\Delta t \to 0} \frac{1}{\Delta t} \int_{-\infty}^{\infty} (y - x)^2 f(y, t + \Delta t)|x, t) dy$$

$$= b(x, t)$$

A Markov process that satisfies these three conditions is called a diffusion process. The function $a(x, t)$ is called the *instantaneous* (or infinitesimal) *mean* (or drift) of $X(t)$, and the function $b(x, t)$ is called the *instantaneous* (or infinitesimal) *variance* of $X(t)$. Let the small increment in $X(t)$ over any small interval dt be denoted by $dX(t)$. Then it can be shown that if $B(t)$ is a standard Brownian motion we can incorporate the above properties into the following stochastic differential equation:

$$dX(t) = a(x, t)dt + [b(x, t)]^{1/2}dB(t) \tag{10.4}$$

where $dB(t)$ is the increment of $B(t)$ over the small interval $(t, t + \Delta t)$.

10.3 Models of Diffusion

The diffusion equation is a partial differential equation that describes the density fluctuations in a material undergoing diffusion. There are different models of the diffusion equation. In this section, we consider a few of them, including the Fokker–Planck equation, the Langevin equation, and the Fick's equations.

10.3.1 Diffusion as a Limit of Random Walk: The Fokker–Planck Equation

Diffusion can be obtained as a limit of the random walk. Consider a one-dimensional random walk where in each interval of length Δt the process makes a movement of length Δx with probability p or a movement of length $-\Delta x$ with probability $q = 1 - p$. Let $P[x, t|x_0, t_0]$ denote the probability that the process is at x at time t, given that it was at x_0 at time t_0 where $\Delta x = x - x_0$ and $\Delta t = t - t_0$. Assume that p and q are independent of x and t. Then we have that

$$P[x, t|x_0, t_0] = pP[x - \Delta x, t - \Delta t|x_0, t_0] + qP[x + \Delta x, t - \Delta t|x_0, t_0]$$

Since $P[x, t|x_0, t_0] = f(x, t|x_0, t_0)\Delta x$, we have that

$$f(x, t|x_0, t_0)\Delta x = pf(x - \Delta x, t - \Delta t|x_0, t_0)\Delta x + qf(x + \Delta x, t - \Delta t|x_0, t_0)\Delta x$$

Thus, we obtain

$$f(x, t|x_0, t_0) = pf(x - \Delta x, t - \Delta t|x_0, t_0) + qf(x + \Delta x, t - \Delta t|x_0, t_0)$$

We assume that $f(x, t; x_0, t_0)$ is a smooth function of its arguments so that we can expand it in a Taylor series in x and t. Thus, we have that:

$$
\begin{aligned}
f(x,t|x_0,t_0) = &\, p\left\{ f(x,t|x_0,t_0) - \Delta x \frac{\partial f}{\partial x} - \Delta t \frac{\partial f}{\partial t} + \frac{(\Delta x)^2}{2}\frac{\partial^2 f}{\partial x^2} + \frac{(\Delta t)^2}{2}\frac{\partial^2 f}{\partial t^2} + (\Delta x)(\Delta t)\frac{\partial^2 f}{\partial x \partial t} \right\} \\
&+ q\left\{ f(x,t|x_0,t_0) + \Delta x \frac{\partial f}{\partial x} - \Delta t \frac{\partial f}{\partial t} + \frac{(\Delta x)^2}{2}\frac{\partial^2 f}{\partial x^2} + \frac{(\Delta t)^2}{2}\frac{\partial^2 f}{\partial t^2} - (\Delta x)(\Delta t)\frac{\partial^2 f}{\partial x \partial t} \right\} \\
&+ o(\{\Delta x\}^3) + o(\{\Delta t\}^3) \\
= &\, f(x,t|x_0,t_0) - \Delta t \frac{\partial f}{\partial t} - (p-q)\Delta x \frac{\partial f}{\partial x} + \frac{(\Delta x)^2}{2}\frac{\partial^2 f}{\partial x^2} + \frac{(\Delta t)^2}{2}\frac{\partial^2 f}{\partial t^2} \\
&+ (p-q)(\Delta x)(\Delta t)\frac{\partial^2 f}{\partial x \partial t} + o(\{\Delta x\}^3) + o(\{\Delta t\}^3)
\end{aligned}
$$

From this we obtain

$$
\begin{aligned}
\frac{\partial f}{\partial t} = &\, -\frac{(p-q)\Delta x}{\Delta t}\frac{\partial f}{\partial x} + \frac{(\Delta x)^2}{2\Delta t}\frac{\partial^2 f}{\partial x^2} + \frac{\Delta t}{2}\frac{\partial^2 f}{\partial t^2} \\
&+ (p-q)\Delta x \frac{\partial^2 f}{\partial x \partial t} + o\left(\frac{\{\Delta x\}^3}{\Delta t}\right) + o(\{\Delta t\}^2)
\end{aligned}
$$

Assume that

$$
\mu = \lim_{\Delta t, \Delta x \to 0} \frac{(p-q)\Delta x}{\Delta t}
$$

$$
D_0 = \lim_{\Delta t, \Delta x \to 0} \left\{ \frac{(\Delta x)^2}{2\Delta t} \right\}
$$

(10.5)

Then in the limit as $\Delta x, \Delta t \to 0$ we have that

$$
\frac{\partial f}{\partial t} = -\mu \frac{\partial f}{\partial x} + D_0 \frac{\partial^2 f}{\partial x^2}
$$

(10.6)

This is called the *forward diffusion equation* because it involves differentiation in x. This equation is commonly called the *Fokker–Planck equation*. The constant D_0 is called the *diffusion constant* or *diffusivity*; it describes how fast or slow an object diffuses. The equation obtained when $\mu = 0$ (i.e., when $p = q = 1/2$) is known as the *heat equation*, which is

$$
\frac{\partial f}{\partial t} = D_0 \frac{\partial^2 f}{\partial x^2}
$$

(10.7)

Similarly, when diffusion is absent (i.e., $D_0 = 0$), we obtain the *advection* equation:

$$\frac{\partial f}{\partial t} + \mu \frac{\partial f}{\partial x} = 0 \tag{10.8}$$

which describes the evolution of $f(x, t; x_0, t_0)$ as it is carried along by a constant flow process.

10.3.1.1 Forward Versus Backward Diffusion Equations

There are two partial differential equations that characterize the dynamics of the distribution of the diffusion process. These are the forward diffusion equation that we derived earlier and the backward diffusion equation. The forward diffusion equation addresses the following issue: If at time t the state of the system is x_0, what can we say about the distribution of the state at a future time $s > t$ (hence the term "forward")? The backward equation, on the other hand, is useful when we address the question that given that the state of the system at a future time s has a particular value x, what can we say about the state of the system at an earlier time $t < s$? This imposes a terminal condition on the partial differential equation, which is integrated backward in time, from s to t (hence the term "backward" is associated with this). For example, if the forward diffusion equation models the evolution of the temperature from its initial values, the backward equation attempts to determine the initial temperature distribution from which the current temperature could have evolved.

Thus, the *backward diffusion equation*, which involves differentiation in x_0, can be obtained as follows:

$$\frac{\partial f}{\partial t} = \mu \frac{\partial f}{\partial x_0} + D_0 \frac{\partial^2 f}{\partial x_0^2} \tag{10.9}$$

The term $\mu \partial f / \partial x$ is called a *convection term* while the term $D_0 \partial^2 f / \partial x^2$ is called the *diffusion term*. Thus, Eq. (10.9) is often called a *diffusion equation with convection*, or a *convection–diffusion equation*.

10.3.2 The Langevin Equation

Brownian motion can be modeled by the Langevin equation. Consider a particle with mass m performing a random walk inside a fluid due to the bombardment by the fluid molecules that obey an equilibrium distribution. Pierre Langevin described this motion with a simple stochastic differential equation, which in the one-dimensional case is given by

$$m\ddot{x} = -\alpha \dot{x} + F(t) \tag{10.10}$$

where the term $\alpha \dot{x}$ represents the friction force, $\dot{x}$ is the particle velocity, α is the damping rate that depends on the radius of the particle and the viscosity of the fluid, and $F(t)$ is a zero-mean random fluctuating force due to the random bombardment of the particle by the fluid molecules. The random fluctuating force at one time is uncorrelated with the force at another time; thus we have that

$$E[F(t)] = 0 \quad E[F(t)F(u)] = \delta(t - u)$$

Multiplying Eq. (10.10) by x, we have

$$m x \ddot{x} = m x \frac{d \dot{x}}{dt} = m \left[\frac{d(x \dot{x})}{dt} - \dot{x}^2 \right] = -\alpha x \dot{x} + x F(t)$$

Taking averages over a large number of particles and recognizing that $E[x F(t)] = 0$ due to the random nature of the force $F(t)$, we obtain

$$m E \left[\frac{d}{dt}(x \dot{x}) \right] = m \frac{d}{dt} E[x \dot{x}] = m E \left[\dot{x}^2 \right] - \alpha E[x \dot{x}] \tag{10.11}$$

Since the background fluid is in equilibrium, the kinetic energy of the particle is proportional to the fluid temperature; that is, $m E[\dot{x}^2]/2 = kT/2$, where k is the Boltzmann constant and T is the temperature of the fluid. Thus, Eq. (10.11) now takes the form

$$\frac{d}{dt} E[x \dot{x}] = \frac{kT}{m} - \gamma E[x \dot{x}]$$

where $\gamma = \alpha/m$, which has the solution

$$E[x \dot{x}] = C e^{-\gamma t} + \frac{kT}{\alpha} \tag{10.12}$$

where C is an arbitrary constant of integration. Note that $\gamma^{-1} = \tau$ acts like a time constant of the system. Now, at $t = 0$, the mean square displacement is zero, so that $0 = C + kT/\alpha \Rightarrow C = -kT/\alpha$, and Eq. (10.12) becomes

$$E[x \dot{x}] \equiv \frac{1}{2} \frac{d}{dt} E \left[x^2 \right] = \frac{kT}{\alpha} (1 - e^{-\gamma t}) \tag{10.13}$$

This means that

$$\frac{d}{dt} E \left[x^2 \right] = \frac{2kT}{\alpha} (1 - e^{-\gamma t})$$

On integrating the above equation we find the solution

$$E \left[x^2 \right] = \frac{kT}{\alpha} \{ t - \gamma^{-1} (1 - e^{-\gamma t}) \} \tag{10.14}$$

If $t \ll \gamma^{-1}$ (that is, time is much shorter than the collision time), we use the approximation

$$e^{-\gamma t} = 1 - \gamma t + \frac{1}{2}\gamma^2 t^2 - \cdots$$

and obtain the solution

$$E[x^2] = \frac{kT}{m}t^2 \tag{10.15}$$

which is called "ballistic" diffusion and means that at small times particles are not hindered by collisions yet and diffuse very fast. On the other hand, if $t \gg \gamma^{-1}$, then $e^{-\gamma t} \to 0$ and the solution has the form

$$E[x^2] = \left\{\frac{kT}{\alpha}\right\}t = D_0 t \tag{10.16}$$

which is the expression of the diffusion approximation of a random walk. In this case, the mean-squared displacement grows linearly with time and D_0 is the diffusion constant.

10.3.3   The Fick's Equations

A physically motivated derivation of the diffusion equation starts with the Fick's laws. As discussed earlier, diffusion is essentially the movement of particles from a region of high concentration to a region of low concentration. Given enough time, this movement will eventually result in homogeneity within the entire region causing the movement due to a concentration gradient to stop. The first Fick's law states that the magnitude of the diffusive flux of particles across a given plane from the region of high concentration to the region of low concentration is proportional to the concentration gradient across the plane. That is, if φ is the concentration of the diffusive substance and J is the diffusion flux (or the amount of substance per unit area per unit time), then

$$J = -D_0 \frac{\partial \phi}{\partial x} \tag{10.17}$$

where D_0 is the diffusion constant discussed earlier. The negative sign of the right-hand side of the equation indicates that the substance is flowing in the direction of lower concentration. Note that this law does not consider the fact that the gradient and local concentration of the particles decrease as time increases. The flux of particles entering a section of a bar with a concentration gradient is different from the flux leaving the same section. From the law of conservation of matter, the difference between the two fluxes must lead to a change in the concentration of particles within the section. This is precisely *Fick's second law* which states that the change in particle concentration over time is equal to the change in local diffusion flux. That is,

$$\frac{\partial \phi}{\partial t} = -\frac{\partial J}{\partial x}$$

Combining this with Fick's first law we obtain

$$\frac{\partial \phi}{\partial t} = -\frac{\partial J}{\partial x} = \frac{\partial}{\partial x}\left(D_0 \frac{\partial \phi}{\partial x}\right) = D_0 \frac{\partial^2 \phi}{\partial x^2} \tag{10.18}$$

10.4 Examples of Diffusion Processes

The following are some examples of the diffusion process. These processes differ only in their values of the infinitesimal mean and infinitesimal variance. In particular, we consider the Brownian motion and the Brownian motion with drift.

10.4.1 Brownian Motion

The Brownian motion is a diffusion process on the interval $(-\infty, \infty)$ with zero mean and constant variance. That is, for the standard Brownian motion, $\mu = 0$ and $D_0 = \sigma^2/2$, where $\sigma^2 > 0$ is the variance. Thus, the forward diffusion equation becomes

$$\frac{\partial f}{\partial t} = \frac{\sigma^2}{2}\frac{\partial^2 f}{\partial x^2} \tag{10.19}$$

As stated earlier, this equation is an example of a heat equation, which describes the variation in temperature as heat diffuses through an isotropic and homogeneous physical medium in one-dimensional space. Similarly, the backward diffusion equation becomes

$$\frac{\partial f}{\partial t} = -\frac{\sigma^2}{2}\frac{\partial^2 f}{\partial x_0^2} \tag{10.20}$$

We can solve the forward diffusion equation as follows. If we define $\lambda^2 = \sigma^2/2$, the forward diffusion equation becomes

$$\frac{\partial}{\partial t}f(x,t) = \lambda^2 \frac{\partial^2}{\partial x^2}f(x,t) \quad 0 \le x \le L; 0 \le t \le T$$

where

$$f(x,t) = \frac{\partial}{\partial x}P[X(t) \le x]$$

Assume that the initial condition is $f(x,0) = \phi(x), 0 \le x \le L$, and the boundary conditions are $f(0,t) = f(L,t) = 0, 0 \le t \le T$. If we assume that $f(x,t)$ is a separable function of t and x, then we can write

$$f(x,t) = g(t)h(x)$$

with the boundary conditions $h(0) = h(L) = 0$. Thus, the differential equation becomes

$$\frac{dg(t)}{dt} h(x) = \lambda^2 g(t) \frac{d^2 h(x)}{dx^2}$$

which gives

$$\frac{1}{\lambda^2} \frac{dg(t)}{dt} \frac{1}{g(t)} = \frac{d^2 h(x)}{dx^2} \frac{1}{h(x)}$$

Because the left side of the equation is a function of t alone and the right side is a function of x alone, the equality can be satisfied only if the two sides are equal to some constant m; that is,

$$\frac{dg(t)}{dt} = m\lambda^2 g(t) \quad 0 \leq t \leq T$$

$$\frac{d^2 h(x)}{dx^2} = mh(x) \quad 0 \leq x \leq L$$

Using the method of characteristic equation, the solutions to these equations are given by

$$g(t) = C_0 e^{m\lambda^2 t}$$
$$h(x) = C_1 e^{x\sqrt{m}} + C_2 e^{-x\sqrt{m}}$$

where C_0, C_1, and C_2 are constants. To avoid a trivial solution obtained when $m = 0$, and to obtain a solution that does not increase exponentially with t, we require that $m < 0$. Thus, if we define $m = -p^2$, the solutions become

$$g(t) = C_0 e^{-\lambda^2 p^2 t}$$
$$h(x) = C_1 e^{jpx} - C_2 e^{-jpx} = B_1 \sin(px) + B_2 \cos(px)$$
$$f(x,t) = C_0 e^{-\lambda^2 p^2 t} \{B_1 \sin(px) + B_2 \cos(px)\}$$

From the boundary condition $f(0,t) = 0$, we obtain $B_2 = 0$. Similarly, from the boundary condition $f(L,t) = 0$, we obtain $B_1 \sin(pL) = 0$, which gives $pL = k\pi$, $k = 1, 2, \ldots$. Thus,

$$p = \frac{k\pi}{L} \quad k = 1, 2, \ldots$$

We can then define the following functions:

$$g_k(t) = C_{0k} \exp\left\{ -\frac{\lambda^2 k^2 \pi^2 t}{L^2} \right\}$$

$$h_k(x) = B_{1k} \sin\left\{ \frac{k\pi x}{L} \right\}$$

$$f_k(x,t) = C_k \exp\left\{ -\frac{\lambda^2 k^2 \pi^2 t}{L^2} \right\} \sin\left\{ \frac{k\pi x}{L} \right\}$$

$$f(x,t) = \sum_{k=1}^{\infty} f_k(x,t) = \sum_{k=1}^{\infty} C_k \exp\left\{ -\frac{\lambda^2 k^2 \pi^2 t}{L^2} \right\} \sin\left\{ \frac{k\pi x}{L} \right\}$$

where $C_k = C_{0k} \times B_{1k}$. From the initial condition, we obtain

$$f(x,0) = \phi(x) = \sum_{k=1}^{\infty} C_k \sin\left(\frac{k\pi x}{L}\right)$$

Because

$$\int_0^L \sin\left(\frac{m\pi u}{L}\right) \sin\left(\frac{k\pi u}{L}\right) du = \begin{cases} 0 & \text{if } k \neq m \\ \frac{L}{2} & \text{if } k = m \end{cases}$$

that is, the functions $\{\sin(k\pi u)/L\}$ are orthogonal on the interval $0 \leq u \leq L$, we obtain

$$C_k = \frac{2}{L}\int_0^L \phi(u) \sin\left(\frac{k\pi u}{L}\right) du$$

This means that

$$f(x,t) = \frac{2}{L}\sum_{k=1}^{\infty} \left\{ \int_0^L \phi(u) \sin\left(\frac{k\pi u}{L}\right) du \right\} \exp\left\{ -\frac{\lambda^2 k^2 \pi^2 t}{L^2} \right\} \sin\left\{ \frac{k\pi x}{L} \right\} \qquad (10.21)$$

10.4.2 Brownian Motion with Drift

In this process, $a(x,t) = \mu$ and $b(x,t) = \sigma^2$, where μ is the drift rate. The forward diffusion equation becomes

$$\frac{\partial f}{\partial t} = -\mu \frac{\partial f}{\partial x} + \frac{\sigma^2}{2} \frac{\partial^2 f}{\partial x^2} \tag{10.22}$$

Similarly, the backward diffusion equation becomes

$$\frac{\partial f}{\partial t} = -\mu \frac{\partial f}{\partial x_0} - \frac{\sigma^2}{2} \frac{\partial^2 f}{\partial x_0^2} \tag{10.23}$$

10.5 Correlated Random Walk and the Telegraph Equation

The telegraph equation was originally used to describe the propagation of electric current in a conducting wire with leakage. Specifically, let $\varphi(x,t)$ denote the current at point x in the wire at time t. It has been found that $\varphi(x,t)$ satisfies the following partial differential equation:

$$\frac{\partial^2 \varphi(x,t)}{\partial x^2} = C \frac{\partial^2 \varphi(x,t)}{\partial t^2} + (RC + SL) \frac{\partial \varphi(x,t)}{\partial t} + RS\varphi(x,t)$$

where C is the capacitance, R is the resistance, L is the inductance, and S is the leakage of the wire. Because of this historical fact, a partial differential equation of the form

$$\frac{\partial^2 \varphi(x,t)}{\partial t^2} + 2b \frac{\partial \varphi(x,t)}{\partial t} + a\varphi(x,t) = c^2 \frac{\partial^2 \varphi(x,t)}{\partial x^2}$$

is now referred to as the telegraph equation, where $a \geq 0, b \geq 0$, and $c > 0$. This equation includes the *wave equation* ($c^2\varphi_{xx} = \varphi_{tt}$) and the heat equation ($c^2\varphi_{xx} = \varphi_t$), where $\varphi_{xx} = \partial^2\varphi/\partial x^2, \varphi_{tt} = \partial^2\varphi/\partial t^2$, and $\varphi_t = \partial\varphi/\partial t$. In this section, we show how the telegraph equation can be derived from the correlated random walk.

Let $P(x,t)$ denote the probability that a walker performing a one-dimensional correlated random walk reaches x from 0 at a time t. Let $R(x,t)$ denote the probability that the walker arrives at x from the right, and let $L(x,t)$ denote the probability that the walker reaches x from the left. Note that $R(x,t)$ is the probability that at time $t - \Delta t$ the walker was at $x - \Delta x$ and moved to x by going right. Similarly, $L(x,t)$ is the probability that at time $t - \Delta t$ the walker was at $x + \Delta x$ and moved to x by going left. Then we have that

$$P(x,t) = R(x,t) + L(x,t) \tag{10.24}$$

Let p_1 denote the probability that the next movement is to the right, given that the walker's last move was to the right; let $q_1 = 1 - p_1$ denote the probability that the next movement is to the left, given that walker's last move was to the right; let p_2 denote the probability that the next movement is to the left, given that the walker's last move was to the left; and let $q_2 = 1 - p_2$ denote the probability that the next movement is to the right, given that the walker's last move was to the left. We analyze the problem using a method used in Jain (1971). Assume that the walker takes steps of length Δx and the time between two consecutive steps is Δt. From these we obtain the following system of partial difference equations:

$$
\begin{aligned}
R(x, t + \Delta t) &= p_1 R(x - \Delta x, t) + q_2 L(x - \Delta x, t) \\
&= p_1 \{P(x - \Delta x, t) - L(x - \Delta x, t)\} + q_2 L(x - \Delta x, t) \\
&= p_1 P(x - \Delta x, t) - (p_1 - q_2) L(x - \Delta x, t) \\
L(x, t + \Delta t) &= p_2 L(x + \Delta x, t) + q_1 R(x + \Delta x, t) \\
&= p_2 \{P(x + \Delta x, t) - R(x + \Delta x, t)\} + q_1 R(x + \Delta x, t) \\
&= p_2 P(x + \Delta x, t) - (p_1 - q_2) R(x + \Delta x, t)
\end{aligned}
\tag{10.25}
$$

where the last equality follows from the fact that $p_2 - q_1 = (1 - q_2) - (1 - p_1) = p_1 - q_2$. Also,

$$
\begin{aligned}
R(x + \Delta x, t) + L(x - \Delta x, t) &= \{p_1 R(x, t - \Delta t) + q_2 L(x, t - \Delta t)\} \\
&\quad + \{p_2 L(x, t - \Delta t) + q_1 R(x, t - \Delta t)\} \\
&= R(x, t - \Delta t) + L(x, t - \Delta t) \\
&= P(x, t - \Delta t)
\end{aligned}
\tag{10.26}
$$

From Eqs. (10.24), (10.25), and (10.26) we have that

$$
\begin{aligned}
P(x, t + \Delta t) &= p_1 P(x - \Delta x, t) + p_2 P(x + \Delta x, t) \\
&\quad - (p_1 - q_2)\{R(x + \Delta x, t) + L(x - \Delta x, t)\} \\
&= p_1 P(x - \Delta x, t) + p_2 P(x + \Delta x, t) - (p_1 - q_2) P(x, t - \Delta t)
\end{aligned}
\tag{10.27}
$$

Using the Taylor series expansion we have that

$$
P(x, t + \Delta t) = P(x, t) + \Delta t \frac{\partial P}{\partial t} + \frac{1}{2}(\Delta t)^2 \frac{\partial^2 P}{\partial t^2} + o(\{\Delta t\}^3)
$$

$$
P(x, t - \Delta t) = P(x, t) - \Delta t \frac{\partial P}{\partial t} + \frac{1}{2}(\Delta t)^2 \frac{\partial^2 P}{\partial t^2} + o(\{\Delta t\}^3)
$$

$$
P(x + \Delta x, t) = P(x, t) + \Delta x \frac{\partial P}{\partial x} + \frac{1}{2}(\Delta x)^2 \frac{\partial^2 P}{\partial x^2} + o(\{\Delta x\}^3)
$$

$$
P(x - \Delta x, t) = P(x, t) - \Delta x \frac{\partial P}{\partial x} + \frac{1}{2}(\Delta x)^2 \frac{\partial^2 P}{\partial x^2} + o(\{\Delta x\}^3)
$$

Thus, substituting in Eq. (10.27), we obtain

$$P(x,t) + \Delta t \frac{\partial P}{\partial t} + \frac{1}{2}(\Delta t)^2 \frac{\partial^2 P}{\partial t^2} = p_1 \left\{ P(x,t) - \Delta x \frac{\partial P}{\partial x} + \frac{1}{2}(\Delta x)^2 \frac{\partial^2 P}{\partial x^2} \right\}$$

$$+ p_2 \left\{ P(x,t) + \Delta x \frac{\partial P}{\partial x} + \frac{1}{2}(\Delta x)^2 \frac{\partial^2 P}{\partial x^2} \right\}$$

$$- (p_1 - q_2) \left\{ P(x,t) - \Delta t \frac{\partial P}{\partial t} + \frac{1}{2}(\Delta t)^2 \frac{\partial^2 P}{\partial t^2} \right\}$$

$$+ o(\{\Delta t\}^3) + o(\{\Delta x\}^3)$$

This gives

$$\Delta t \frac{\partial P}{\partial t}(1 - p_1 + q_2) + \frac{1}{2}(\Delta t)^2 \frac{\partial^2 P}{\partial t^2}(1 + p_1 - q_2)$$

$$= (p_1 + p_2)\frac{1}{2}(\Delta x)^2 \frac{\partial^2 P}{\partial x^2} - (p_1 - p_2)\Delta x \frac{\partial P}{\partial x} + o(\{\Delta x\}^3) + o(\{\Delta t\}^3)$$

That is,

$$\frac{\partial^2 P}{\partial t^2} + \frac{2(q_1 + q_2)}{(p_1 + p_2)\Delta t} \frac{\partial P}{\partial t} = \frac{(p_1 + p_2)(\Delta x)^2}{(1 + p_1 - q_2)(\Delta t)^2} \frac{\partial^2 P}{\partial x^2} - \frac{2(p_1 - p_2)\Delta x}{(1 + p_1 - q_2)(\Delta t)^2} \frac{\partial P}{\partial x}$$

$$+ o(\{\Delta x\}^3) + o(\{\Delta t\}^3)$$

$$= \frac{(p_1 + p_2)}{(1 + p_1 - q_2)} \left(\frac{\Delta x}{\Delta t}\right)^2 \frac{\partial^2 P}{\partial x^2} - \frac{2(p_1 - p_2)}{(1 + p_1 - q_2)\Delta x} \left(\frac{\Delta x}{\Delta t}\right)^2 \frac{\partial P}{\partial x}$$

$$+ o(\{\Delta t\}) + o\left(\frac{\{\Delta x\}^3}{\{\Delta t\}^2}\right)$$

$$= \left(\frac{\Delta x}{\Delta t}\right)^2 \frac{\partial^2 P}{\partial x^2} - \frac{2(p_1 - p_2)}{(p_1 + p_2)\Delta x} \left(\frac{\Delta x}{\Delta t}\right)^2 \frac{\partial P}{\partial x} + o(\{\Delta t\}) + o\left(\frac{\{\Delta x\}^3}{\{\Delta t\}^2}\right)$$

Let $v = \lim_{\Delta x, \Delta t \to 0}\{\Delta x/\Delta t\}$ be the velocity of the walker and let the constants C and D be defined as follows:

$$T = \lim_{\Delta t \to 0} \frac{\Delta t(p_1 + p_2)}{(q_1 + q_2)}$$

$$D_1 = \lim_{\Delta x \to 0} \frac{p_1 - p_2}{(p_2 + p_2)\Delta x}$$

(10.28)

Then, in the limit $\Delta t, \Delta x \to 0$, we obtain

$$\frac{\partial^2 P}{\partial t^2} + \frac{2}{T}\frac{\partial P}{\partial t} = v^2\left\{\frac{\partial^2 P}{\partial x^2} - 2D_1\frac{\partial P}{\partial x}\right\} \tag{10.29}$$

which is the telegraph equation. When $p_1 = p_2$, we have that $D_1 = 0$, and we obtain the telegraph equation without leakage, namely

$$\frac{\partial^2 P}{\partial t^2} + \frac{2}{T}\frac{\partial P}{\partial t} = v^2\frac{\partial^2 P}{\partial x^2} \tag{10.30}$$

which is the equation of wave motion with speed v. Thus, the telegraph equation features the characteristics of both the diffusion process and wave motion. The physical difference between the telegraph equation and the diffusion equation is that the velocity of dispersion is finite in the telegraph equation but infinite in the diffusion equation.

10.6 Introduction to Fractional Calculus

For the function $f(x)$, the notations $df(x)/dx \equiv D^1 f(x)$ and $d^2 f(x)/dx^2 \equiv D^2 f(x)$ are well understood to mean the first derivative and second derivative, respectively. But a notation like $d^{1/2}f(x)/dx^{1/2}$ or $D^{1/2}f(x)$ is not familiar to many people. This is what fractional calculus deals with. That is, fractional calculus is the branch of calculus that generalizes ordinary calculus by letting differentiation and integration be of any arbitrary real or complex order. It has become popular due to its demonstrated applications in many diverse fields of science and engineering.

The increase in its use in applications arises from the fact that it has been observed that while many analytical models assume that there is no memory in the system under investigation, many of these systems actually exhibit some form of memory. Fractional calculus has proved to be a natural tool used to model memory-dependent phenomena, which accounts for its increasing popularity in memory-dependent systems modeling.

The concept of factional calculus is believed to have stemmed from a question that L'Hospital asked Leibniz in 1695 about the meaning of the derivative of order n, $d^n y/dx^n$, when $n = 1/2$. Leibniz is said to have responded as follows: "This is an apparent paradox from which, one day, useful consequences will be drawn." Today, fractional calculus has found application in such areas as fluid flow, dynamic processes in self-similar and porous structures, diffusive transport, electrical networks, chemical physics, and signal processing. In this chapter, we review the basics of fractional calculus and discuss some of its applications. We will use the term "classical calculus" to refer to the traditional (or ordinary) calculus. More detailed discussion on fractional calculus can be found in such texts as Oldham and Spanier (1974), Podlubny (1999), Kilbas et al. (2006), Diethelm (2010), and Herrmann (2011).

10.6.1 Gamma Function

The factorial of an integer n, denoted by $n!$, is defined by

$$n! = n \times (n-1) \times \cdots \times 3 \times 2 \times 1 = \prod_{k=1}^{n} k \qquad (10.31)$$

Its properties include the following:

$$1! = 1$$
$$n! = n(n-1)!$$

By definition, $0! = 1$. The gamma function, which was introduced by Euler, is a generalization of the factorial to noninteger values. For any $\alpha > 0$, the gamma function is defined by

$$\Gamma(\alpha) = \int_{0}^{\infty} e^{-t} t^{\alpha-1} dt = 2 \int_{0}^{\infty} e^{-t^2} t^{2\alpha-1} dt \qquad (10.32)$$

It can be shown that

$$\Gamma\left(\frac{1}{2}\right) = \sqrt{\pi}$$
$$\Gamma(\alpha+1) = \alpha\Gamma(\alpha)$$

Thus, we have that

$$\Gamma\left(\frac{3}{2}\right) = \frac{1}{2}\Gamma\left(\frac{1}{2}\right) = \frac{1}{2}\sqrt{\pi}$$

$$\Gamma\left(\frac{5}{2}\right) = \frac{3}{2}\Gamma\left(\frac{3}{2}\right) = \frac{3}{4}\sqrt{\pi}$$

Also, when α is an integer,

$$\Gamma(\alpha) = (\alpha-1)!$$

The gamma function is related to the *beta function* $B(x,y)$, $x > 0$, $y > 0$, as follows:

$$B(x,y) = \frac{\Gamma(x)\Gamma(y)}{\Gamma(x+y)} \qquad (10.33)$$

where

$$B(x, y) = \int_0^1 u^{x-1}(1-u)^{y-1}du \tag{10.34}$$

10.6.2 Mittag-Leffler Functions

The exponential function e^x is defined as follows:

$$e^x = \sum_{n=0}^{\infty} \frac{x^n}{n!} = \sum_{n=0}^{\infty} \frac{x^n}{\Gamma(n+1)} \tag{10.35}$$

The Mittag-Leffler function is an extension of the exponential function to arbitrary real numbers $\alpha > 0$, and defined as follows:

$$E_\alpha(x) = \sum_{n=0}^{\infty} \frac{x^n}{\Gamma(n\alpha + 1)} \tag{10.36}$$

The Mittag-Leffler function for special values for some integer values of α are as follows:

$$E_0(x) = \sum_{n=0}^{\infty} \frac{x^n}{\Gamma(1)} = \sum_{n=0}^{\infty} x^n = \frac{1}{1-x}$$

$$E_1(x) = \sum_{n=0}^{\infty} \frac{x^n}{\Gamma(n+1)} = e^x$$

$$E_2(x) = \sum_{n=0}^{\infty} \frac{x^n}{\Gamma(2n+1)} = \cosh(\sqrt{x})$$

The generalized Mittag-Leffler function can be defined for $\alpha, \beta > 0$ as follows:

$$E_{\alpha,\beta}(x) = \sum_{n=0}^{\infty} \frac{x^n}{\Gamma(n\alpha + \beta)} \tag{10.37}$$

This means that

$$E_\alpha(x) = E_{\alpha,1}(x) \tag{10.38}$$

The following are some of the properties of $E_{\alpha,\beta}(x)$:

a. $E_{\alpha,\beta}(x) = xE_{\alpha,\alpha+\beta}(x) + \frac{1}{\Gamma(\beta)}$

b. $E_{\alpha,\alpha+\beta}(x) = \frac{1}{x}\left\{E_{\alpha,\beta}(x) - \frac{1}{\Gamma(\beta)}\right\}$

c. $E_{\alpha,\beta}(x) = \beta E_{\alpha,\beta+1}(x) + \alpha x \frac{d}{dx} E_{\alpha,\beta+1}(x)$

d. $E_{\alpha,\beta}^{(r)}(x) = \frac{d^r}{dx^r} E_{\alpha,\beta}(x) = \sum_{k=0}^{\infty} \frac{(k+r)! x^k}{k! \Gamma(k\alpha + r\alpha + \beta)} = r! \sum_{k=r}^{\infty} \binom{k}{r} \frac{x^{k-r}}{\Gamma(k\alpha + \beta)} \quad r = 0,1,2,\ldots$

e. $E'_{\alpha,\beta}(x) = \frac{d}{dx} E_{\alpha,\beta}(x) = \frac{E_{\alpha,\beta-1}(x) - (\beta-1)E_{\alpha,\beta}(x)}{\alpha x}$, where $\beta \neq 1$

f. $\frac{d^m}{dx^m}\left\{x^{\beta-1}E_{\alpha,\beta}(x^\alpha)\right\} = x^{\beta-m-1}E_{\alpha,\beta-m}(x^\alpha) \quad \text{Re}(\beta - m) > 0, \quad m = 0,1,\ldots$

g. $\int_0^t E_{\alpha,\beta}(\lambda x^\alpha)x^{\beta-1}dx = t^\beta E_{\alpha,\beta+1}(\lambda t^\alpha) \quad \beta > 0$

Examples of simple functions that can be expressed in terms of the generalized Mittag-Leffler function include:

$$E_{1,2}(x) = \sum_{n=0}^{\infty} \frac{x^n}{\Gamma(n+2)} = \sum_{n=0}^{\infty} \frac{x^n}{(n+1)!} = \frac{1}{x}\sum_{n=0}^{\infty} \frac{x^{n+1}}{(n+1)!} = \frac{e^x - 1}{x}$$

$$E_{1,3}(x) = \sum_{n=0}^{\infty} \frac{x^n}{\Gamma(n+3)} = \sum_{n=0}^{\infty} \frac{x^n}{(n+2)!} = \frac{1}{x^2}\sum_{n=0}^{\infty} \frac{x^{n+2}}{(n+2)!} = \frac{e^x - 1 - x}{x^2}$$

In general,

$$E_{1,m}(x) = \frac{1}{x^{m-1}}\left\{e^x - \sum_{n=0}^{m-2} \frac{x^n}{\Gamma(n+1)}\right\}$$

Also,

$$E_{2,1}(-x^2) = \sum_{n=0}^{\infty} \frac{(-1)^n x^{2n}}{\Gamma(2n+1)} = \sum_{n=0}^{\infty} \frac{(-1)^n x^{2n}}{(2n)!} = \cos(x)$$

$$E_{2,2}(-x^2) = \sum_{n=0}^{\infty} \frac{(-1)^n x^{2n}}{\Gamma(2n+2)} = \frac{1}{x}\sum_{n=0}^{\infty} \frac{(-1)^n x^{2n+1}}{(2n+1)!} = \frac{\sin(x)}{x}$$

10.6.3 Laplace Transform

The Laplace transform of a function $f(t)$, denoted by $L\{f(t)\}$ or $F(s)$, is defined by the following equation:

$$F(s) = L\{f(t)\} = \int_0^\infty e^{-st}f(t)dt \tag{10.39}$$

The inverse Laplace transform enables us to recover $f(t)$, and is given by

$$f(t) = L^{-1}\{F(s)\} \tag{10.40}$$

The Laplace transform is commonly used in the solution of differential equations. Some of its properties include the following:

a. *Linearity property*: If $f(t)$ and $g(t)$ are two functions and a and b are two real numbers, then $L\{af(t) + bg(t)\} = aF(s) + bG(s)$

b. *Frequency differentiation*: $L\{t^n f(t)\} = (-1)^n F^{(n)}(s)$, where $F^{(n)}(s)$ is the nth derivative of $F(s)$.

c. *Differentiation*: If $f(t)$ is n times differentiable, then

$$L\left\{\frac{d^n f(t)}{dt^n}\right\} = L\{f^{(n)}(t)\} = s^n F(s) - \sum_{k=0}^{n-1} s^{n-k-1} f^{(k)}(0)$$

$$\tag{10.41}$$

$$= s^n F(s) - \sum_{k=0}^{n-1} s^k f^{(n-k-1)}(0)$$

In particular,

$$L\left\{\frac{df(t)}{dt}\right\} = L\{f'(t)\} = sF(s) - f(0)$$

d. *Integration*:

    i. $L\left\{\int_0^t f(u)du\right\} = \frac{1}{s}F(s) = s^{-1}F(s)$

    ii. $L\left\{\int_0^t \cdots \int_0^t f(u)du^n\right\} = L\left\{\int_0^t \frac{(t-u)^{n-1}}{(n-1)!}f(u)du\right\} = \frac{1}{s^n}F(s) = s^{-n}F(s)$

e. *Time scaling*: $L\{f(at)\} = (1/|a|)F(s/a)$

f. *Convolution*: $L\{f(t) * g(t)\} = F(s)G(s)$

g. *Frequency shift*: $L\{e^{at}f(t)\} = F(s-a)$

h. *Frequency integration*: $L\{f(t)/t\} = \int_s^\infty F(u)du$

i. *Constant term*: $L\{a\} = a/s$

j. *Time shifting*: For $t \geq a$, $L\{f(t-a)\} = e^{-as}F(s)$

k. *Power*: $L\{t^n\} = (n!/s^{n+1})$, $n = 0, 1, 2, \ldots$

l. *Mittag-Leffler functions*:

    i. $L\{x^{\beta-1}E_{\alpha,\beta}(bx^\alpha)\} = \frac{s^{\alpha-\beta}}{s^\alpha - b}$

    ii. $L\{x^{\beta-1}E_{\alpha,\beta}(-bx^\alpha)\} = \frac{s^{\alpha-\beta}}{s^\alpha + b}$

    iii. $L\{E_\alpha(bx^\alpha)\} = \frac{s^{\alpha-1}}{s^\alpha - b}$

    iv. $L\{E_\alpha(-bx^\alpha)\} = \frac{s^{\alpha-1}}{s^\alpha + b}$

    v. $L\{x^{\alpha-1}E_{\alpha,\alpha}(-bx^\alpha)\} = \frac{1}{s^\alpha + b}$

    vi. $L\{x^{\alpha k}E_\alpha^{(k)}(-x^\alpha)\} = L\left\{x^{\alpha k}\frac{d^k}{dx^k}E(-x^\varepsilon)\right\} = \frac{k!s^{\alpha-1}}{(s^\alpha+1)^{k+1}}$ $k = 0, 1, 2, \ldots$

10.6.4 Fractional Derivatives

Consider the derivatives of $f(x) = x^k, k \in N$, in classical calculus. The nth derivative of $f(x), n \in N$, is given by

$$\frac{d^n f}{dx^n} = D^n x^k = k(k-1)\cdots(k-n-1)x^{k-n} = \frac{k!}{(k-n)!}x^{k-n}$$

$$= \frac{\Gamma(k+1)}{\Gamma(k-n+1)}x^{k-n} \tag{10.42}$$

We can extend the derivatives of $f(x) = x^k$ to the case when the positive integer n is replaced by the arbitrary number α in terms of the gamma function as follows:

$$\frac{d^\alpha f}{dx^\alpha} = D^\alpha x^k = \frac{\Gamma(k+1)}{\Gamma(k-\alpha+1)}x^{k-\alpha} \quad \alpha > 0 \tag{10.43}$$

This result allows us to extend the concept of a fractional derivative to a large number of functions. We can extend the result to any function that can be expanded in a Taylor series in powers of x as follows:

$$f(x) = \sum_{k=0}^{\infty} a_k x^k$$

$$D^\alpha f(x) = \sum_{k=0}^{\infty} a_k D^\alpha x^k = \sum_{k=0}^{\infty} a_k \frac{\Gamma(k+1)}{\Gamma(k-\alpha+1)}x^{k-\alpha} \tag{10.44}$$

This result enables us to evaluate the fractional derivative of a function such as $f(x) = e^x$. From the Taylor series, we have that

$$e^x = \sum_{k=0}^{\infty} \frac{1}{k!}x^k \Rightarrow D^\alpha e^x = \sum_{k=0}^{\infty} \frac{1}{\Gamma(k+1)}\left\{\frac{\Gamma(k+1)}{\Gamma(k-\alpha+1)}x^{k-\alpha}\right\}$$

$$= \sum_{k=0}^{\infty} \frac{1}{\Gamma(k-\alpha+1)}x^{k-\alpha} \tag{10.45}$$

From Eq. (10.43), we have that

$$\frac{d^{1/2}}{dx^{1/2}}x = D^{1/2}x = \frac{\Gamma(1+1)}{\Gamma(1-1/2+1)}x^{1-1/2} = \frac{\Gamma(2)}{\Gamma(3/2)}x^{1/2} = \frac{2x^{1/2}}{\sqrt{\pi}}$$

Also, the derivative of a constant, say M, is given by

$$\frac{d^\alpha}{dx^\alpha}M = D^\alpha M x^0 = M D^\alpha x^0 = M \frac{\Gamma(0+1)}{\Gamma(0-\alpha+1)}x^{0-\alpha} = \frac{Mx^{-\alpha}}{\Gamma(1-\alpha)}$$

Thus, while the regular derivative of a constant is zero, the fractional derivative is nonzero.

10.6.5 Fractional Integrals

Because integration and differentiation are inverse operations, we may write

$$D^{-1}f(x) = If(x) = \int_0^x f(u)du$$

Similarly, we have that

$$D^{-2}f(x) = \int_0^x \int_{u_1}^x f(u_1)du_2 du_1 = \int_0^x f(u_1)\int_{u_1}^x du_2 du_1 = \int_0^x f(u_1)(x-u_1)du_1$$

$$\equiv \int_0^x f(u)(x-u)du$$

$$D^{-3}f(x) = \int_0^x \int_{u_1}^x \int_{u_2}^x f(u_1)du_3 du_2 du_1 = \int_0^x f(u_1)\int_{u_1}^x \int_{u_2}^x du_3 du_2 du_1$$

$$= \int_0^x f(u_1)\int_{u_1}^x (x-u_2)du_2 du_1 = \frac{1}{2}\int_0^x f(u_1)(x-u_1)^2 du_1$$

$$\equiv \frac{1}{2}\int_0^x f(u)(x-u)^2 du$$

In the same way it can be shown that

$$D^{-4}f(x) = \frac{1}{3\times 2}\int_0^x f(u)(x-u)^3 du$$

In general, we have that

$$D^{-n}f(x) = \frac{1}{(n-1)!}\int_0^x f(u)(x-u)^{n-1}du = \frac{1}{\Gamma(n)}\int_0^x f(u)(x-u)^{n-1}du \qquad (10.46)$$

Thus, the fractional integral of order α is defined, for a function $f(x)$, by

$$I_x^\alpha f(x) = \frac{1}{\Gamma(\alpha)}\int_0^x f(u)(x-u)^{\alpha-1}du = \frac{1}{\Gamma(\alpha)}\int_0^x \frac{f(u)}{(x-u)^{1-\alpha}}du \qquad \alpha > 0 \qquad (10.47)$$

10.6.6   Definitions of Fractional Integro-Differentials

There are several definitions of the fractional integro-differentials. Each of the definitions has its advantages and drawbacks and the choice depends mainly on the purpose and the area of application. The Riemann–Liouville (RL) derivative was the first fractional derivative to be developed and has a well-established mathematical theory. However, it has certain features that lead to difficulties when it is applied to real-world problems. Because of these limitations, many other models were developed. One of these models is the Caputo derivative, which has a close relationship to the Riemann–Liouville model but with certain modifications that attempt to avoid the limitations of the Riemann–Liouville model. We give a brief introduction to these two definitions.

10.6.7   Riemann–Liouville Fractional Derivative

We can generalize the derivative to noninteger values by replacing the $-n$ in Eq. (10.46) with arbitrary α and the factorial with the gamma function to obtain the so-called *Riemann–Liouville derivative*:

$$D^\alpha f(x) = \frac{1}{\Gamma(-\alpha)} \int_0^x \frac{f(u)}{(x-u)^{\alpha+1}} \, du \tag{10.48}$$

In the preceding equation, the lower limit of integration is 0. The limits of integration are generally indicated with subscripts as follows:

$$_b D_x^\alpha f(x) = \frac{1}{\Gamma(-\alpha)} \int_b^x \frac{f(u)}{(x-u)^{\alpha+1}} \, du \tag{10.49}$$

That is, the left subscript is the lower limit, the right subscript is the upper limit of the integration, and the superscript is the order of differentiation. Similarly, we may write

$$_b D_x^{-\alpha} f(x) \quad = {}_b I_x^\alpha f(x) = \frac{1}{\Gamma(\alpha)} \int_b^x f(u)(x-u)^{\alpha-1} du = \frac{1}{\Gamma(\alpha)} \int_b^x \frac{f(u)}{(x-u)^{1-\alpha}} \, du \quad \alpha > 0 \tag{10.50}$$

The following law of exponents holds whenever α and β are natural numbers:

$$D^\alpha D^\beta = D^{\alpha+\beta} \tag{10.51}$$

In particular, let $m = \lceil \alpha \rceil$; that is, m is the smallest integer that is greater than α. Also, let $v = m - \alpha \Rightarrow \alpha = m - v$. Then

$$_0 D_x^\alpha f(x) = D^m \{ {}_0 D_x^{-v} f(x) \} \tag{10.52}$$

Thus, the fractional derivative involves integer-order differentiation of an integral. For example, let $\alpha = 0.4$. Then $m = \lceil 0.4 \rceil = 1$ and $v = m - \alpha = 0.6$. Thus,

$$_0D_x^{0.4}f(x) = D\{_0D_x^{-0.6}f(x)\} = D\left\{\frac{1}{\Gamma(0.6)}\int_0^x (x-u)^{-0.4}f(u)du\right\}$$

In general,

$$_0D_x^{\alpha}f(x) = D^m\{_0D_x^{-v}f(x)\} = D^m\left\{\frac{1}{\Gamma(v)}\int_0^x (x-u)^{v-1}f(u)du\right\} \tag{10.53}$$

Substituting $v = m - \alpha$ and taking $1/\Gamma(v) = 1/\Gamma(m-\alpha)$ outside the differential gives the following equation:

$$_0D_x^{\alpha}f(x) = \frac{1}{\Gamma(m-\alpha)}D^m\left\{\int_0^x (x-u)^{m-\alpha-1}f(u)du\right\} \tag{10.54}$$

10.6.8 Caputo Fractional Derivative

Another type of fractional derivatives is the Caputo fractional derivative of a causal function $f(x)$ (i.e., $f(x) = 0$ for $x < 0$), which is defined as follows:

$$_b^CD_x^{\alpha}f(x) = {}_bI_x^{n-\alpha}\frac{d^n}{dx^n}f(x) = {}_bD_x^{-(n-\alpha)}f^{(n)}(x) = \frac{1}{\Gamma(n-\alpha)}\int_b^x \frac{f^{(n)}(u)}{(x-u)^{\alpha+1-n}}du \tag{10.55}$$

where n is the smallest integer bigger than α. The operator $_b^CD_x^{\alpha}$ is sometimes denoted by $_bD_{*x}^{\alpha}$ and is called the *Caputo differential operator* of order α. From Eq. (10.55) we have that the Caputo fractional derivative of a constant M is given by

$$_bD_{*x}^{\alpha}f(x) = {}_b^CD_x^{\alpha}f(x) = {}_bD_x^{-(n-\alpha)}f^{(n)}(x) = {}_bD_x^{-(n-\alpha)}\frac{d^nM}{dx^n} = 0$$

Recall that the Riemann–Liouville fractional derivative of a constant M is given by

$$\frac{d^{\alpha}}{dx^{\alpha}}M = {}_0D_x^{\alpha}Mx = \frac{Mx^{-\alpha}}{\Gamma(1-\alpha)}$$

Thus, one of the advantages of the Caputo derivative is its ability to give the derivative of a constant as zero, as in ordinary derivative.

To compare the Riemann–Liouville and Caputo fractional derivatives side by side, we reproduce their values, respectively, as follows. For the Riemann–Liouville derivative, we have

$$
D_x^\alpha f(x) = \begin{cases} \dfrac{1}{\Gamma(m-\alpha)} \dfrac{d^m}{dx^m} \left[\displaystyle\int_0^x \dfrac{f(u)du}{(x-u)^{\alpha+1-m}} \right] & m-1 < \alpha < m \\[2em] \dfrac{d^m}{dx^m} f(x) & \alpha = m \end{cases}
$$

Similarly, for the Caputo derivative we have

$$
D_{*x}^\alpha f(x) = \begin{cases} \dfrac{1}{\Gamma(m-\alpha)} \displaystyle\int_0^x \dfrac{f^{(m)}(u)du}{(x-u)^{\alpha+1-m}} & m-1 < \alpha < m \\[2em] \dfrac{d^m}{dx^m} f(x) & \alpha = m \end{cases}
$$

In summary, two of the limitations of the Riemann–Liouville derivative are as follows:

1. The derivative of a constant is not zero, which might be a problem when using RL operators to write evolution equations. Specifically, $_bD_x^\alpha A = (A(x-b)^{-\alpha}/\Gamma(1-\alpha))$.
2. The Laplace transform of the RL derivative depends on the fractional derivative at zero, which is usually an issue when solving initial-value problems:

$$
L\{_0 D_x^\alpha f\} = s^\alpha F(s) - \sum_{k=0}^{n-1} s^k [_0 D_x^{\alpha-k-1} f]_{x=0} \quad n-1 \le \alpha < n \tag{10.56}
$$

10.6.9 Fractional Differential Equations

Fractional differential equations (FDEs) involve fractional derivatives of the form (d^α/dx^α), which are defined for $\alpha > 0$, where α is not necessarily an integer. They are generalizations of the ordinary differential equations to a random (noninteger) order. They have attracted considerable interest due to their ability to model complex phenomena. The equation

$$
D^\alpha f(x) = u(x, f(x)) \tag{10.57}
$$

is called an *FDE of the Riemann–Liouville type*. The initial conditions for this type of FDE are of the form

$$
D^{\alpha-k} f(0) = b_k \quad k = 1, 2, \ldots, n-1
$$
$$
I^{n-\alpha} f(0) = b_n
$$

Similarly, the equation

$$D_*^\alpha f(x) = u(x, f(x)) \tag{10.58}$$

is called an FDE of the Caputo type, and the initial conditions are of the form

$$D^k f(0) = b_k \quad k = 1, 2, \ldots, n - 1$$

Consider the FDE based on the Riemann–Liouville derivative:

$$_0 D_t^\beta f = \frac{1}{\Gamma(1 - \beta)} \frac{\partial}{\partial t} \int_0^t \frac{f(\tau)}{(t - \tau)^\beta} d\tau$$

Its Laplace transform is

$$L[_0 D_t^\beta f] = s^\beta \tilde{f}(s) - [_0 D_t^{-(1 - \beta)} f](0)$$

The Laplace transform depends on the initial value of the fractional integral of f rather than the initial value of f, which is typically given in physical applications. It is well known that to solve classical and FDEs we need to specify additional conditions in order to produce a unique solution. For Riemann–Liouville FDEs, these additional conditions constitute certain fractional derivatives and/or integrals of the unknown solution at the initial point $x = 0$, which are functions of x. Unfortunately, these initial conditions are not physical and cannot be generally measured, and this is a major drawback of this type of fractional derivative. A solution of this problem is provided by using the Caputo derivative of the fractional derivative where these additional conditions are essentially the traditional conditions that are similar to those of classical differential equations that we are familiar with. Thus, the equation of choice in most cases is that based on the Caputo derivative that incorporates the initial values of the function and of its integer derivatives of lower order. A fractional derivative in the Caputo sense is defined as follows:

$$D_*^\alpha f(x) = I^{m-\alpha} D^m f(x) = \frac{1}{\Gamma(m - \alpha)} \int_0^x (x - u)^{m-\alpha-1} f^{(m)}(u) du \tag{10.59}$$

where $m - 1 < \alpha \le m, m \in N, x > 0$. The following properties apply:

$$D_*^\alpha I^\alpha f(x) = f(x)$$

$$I^\alpha D_*^\alpha f(x) = f(x) - \sum_{k=0}^{m-1} f^{(k)}(0^+) \frac{x^k}{k!} \quad x > 0$$

Also, most of the applications of FDEs involve relaxation and oscillation models. We begin by reviewing the regular differential equations of these two models, which involve integer orders.

10.6.10   Relaxation Differential Equation of Integer Order

A relaxation differential equation is an initial-value differential equation of the form:

$$\frac{df(t)}{dt} = -cf(t) + u(t) \quad t > 0, f(0^+) = f_0 \tag{10.60}$$

where c is a constant. The solution to this equation is

$$f(t) = f_0 e^{-ct} + \int_0^t u(t-\tau)e^{-c\tau}d\tau \tag{10.61}$$

10.6.11   Oscillation Differential Equation of Inter Order

An oscillation differential equation is an initial-value differential equation of the form:

$$\frac{d^2f(t)}{dt^2} = -f(t) + u(t) \quad t > 0, f(0^+) = f_0, f'(0^+) = f_1 \tag{10.62}$$

It is called an oscillation differential equation because it has an oscillating sinusoidal solution as follows:

$$f(t) = f_0 \cos(t) + f_1 \sin(t) + \int_0^t u(t-\tau)\sin(\tau)d\tau \tag{10.63}$$

10.6.12   Relaxation and Oscillation FDEs

The relaxation FDE has order $0 < \alpha \le 1$, while the oscillation FDE has an order $1 < \alpha \le 2$. However, unlike in the integer-order ordinary differential equation there is no need to make a distinction between the two forms of FDE. Using the Caputo FDE, we can represent both forms as follows:

$$D_*^\alpha f(t) = D^\alpha \left\{ f(t) - \sum_{k=0}^{m-1} \frac{t^k}{k!} f^{(k)}(0) \right\} = -f(t) + u(t) \quad m - 1 < \alpha \le m \tag{10.64}$$

where m is an integer. Since $I^\alpha D^\alpha f(t) = f(t)$, performing the I^α operation on the equation gives:

$$f(t) = \sum_{k=0}^{m-1} \frac{t^k}{k!} - I^\alpha f(t) + I^\alpha u(t) \quad m - 1 < \alpha \le m \tag{10.65}$$

The Laplace transform of the Mittag-Leffler function helps us to write the solution of FDEs in terms of this function. It can be shown that

$$L\{t^{bk+\beta-1}E_{\alpha,\beta}^{(k)}(bt^{\alpha})\} = \frac{k!s^{\alpha-\beta}}{(s^{\alpha}-b)^{k+1}}$$

$$L\{t^{bk+\beta-1}E_{\alpha,\beta}^{(k)}(-bt^{\alpha})\} = \frac{k!s^{\alpha-\beta}}{(s^{\alpha}+b)^{k+1}}$$

where $E_{\alpha,\beta}^{(k)}(x) = (d^k/dx^k)E_{\alpha,\beta}(x)$.

We can solve Eq. (10.64) using the Laplace transform method, which yields

$$F(s) = \sum_{k=0}^{m-1}\frac{1}{s^{k+1}}f^{(k)}(0) - \frac{1}{s^{\alpha}}F(s) + \frac{1}{s^{\alpha}}U(s)$$

This gives

$$F(s) = \sum_{k=0}^{m-1}\frac{s^{\alpha-k-1}}{s^{\alpha}+1}f^{(k)}(0) + \frac{1}{s^{\alpha}+1}U(s) = \sum_{k=0}^{m-1}\frac{1}{s^k}\left\{\frac{s^{\alpha-1}}{s^{\alpha}+1}\right\}f^{(k)}(0) + \frac{1}{s^{\alpha}+1}U(s)$$

$$(10.66)$$

From the properties of the Laplace transform in Section 10.6.3, we have that

$$\frac{1}{s^k}\left\{\frac{s^{\alpha-1}}{s^{\alpha}+1}\right\} = L\{I^k E_{\alpha}(-t^{\alpha})\}$$

Also, from Section 10.6.3 we have that

$$\frac{1}{s^{\alpha}+1} = L\left\{\frac{d}{dt}E_{\alpha}(-t^{\alpha})\right\}$$

Thus, we have the solution to the FDE as:

$$f(t) = \sum_{k=0}^{m-1}I^k E_{\alpha}(-t^{\alpha})f^{(k)}(0) - u(t) * E_{\alpha}'(-t^{\alpha})$$

$$(10.67)$$

where $*$ is the convolution operator.

10.7 Anomalous (or Fractional) Diffusion

As stated earlier, diffusion is macroscopically associated with a gradient of concentration: It can be considered as the movement of molecules from a higher concentration region to a lower concentration region. The basic property of the normal diffusion is that the second-order moment of the space coordinate of the particle

$E[X^2(t)] = \int x^2 f(x,t) \mathrm{d}x$ is proportional to time, t, where $f(x,t)$ is the probability density function (PDF) of finding the particle at x at time t; that is,

$$E[X^2(t)] \propto t^\alpha$$

where $\alpha = 1$. A diffusion process in which $\alpha \neq 1$ is called *anomalous diffusion*. Anomalous diffusion is also referred to as *fractional diffusion*. The one-dimensional normal diffusion equation and its corresponding solution are given by:

$$\frac{\partial u(x,t)}{\partial t} = D \frac{\partial^2 u(x,t)}{\partial x^2} \tag{10.68}$$

$$u(x,t) = \frac{1}{\sqrt{4\pi Dt}} \exp\left\{ -\frac{x^2}{4Dt} \right\} \tag{10.69}$$

If either one or both of the two derivatives are replaced by fractional-order derivatives, the resulting equation is called a *fractional diffusion equation*. We call the two types of fractional diffusion equation *time-fractional diffusion equation* and *space-fractional diffusion equation*. The time-fractional diffusion equation implies that the process has memory, while the space-fractional diffusion equation describes processes that are nonlocal. The *space-time fractional equation* considers both processes that have memory as well as being nonlocal.

The main physical purpose for studying diffusion equations of fractional order is to describe phenomena of anomalous diffusion usually met in transport processes through complex and/or disordered systems. Anomalous diffusion is classified through the scaling index α. When $\alpha = 1$, we obtain normal diffusion. When $\alpha > 1$, we obtain *superdiffusion* that includes the case when $\alpha = 2$, which is called *ballistic diffusion*. Superdiffusion is also called *enhanced diffusion*. When $0 < \alpha < 1$, we obtain *subdiffusion*, which is also called *suppressed diffusion*. Thus, an anomalous diffusion is a diffusion that is characterized by a mean square displacement that is proportional to a power law in time. This type of diffusion is found in many environments including diffusion in porous media, disordered solids, biological media, atmospheric turbulence, transport in polymers, and Levy flights. Fractional diffusion has been the subject of many reviews such as Haus and Kehr (1987), Bouchaud and Georges (1990), Havlin and Ben-Avraham (2002), Zaslavsky (2002), and Metzler and Klafter (2000,2004), and other publications, such as Vainstein (2008).

In subdiffusion, the travel times of particles are much longer than those expected from classical diffusion because particles tend to halt between steps. Thus, particles diffuse slower than in regular diffusion. On the other hand, in superdiffusion, the particles spread faster than in regular diffusion. Superdiffusion occurs in biological systems while subdiffusion can be found in transport processes.

Fractional diffusion has been successfully modeled by the time-fractional diffusion equation that is obtained by replacing the integer-order time derivative in the diffusion equation with a derivative of noninteger order, which gives

$$_0D_x^\alpha u(x,t) = \frac{\partial^\alpha u(x,t)}{\partial t^\alpha} = C_\alpha \frac{\partial^2 u(x,t)}{\partial x^2} \tag{10.70}$$

where $_0D_x^\alpha$ is the fractional Riemann−Liouville derivative operator of order α.

10.7.1   Fractional Diffusion and Continuous-Time Random Walk

Fractional diffusion equation can be derived from the continuous-time random walk (CTRW). Recall from Chapter 8 that CTRW is a random walk that permits intervals between successive walks to be independent and identically distributed. Thus, the walker starts at the point zero at time $T_0 = 0$ and waits until time T_1 when he makes a jump of size x_1, which is not necessarily positive. The walker then waits until time T_2 when he makes another jump of size x_2, and so on. The jump sizes x_i are also assumed to be independent and identically distributed. Thus, we assume that the times $T_1, T_2, \ldots$ are the instants when the walker makes jumps. The intervals $\tau_i = T_i - T_{i-1}$, $i = 1, 2, \ldots$, are called the *waiting times* (or *pausing times*) and are assumed to be independent and identically distributed.

Let T denote the waiting time and let X denote the jump size. Similarly, let $f_X(x)$ denote the PDF of X, let $f_T(t)$ denote the PDF of T, and let $P(x, t)$ denote the probability that the position of the walker at time t is x, given that it was in position 0 at time $t = 0$; that is,

$$P(x, t) = P[X(t) = x | X(0) = 0] \qquad (10.71)$$

We consider an uncoupled CTRW in which the waiting time and the jump size are independent so that the master equation is given by:

$$P(x, t) = \delta(x)R_T(t) + \int_0^t \int_{-\infty}^{\infty} P(u, \tau)f_X(x - u)f_T(t - \tau)\,\mathrm{d}u\,\mathrm{d}\tau \qquad (10.72)$$

where $\delta(x)$ is the Dirac delta function and $R_T(t) = P[T > t]$ is called the *survival probability*, which is the probability that the waiting time when the process is in a given state is greater than t. The master equation states that the probability that $X(t) = x$ is equal to the probability that the process was in state 0 up to time t, plus the probability that the process was at some state u at time τ, where $0 < \tau \le t$, and within the waiting time $t - \tau$, a jump of size $x - u$ took place. Note that

$$R_T(t) = \int_t^{\infty} f_T(v)\,\mathrm{d}v = 1 - \int_0^t f_T(v)\,\mathrm{d}v$$

$$f_T(t) = -\frac{\mathrm{d}R_T(t)}{\mathrm{d}t}$$

When the waiting times are exponentially distributed such that $f_T(t) = \lambda e^{-\lambda t}$ $t > 0$, the survival probability is $R_T(t) = e^{-\lambda t}$ and satisfies the following relaxation ordinary differential equation:

$$\frac{\mathrm{d}}{\mathrm{d}t}R_T(t) = -\lambda R_T(t) \quad t > 0, R_T(0^+) = 1 \qquad (10.73)$$

The simplest fractional generalization of Eq. (10.73) that gives rise to anomalous relaxation and power-law tails in the waiting time PDF can be written as follows:

$$\frac{\mathrm{d}^{\alpha}}{\mathrm{d}t^{\alpha}}R_T(t) = -R_T(t) \quad t > 0, \ 0 < \alpha < 1; \ R_T(0^+) = 1 \qquad (10.74)$$

where d^α/dt^α is the Caputo fractional derivative; that is,

$$\frac{d^\alpha}{dt^\alpha} R_T(t) = \frac{1}{\Gamma(1-\alpha)} \frac{d}{dt} \int_0^t \frac{R_T(u)}{(t-u)^\alpha} du - \frac{t^{-\alpha}}{\Gamma(1-\alpha)} R_T(0^+) \qquad (10.75)$$

Taking the Laplace transform of Eq. (10.74) incorporating Eq. (10.75) we obtain

$$s^\alpha \tilde{R}_T(s) - s^{\alpha-1} R_T(0^+) = -\tilde{R}_T(s) \Rightarrow \tilde{R}_T(s) = \frac{s^{\alpha-1} R_T(0^+)}{1+s^\alpha} = \frac{s^{\alpha-1}}{1+s^\alpha} \qquad (10.76)$$

From Section 10.9.3 we have that the inverse Laplace transform of $\tilde{R}_T(s)$ is

$$R_T(t) = L^{-1}\left\{ \frac{s^{\alpha-1}}{1+s^\alpha} \right\} = E_\alpha(-t^\alpha) \qquad (10.77)$$

Thus, the corresponding PDF of the waiting time is

$$f_T(t) = -\frac{dR_T(t)}{dt} = -\frac{dE_\alpha(-t^\alpha)}{dt} = t^{\alpha-1} E_{\alpha,\alpha}(-t^\alpha)$$

The Laplace transform is:

$$\tilde{F}_T(s) = \frac{1}{s^\alpha + 1} \qquad 0 < \alpha \le 1$$

The asymptotic behavior of the PDF of the waiting time is as follows:

$$f_T(t) \sim \begin{cases} t^{\alpha-1} & t \to 0 \\ \dfrac{1}{t^{\alpha+1}} & t \to \infty \end{cases}$$

From this, we observe that for $0 < \alpha < 1$ at large t, the function $f_T(t)$ does not decay exponentially anymore; instead, it decays according to a power law. This means that because of the power-law asymptotic behavior of the process, it is no longer Markovian but of the long-memory type. Thus, the kind of diffusion that is associated with the CTRW depends on the distribution of the step increments. If the increments are small, we obtain normal diffusion. In this case, for jump sizes (or displacements) with finite variance σ^2 and waiting times with finite mean τ we have that

$$\frac{\partial P(x,t)}{\partial t} = D \frac{\partial^2 P(x,t)}{\partial x^2}$$

with a diffusion coefficient $D = \sigma^2/\tau$. Assume that both $f_X(x)$ and $f_T(t)$ exhibit algebraic tail such that

$$f_X(x) \sim |x|^{-(1+\beta)} \text{ and } f_T(t) \sim t^{-(1+\alpha)}$$

for which σ^2 and τ are infinite. In this case we can derive a space-time fractional diffusion equation for the dynamics of $P(x, t)$:

$$\frac{\partial^\alpha P(x, t)}{\partial t^\alpha} = D_{\alpha, \beta} \frac{\partial^\beta P(x, t)}{\partial |x|^\beta}$$

where the constant $D_{\alpha, \beta}$ is a generalized diffusion coefficient. Thus, the space-time fractional diffusion equation is obtained by replacing the first-order time derivative and second-order space derivative in the standard diffusion equation by a fractional derivative of order α and β, respectively.

As stated earlier, if we limit the power-law distribution to the waiting time, we obtain the time-fractional diffusion equation of the form:

$$\frac{\partial^\alpha P(x, t)}{\partial t^\alpha} = D_\alpha \frac{\partial^2 P(x, t)}{\partial x^2}$$

When $\alpha = 1$, we obtain the classical diffusion equation. Similarly, if only the jump size (or displacement) has power-law distribution, we obtain the space-fractional diffusion equation of the form:

$$\frac{\partial P(x, t)}{\partial t} = D_\beta \frac{\partial^\beta P(x, t)}{\partial x^\beta}$$

In this case, when $\beta = 2$, we obtain the classical diffusion equation. As stated earlier, the Caputo fractional derivative is generally used to solve FDEs because it allows traditional initial and boundary conditions to be included in the formulation in a standard way, whereas models based on other fractional derivatives may require the values of the fractional derivative terms at the initial time.

10.7.2   Solution of the Fractional Diffusion Equation

Consider the time-fractional diffusion equation:

$$\frac{\partial^\alpha}{\partial t^\alpha} P(x, t) = D \frac{\partial^2}{\partial x^2} P(x, t)$$

$$P(x, 0) = \delta(x) \tag{10.78}$$

$$\lim_{x \to \pm \infty} P(x, t) = 0$$

Let $\tilde{a}$ denote the Laplace transform of a, and let $\hat{b}$ denote the Fourier transform of b. Recall that for a function $g(t)$, the Laplace transform of the Caputo derivative is:

$$L\left\{\frac{d^\alpha g(t)}{dt^\alpha}\right\} = s^\alpha \tilde{g}(s) - s^{\alpha-1} g(0^+)$$

Thus, applying the Laplace transform with respect to the time variable t and the Fourier transform with respect to the space variable x in Eq. (10.77) we obtain:

$$s^\alpha \hat{\tilde{P}}(w, s) - s^{\alpha-1} = -Dw^2 \hat{\tilde{P}}(w, s) \Rightarrow \hat{\tilde{P}}(w, s) = \frac{s^{\alpha-1}}{s^\alpha + Dw^2}$$

Inverting the Laplace transform, we obtain

$$\hat{P}(w, t) = L^{-1} \left\{ \frac{s^{\alpha-1}}{s^\alpha + Dw^2} \right\} = E_\alpha(-Dw^2 t^\alpha)$$

where $E_\alpha(\cdot)$ is the Mittag-Leffler function. Finally, inverting the Fourier transform we obtain the probability $P(x, t)$ of having traversed a distance x at time t as follows:

$$P(x, t) = F^{-1} \left\{ E_\alpha(-Dw^2 t^\alpha) \right\} = \frac{1}{2\pi} \int_{-\infty}^{\infty} e^{-iwx} E_\alpha(-Dw^2 t^\alpha) dw$$

We discuss the space-fractional diffusion equation through the Levy flights, which are random walk processes that have infinite second-order moments $\langle X^2(t) \rangle$ and PDFs with heavy tails:

$$f_X(x) \sim \frac{1}{|x|^{1+\alpha}} \quad 0 < \alpha < 2, x \to \infty$$

Levy flights are discussed in Chapter 11. Since the moments are infinite, we consider such processes as anomalous diffusion. Consider a CTRW with exponentially distributed waiting time, but a Levy step length whose asymptotic PDF has a power law:

$$f_X(x) = c|x|^{-1-\alpha} \quad 1 < \alpha < 2$$

where c is a normalizing constant. The Fourier transform of the PDF is

$$\hat{F}_X(w) = \exp(-c|w|^\alpha) \sim 1 - c|w|^\alpha \tag{10.79}$$

Using the Laplace–Fourier transform of Eq. (10.77), we have

$$\hat{\tilde{P}}(w, s) = \frac{1 - \tilde{F}_T(s)}{s} \frac{1}{1 - \tilde{F}_T(s)\hat{F}_X(w)} = \frac{1}{s + \lambda[1 - \exp(-c|w|^\alpha]}$$

Taking the inverse Laplace transform, we obtain

$$
\begin{aligned}
\hat{P}(w,t) &= \exp\{-\lambda t[1 - \exp(-c|w|^\alpha)]\} \sim 1 - \lambda t[1 - \exp(-c|w|^\alpha)] \\
&\sim 1 - \lambda t[1 - \{1 - c|w|^\alpha\}] = 1 - \lambda c t|w|^\alpha
\end{aligned}
$$

Finally, taking the inverse Fourier transform and recognizing that the equation is similar to Eq. (10.79), we obtain the asymptotic long-time behavior as

$$
P(x,t) \sim \frac{\lambda c t}{|x|^{1+\alpha}} \quad 1 < \alpha < 2
$$

The solution of the space-time fractional diffusion equation is more involved than that of either the time-fraction or space-fractional version. But the solution principles are the same.

10.8 Problems

10.1 Show that the Fokker–Planck equation

$$
\frac{\partial P}{\partial t} = -a\frac{\partial P}{\partial x} + \frac{D}{2}\frac{\partial^2 P}{\partial x^2}
$$

has the solution

$$
P(x,t) = \frac{1}{\sqrt{2\pi Dt}} e^{-(x-a)^2/2Dt}
$$

10.2 Consider a particle whose position, $x(t)$, undergoes the diffusion and damping process

$$
dx = -\mu x dt + (1-x^2)\sigma\, dB
$$

What is the steady-state PDF of x?

10.3 Let $\{X(t), t \geq 0\}$ be a continuous-time continuous-state Markov process whose transition PDF $f(x,t;x_0,t_0)$ satisfies the following forward Kolmogorov equation:

$$
\frac{\partial f}{\partial t} = -\frac{\partial}{\partial x}(a(t,x)f) + \frac{1}{2}\frac{\partial^2}{\partial x^2}(b(t,x)f)
$$

Assume that $a(t,x) = a(t)$ and $b(t,x) = b(t)$. Show that the forward Kolmogorov equation can be reduced to

$$
\frac{\partial f}{\partial t} = \frac{1}{2}\frac{\partial^2 f}{\partial x^2}
$$

Also, show that the corresponding distribution is Gaussian.

10.4 Another way to define a diffusion process is as follows. Let $\mu(t,x)$ and $\sigma(t,x)$ be continuous functions of t and x, where

$$\int_0^t E[\sigma^2(u, B(u))]du < \infty$$

Define

$$X(t) = X(0) + \int_0^t \mu(u, B(u))du + \int_0^t \sigma(u, B(u))dB(u) \quad t \geq 0$$

Then, $\{X(t), t \geq 0\}$ is a diffusion process with μ as its drift function and σ the diffusion function. If $X(0) = x_0$, $\mu(t, x) = \mu$ and $\sigma(t, x) = \sigma$, show that $\{X(t)\}$ is a Brownian motion with drift, where the drift is the initial state x_0.

11 Levy Processes

11.1 Introduction

Both the Poisson process and the Brownian motion have stationary and independent increments. However, they have different sample paths: the Brownian motion has continuous sample paths, while the Poisson process has discontinuities (or jumps) of size 1. At a high level, Levy processes are stochastic processes with both stationary and independent increments. They constitute a wide class of stochastic processes whose sample paths can be continuous, continuous with occasional discontinuities, and purely discontinuous. In this chapter, we discuss Levy process as well as power-law and stable distributions, which will enable us to study the Levy distribution. Finally, we discuss jump-diffusion processes that are used in energy finance, option pricing, and other financial instruments.

11.2 Generalized Central Limit Theorem

The central limit theorem describes the behavior of sums of random variables. Specifically, let $\{X_1, X_2, \ldots, X_n\}$ be a sequence of independent random variables with a common probability density function (PDF) $f_X(x)$. Let the random variable $Y_n = X_1 + X_2 + \cdots + X_n$ be the sum of these random variables. Then, according to the central limit theorem, as n becomes large in the limit, the PDF of Y_n is normal, that is, $\lim_{n \to \infty} Y_n \sim N(\mu, \sigma^2)$, where $\mu = n\mu_X$ and $\sigma^2 = n\sigma_X^2$. Thus, we have that

$$f_{Y_n}(y) = \frac{1}{\sqrt{2\pi n \sigma_X^2}} \exp(-[y - n\mu_X]^2 / 2n\sigma_X^2) \quad -\infty < y < \infty$$

The central limit theorem does not depend on the PDF or probability mass function (PMF) of the X_i, and this makes the normal distribution act as a "black hole of statistics." Thus, we say that the PDF $f_X(x)$ belongs to the domain of attraction of the Gaussian if the variance σ_X^2 is finite. The requirements for the central limit theorem to be applicable are as follows:

- The random variables summed must be independent.
- All random variables must have finite mean and finite variance.
- No variable can make an excessively large contribution to the sum; their contributions to the sum are essentially identical.

Markov Processes for Stochastic Modeling. DOI: http://dx.doi.org/10.1016/B978-0-12-407795-9.00011-6

Consider a process that can be described by the power-law PDF $f_X(x) \sim 1/x^{\gamma}$. This is a probability distribution in which the probability of a given value of the random variable occurring falls off very slowly with increasing value of the random variable, unlike ordinary distributions, where the rate of falloff is much faster. In this case, the mean and the second moment are given respectively by

$$E[X] = \int_{-\infty}^{\infty} x f_X(x) dx \sim \int_{-\infty}^{\infty} x^{1-\gamma} dx = \left. \frac{x^{2-\gamma}}{2-\gamma} \right|_{x=-\infty}^{\infty}$$

$$E[X^2] = \int_{-\infty}^{\infty} x^2 f_X(x) dx \sim \int_{-\infty}^{\infty} x^{2-\gamma} dx = \left. \frac{x^{3-\gamma}}{3-\gamma} \right|_{x=-\infty}^{\infty}$$

From this, we observe that the mean diverges when $\gamma < 2$, and the second moment diverges when $\gamma < 3$. This means that if we allow the jump sizes in a random walk to have a power-law PDF $f_X(x) \sim x^{-\gamma}$ such that $1 < \gamma < 3$, then $E[X^2] = \infty$, and the variance is also infinite. It is a common practice to write $\gamma = 1 + \alpha$. Thus, we have that $f_X(x) \sim x^{-(1+\alpha)}$, where $0 < \alpha < 2$.

The physical significance of infinite variance can be understood by recalling that variance is a measure of central tendency. When the variance is finite, values are known to be clustered around the mean; the smaller the variance, the more the values are clustered around the mean. Generally, the probability of large deviations from the mean is very small. However, in the case of infinite variance, there is no such clustering, and regardless of the scale on which measurements are made, there is no change in their central tendency; thus, all scales look the same. Physically, this means that it is difficult or impossible to put limits on the values of the random variable that one may observe. Such values can become arbitrarily large in absolute value with a much higher frequency than in the case with better-behaved distributions such as the normal distribution.

Another property of power-law distributions is that very few number of terms dominate all the others, which means that contributions of other terms to the sum are very negligible. This is contrary to Gaussian sums where, as we stated earlier, each term contributes essentially equally to the sum. Thus, the central limit theorem cannot be applied to power-law distributions.

The generalized central limit theorem is an extension of the classical central limit theorem, which was developed to deal with sums of power-law random variables whose variances are infinite. Specifically, it is an answer to the following question: Can there be a limiting distribution for the sum of an infinite number of independent and identically distributed random variables with infinite variance, which precludes their convergence to the normal distribution? The theorem states that the answer to the question is "yes." Specifically, the sum of independent and identically distributed random variables converges to a distribution that is a member of the family of random variables that have a *stable distribution*. Thus, stable distributions are a generalization of the normal distribution.

11.3   Stable Distribution

A random variable is defined to be stable (also called α-*stable* or *Levy stable*) if a linear combination of two independent copies of the random variable has the same distribution as the random variable. Specifically, X is defined to be stable if for any positive numbers a and b, there exists a positive number c and a real number d such that

$$aX_1 + bX_2 \sim cX + d \tag{11.1}$$

where X_1 and X_2 are independent copies of X and $u \sim v$ means that u and v are identical in distribution. If $d = 0$ for all choices of a and b, then X is said to be *strictly stable*. If the distribution of X is symmetric (i.e., if X and $-X$ have the same distribution), then X is called a *symmetric* stable random variable. For any stable random variable X that satisfies Eq. (11.1), there exists a number $\alpha \in (0, 2]$ such that

$$c^\alpha = a^\alpha + b^\alpha \tag{11.2}$$

There are other ways to define a stable distribution. One such way is as follows: A random variable X has a stable distribution if for $n \geq 2$, there exist a positive number c_n and a real number d_n such that if the random variables $X_1, X_2, \ldots, X_n$ are independent copies of X, then

$$X_1 + X_2 + \cdots + X_n = c_n X + d_n \tag{11.3}$$

Another definition is as follows: A random variable X is defined to have a Levy stable distribution if there are parameters $0 < \alpha \leq 2$, $-1 \leq \beta \leq 1$, and a real value μ such that its characteristic function has the form

$$\Phi_X(w) = E\left[e^{iwX}\right] = \begin{cases} \exp\left\{-\sigma^\alpha |w|^\alpha \left(1 - i\beta(\mathrm{sgn}\, w)\tan\left\{\dfrac{\pi\alpha}{2}\right\}\right) + i\mu w\right\} & \alpha \neq 1 \\[4mm] \exp\left\{-\sigma |w| \left(1 + i\beta\dfrac{2}{\pi}(\mathrm{sgn}\, w)\ln|w|\right) + i\mu w\right\} & \alpha = 1 \end{cases}$$

$$\tag{11.4}$$

The parameter α is called the *stability index* and

$$\mathrm{sgn}\, w = \begin{cases} 1 & \text{if } w > 0 \\ 0 & \text{if } w = 0 \\ -1 & \text{if } w < 0 \end{cases}$$

Stable distributions are generally characterized by four parameters:

1. A *stability index* $\alpha \in (0, 2]$, which is also called the *tail index, tail exponent*, or *character-istic exponent*, determines the rate at which the tails of the distribution taper off. In partic-ular, the smaller the value of α, the greater the frequency and size of the extreme events. When $\alpha > 1$, the mean of the distribution exists and is equal to μ.
2. A *skewness parameter* $\beta \in [-1, 1]$, which has the following property: when it is positive, the distribution is skewed to the right, which means that the right tail is thicker; and when it is negative, it is skewed to the left. When $\beta = 0$, the distribution is symmetric about μ. As α approaches 2, β loses its effect, and the distribution approaches the Gaussian distribution regardless of β.
3. A *scale parameter* $\sigma > 0$, which determines the width and thus dispersion of the PDF.
4. A *location* or *shift parameter* $\mu \in \Re$, which measures the shift of the mode (i.e., the peak) of the distribution and plays the role that the mean plays in a normal distribution.

Thus, if X is a stable random variable, it is generally expressed as $X \sim S(\alpha, \beta, \sigma, \mu)$. One major drawback of the stable distribution is that, with the exception of three spe-cial cases, its PDF and cumulative distribution function (CDF) do not have closed form expressions. Thus, the stable distribution is usually described by its characteristic function. The PDF of X is generally written in the form $f_X(x; \alpha, \beta, \sigma, \mu)$. The stability index and the skewness parameter play a more important role than the scale and shift parameters. The three special cases are as follows:

1. The Gaussian distribution in which $\alpha = 2$ and the mean is μ. The skewness parameter has no effect and so we use $\beta = 0$. Thus, the PDF is $f_X(x; 2, 0, \sigma, \mu)$, and the variance is given by $\sigma_X^2 = 2\sigma^2$. Thus, the PDF and CDF are given by

$$f_X(x; 2, 0, \sigma, \mu) = \frac{1}{\sqrt{4\pi\sigma^2}} \exp\left\{ -\frac{(x-\mu)^2}{4\sigma^2} \right\} \quad -\infty < x < \infty$$

$$F_X(x; 2, 0, \sigma, \mu) = \Phi\left(\frac{x - \mu}{\sigma\sqrt{2}} \right)$$

(11.5)

2. The Cauchy distribution in which $\alpha = 1$ and $\beta = 0$. Thus, the PDF and CDF are

$$f_X(x; 1, 0, \sigma, \mu) = \frac{1}{\pi\sigma[1 + ((x-\mu)/\sigma)^2]} = \frac{\sigma}{\pi[\sigma^2 + (x-\mu)^2]} \quad -\infty < x < \infty \qquad (11.6a)$$

$$F_X(x; 1, 0, \sigma, \mu) = \frac{1}{2} + \frac{1}{\pi} \text{arc } tan\left(\frac{x - \mu}{\sigma} \right) \qquad (11.6b)$$

The case where $\mu = 0$ and $\sigma = 1$ is called the standard Cauchy distribution whose PDF and CDF become

$$f_X(x; 1, 0, 1, 0) = f_X(x; 1) = \frac{1}{\pi[1 + x^2]} \quad -\infty < x < \infty \qquad (11.7a)$$

$$F_X(x; 1, 0, 1, 0) = \frac{1}{2} + \frac{1}{\pi} \text{arc } tan(x) \qquad (11.7b)$$

From this, we observe that the general Cauchy distribution is related to the standard Cauchy distribution as follows:

$$f_X(x; 1, 0, \sigma, \mu) = \frac{1}{\sigma} f_X\left(\frac{x-\mu}{\sigma}; 1\right)$$

3. The Levy distribution in which $\alpha = 0.5$ and $\beta = 1$. Thus, the PDF and CDF are

$$f_X\left(x; \frac{1}{2}, 1, \sigma, \mu\right) = \left(\frac{\sigma}{2\pi}\right)^{1/2} \frac{1}{(x-\mu)^{3/2}} \exp\left\{-\frac{\sigma}{2(x-\mu)}\right\} \quad \mu < x < \infty$$

$$F_X\left(x; \frac{1}{2}, 0, \sigma, \mu\right) = 2\left\{1 - \Phi\left(\frac{\sigma}{x-\mu}\right)\right\}$$

(11.8)

It must be emphasized that with the exception of the Gaussian distribution (i.e., $\alpha = 2$), all stable distributions are *leptokurtic* and heavy-tailed distributions. Leptokurtic distributions have higher peaks around the mean compared to the normal distribution, which leads to thick tails on both sides.

We can obtain the parameters of the equation $aX_1 + bX_2 = cX + d$ for the normal random variable $X \sim N(\mu_X, \sigma_X^2)$ as follows. For a normal random variable, $\alpha = 2$. Since X_1 and X_2 are two independent copies of X, we have

$$E[aX_1 + bX_2] = (a+b)\mu_X, \quad \text{Var}(aX_1 + bX_2) = (a^2 + b^2)\sigma_X^2$$
$$E[cX + d] = c\mu_X + d, \quad \text{Var}(cX + d) = c^2\sigma_X^2$$

Thus, we have

$$aX_1 + bX_2 \sim N(\{a+b\}\mu_X, \{a^2 + b^2\}\sigma_X^2)$$
$$cX + d \sim N(c\mu_X + d, c^2\sigma_X^2)$$

Since the two are equal in distribution, we have

$$c^2 = a^2 + b^2 \Rightarrow c = \sqrt{a^2 + b^2}$$
$$d = (a + b - c)\mu_X$$

Note that $c^2 = a^2 + b^2$, which is a property of the stable distribution we discussed earlier, that is, $c^\alpha = a^\alpha + b^\alpha$, where $\alpha = 2$.

Restatement of the generalized central limit theorem: In light of our discussion on stable distributions, we can restate the generalized central limit theorem as follows. Let $X_1, X_2, \ldots, X_n$ be a sequence of independent and identically distributed random variables with a common PDF $f_X(x)$. Let $S_n = X_1 + X_2 + \cdots + X_n$. Let Z_α be an α-stable distribution. The random variable X belongs to the domain of attraction for Z_α (i.e., PDF of an infinite sum of X_i converges to the PDF of Z_α) if there exist constants $b_n \in \Re$ such that

$$\frac{S_n - b_n}{n^{1/\alpha}} \sim Z_\alpha$$

That is, $(S_n - b_n)/n^{1/\alpha}$ converges in distribution to Z_α.

Figure 11.1 shows the plots of the PDFs of the three α-stable distributions for various α, β, σ, and μ values.

Stable laws are of interest for their asymptotic behavior. The asymptotic behavior of the stable distributions for $\alpha < 2$ and skewness parameter β as $x \to \infty$ is of the form

$$\lim_{x \to \pm\infty} f_X(x; \alpha, \beta, \sigma, \mu) \sim \frac{\alpha(1 + \beta)C(\alpha)}{|x|^{1+\alpha}}$$

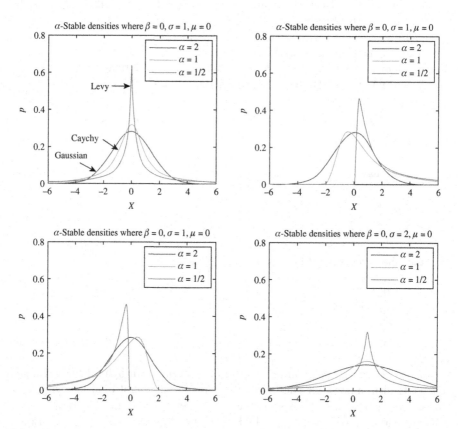

Figure 11.1 α-Stable density functions for various α, β, σ, and μ values.

where

$$C(\alpha) = \frac{1}{\pi}\Gamma(\alpha)\sin\left(\frac{\pi\alpha}{2}\right)$$

$\Gamma(u)$ is the gamma function of u. Thus, for the Cauchy distribution, $C(\alpha) = C(1) = 1/\pi$; for the Levy distribution, $C(\alpha) = C(0.5) = 1/\sqrt{2\pi}$; and for the normal distribution, $C(\alpha) = C(2) = 0$, which is consistent with the fact that the PDF of the normal distribution goes to zero as $x \to \infty$.

Recall that the characteristic function of the PDF $f_X(x; \alpha, \beta, \sigma, \mu)$ has the form

$$\Phi_X(w) = E[e^{iwX}] = \int_{-\infty}^{\infty} f_X(x)e^{iwx}\, dx$$

$$= \begin{cases} \exp\left\{ -\sigma^\alpha|w|^\alpha\left(1 - i\beta(\text{sgn } w)\tan\frac{\pi\alpha}{2}\right) + i\mu w \right\} & \alpha \neq 1 \\[3mm] \exp\left\{ -\sigma|w|\left(1 + i\beta\frac{2}{\pi}(\text{sgn } w)\log|w|\right) + i\mu w \right\} & \alpha = 1 \end{cases}$$

$$f_X(x; \alpha) = \frac{1}{2\pi}\int_{-\infty}^{\infty} \Phi_X(w)e^{-iwx}\, dw$$

Thus, the characteristic functions of the three special cases are as follows:

1. For the normal distribution where $\alpha = 2$ and $\beta = 0$, the characteristic function is given by $\Phi_X(w) = \exp\{i\mu w - \sigma^2 w^2\}$.
2. For the Cauchy distribution where $\alpha = 1$ and $\beta = 0$, the characteristic function is given by $\Phi_X(w) = \exp\{i\mu w - \sigma|w|\}$.
3. For the Levy distribution where $\alpha = 1/2$ and $\beta = 1$, the characteristic function is given by $\Phi_X(w) = \exp\{i\mu w - |\sigma w|^{1/2}(1 - \text{sgn } w)\}$.

11.4   Levy Distribution

The Levy distribution, named after Paul Levy, is a stable distribution with $\alpha = 1/2$ and $\beta = 1$. Thus, the PDF of X is given by

$$f_X(x; \tfrac{1}{2}, 1, \sigma, \mu) = \left(\frac{\sigma}{2\pi}\right)^{1/2}\frac{1}{(x-\mu)^{3/2}}\exp\left\{-\frac{\sigma}{2(x-\mu)}\right\} \qquad \mu < x < \infty \qquad (11.9)$$

The PDF is leptokurtic, which, as we stated earlier, means that it has a fat tail. This gives it the advantage over the Gaussian PDF that its fat tail accounts for a higher probability of extreme events. Both the mean and the variance are infinite. As discussed earlier, the characteristic function is given by

$$\Phi_X(w) = \exp\{i\mu w - |\sigma w|^{1/2}(1 - \text{sgn}w)\} = \begin{cases} \exp\{i\mu w\} & w > 0 \\ \exp\{i\mu w - |\sigma w|^{1/2}\} & w = 0 \\ \exp\{i\mu w - 2|\sigma w|^{1/2}\} & w < 0 \end{cases}$$

Note that for very large values of x, the exponential term becomes nearly 1. This causes the PDF to be approximately $f_X(x) \sim 1/x^{3/2} = 1/x^{(1/2)+1}$, which is a function that decreases as $1/x^{3/2}$, thereby giving it fat tails.

11.5 Levy Processes

As we discussed earlier, Levy processes are stochastic processes with both stationary and independent increments. They constitute a wide class of stochastic processes whose sample paths can be continuous, continuous with occasional discontinuities, and purely discontinuous. Examples of Levy processes include the Brownian motion with drift, the Poisson process, and the compound Poisson process. Among the Levy family of processes, the Brownian process with drift is the only member with continuous sample paths. All the other Levy-type processes have discontinuous trajectories and exhibit jumps.

Levy processes are widely used in the field of quantitative finance to model asset prices and in physics. More formally, a stochastic process $\{X(t), t \geq 0\}$ is a Levy process if the following conditions are satisfied:

- $X(0) = 0$.
- $X(t)$ has independent increments.
- $X(t)$ has stationary increments.
- $X(t)$ is continuous in probability; that is, for all $t \geq 0$ and $\varepsilon > 0$,

$$\lim_{u \to t} P[|X(t) - X(u)| > \varepsilon] = 0$$

11.6 Infinite Divisibility

An interesting property of the Levy process is the *infinite divisibility* property. A random variable Y is said to be infinitely divisible if for every integer $n \geq 2$, there are n independent random variables $Y_{1,n}, \ldots, Y_{n,n}$ such that the sum $Y_n = Y_{1,n} + \cdots + Y_{n,n}$ has the same distribution as Y. Because the cumulative distribution function of the sum of n independent random variables is the n-way convolution of their CDFs, we have that

$$F_Y(y) = F_{Y_n}(y) = F_{Y_{1,n}}(y) * \cdots * F_{Y_{n,n}}(y) \tag{11.10}$$

In terms of the characteristic function, because the $Y_{k,n}$ are also identically distributed, we have that

$$\Phi_Y(w) = [\Phi_{Y_{1,n}}(w)]^n \tag{11.11}$$

$$\Phi_{Y_{1,n}}(w) = [\Phi_Y(w)]^{1/n} \tag{11.12}$$

Because $X(0) = 0$, we can write

$$X(t) = \left\{ X\left(\frac{t}{n}\right) - X\left(\frac{0t}{n}\right) \right\} + \left\{ X\left(\frac{2t}{n}\right) - X\left(\frac{t}{n}\right) \right\} + \cdots$$

$$+ \left\{ X\left(\frac{(n-1)t}{n}\right) - X\left(\frac{(n-2)t}{n}\right) \right\} + \left\{ X\left(\frac{nt}{n}\right) - X\left(\frac{(n-1)t}{n}\right) \right\}$$

$$= \sum_{k=1}^{n} \left\{ X\left(\frac{kt}{n}\right) - X\left(\frac{(k-1)t}{n}\right) \right\}$$

Thus, $X(t)$ is the sum of n independent random variables, all of which have the same distribution as $X(t/n)$. Because the condition is true for all $n \geq 1$, we conclude that $X(t)$ has an infinitely divisible distribution.

11.6.1 Infinite Divisibility of the Poisson Process

As stated earlier, the Poisson process is a Levy process. The characteristic function of the Poisson random variable $X(t)$ with mean λt is given by

$$\Phi_{X(t)}(w) = E\left[e^{iwX(t)}\right] = \sum_{k=-\infty}^{\infty} e^{iwk} p_{X(t)}(k) = \sum_{k=0}^{\infty} e^{iwk} \left\{ \frac{e^{-\lambda t}(\lambda t)^k}{k!} \right\}$$

$$= e^{-\lambda t} \sum_{k=0}^{\infty} \frac{(\lambda t e^{iw})^k}{k!} = e^{-\lambda t(1 - e^{iw})} = e^{-\frac{\lambda t}{n}(1 - e^{iw})n} = \left[e^{-\frac{\lambda t}{n}(1 - e^{iw})} \right]^n$$

$$= [\Phi_{Y(t)}(w)]^n$$

where $Y(t)$ is a Poisson random variable with rate λ/n. Thus, the Poisson random variable has the infinite divisibility property.

11.6.2 Infinite Divisibility of the Compound Poisson Process

The compound Poisson process $X(t)$ is another example of a Levy process. Let $\Phi_Y(w)$ denote the characteristic function of the jump size density. It can be shown, using the random sum of random variable method used in Ibe (2005), that the characteristic function of the compound Poisson process is given by

$$
\Phi_{X(t)}(w) = E\left[e^{iwX(t)}\right] = e^{-\lambda t(1-\Phi_Y(w))} = \exp\left\{-\frac{\lambda t}{n}[1-\Phi_Y(w)]n\right\}
$$

$$
= \left[\exp\left\{-\frac{\lambda t}{n}[1-\Phi_Y(w)]\right\}\right]^n = [\Phi_{V(t)}(w)]^n
$$

where $V(t)$ is a Poisson random variable with rate λ/n. Thus, the compound Poisson random process has the infinite divisibility property.

11.6.3 Infinite Divisibility of the Brownian Motion with Drift

The Brownian motion with drift (or the arithmetic Brownian motion) $X(t)$ is a Levy process. We know that

$$
X(t) = \mu t + \sigma B(t) \sim N(\mu t, \sigma^2 t)
$$

Thus, its characteristic function is given by

$$
\Phi_{X(t)}(w) = E\left[e^{iwX(t)}\right] = \int_{-\infty}^{\infty} e^{iwx} f_{X(t)}(x)dx = \int_{-\infty}^{\infty} e^{iwx}\frac{1}{\sqrt{2\pi\sigma^2 t}}\exp\left\{-\frac{(x-\mu)^2}{2\sigma^2 t}\right\}dx
$$

$$
= \exp(i\mu wt - \frac{1}{2}\sigma^2 w^2 t) = \exp\left\{\left(\frac{i\mu wt}{n} - \frac{1}{2}\frac{\sigma^2 w^2 t}{n}\right)n\right\}
$$

$$
= \left[\exp\left(\frac{i\mu wt}{n} - \frac{1}{2}\frac{\sigma^2 w^2 t}{n}\right)\right]^n = [\Phi_{U(t)}(w)]^n
$$

where $U(t) = \dfrac{\mu t}{n} + \dfrac{\sigma}{\sqrt{n}}B(t) \sim N\left(\dfrac{\mu t}{n}, \dfrac{\sigma^2 t}{n}\right)$.

11.7 Jump-Diffusion Processes

As discussed earlier, a jump process is a process that makes transitions between discrete states at times that may be fixed or random. Thus, it is a process with a piecewise constant trajectory. This is in contrast to many stochastic processes that move continuously between their possible states. For example, both discrete-time and continuous-time Markov chains are jump processes where jumps are made to new states at times $1, 2, \ldots$ in discrete-time Markov chains, but in continuous-time Markov chains, the jump to a new state can occur at any time $t \geq 0$. Jump processes can be classified as *pure* (or *regular* or *nonexplosive*) jump processes and *explosive* jump processes.

A Markov process $\{X(t), t \geq 0\}$ is called a pure jump process if there are a finite number of jumps in every finite time interval. In an explosive jump process, there are infinitely many jumps within a finite interval. For a pure jump process, the expected waiting time in each state n, $E[W_n] = 1/\lambda(n)$, will be finite, where $\lambda(n)$ is the rate of transition out of state n. However, if $\lambda(n) \to \infty$, so that $1/\lambda(n) \to 0$ for any n, the process can move from state to state spending only extremely short waiting time at each state, which leads to infinitely many jumps within a short time interval. Thus, a Markov process $\{X(t), t \geq 0\}$ is a pure jump process only if

$$\sum_{n=1}^{\infty} \frac{1}{\lambda_n} = \infty$$

On the other hand, an explosive Markov jump process is one in which

$$\sum_{n=1}^{\infty} \frac{1}{\lambda_n} < \infty$$

11.7.1 Models of Jump-Diffusion Process

Jump-diffusion models are widely used in energy finance to describe the behavior of spot electricity prices. These jumps are due to limited generation supply and inelastic and volatile demand that result in electricity prices that fluctuate violently over short periods of time.

Jump-diffusion models are also used in option pricing and other financial instruments. Various stochastic models are used in finance to model the price movements of financial instruments. For example, the Black–Scholes model (Black and Scholes, 1973) for pricing options assumes that the underlying instrument follows a traditional diffusion process with small and continuous random movements. Thus, they are modeled by a geometric Brownian motion (GBM). Specifically, let S_t be the stock price at time t and let $\{B_t, t \geq 0\}$ be a standard Brownian motion. Then, the GBM is defined by

$$S_t = S_0 \, e^{\mu t + \sigma B_t} \tag{11.13}$$

where $\mu \in \Re$ is called the *drift rate* because it represents the average rate at which the value increases, $\sigma \in \Re$ is called the volatility because it characterizes the degree of variability, and $\{B(t), t \geq 0\}$ is the standard Brownian motion. We can express Eq. (11.13) as $\log(S_t/S_0) = \mu t + \sigma B_t$. If we define $X_t = \log(S_t/S_0)$, we obtain

$$X_t = \log(S_t/S_0) = \mu t + \sigma B_t$$

Alternatively, we can define the process through the stochastic differential equation:

$$dX_t = \mu \, dt + \sigma \, dB_t \tag{11.14}$$

The first term is used to model deterministic trends and is the expected value of the return provided by the stock for a time period dt, while the second term is used to model a set of unpredictable events occurring during this motion and thus represents the stochastic component of the return.

While asset price evolution can be adequately described by a GBM for most of the time, it was found that from time to time a large jump may occur, and this cannot be adequately captured by a GBM. Also, the model fails to capture the leptokurtic, heavy-tailed distributions that are typically met in finance. An early attempt to address this limitation of the GBM was made by Merton (1976) who superimposed a jump process on a GBM. Thus, the resultant jump-diffusion process provides a simple way of replacing the Gaussian return distributions that arise in GBM models by Gaussian mixture distributions, yielding more appropriate and meaningful models for the leptokurtic, heavy-tailed distributions.

A simple jump-diffusion model is of the form

$$S_t = S_0 \, e^{\mu t + \sigma B_t + J_t} = S_0 \, e^{X_t}$$

where

$$X_t = \mu t + \sigma B_t + J_t = \mu t + \sigma B_t + \sum_{i=1}^{N_t} Y_i \quad X_0 = 0 \tag{11.15}$$

In Eq. (11.15), $\{B_t, t \geq 0\}$ is a standard Brownian motion with $B_0 = 0$, $\{N_t, t \geq 0\}$ is a Poisson process with rate λ, μ is the drift of the diffusion part, $\sigma > 0$ is the volatility of the diffusion part, and the random variables $\{Y_1, Y_2, \ldots\}$ are independent and identically distributed. Thus, there are two parts in the model: a pure diffusion-type process, $\mu t + \sigma B_t$, and a jump process, $J_t = \sum_{i=1}^{N_t} Y_i$, which is a compound Poisson process. Recall that the compound Poisson process is a Levy process, which means that jump-diffusion processes are essentially Levy processes if the jumps are compound Poisson processes.

The corresponding stochastic differential equation for the jump-diffusion model is

$$dX_t = \mu \, dt + \sigma \, dB_t + dJ_t \tag{11.16}$$

Jump-diffusion models have the intuitive appeal that they allow prices and interest rates to change continuously most of the time, but they also take into account the fact that from time to time, larger jumps may occur, which cannot be adequately modeled by pure diffusion-type processes. We can interpret the jump part of the model as the market's response to outside news. Specifically, in the absence of outside news, the asset price simply follows a GBM. Good or bad news arrives according to a Poisson process, and the asset price changes in response to the news. The jump size of the process is proportional to the "severity" of the news.

Jump-diffusion process models are also well suited for the energy market where electricity prices are characterized by abrupt and unanticipated large changes. Temporary price spikes are the result of supply shocks such as generating or transmission constraints and account for a large part of the total variation of changes in spot prices.

In the remainder of this section, we discuss some of the models that have been used in jump-diffusion processes in the financial market. Specifically, we discuss the Merton's normal jump-diffusion model, the Bernoulli jump model, and the double exponential model.

11.7.2 Normal Jump-Diffusion Model

The GBM on which the Black–Scholes option-pricing model is based is the simplest and probably most popular specification in financial models. GBM specifies that the instantaneous percentage change in the exchange rate has a constant drift, μ_B, and volatility, σ_B, so that the exchange rate S_t evolves according to the equation

$$\frac{dS_t}{S_t} = \mu_B \, dt + \sigma_B \, dB_t$$

The error, dB_t, is a standard Brownian motion. This can also be written in the form

$$S_t = S_0 \exp(\mu_B t + \sigma_B B_t)$$

Let $\tau > 0$ denote the interval between observations. Then, the τ-period logarithmic return

$$\ln\left(\frac{S_{t+\tau}}{S_t}\right) \equiv x(\tau) = \mu_B \tau + \sigma_B(B_{t+\tau} - B_t)$$

is normally distributed, that is,

$$x(\tau) \sim N(\mu_B \tau, \sigma_B^2 \tau)$$

with a mean and variance proportional to the observation interval. This follows because the difference $B_{t+\tau} - B_t$ in the Brownian motion is normally distributed with mean zero and variance $\sigma_B^2 \tau$.

As discussed earlier, Merton (1976) added Poisson jumps to a standard GBM process to approximate the movement of stock prices subject to occasional discontinuous breaks. Thus, changes in asset price consist of a "normal" component that is modeled by a Brownian motion with drift and an "abnormal" component that is modeled by a compound Poisson process. The "normal" vibrations in price are usually due to a temporary imbalance between supply and demand, changes in capitalization rates, changes in the economic outlook, or other new information that causes marginal changes in an asset's value. Similarly, the "abnormal" vibrations in price are due to the arrival of important new information about the asset that has more than a marginal effect on price. Typically, such information will be specific to the firm or possibly its industry; occasionally, general economic information could be the source. It is assumed that this important information arrives only at discrete points in time.

The asset prices are assumed to occur independently with identical distribution. Because the jumps occur according to a Poisson process dN_t, the probability that an asset jump occurs during a small time interval dt can be written as follows:

$$P[dN_t = 0] \cong 1 - \lambda \, dt$$
$$P[dN_t = 1] \cong \lambda \, dt$$
$$P[dN_t \geq 2] \cong 0$$

where $\lambda > 0$ is the mean number of jumps per unit time, which is also called the intensity of the jump process. In a small time interval dt, let the asset price jump from S_t to $y_t S_t$. Thus, the relative price jump is

$$\frac{dS_t}{S_t} = \frac{Y_t S_t - S_t}{S_t} = Y_t - 1$$

Merton assumed that the price jump size y_t is a nonnegative random variable that has a log-normal distribution, that is,

$$\log(y_t) \sim N(\mu, \delta^2) \Rightarrow y \sim \log \text{normal}(e^{\mu + \frac{1}{2}\delta^2}, e^{2\mu + \delta^2}(e^{\delta^2} - 1))$$

which implies that

$$E[y_t] = \exp\left\{ \mu + \frac{1}{2}\delta^2 \right\}$$
$$\sigma_{y_t}^2 = e^{2\mu + \delta^2}(e^{\delta^2} - 1)$$

The relative price jump size of $S_t, y_t - 1$ is log-normally distributed with mean

$$E[y_t - 1] = e^{\mu + \frac{1}{2}\delta^2} - 1 \equiv k$$

and the variance is

$$\sigma^2_{y_t-1} = E[(y_t - 1 - E[y_t - 1])^2] = e^{2\mu+\delta^2}(e^{\delta^2} - 1)$$

The stochastic differential equation of the dynamics of asset price that incorporates these properties is of the form

$$\frac{dS_t}{S_t} = (\alpha - \frac{\sigma^2}{2} - \lambda k)dt + \sigma \, dB_t + (y_t - 1)dN_t \tag{11.17}$$

where α is the instantaneous expected return on the asset, σ is the instantaneous volatility of the asset return conditional on the fact that a jump does not occur, k is the expected value of the relative jump size of S_t, B_t is a standard Brownian motion process, and N_t is a Poisson process with intensity λ. It is assumed that $\{B_t\}$, $\{N_t\}$, and $\{y_t\}$ are independent. Equation (11.17) can be rewritten in the following form:

$$\frac{dS_t}{S_t} = \begin{cases} (\alpha - \dfrac{\sigma^2}{2} - \lambda k)dt + \sigma \, dB_t & \text{if the Poisson event does not occur} \\[2ex] (\alpha - \dfrac{\sigma^2}{2} - \lambda k)dt + \sigma \, dB_t + (y_t - 1) & \text{if the Poisson event occurs} \end{cases}$$

It can be shown that the solution to Eq. (11.17) is

$$S_t = S_0 \exp\left\{ \left(\alpha - \frac{\sigma^2}{2} - \lambda k \right) t + \sigma B_t + \sum_{i=1}^{N_t} \ln(y_i) \right\}$$

$$= S_0 \exp\left\{ \left(\alpha - \frac{\sigma^2}{2} - \lambda k \right) t + \sigma B_t + \sum_{i=1}^{N_t} Y_i \right\}$$

where $Y_i = \ln(y_i)$. This result can also be written in the form

$$S(t) = S_0 \exp\left\{ \left(\mu - \frac{\sigma^2}{2} \right) t + \sigma B_t \right\} \prod_{i=1}^{N_t} y_i$$

$$= S_0 \exp\left\{ \left(\mu - \frac{\sigma^2}{2} \right) t + \sigma B_t + \sum_{i=1}^{N_t} Y_i \right\}$$

It must be mentioned that the solution to the Black–Scholes model is

$$S_t = S_0 \exp\left\{ \left(\alpha - \frac{\sigma^2}{2} \right) t + \sigma B_t \right\} \Rightarrow \log\left\{ \frac{S_t}{S_0} \right\} \sim N\left(\left(\alpha - \frac{\sigma^2}{2} \right) t, \sigma^2 t \right)$$

The existence of the compound Poisson jump process $\sum_{i=1}^{N_t} Y_i$ makes the Merton's model nonnormal.

11.7.3 Bernoulli Jump Process

The Bernoulli jump process was proposed by Ball and Torous (1983). Let the random variable N denote the number of Poisson events that occur in a time interval of length t. Define $h = t/n$ for an arbitrary integer n and subdivide the interval $(0, t)$ into n equal subintervals each of length h. Let X_i denote the number of events that occur in subinterval i. Then, by the stationary independent increment property of the Poisson process,

$$N = \sum_{i=1}^{n} X_i$$

is the sum of n independent identically distributed random variables such that

$$P[X_i = 0] = 1 - \lambda h + o(h)$$
$$P[X_i = 1] = \lambda h + o(h) \quad i = 1, 2, \ldots, n$$
$$P[X_i > 1] = o(h)$$

For large n, each X_i has approximately the Bernoulli distribution with probability of success $p = \lambda h = \lambda t/n$. Thus, the random variable N is approximately a binomially distributed random variable, which means that

$$P[N = k] \cong \binom{n}{k} \left(\frac{\lambda t}{n} \right)^k \left(1 - \frac{\lambda t}{n} \right)^{n-k} \quad k = 0, 1, \ldots, n$$

It is well known that

$$\lim_{n \to \infty} \binom{n}{k} \left(\frac{\lambda t}{n} \right)^k \left(1 - \frac{\lambda t}{n} \right)^{n-k} = \frac{(\lambda t)^k \exp(-\lambda t)}{k!} \quad k = 0, 1, \ldots$$

which is a standard construction of the Poisson process. If we assume that t is very small, then we can satisfactorily approximate N by the Bernoulli random variable X defined by

$$P[X = 0] = 1 - \lambda t$$
$$P[X = 1] = \lambda t$$

The distinguishing feature of the Bernoulli jump process is that over a fixed period of time, t, either no information impacts the stock price or one relevant information arrival occurs with probability λt, where λ is the rate of the process. No further information arrivals over this period of time are allowed.

If jumps in stock prices correspond to the arrival of "abnormal" information, then by definition, the number of such information arrivals ought not to be very large. From the assumptions we made in the preceding analysis, if t corresponds to one trading day, no more than one "abnormal" information arrival is to be expected on average. Also, if returns were computed for finer time intervals, the Bernoulli model would converge to the Poisson model.

11.7.4 Double Exponential Jump-Diffusion Model

The double exponential jump-diffusion process was proposed by Kou (2002, 2008). In this model, the process Y has a double exponential distribution, that is, the PDF of Y is

$$f_Y(y) = p\eta_1 \, e^{-\eta_1 y} 1_{\{y \geq 0\}} + q\eta_2 \, e^{-\eta_2 y} 1_{\{y < 0\}} \tag{11.18}$$

where $p, q \geq 0$ are constants, $p + q = 1$, and $\eta_1, \eta_2 > 0$. The means of the two exponential distributions are $1/\eta_1$ and $1/\eta_2$, respectively. Thus,

$$E[Y] = \frac{p}{\eta_1} + \frac{1-p}{\eta_2}$$

A major attraction of the double exponential jump-diffusion model is its simplicity, particularly its analytical tractability for path-dependent options and interest rate derivatives. The model is superior to the Black–Scholes model in fitting historical stock data while being more tractable than rival jump process models for the purpose of option pricing because for several important types of options, explicit formulas can be given for the option price in the model but not for the others. This property of the model is due to the "memorylessness" property of the exponential distribution.

As with all Levy processes, the distribution of values of the process at a given moment in time is infinitely divisible, but in this case, it is not closed under convolution. This means that the distribution of returns depends on the timescale over which the data is sampled (i.e., the size of increment Δt). When Δt is small, an approximate formula for the PDF can be obtained by means of the Taylor expansion.

11.7.5 Jump Diffusions and Levy Processes

As discussed earlier, the simplest case of a jump diffusion is obtained by combining a Brownian motion with drift and a compound Poisson process. In this way, we obtain a process that sometimes jumps and has a continuous but random evolution between the jump times, that is,

$$X_t = \mu t + \sigma B_t + \sum_{i=1}^{N_t} Y_i \tag{11.19}$$

A compound Poisson process is a pure jump Levy process with paths that are constant apart from a finite number of jumps at a finite time. Thus, the process in Eq. (11.19) is a Levy process called the Levy jump-diffusion process, and its characteristic function can be computed by multiplying the characteristic function of the Brownian motion and that of the compound Poisson process (since the two parts are independent):

$$E[e^{iuX_t}] = \exp\left\{\left[i\mu u - \frac{\sigma^2 u^2}{2} + \lambda \int_R (e^{iux} - 1)f(dx)\right]t\right\}$$

We calculate the characteristic function of the Levy jump diffusion, since it offers significant insight into the structure of the characteristic function of general Levy processes. Assume that the process $X = \{X_t, t \geq 0\}$ is a Levy jump diffusion that consists of a linear deterministic process, plus a Brownian motion, plus a compensated compound Poisson process. The paths of this process are described by

$$X_t = \mu t + \sigma B_t + \left(\sum_{k=1}^{N_t} J_k - \beta \lambda t\right)$$

where $B = \{B_t, t \geq 0\}$ is a standard Brownian motion, $N = \{N_t, t \geq 0\}$ is a Poisson process with intensity $\lambda > 0$ (i.e., $E[N_t] = \lambda t$), and $J = \{J_k, k \geq 1\}$ is a sequence of independent and identically distributed random variables with probability distribution F and $E[J_t] = \beta < \infty$. Here, F describes the distribution of the jumps, which arrive according to the Poisson process N. All sources of randomness are assumed to be *mutually independent*.

The characteristic function of X_t, taking into account that all sources of randomness are independent, is

$$\begin{aligned}
E[e^{iuX_t}] &= E\left[\exp\left(iu\left\{\mu t + \sigma B_t + \sum_{k=1}^{N_t} J_k - \beta \lambda t\right\}\right)\right] \\
&= \exp[iu\mu t]E[\exp(iu\sigma B_t)]E\left[\exp\left\{iu\sum_{k=1}^{N_t} J_k - iu\beta \lambda t\right\}\right]
\end{aligned} \tag{11.20}$$

But the characteristic functions of the Brownian motion and compound Poisson distributions are given respectively by

$$E[e^{iu\sigma B_t}] = \exp\left\{-\frac{1}{2}\sigma^2 u^2 t\right\}$$

$$E\left[e^{iu\sum_{k=1}^{N_t} J_k}\right] = \exp\{\lambda t(E[iuJ_k - 1])\}$$

Thus, from Eq. (11.20), we have that

$$
\begin{aligned}
E[e^{iuX_t}] &= \exp[iu\mu t]E[\exp(iu\sigma B_t)]E\left[\exp\left\{\sum_{k=1}^{N_t} J_k - \beta\lambda t\right\}\right] \\
&= \exp[iu\mu t]\exp\left[-\frac{1}{2}u^2\sigma^2 t\right]\exp\left[\lambda t\left(E[e^{iuJ_k}] - 1 - iuE[J_k]\right)\right] \\
&= \exp[iu\mu t]\exp\left[-\frac{1}{2}u^2\sigma^2 t\right]\exp\left[\lambda t\left(E\left[e^{iuJ_k}\right] - 1 - iuJ_k\right]\right)\right] \\
&= \exp[iu\mu t]\exp\left[-\frac{1}{2}u^2\sigma^2 t\right]\exp\left[\lambda t\left(\int_R [e^{iux} - 1 - iux]f(x)dx\right)\right] \\
&= \exp\left[\left(iu\mu - \frac{1}{2}u^2\sigma^2 + \lambda\int_R [e^{iux} - 1 - iux]f(x)dx\right)t\right]
\end{aligned}
$$

We can make the following observations based on the structure of the characteristic function of the random variable X_t from the Levy jump diffusion:

- $E[e^{iuX_t}]$ time and space factorize.
- The drift, the diffusion, and the jump parts are separated.

One would naturally ask if these observations are true for any Levy process. The answer is yes. Since the characteristic function of a random variable determines its distribution, the preceding equation provides a characterization of the distribution of the random variables X_t from the Levy jump diffusion X. In general, every Levy process can be represented in the form

$$
X_t = \gamma t + \sigma B_t + Z_t
$$

where Z_t is a jump process with (possibly) infinitely many jumps. This result is an example of the so-called *Levy–Ito decomposition* (Applebaum 2004a,b).

12 Markovian Arrival Processes

12.1 Introduction

Modern telecommunication networks are designed to support multimedia traffic including voice, data, and video. One major feature of these networks is that the input traffic is usually highly bursty. Also, because these networks operate in a packet switching mode, there is usually a strong correlation between packet arrivals. Thus, for these networks, the traditional Poisson traffic model cannot be applied because the presence of correlation between traffic arrivals violates the independence assumption associated with the Poisson process. For example, in an *asynchronous transfer mode* (ATM) network, different types of traffic from different sources arrive at an ATM switch that statistically multiplexes the traffic and transmits them as fixed length packets called cells, as discussed in several books such as Ibe (1997). Thus, there is a high degree of correlation between the individual user traffic and the aggregate arrival process that cannot be captured by traditional Poisson models.

In fact, traffic measurement studies reported by Leland et al. (1994) and Crovella and Bestavros (1997) indicate the Ethernet and Internet traffic display behavior that is associated with long-range dependence (LRD) and self-similarity. As discussed in Chapter 9, self-similarity is a feature whereby parts of an object show the same statistical properties as the object at many scales. For example, in the case of Internet Protocol (IP) traffic, self-similarity means that similar looking traffic bursts can be seen at every time scale ranging from a few milliseconds to minutes and even hours. Similarly, LRD means that values at any instant tend to be positively correlated with values at several future instants. As discussed in Chapter 9, self-similar processes exhibit LRD.

The Internet has become a multiservice network, and one of the features of such networks is the burstiness exhibited by the different services, such as voice, compressed video, and file transfer, that use the network. A traffic process is defined to be bursty if the traffic arrival points $\{t_n\}$ tend to form visual clusters. This means the $\{t_n\}$ tend to consist of a bunch of several relatively short interarrival times followed by a relatively long one.

To deal with the new multiservice network traffic pattern, teletraffic systems analysts have developed a set of traffic models that have been shown to be analytically tractable while capturing the true nature of the traffic better than the traditional Poisson model does. These models are characterized by the fact that they are doubly

Markov Processes for Stochastic Modeling. DOI: http://dx.doi.org/10.1016/B978-0-12-407795-9.00012-8

stochastic Poisson processes that are obtained as a natural extension of the homogeneous Poisson process by allowing the arrival rate to be a stochastic process.

This chapter deals with some of these traffic models, which include the Markovian arrival process (MAP), the batch Markovian arrival process (BMAP), the Markov-modulated Poisson process (MMPP), and the Markov-modulated Bernoulli process (MMBP).

Most of the models discussed in this chapter are usually analyzed via the matrix-analytic method proposed by Neuts (1981) and discussed in great detail by Latouche and Ramaswami (1999). Thus, we begin by providing a brief discussion on quasi-birth-and-death (QBD) processes and matrix-analytic methods.

12.2 Overview of Matrix-Analytic Methods

Recall from Chapter 5 that the infinitesimal generator (or intensity) matrix Q for a continuous-time Markov chain (CTMC) is given by

$$
Q = \begin{bmatrix}
-q_1 & q_{12} & q_{13} & q_{14} & \cdots \\
q_{21} & -q_2 & q_{23} & q_{24} & \cdots \\
q_{31} & q_{32} & -q_3 & q_{34} & \cdots \\
q_{41} & q_{42} & q_{43} & -q_4 & \cdots \\
\cdots & \cdots & \cdots & \cdots & \cdots
\end{bmatrix}
$$

where

$$
q_i = \sum_{j \neq i} q_{ij} \quad i = 1, 2, \ldots
$$

For the special case of a birth-and-death process, it has been shown in Chapter 5 that the intensity matrix is given by

$$
Q_{BD} = \begin{bmatrix}
-\lambda_0 & \lambda_0 & 0 & 0 & 0 & \cdots \\
\mu_1 & -(\lambda_1 + \mu_1) & \lambda_1 & 0 & 0 & \cdots \\
0 & \mu_2 & -(\lambda_2 + \mu_2) & \lambda_2 & 0 & \cdots \\
0 & 0 & \mu_3 & -(\lambda_3 + \mu_3) & \lambda_3 & \cdots \\
\cdots & \cdots & \cdots & \cdots & \cdots & \cdots
\end{bmatrix}
$$

From this, we obtain the intensity matrix for the M/M/1 queue as follows:

$$
Q_{M/M/1} = \begin{bmatrix}
-\lambda & \lambda & 0 & 0 & 0 & \cdots \\
\mu & -(\lambda + \mu) & \lambda & 0 & 0 & \cdots \\
0 & \mu & -(\lambda + \mu) & \lambda & 0 & \cdots \\
0 & 0 & \mu & -(\lambda + \mu) & \lambda & \cdots \\
\cdots & \cdots & \cdots & \cdots & \cdots & \cdots
\end{bmatrix}
$$

We observe that $Q_{M/M/1}$ is a tridiagonal matrix in which all elements above the main diagonal are equal and all elements below the main diagonal are equal. With the exception of the topmost element, all elements of the main diagonal are also equal and are negative sums. The topmost element of the main diagonal is different because that state is a boundary state that has no transition to a lower state. The importance of Q can be explained as follows. Let $P(t)$ be an $n \times n$ matrix such that $p_{ij}(t) \in P(t)$ is defined by

$$p_{ij}(t) = P[X(t) = j | X(0) = i]$$

Then from the forward Kolmogorov equation, we obtain

$$\frac{dP(t)}{dt} = P(t)Q$$

The solution to this matrix equation is

$$P(t) = e^{Qt} = \sum_{k=0}^{\infty} \frac{t^n}{n!} Q^n \quad t \geq 0$$

Let $p_i(t) = P[X(t) = i]$ and $p(t) = [p_1(t), p_2(t), \ldots, p_n(t)]$. Then,

$$p(t) = p(0)P(t)$$

Thus, while the solution to the matrix equation may not be easy to obtain in practice, knowledge of Q enables us to obtain the $p(t)$.

A QBD process is a special case of an infinite-state CTMC that provides a two-dimensional state space version of the birth-and-death process. In this process, states are grouped into *levels*, and transitions are allowed only between levels and within a level. A level l_i consists of m_i *phases*, where m_i can be finite or infinite. An example of the QBD is shown in Figure 12.1.

Similar to the CTMC, all levels are alike except for the first level, which can be different because it is the *boundary level* while others are *repeating levels* that usually have the same transition structure. Sometimes, the second level is referred to as the *border level* that can have some slightly different structure from the other repeating levels. In Figure 12.1, level 0 is the boundary level and level 1 is the border level. Also, for a homogeneous process, $m_i = m$, for $i \geq 1$, if we assume that l_0 is the boundary level. Thus, the Q-matrix for a homogeneous QBD process is of the following tridiagonal form:

$$Q_{QBD} = \begin{bmatrix} D_0 & D_1 & 0 & 0 & 0 & 0 & \cdots \\ D_2 & A_0 & A_1 & 0 & 0 & 0 & \cdots \\ 0 & A_2 & A_0 & A_1 & 0 & 0 & \cdots \\ 0 & 0 & A_2 & A_0 & A_1 & 0 & \cdots \\ \cdots & \cdots & \cdots & \cdots & \cdots & \cdots & \cdots \end{bmatrix}$$

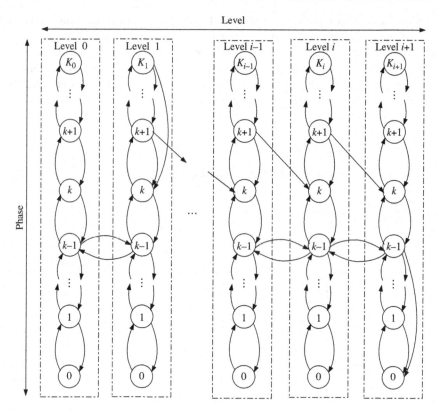

Figure 12.1 Example of the QBD process.

where A_0, A_1, and A_2 are $m \times m$ matrices, where m is the number of *phases* in a level that is not the boundary level; D_0 is an $n \times n$ submatrix, where n is the number of phases in the boundary level; D_1 is an $n \times m$ submatrix; and D_2 is an $m \times n$ submatrix. The states in a given level are called the phases of the level. In general, A_0 and D_0 have nonnegative off-diagonal elements and strictly negative diagonal elements while A_1, A_2, D_1, and D_2 are nonnegative matrices.

One way to visualize these matrices is from Figure 12.2, where the submatrix D_1 deals with the transition rates of the transitions from the boundary level to the border level, D_2 deals with the transition rates of transitions from the border level to the boundary level, A_1 deals with the transition rates of transitions from a repeating level to the next higher repeating level, and A_2 deals with the transition rates of transitions from a repeating level to a preceding repeating level. D_0 and A_0 can be likened to self-loops that deal with intralevel transitions: D_0 for the boundary level and A_0 for the repeating levels.

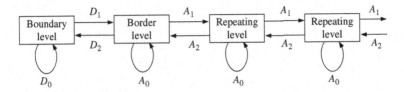

Figure 12.2 Visualization of the roles of the QBD submatrices.

Example 12.1

Consider an M/H_2/1 queue, which is a single-server queue to which customers arrive according to a Poisson process with rate λ, and each customer requires two exponential stages of service with service rates μ_1 and μ_2. Each state is usually represented by (k, s), where k is the number of customers in the system and s is the stage of service of the customer who is currently receiving service. The state transition rate diagram for the queue is shown in Figure 12.3.

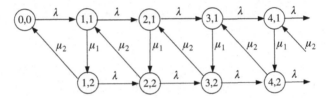

Figure 12.3 State transition rate diagram of the M/H_2/1 queue.

If we lexicographically order the states such that (k_1, s_1) precedes (k_2, s_2) if and only if $k_1 < k_2$ or $\{k_1 = k_2$ and $s_1 < s_2\}$, we obtain the states

$$S = \{(0, 0), (1, 1), (1, 2), (2, 1), (2, 2), (3, 1), (3, 2), \ldots\}$$

and the Q-matrix for the process becomes

$$Q_{M/H_2/1} = \begin{bmatrix}
-\lambda & \lambda & 0 & 0 & 0 & 0 & 0 & 0 & 0 & \cdots \\
0 & -(\lambda+\mu_1) & \mu_1 & \lambda & 0 & 0 & 0 & 0 & 0 & \cdots \\
\mu_2 & 0 & -(\lambda+\mu_2) & 0 & \lambda & 0 & 0 & 0 & 0 & \cdots \\
0 & 0 & 0 & -(\lambda+\mu_1) & \mu_1 & \lambda & 0 & 0 & 0 & \cdots \\
0 & \mu_2 & 0 & 0 & -(\lambda+\mu_2) & 0 & \lambda & 0 & 0 & \cdots \\
0 & 0 & 0 & 0 & 0 & -(\lambda+\mu_1) & \mu_1 & \lambda & 0 & \cdots \\
\cdots & \cdots & \cdots & \cdots & \cdots & \cdots & \cdots & \cdots & \cdots & \cdots
\end{bmatrix}$$

$$= \begin{bmatrix}
D_0 & D_1 & 0 & 0 & 0 & 0 & \cdots \\
D_2 & A_0 & A_1 & 0 & 0 & 0 & \cdots \\
0 & A_2 & A_0 & A_1 & 0 & 0 & \cdots \\
0 & 0 & A_2 & A_0 & A_1 & 0 & \cdots \\
\cdots & \cdots & \cdots & \cdots & \cdots & \cdots & \cdots
\end{bmatrix}$$

where

$$D_0 = \begin{bmatrix} -\lambda & \lambda & 0 \\ 0 & -(\lambda+\mu_1) & \mu_1 \\ \mu_2 & 0 & -(\lambda+\mu_2) \end{bmatrix} \quad D_1 = \begin{bmatrix} 0 & 0 \\ \lambda & 0 \\ 0 & \lambda \end{bmatrix}$$

$$D_2 = \begin{bmatrix} 0 & 0 & 0 \\ 0 & \mu_2 & 0 \end{bmatrix} \quad A_0 = \begin{bmatrix} -(\lambda+\mu_1) & \mu_1 \\ 0 & -(\lambda+\mu_2) \end{bmatrix}$$

$$A_1 = \begin{bmatrix} \lambda & 0 \\ 0 & \lambda \end{bmatrix} \quad A_2 = \begin{bmatrix} 0 & 0 \\ \mu_2 & 0 \end{bmatrix}$$

Thus, for this example, $m = 2$ and $n = 3$.

QBDs are usually analyzed via matrix-analytic methods whose fundamental premise is that the intensity matrix Q of many complex Markov processes has an internal structure that can be exploited to simplify their analysis. In particular, the matrix Q for QBDs can be written as a block-tridiagonal matrix that is similar to the scalar-tridiagonal Q-matrix of the CTMC.

Recall that for an M/M/1 queue, the limiting state probabilities are the geometric distribution obtained as follows:

$$\pi_k = \rho\pi_{k-1} = \rho^k\pi_0 \quad k = 0, 1, 2, \ldots$$

where $\rho = \lambda/\mu$. From the law of total probability, we obtain

$$1 = \sum_{k=0}^{\infty} \pi_k = \pi_0 \sum_{k=0}^{\infty} \rho^k = \frac{\pi_0}{1-\rho}$$

Thus,

$$\pi_0 = 1 - \rho$$
$$\pi_k = \rho^k\pi_0 = (1-\rho)\rho^k \quad k = 0, 1, 2, \ldots$$

Observe that the Q-matrix for the homogeneous QBD is similar to that of the M/M/1 queue. Let the stationary probability vector be π that is partitioned into subvectors π_k, where $\pi_0 = \{\pi(0,0), \pi(0,1), \pi(0,2)\}$, and for $k \geq 1$, $\pi_k = \{\pi(k,1), \pi(k,2)\}$; $\pi(i,j)$ is the stationary probability of being in state (i,j). The key to obtaining π is the fact that a geometric relationship exists among the π_k, which is

$$\pi_k = \pi_{k-1}R \quad k \geq 2$$

This solution is called a *matrix geometric solution,* and the matrix R is called the *geometric coefficient.* Applying successive substitution, we obtain

$$\pi_k = \pi_1 R^{k-1} \quad k \geq 1$$

The balance equations of the QBD process are given by $\pi Q = 0$, which means that

$$\pi_0 D_0 + \pi_1 D_2 = 0$$
$$\pi_0 D_1 + \pi_1 A_0 + \pi_2 A_2 = 0$$
$$\pi_{k-1} A_1 + \pi_k A_0 + \pi_{k+1} A_2 = 0, \quad k \geq 2$$

Substituting $\pi_k = \pi_1 R^{k-1}$, we obtain

$$\pi_0 D_0 + \pi_1 D_2 = 0$$
$$\pi_0 D_1 + \pi_1 A_0 + \pi_2 A_2 = 0$$
$$A_1 + R A_0 + R^2 A_2 = 0$$

where the last equation follows from the fact that π_1 cannot be identically zero. If we can find a matrix R that solves these equations, then the proposition that $\pi_k = \pi_1 R^{k-1}$ is correct. A number of iterative methods have been proposed for solving these quadratic matrix equations. These are given in Latouche and Ramaswami (1999) and will not be repeated here. Another method of solution is proposed in Servi (2002).

12.3 Markovian Arrival Process

The Markovian arrival process was introduced by Lucantoni et al. (1990) as a simpler version of an earlier model proposed by Neuts (1989). It is a generalization of the Markov process where arrivals are governed by an underlying m-state Markov chain. MAP includes phase-type renewal processes and the Markov-modulated Poisson process. The discrete-time version of the process is called DMAP, and a version that includes batch arrivals is called BMAP, which is discussed later in this chapter. One important property of both MAP and BMAP is that the superpositions of independent processes of these types are also processes of the same type.

MAP generalizes the Poisson process by permitting interarrival times that are not exponential while maintaining its Markovian structure. Consider a Poisson process $\{N(t)\}$ with rate λ, where $N(t)$ is the number of arrivals in $(0, t]$ and thus takes nonnegative integer values. The state space of $N(t)$ is $\{0, 1, 2, \ldots\}$, and the state transition rate diagram of the process is shown in Figure 12.4.

Thus, the Q-matrix for a Poisson process is given by

$$
Q_{\text{Poisson}} =
\begin{bmatrix}
-\lambda & \lambda & 0 & 0 & 0 & 0 & \cdots \\
0 & -\lambda & \lambda & 0 & 0 & 0 & \cdots \\
0 & 0 & -\lambda & \lambda & 0 & 0 & \cdots \\
0 & 0 & o & -\lambda & \lambda & 0 & \cdots \\
\cdots & \cdots & \cdots & \cdots & \cdots & \cdots & \cdots
\end{bmatrix}
=
\begin{bmatrix}
d_0 & d_1 & 0 & 0 & 0 & 0 & \cdots \\
0 & d_0 & d_1 & 0 & 0 & 0 & \cdots \\
0 & 0 & d_0 & d_1 & 0 & 0 & \cdots \\
0 & 0 & 0 & d_0 & d_1 & 0 & \cdots \\
\cdots & \cdots & \cdots & \cdots & \cdots & \cdots & \cdots
\end{bmatrix}
$$

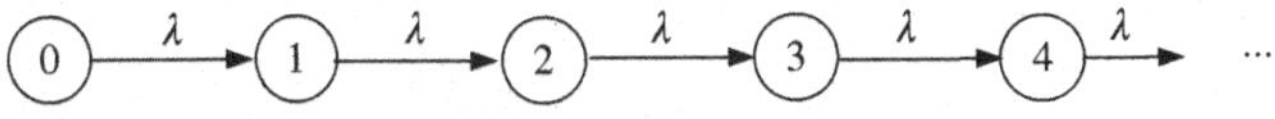

Figure 12.4 State transition rate diagram of a Poisson process.

where $d_0 = -\lambda$ and $d_1 = \lambda$. Let $\{J(t)\}$ be an additional process, called the phase process, that takes values in $\{1, 2, \ldots, m\}$ such that when the process is in state $j \in J(t)$ the Poisson arrival rate is λ_j. Additionally, the state transition rate from state j to k is α_{jk}, where $j, k \in J(t)$. Thus, $\{J(t)\}$ is an irreducible CTMC. The two-dimensional process $\{N(t), J(t)\}$ is an MAP, which represents a Markov process on the state space $\{(i, j) | i = 0, 1, \ldots; 1 \leq j \leq m\}$. $N(t)$ counts the number of arrivals during $(0, t]$, and $J(t)$ represents the phase of the arrival process. The value of m defines the *order* of an MAP. For example, an MAP that has $m = 2$ is called an MAP of order 2 and is denoted by MAP(2). The state transition rate diagram for MAP(4) is shown in Figure 12.5.

If we arrange the states in a lexicographical order, then the infinitesimal generator matrix Q is given by

$$Q_{\text{MAP}} = \begin{bmatrix} -(\lambda_1 + \alpha_1) & \alpha_{12} & \alpha_{13} & \cdots & \alpha_{1m} & \lambda_{11} & \lambda_{12} & \lambda_{13} & \cdots & \lambda_{1m} & 0 & \cdots \\ \alpha_{21} & -(\lambda_2 + \alpha_2) & \alpha_{23} & \cdots & \alpha_{2m} & \lambda_{21} & \lambda_{22} & \lambda_{23} & \cdots & \lambda_{2m} & 0 & \cdots \\ \cdots & \cdots & \cdots & \cdots & & \cdots & \cdots & \cdots & \cdots & \cdots & \cdots & \cdots \\ \alpha_{m1} & \alpha_{m2} & \alpha_{m3} & \cdots & -(\lambda_m + \alpha_m) & \lambda_{m1} & \lambda_{m2} & \lambda_{m3} & \cdots & \lambda_{mm} & 0 & \cdots \\ 0 & 0 & 0 & 0 & 0 & -(\lambda_1 + \alpha_1) & \alpha_{12} & \alpha_{13} & \cdots & \alpha_{1m} & \lambda_{11} & \cdots \\ \cdots & \cdots & \cdots & \cdots & & \cdots & \cdots & \cdots & \cdots & \cdots & \cdots & \cdots \end{bmatrix}$$

where, for $i = 1, 2, \ldots, m$,

$$\lambda_i = \sum_{k=1}^{m} \lambda_{ik}$$

$$\alpha_i = \sum_{k=1}^{m} \alpha_{ik}$$

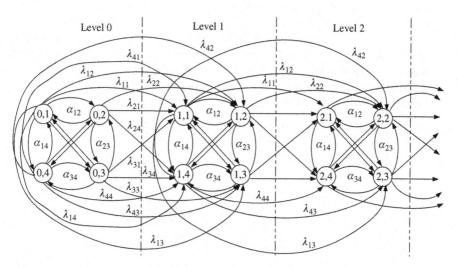

Figure 12.5 State transition rate diagram of MAP(4).

If we define D_0 and D_1 by the following $m \times m$ matrices,

$$D_0 = \begin{bmatrix} -(\lambda_1 + \alpha_1) & \alpha_{12} & \alpha_{13} & \cdots & \alpha_{1m} \\ \alpha_{21} & -(\lambda_2 + \alpha_2) & \alpha_{23} & \cdots & \alpha_{2m} \\ \alpha_{31} & \alpha_{32} & -(\lambda_3 + \alpha_3) & \cdots & \alpha_{3m} \\ \vdots & \vdots & \vdots & \ddots & \vdots \\ \alpha_{m1} & \alpha_{m2} & \alpha_{m3} & \cdots & -(\lambda_m + \alpha_m) \end{bmatrix}$$

$$D_1 = \begin{bmatrix} \lambda_{11} & \lambda_{12} & \lambda_{13} & \cdots & \lambda_{1m} \\ \lambda_{21} & \lambda_{22} & \lambda_{23} & \cdots & \lambda_{2m} \\ \lambda_{31} & \lambda_{32} & \lambda_{33} & \cdots & \lambda_{3m} \\ \vdots & \vdots & \vdots & \ddots & \vdots \\ \lambda_{m1} & \lambda_{m2} & \lambda_{m3} & \cdots & \lambda_{mm} \end{bmatrix}$$

then Q_{MAP} can be represented in the following block form

$$Q_{\mathrm{MAP}} = \begin{bmatrix} D_0 & D_1 & 0 & 0 & 0 & \cdots \\ 0 & D_0 & D_1 & 0 & 0 & \cdots \\ 0 & 0 & D_0 & D_1 & 0 & \cdots \\ \cdots & \cdots & \cdots & \cdots & \cdots & \cdots \end{bmatrix}$$

that has a structure similar to Q_{Poisson}. As can be observed, D_0 has negative diagonal elements and nonnegative off-diagonal elements, and its elements correspond to state transitions without an arrival; that is, they are phase transitions. Similarly, D_1 is a non-negative matrix whose elements represent state transitions with one arrival. Because the Poisson process is a pure birth process, the structure of Q_{MAP} suggests that MAP behaves like a quasi-birth process. Sometimes, MAP is denoted by $\mathrm{MAP}(D_0, D_1)$ to stress the fact that it is completely characterized by these two matrices.

12.3.1 Properties of MAP

In this section, we discuss some of the properties of MAP. The first property of MAP is that the process is so broad that any stationary point process can be approximated arbitrarily closely by a MAP. The second property of MAP is that the superposition of two independent MAPs, say $\mathrm{MAP}(C_0, C_1)$ and $\mathrm{MAP}(D_0, D_1)$, is another MAP, $\mathrm{MAP}(E_0, E_1)$, where

$$E_0 = C_0 \oplus D_0$$
$$E_1 = C_1 \oplus D_1$$

where $\oplus$ represents the *Kronecker sum*, which is defined as follows. Let A be a $k \times k$ matrix and B an $n \times n$ matrix. Let I_k and I_n be identity matrices of order k and n, respectively. Then,

$$A \oplus B = (A \otimes I_n) + (I_k \otimes B)$$

where $\otimes$ represents the *Kronecker product*, which is given by

$$G \otimes F = \begin{bmatrix} g_{11}F & g_{12}F & \cdots & g_{1m}F \\ g_{21}F & g_{22}F & \cdots & g_{2m}F \\ \vdots & \vdots & \ddots & \vdots \\ g_{n1}F & g_{n2}F & \cdots & g_{nm}F \end{bmatrix}$$

where G is an $n \times m$ matrix and F is a $p \times q$ matrix. Thus, $G \otimes F$ is an $np \times mq$ matrix. This construction can be extended to the superpositions of $n > 2$ MAPs.

Let the matrix D be defined as follows:

$$D = D_1 + D_0$$

Then D is the irreducible infinitesimal generator of the underlying Markov chain $\{J(t)\}$. Let π be the stationary probability vector in the Markov chain with infinitesimal generator D. We know that if we define e as the column vector of 1s (i.e., $e = [1, 1, \ldots, 1]^T$ of length m), then

$$\pi D = 0$$
$$\pi e = 1$$

The average rate of events in a MAP, which is called the *fundamental rate* of the MAP, is given by

$$\lambda = \pi D_1 \mathbf{e}$$

Let X_n denote the time between the nth arrival and the $(n + 1)$th arrival, and let $J_n, n \geq 1$, denote the state of the Markov chain with infinitesimal generator D. Then $\{(J_n, X_n), n \geq 1\}$ is a Markov renewal sequence with the transition probability matrix $F(x)$ whose (i,j)th element is

$$F_{ij}(x) = P[X_n \leq x, J_n = j | J_{n-1} = i]$$

In Neuts (1992) it is shown that, for $x \geq 0$,

$$F(x) = \int_0^x \exp(D_0 u) du \, D_1 = \{I - \exp(D_0 x)\}(-D_0)^{-1} D_1$$

The sequence $\{J_n, n \geq 1\}$ forms a Markov chain with state transition probability matrix P given by

$$P = F(\infty) = (-D_0)^{-1} D_1$$

Let p denote the stationary distribution vector of P. The relationship between p and π is

$$p = \frac{1}{\lambda} \pi D_1$$

Also, as defined earlier, $N(t)$ is the number of arrivals in $(0, t]$. Let J_t denote the phase of the MAP at time t. Let the (i, j)th element of the $m \times m$ matrix $P(n, t)$ be defined as follows:

$$P_{ij}(n, t) = P[N(t) = n, J_t = j | N(0) = 0, J_0 = i]$$

In Lucantoni (1993), it is shown that the matrix generating function of $P(n, t)$ is given by

$$G_{P(n,t)}(z, t) = \sum_{n=0}^{\infty} z^n P(n, t) = \exp\{(D_0 + zD_1)t\} \quad t \geq 0$$

Finally, the generating function of the PMF of $N(t)$ is given by

$$G_{N(t)}(z, t) = \pi G_{P(n,t)}(z, t)\mathbf{e}$$

12.4   Batch Markovian Arrival Process

Batch Markovian arrival process was proposed by Lucantoni (1991) as an extension of MAP that provides a far more accurate view of the IP traffic because it captures two important statistical properties of the IP traffic, namely, self-similarity and bur-stiness. As stated earlier in the chapter, Crovella and Bestavros (1997) show that the World Wide Web traffic exhibits self-similarity, which means that the Poisson process cannot be used to model such traffic because it cannot effectively capture the dependence and correlation of the traffic arrival process in the Internet.

BMAP is particularly useful in modeling interactive data transfer in ATM networks. A bulk data transmission in an ATM network results in a large number of ATM cells. For example, a single file transfer request results in a batch arrival of several cells. Thus, BMAP jointly characterizes the traffic arrival process and batch size distribution.

We motivate our discussion on BMAP by considering a batch Poisson process (BPP) in which batches arrive with rate λ. Assume that the batch size B is a discrete random variable with the PMF

$$p_B(k) = P[B = k] \equiv p_k \quad 1 \leq k \leq m$$

$$\sum_{k=1}^{m} p_k = 1$$

where m can be infinite. Furthermore, let $N(t)$ denote the number of arrivals in the interval $(0, t]$. Then $\{N(t), t \geq 0\}$ is a CTMC with state space $\{0, 1, 2, \ldots\}$ whose state transition rate diagram is shown in Figure 12.6.

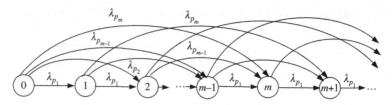

Figure 12.6 State transition rate diagram of batch Poisson process.

The infinitesimal generator is given by

$$Q_{BPP} = \begin{bmatrix} -\lambda & \lambda p_1 & \lambda p_2 & \cdots & \lambda p_m & 0 & 0 & 0 & 0 & \cdots \\ 0 & -\lambda & \lambda p_1 & \lambda p_2 & \cdots & \lambda p_m & 0 & 0 & 0 & \cdots \\ 0 & 0 & -\lambda & \lambda p_1 & \lambda p_2 & \cdots & \lambda p_m & 0 & 0 & \cdots \\ 0 & 0 & 0 & -\lambda & \lambda p_1 & \lambda p_2 & \cdots & \lambda p_m & 0 & \cdots \\ \cdots & \cdots & \cdots & \cdots & \cdots & \cdots & \cdots & \cdots & \cdots & \cdots \end{bmatrix}$$

$$= \begin{bmatrix} d_0 & d_1 & d_2 & \cdots & d_m & 0 & 0 & 0 & 0 & \cdots \\ 0 & d_0 & d_1 & d_2 & \cdots & d_m & 0 & 0 & 0 & \cdots \\ 0 & 0 & d_0 & d_1 & d_2 & \cdots & d_m & 0 & 0 & \cdots \\ 0 & 0 & 0 & d_0 & d_1 & d_2 & \cdots & d_m & 0 & \cdots \\ \cdots & \cdots & \cdots & \cdots & \cdots & \cdots & \cdots & \cdots & \cdots & \cdots \end{bmatrix}$$

where $d_0 = -\lambda$ and $d_k = \lambda p_k, k = 1, 2, \ldots, m$.

As defined by Lucantoni (1991), a BMAP is a doubly stochastic process that operates as follows. There is an ergodic CTMC with a finite state space $\{0, 1, 2, \ldots, m\}$. When the process is in state i, the sojourn time of the process is exponentially distributed with mean $1/\lambda_i$. At the end of the sojourn time, a batch of size $l \geq 1$ can arrive with probability $p_{ij}(l)$, and the Markov chain moves to state $j \neq i$. Thus, the BMAP is a two-dimensional Markov process $X(t) = \{N(t), J(t)\}$ on the state space $\{(i, j)|i \geq 0, 1 \leq j \leq m\}$, where $N(t)$ defines the CTMC and $J(t)$ is the phase process. Figure 12.7 illustrates the state transition rate diagram of BMAP for the case of $m = 4$; α_{jk} has the same notation that we used for MAP.

Arranging the states in a lexicographic order, we obtain the infinitesimal generator of the process as follows:

$$G_{BMAP} = \begin{bmatrix} -(\lambda_1 + \alpha_1) & \alpha_{12} & \alpha_{13} & \cdots & \alpha_{1m} & p_{11}(1)\lambda_1 & p_{12}(1)\lambda_1 & \cdots & p_{1m}(1)\lambda_1 & p_{11}(2)\lambda_1 & \cdots \\ \alpha_{21} & -(\lambda_2 + \alpha_2) & \alpha_{23} & \cdots & \alpha_{2m} & p_{21}(1)\lambda_2 & p_{22}(1)\lambda_2 & \cdots & p_{2m}(1)\lambda_2 & p_{21}(2)\lambda_2 & \cdots \\ \alpha_{31} & \alpha_{32} & -(\lambda_3 + \alpha_3) & \cdots & \alpha_{3m} & p_{31}(1)\lambda_3 & p_{32}(1)\lambda_3 & \cdots & p_{3m}(1)\lambda_3 & p_{31}(2)\lambda_3 & \cdots \\ \vdots & \vdots & \vdots & \cdots & \vdots & \vdots & \vdots & \cdots & \vdots & \vdots & \\ \alpha_{m1} & \alpha_{m2} & \alpha_{m3} & \cdots & -(\lambda_m + \alpha_m) & p_{m1}(1)\lambda_m & p_{m2}(1)\lambda_m & \cdots & p_{mm}(1)\lambda_m & & \cdots \\ 0 & 0 & 0 & \cdots & 0 & -(\lambda_1 + \alpha_1) & \alpha_{12} & \cdots & \alpha_{1m} & p_{11}(1)\lambda_1 & \cdots \\ 0 & 0 & 0 & \cdots & 0 & \alpha_{21} & -(\lambda_2 + \alpha_2) & \cdots & \alpha_{2m} & p_{21}(1)\lambda_2 & \cdots \\ \cdots & \cdots & \cdots & \cdots & \cdots & \cdots & \cdots & & \cdots & \cdots & \\ 0 & 0 & 0 & \cdots & 0 & \alpha_{m1} & \alpha_{m2} & \cdots & -(\lambda_m + \alpha_m) & p_{m1}(1)\lambda_{m1} & \cdots \\ 0 & 0 & 0 & \cdots & 0 & 0 & 0 & \cdots & 0 & -(\lambda_1 + \alpha_1) & \cdots \\ \cdots & \cdots & \cdots & \cdots & \cdots & \cdots & \cdots & & \cdots & \cdots & \end{bmatrix}$$

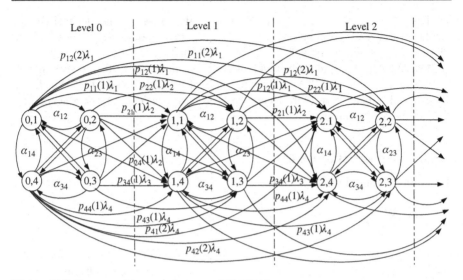

Figure 12.7 State transition rate diagram of BMAP(4).

If we define D_0 and $D_i, i = 1, 2, \ldots,$ by the following $m \times m$ matrices

$$
D_0 = \begin{bmatrix}
-(\lambda_1 + \alpha_1) & \alpha_{12} & \alpha_{13} & \cdots & \alpha_{1m} \\
\alpha_{21} & -(\lambda_2 + \alpha_2) & \alpha_{23} & \cdots & \alpha_{2m} \\
\alpha_{31} & \alpha_{32} & -(\lambda_3 + \alpha_3) & \cdots & \alpha_{3m} \\
\vdots & \vdots & \vdots & \ddots & \vdots \\
\alpha_{m1} & \alpha_{m2} & \alpha_{m3} & \cdots & -(\lambda_m + \alpha_m)
\end{bmatrix}
$$

$$
D_i = \begin{bmatrix}
p_{11}(i)\lambda_1 & p_{12}(i)\lambda_1 & p_{13}(i)\lambda_1 & \cdots & p_{1m}(i)\lambda_1 \\
p_{21}(i)\lambda_2 & p_{21}(i)\lambda_2 & p_{23}(i)\lambda_2 & \cdots & p_{2m}(i)\lambda_2 \\
p_{31}(i)\lambda_3 & p_{32}(i)\lambda_3 & p_{33}(i)\lambda_3 & \cdots & p_{3m}(i)\lambda_3 \\
\vdots & \vdots & \vdots & \ddots & \vdots \\
p_{m1}(i)\lambda_m & p_{m2}(i)\lambda_m & p_{m3}(i)\lambda_m & \cdots & p_{mm}(i)\lambda_m
\end{bmatrix} \quad i = 1, 2, \ldots, m
$$

then Q_{BMAP} can be represented in the following block form

$$
Q_{\text{BMAP}} = \begin{bmatrix}
D_0 & D_1 & D_2 & D_3 & D_4 & \cdots & D_m & 0 & 0 & \cdots \\
0 & D_0 & D_1 & D_2 & D_3 & \cdots & D_{m-1} & D_m & 0 & \cdots \\
0 & 0 & D_0 & D_1 & D_2 & \cdots & D_{m-2} & D_{m-1} & D_m & \cdots \\
0 & 0 & 0 & D_0 & D_1 & \cdots & D_{m-3} & D_{m-2} & D_{m-1} & \cdots \\
0 & 0 & 0 & 0 & D_0 & \cdots & D_{m-4} & D_{m-3} & D_{m-2} & \cdots \\
\cdots & \cdots & \cdots & \cdots & \cdots & \cdots & \cdots & \cdots & \cdots & \cdots
\end{bmatrix}
$$

Thus, Q_{BMAP} has a structure that is similar to that of Q_{BPP}. Observe that D_0 has negative entries in the main diagonal, and all other entries are nonnegative. Each

D_i has only nonnegative entries. To ensure that D_0 is a nondegenerate and stable matrix and thus invertible, we require that Q_{BMAP} be irreducible and $Q_{\text{BMAP}} \neq D_0$, which ensures that arrivals will occur.

BMAP has many applications and variants. Special cases of the process include the Poisson process, MMPP, phase-type renewal processes, and MAP, which is a BMAP with a batch size of 1. Thus, BMAP can be considered to be a generalization of the Poisson process.

12.4.1 Properties of BMAP

Let D be the sum of the D_k; that is,

$$D = \sum_{k=0}^{\infty} D_k$$

We observe that D is the infinitesimal generator for the phase process. Let π_{BMAP} be the stationary probability vector in the Markov chain with infinitesimal generator D.

$$\pi_{\text{BMAP}}D = 0$$

$$\pi_{\text{BMAP}}e = 1$$

The fundamental arrival rate of the BMAP is given by

$$\lambda_{\text{BMAP}} = \pi_{\text{BMAP}} \sum_{k=1}^{\infty} kD_k e$$

Performance measures that are related to the interarrival times of batches are usually obtained from a MAP that is derived from BMAP by setting all nonzero batches to size 1. Thus, the batch arrival rate is given by

$$\lambda_{\text{B}} = \pi_{\text{BMAP}}(-D_0)e$$

Similarly, the squared coefficient of variation of the interbatch arrival time X is given by

$$c_{\text{BMAP}}^2 = \frac{E[X^2]}{(E[X])^2} - 1 = 2\lambda_{\text{B}}\pi_{\text{BMAP}}(-D_0)^{-1}e - 1$$

Let X_0 and X_k be two interbatch times that are k lag times apart, where $k > 0$. In Neuts (1995), the lag-k coefficients of correlation are obtained as

$$\text{corr}[X_0, X_k] = \frac{E[X_0 - E[X]]E[X_k - X[X]]}{\text{Var}[X]}$$

$$= \frac{\lambda_{\text{B}}\pi_{\text{BMAP}}[(-D_0)^{-1}(D-D_0)]^k(-D_0)^{-1}e - 1}{2\lambda_{\text{B}}\pi_{\text{BMAP}}(-D_0)^{-1}e - 1}$$

Also, let $D(z)$ be the z-transform of the matrix $\{D_k, k = 0, 1, \ldots\}$. That is,

$$D(z) = \sum_{k=0}^{\infty} z^k D_k$$

Then the z-transform of $P(n, t)$ is given by

$$G_P(z, t) = \sum_{n=0}^{\infty} z^n P(n, t) = e^{D(z)t} \quad t \geq 0$$

Finally, assume that the BMAP starts with the initial phase distribution π. That is, $\pi = \{\pi_j, j \in J(0)\}$. Let $\mathbf{1}$ be the column vector whose entries are all 1. Then the expected number of arrivals up to time t is given by

$$E_\pi[N_t] = t\pi \sum_{k=1}^{\infty} kD_k \mathbf{1}$$

12.5 Markov-Modulated Poisson Process

The Markov-modulated Poisson process is a doubly stochastic Poisson process whose rate varies according to a CTMC. As with MAP and BMAP, the use of MMPP permits modeling of time-varying systems while keeping the analytical solution tractable. A review of MMPP is given by Fischer and Meier-Hellstern (1992).

MMPP is a variation of MAP and hence of BMAP, and it is a MAP(D_0, D_1) whose D_1 is diagonal. That is, in the Markov chain that defines an MMPP, all the transitions that are associated with events do not change the phase of a state. The state transition rate diagram for MMPP(4) is shown in Figure 12.8.

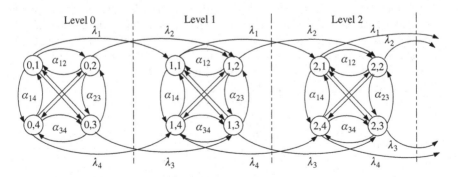

Figure 12.8 State transition rate diagram of MMPP(4).

Thus, we obtain

$$
D_0 = \begin{bmatrix}
-(\lambda_1 + \alpha_1) & \alpha_{12} & \alpha_{13} & \cdots & \alpha_{1m} \\
\alpha_{21} & -(\lambda_2 + \alpha_2) & \alpha_{23} & \cdots & \alpha_{2m} \\
\alpha_{31} & \alpha_{32} & -(\lambda_3 + \alpha_3) & \cdots & \alpha_{3m} \\
\vdots & \vdots & \vdots & \ddots & \vdots \\
\alpha_{m1} & \alpha_{m2} & \alpha_{m3} & \cdots & -(\lambda_m + \alpha_m)
\end{bmatrix}
$$

$$
D_1 = \begin{bmatrix}
\lambda_1 & 0 & 0 & \cdots & 0 \\
0 & \lambda_2 & 0 & \cdots & 0 \\
0 & 0 & \lambda_3 & \cdots & 0 \\
\cdots & \cdots & \cdots & \cdots & \cdots \\
0 & 0 & 0 & \cdots & \lambda_m
\end{bmatrix} = \Lambda
$$

$$
Q_{\mathrm{MMPP}} = \begin{bmatrix}
D_0 & D_1 & 0 & 0 & 0 & \cdots \\
0 & D_0 & D_1 & 0 & 0 & \cdots \\
0 & 0 & D_0 & D_1 & 0 & \cdots \\
\cdots & \cdots & \cdots & \cdots & \cdots & \cdots
\end{bmatrix}
$$

where Λ is the $m \times m$ diagonal matrix whose elements are the arrival rates $\lambda_i, i = 1, 2, \ldots, m$, that is, $\Lambda = \mathrm{diag}(\lambda_1, \lambda_2, \ldots, \lambda_m)$.

12.5.1 The Interrupted Poisson Process

The most basic type of MMPP is a Poisson process that is controlled by a two-state Markov chain, which is typically associated with a voice source that alternates between a talkspurt mode and a silence mode. The generic names of the states are the ON state and the OFF state. When the chain is in the ON state, it is said to be in the talkspurt mode that generates voice traffic. Similarly, when it is in the OFF state, it is said to be in the silence mode and does not generate any traffic. The time spent in the ON state is exponentially distributed with mean $1/\beta$, and the time spent in the OFF state is independent of the time in the ON state and is also exponentially distributed with mean $1/\alpha$. Such a process in which arrivals are blocked in the OFF state is called an *interrupted Poisson process* (IPP). If we denote the ON state by state 0 and the OFF state by state 1, then we can represent IPP by the state transition rate diagram of IPP shown in Figure 12.9.

The infinitesimal generator for IPP is given by

$$
Q_{\mathrm{IPP}} = \begin{bmatrix}
D_0 & D_1 & 0 & 0 & 0 & \cdots \\
0 & D_0 & D_1 & 0 & 0 & \cdots \\
0 & 0 & D_0 & D_1 & 0 & \cdots \\
\cdots & \cdots & \cdots & \cdots & \cdots & \cdots
\end{bmatrix}
$$

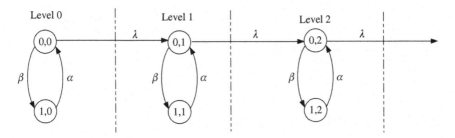

Figure 12.9 State transition rate diagram of IPP.

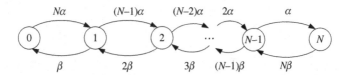

Figure 12.10 State transition rate diagram for number of ON sources.

where

$$D_0 = \begin{bmatrix} -\alpha & \alpha \\ \beta & -(\lambda + \beta) \end{bmatrix}, \quad D_1 = \begin{bmatrix} 0 & 0 \\ 0 & \lambda \end{bmatrix} = \Lambda$$

When N IPP sources are multiplexed, the number of sources in the ON state, n, is represented as a birth-and-death process with birth rate $\lambda(n)$ and death rate $\mu(n)$ given by

$$\lambda(n) = (N - n)\alpha \quad n = 0, 1, 2, \ldots, N$$
$$\mu(n) = n\alpha$$

The state transition rate diagram for the number of ON sources is shown in Figure 12.10. The probability P_n that n of the sources are in the ON state can be obtained using the techniques developed in Chapter 4.

IPP has been used to model overflow systems in Kuczura (1973) and Meier-Hellstern (1989). Such systems operate as follows. Assume we have a queueing system with two facilities labeled primary facility and overflow facility such that arriving customers first try to receive service at the primary facility. If the primary facility is busy (or full), the arrivals are directed to the overflow system. Thus, during the busy period, customers arrive at the overflow facility according to a Poisson process; during nonbusy periods, no customers arrive. This is illustrated in Figure 12.11.

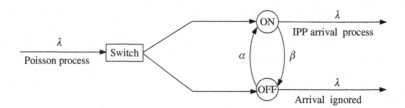

Figure 12.11 Illustration of an interrupted poisson process.

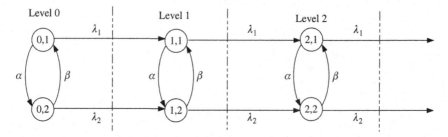

Figure 12.12 State transition rate diagram for SPP.

12.5.2 The Switched Poisson Process

The switched Poisson process (SPP) is very closely related to IPP. Like the latter, it has two states, but unlike IPP, SPP permits traffic to be generated in both states, but with two different rates, λ_1 and λ_2. Thus, it is essentially an MMPP(2), and its state transition rate diagram is shown in Figure 12.12.

The infinitesimal generator for SPP is given by

$$Q_{\text{SPP}} = \begin{bmatrix} D_0 & D_1 & 0 & 0 & 0 & \cdots \\ 0 & D_0 & D_1 & 0 & 0 & \cdots \\ 0 & 0 & D_0 & D_1 & 0 & \cdots \\ \cdots & \cdots & \cdots & \cdots & \cdots & \cdots \end{bmatrix}$$

where

$$D_0 = \begin{bmatrix} -(\lambda_1 + \alpha) & \alpha \\ \beta & -(\lambda_2 + \beta) \end{bmatrix}, \quad D_1 = \begin{bmatrix} \lambda_1 & 0 \\ 0 & \lambda_2 \end{bmatrix} = \Lambda$$

12.5.3 Properties of MMPP

The properties of MMPP are similar to those of MAP and BMAP. First, the super-position of n MMPPs with individual infinitesimal generators D_{0i} and rate matrices Λ_i is a map with infinitesimal generator and rate matrix given by

$$D_0 = D_{01} \oplus D_{02} \oplus \cdots \oplus D_{0n}$$
$$\Lambda = \Lambda_{01} \oplus \Lambda_{02} \oplus \cdots \oplus \Lambda_{0n}$$

Example 12.2

Consider two superposed MMPP(2) systems, MMPP(D_{01}, Λ_1) and MMPP(D_{02}, Λ_2), where

$$D_{01} = \begin{bmatrix} -(\lambda_1 + \alpha_{12}) & \alpha_{12} \\ \alpha_{21} & -(\lambda_2 + \alpha_{21}) \end{bmatrix} \quad \Lambda_1 = \begin{bmatrix} \lambda_1 & 0 \\ 0 & \lambda_2 \end{bmatrix}$$

$$D_{02} = \begin{bmatrix} -(\lambda_3 + \beta_{12}) & \beta_{12} \\ \beta_{21} & -(\lambda_4 + \beta_{21}) \end{bmatrix} \quad \Lambda_1 = \begin{bmatrix} \lambda_3 & 0 \\ 0 & \lambda_4 \end{bmatrix}$$

The resulting process is MMPP(D_0, Λ), where

$$D_0 = D_{01} \oplus D_{02} = (D_{01} \otimes I_2) + (I_2 \otimes D_{02})$$

$$= \begin{bmatrix} -(\lambda_1 + \alpha_{12}) & 0 & \alpha_{12} & 0 \\ 0 & -(\lambda_1 + \alpha_{12}) & 0 & \alpha_{12} \\ \alpha_{21} & 0 & -(\lambda_2 + \alpha_{21}) & 0 \\ 0 & \alpha_{21} & 0 & -(\lambda_2 + \alpha_{21}) \end{bmatrix}$$

$$+ \begin{bmatrix} -(\lambda_3 + \beta_{12}) & \beta_{12} & 0 & 0 \\ \beta_{21} & -(\lambda_4 + \beta_{21}) & 0 & 0 \\ 0 & 0 & -(\lambda_3 + \beta_{12}) & \beta_{12} \\ 0 & 0 & \beta_{21} & -(\lambda_4 + \beta_{21}) \end{bmatrix}$$

$$= \begin{bmatrix} -(\lambda_1 + \lambda_3 + \alpha_{12} + \beta_{12}) & \beta_{12} & \alpha_{12} & 0 \\ \beta_{21} & -(\lambda_1 + \lambda_4 + \alpha_{12} + \beta_{21}) & 0 & \alpha_{12} \\ \alpha_{21} & 0 & -(\lambda_2 + \lambda_3 + \alpha_{21} + \beta_{12}) & \beta_{12} \\ 0 & \alpha_{21} & \beta_{21} & -(\lambda_2 + \lambda_4 + \alpha_{21} + \beta_{21}) \end{bmatrix}$$

$$\Lambda = \Lambda_1 \oplus \Lambda_2 = (\Lambda_1 \otimes I_2) + (I_2 \otimes \Lambda_2)$$

$$= \begin{bmatrix} \lambda_1 & 0 \\ 0 & \lambda_2 \end{bmatrix} \otimes \begin{bmatrix} 1 & 0 \\ 0 & 1 \end{bmatrix} + \begin{bmatrix} \lambda_3 & 0 \\ 0 & \lambda_4 \end{bmatrix} \otimes \begin{bmatrix} 1 & 0 \\ 0 & 1 \end{bmatrix} = \begin{bmatrix} \lambda_1 & 0 & 0 & 0 \\ 0 & \lambda_1 & 0 & 0 \\ 0 & 0 & \lambda_2 & 0 \\ 0 & 0 & 0 & \lambda_2 \end{bmatrix} + \begin{bmatrix} \lambda_3 & 0 & 0 & 0 \\ 0 & \lambda_4 & 0 & 0 \\ 0 & 0 & \lambda_3 & 0 \\ 0 & 0 & 0 & \lambda_4 \end{bmatrix}$$

$$= \begin{bmatrix} \lambda_1 + \lambda_3 & 0 & 0 & 0 \\ 0 & \lambda_1 + \lambda_4 & 0 & 0 \\ 0 & 0 & \lambda_2 + \lambda_3 & 0 \\ 0 & 0 & 0 & \lambda_2 + \lambda_4 \end{bmatrix}$$

12.5.4 The MMPP(2)/M/1 Queue

We illustrate the application of MMPP in performance analysis by considering the MMPP(2)/M/1 queue. MMPP(2) is a process that behaves like a Poisson process with parameter λ_1 for a time that is exponentially distributed with a mean of $1/\alpha$. Then it switches to a Poisson process with parameter λ_2 for a time period that is exponentially distributed with a mean of $1/\beta$. It then switches back to a Poisson process with parameter λ_1 for a time that is exponentially distributed with a mean of $1/\alpha$, and so on. To motivate the discussion on the MMPP(2)/M/1 queue, we consider the following situation.

The weather condition in a certain city is very unpredictable. On any given day, it constantly switches between being sunny and being showery. The rate at which people in the city arrive at a local video rental store to rent videos depends on the weather condition. The duration of a sunny spell is exponentially distributed with a

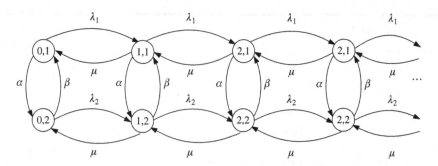

Figure 12.13 State transition rate diagram for MMPP(2)/M/1 queue.

mean of $1/\alpha$, and the duration of a showery spell is exponentially distributed with a mean of $1/\beta$. During a sunny spell, people arrive at the video store according a Poisson process with rate λ_1. Similarly, during a showery spell, people arrive at the video store according to a Poisson process with rate λ_2. Regardless of the prevailing weather condition, the time to serve each customer at the store is exponentially distributed with a mean of $1/\mu$. Thus, we can model the video store by an MMPP(2)/M/1 queue whose state transition rate diagram is shown in Figure 12.13, where the "sunny" condition is state 1 and the "showery" condition is state 2.

From Figure 12.13 we observe that if lexicographic ordering of the states is used, then the infinitesimal generator is given by

$$
Q = \begin{bmatrix}
-(\lambda_1+\alpha) & \alpha & \lambda_1 & 0 & 0 & 0 & 0 & 0 & \cdots \\
\beta & -(\lambda_2+\beta) & 0 & \lambda_2 & 0 & 0 & 0 & 0 & \cdots \\
\mu & 0 & -(\lambda_1+\alpha+\mu) & \alpha & \lambda_1 & 0 & 0 & 0 & \cdots \\
0 & \mu & \beta & -(\lambda_2+\beta+\mu) & 0 & \lambda_2 & 0 & 0 & \cdots \\
0 & 0 & \mu & 0 & -(\lambda_1+\alpha+\mu) & \alpha & \lambda_1 & 0 & \cdots \\
0 & 0 & 0 & \mu & \beta & -(\lambda_2+\beta+\mu) & 0 & \lambda_2 & \cdots \\
\cdots & & \cdots & & \cdots & & \cdots & & \cdots & \cdots & \cdots
\end{bmatrix}
$$

$$
= \begin{bmatrix}
D_0 & D_1 & 0 & 0 & 0 & 0 & \cdots \\
A_2 & A_0 & A_1 & 0 & 0 & 0 & \cdots \\
0 & A_2 & A_0 & A_1 & 0 & 0 & \cdots \\
0 & 0 & A_2 & A_0 & A_1 & 0 & \cdots \\
0 & 0 & 0 & A_2 & A_0 & A_1 & \cdots \\
0 & 0 & 0 & 0 & A_2 & A_0 & \cdots \\
\cdots & \cdots & \cdots & \cdots & \cdots & \cdots & \cdots
\end{bmatrix}
$$

where

$$
D_0 = \begin{bmatrix} -(\lambda_1+\alpha) & \alpha \\ \beta & -(\lambda_2+\beta) \end{bmatrix} \quad D_1 = \begin{bmatrix} \lambda_1 & 0 \\ 0 & \lambda_2 \end{bmatrix}
$$

$$
A_0 = \begin{bmatrix} -(\lambda_1+\alpha+\mu) & \alpha \\ \beta & -(\lambda_2+\beta+\mu) \end{bmatrix} \quad A_1 = \begin{bmatrix} \lambda_1 & 0 \\ 0 & \lambda_2 \end{bmatrix} \quad A_2 = \begin{bmatrix} \mu & 0 \\ 0 & \mu \end{bmatrix}
$$

Let π_{ij} denote the steady-state probability of being in state (i,j) and p_m the probability that the process is in phase m, where $m = 1, 2$. Then we have that

$$\alpha p_1 = \beta p_2$$
$$1 = p_1 + p_2$$

Thus,

$$p_1 = \frac{\beta}{\alpha + \beta} \quad p_2 = \frac{\alpha}{\alpha + \beta}$$

Note that

$$p_j = \sum_{i=0}^{\infty} \pi_{ij} \quad j = 1, 2$$

Also, the average arrival rate is given by

$$\lambda = \lambda_1 p_1 + \lambda_2 p_2 = \frac{\lambda_1 \beta}{\alpha + \beta} + \frac{\lambda_2 \alpha}{\alpha + \beta}$$

Let $\pi_i = [\pi_{i1}, \pi_{i2}]$ denote the vector of the probabilities that the process is in level i, and let the vector π be defined by

$$\pi = [\pi_1, \pi_2, \pi_3, \ldots]$$

Because $\pi Q = 0$, we have that

$$\pi_0 D_0 + \pi_1 A_2 = 0$$
$$\pi_0 D_1 + \pi_1 A_0 + \pi_2 A_2 = 0$$
$$\pi_{k-1} A_1 + \pi_k A_0 + \pi_{k+1} A_2 = 0 \quad k \geq 1$$

As discussed earlier, the analysis of the QBD process is based on the fact that there exists a matrix R such that

$$\pi_k = \pi_{k-1} R \quad k \geq 1$$

Then by successive substitution we have that

$$\pi_k = \pi_0 R^k \quad k \geq 0$$

Thus, we have that

$$D_0 + R A_2 = 0$$
$$A_1 + R A_0 + R^2 A_2 = 0$$

If we can find a matrix R that satisfies these equations, then the proposition that $\pi_{k+1} = \pi_k R$ is correct.

We can also rewrite the preceding equation in the following matrix form:

$$\pi_0 D_0 + \pi_1 A_2 = 0$$
$$\pi_0 D_1 + \pi_1 A_0 + \pi_2 A_2 = \pi_0 A_1 + \pi_1 (A_0 + R A_2) = 0$$

This means that

$$[\pi_0 \quad \pi_1] \begin{bmatrix} D_0 & D_1 \\ A_2 & A_0 + R A_2 \end{bmatrix} = 0$$

This equation can be uniquely solved together with the normalization equation

$$\sum_{k=0}^{\infty} \pi_k \mathbf{e} = \pi_0 \sum_{k=0}^{\infty} R^k \mathbf{e} = \pi_0 [I - R]^{-1} \mathbf{e} = 1$$

where $\mathbf{e}$ is the column vector $e = [1 \ 1]^T$. Note that $[I - R]^{-1}$ is not guaranteed to exist. Thus, when the matrix does not exist, it is a common practice to use the iterative procedure that is derived from the equation

$$\pi_0 A_1 + \pi_1 A_0 + \pi_2 A_2 = \pi_0 [A_1 + R A_0 + R^2 A_2] = 0 \Rightarrow R = -[A_1 + R^2 A_2] A_0^{-1}$$

The recursive solution is given by

$$R(0) = 0$$
$$R(k+1) = -[A_1 + R^2(k) A_2] A_0^{-1}$$

where $R(k)$ is the value of R in the kth iteration. The iteration is repeated until the results of two successive iterations differ by less than a predefined parameter ε, that is,

$$\| R(k+1) - R(k) \| < \varepsilon$$

where $\| \cdot \|$ is a matrix norm. The mean total number of customers in the system is given by

$$E[N] = \sum_{j=1}^{2} \sum_{i=0}^{\infty} i \pi_{ij} = \sum_{j=1}^{2} \sum_{i=0}^{\infty} i \pi_0 R_j^i$$

where R_j^i is the jth column of the matrix R^i. As stated earlier, several techniques have been proposed for computing the matrix R, and these can be found by Latouche and Ramaswami (1999).

12.6 Markov-Modulated Bernoulli Process

The Markov-modulated Bernoulli process is the discrete-time analog of the MMPP. It is particularly used to model traffic in ATM networks. In Ozekici (1997) and Ozekici and Soyer (2003), MMBP is used in reliability modeling where systems and components function in a randomly changing environment. For example, it is used in reliability assessment of power systems that are subject to fluctuating weather conditions over time. Without loss of generality, we consider the use of MMBP in teletraffic applications. However, our formulation of the problem is modeled along the method used in Ozekici (1997) for a more generic system.

Assume that K is an m-state discrete-time Markov chain such that given that the process has just entered state $k \in K, 1 \leq k \leq m$, the time it spends in the state is geometrically distributed with a mean of $1/p_k$. This implies that each time the process enters state k, the probability that it makes a transition to another state that is different from k in the next time slot is p_k. Thus, each transition is a Bernoulli trial with success probability of state k. For the simple case of $m = 3$, the state transition matrix is as follows:

$$P = \begin{bmatrix} 1 - p_1 & p_{12} & p_{13} \\ p_{21} & 1 - p_2 & p_{23} \\ p_{31} & p_{32} & 1 - p_3 \end{bmatrix}$$

where $p_1 = p_{12} + p_{13}, p_2 = p_{21} + p_{23}$ and $p_3 = p_{31} + p_{32}$. The state transition rate diagram for K is shown in Figure 12.14.

Furthermore, given that the process is in state k in the current time slot, the probability that it will generate a packet in the next time slot is $\alpha_k, 1 \leq k \leq m$ Let N_n denote the number of packets that have arrived in the time interval $(0, n]$ and let K_n denote the state of the Markov chain in time slot n. The two-dimensional process $Y = \{N_n, K_n\}$ represents a Markov process on the state space $\{(i, k)|i = 0, 1, \ldots; k = 1, \ldots, m\}$ and is called a Markov-modulated Bernoulli process. Thus, K_n represents the phase of the process. The state transition rate diagram for Y for the case of $m = 3$, which is referred to as the three-state MMBP or MMBP(3), is shown in Figure 12.15.

Figure 12.14 State transition rate diagram for K.

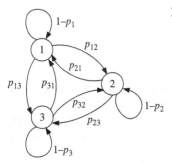

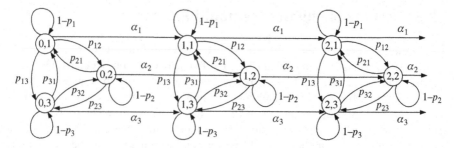

Figure 12.15 State transition rate diagram for MMBP(3).

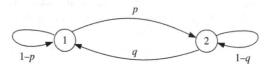

Figure 12.16 State transition rate diagram for K.

Let S_k denote the sojourn time in phase k, which is the number of time slots the Markov chain spends in phase k before making a transition to another phase $l \neq k$. The probability mass function of S_k is the geometric distribution given by

$$p_{S_k}(x) = P[S_k = x] = p_k(1 - p_k)^{x-1} \quad x = 1, 2, \ldots$$

and the number of slots until a packet is generated in phase k is geometrically distributed with mean $1/\alpha_k$.

12.6.1 The MMBP(2)

The special case of $m = 2$, which is the MMBP(2), is used to model cell arrivals in ATM networks. For this case, we assume that $p_1 = p$ and $p_2 = q$. The state transition rate diagram for the two phases is shown in Figure 12.16.

Thus, the state transition matrix is given by

$$P = \begin{bmatrix} 1 - p & p \\ q & 1 - q \end{bmatrix}$$

Also, given that the system is in phase 1 in the current time slot, the probability that it will generate a packet in the next time slot is α. Similarly, given that it is in phase 2 in the current time slot, the probability that it will generate a packet in the next time slot is β. Thus, the state transition rate diagram for Y is shown in Figure 12.17.

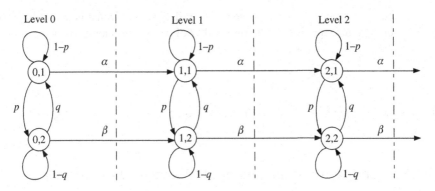

Figure 12.17 State transition rate diagram for Y.

Let π_1 and π_2 denote the steady-state probabilities of being in state 1 and state 2, respectively. Solving the equations $\pi = \pi P$, where $\pi = [\pi_1, \pi_2]$, together with $\pi_1 + \pi_2 = 1$ gives the result

$$\pi_1 = \frac{q}{p+q}, \quad \pi_2 = \frac{p}{p+q}$$

For ATM traffic, it is often assumed that $\alpha = 0$, that is, traffic is only generated when the underlying Markov chain is in phase 2. In Viterbi (1986), the "burstiness" parameter of traffic, γ, is defined as $\gamma = p_{22} - p_{12}$, where p_{12} is the probability that a packet arrives in the current slot given that there was no arrival in the previous slot, and p_{22} is the probability that a packet arrives in the current slot given that there was an arrival in the previous slot. Thus, $p_{12} = p$ and $p_{22} = 1 - q$. which gives $\gamma = 1 - p - q$. From this we can express π_1 and π_2 as

$$\pi_1 = \frac{q}{p+q} = \frac{q}{1-\gamma}$$

$$\pi_2 = \frac{p}{p+q} = \frac{p}{1-\gamma}$$

Alternatively, we have that

$$p = \pi_2(1 - \gamma)$$
$$q = \pi_1(1 - \gamma)$$

The arrival rate of packets, λ, is the probability that the system is in phase 2. That is,

$$\lambda = \frac{p}{1-\gamma}$$

Another parameter of interest is the source utilization, ρ, which is the probability that a slot contains a packet (or the probability of an arrival) and is given by

$$\rho = \pi_1 \alpha + \pi_2 \beta = \pi_2 \beta = \frac{p\beta}{p+q}$$

because we assume that $\alpha = 0$ in ATM traffic modeling.

12.7   Sample Applications of MAP and Its Derivatives

MAP has been widely used for modeling many types of queueing systems. Thus, different types of MAP-based queueing models have been developed, including MAP/M/1 queue, MAP/G/1 queue, and MAP/PH/1 queue.

One major advantage of BMAP over MAP is that batches associated with BMAP add to the modeling power and flexibility of MAP. This fact has been exploited by Klemm et al. (2003) to model IP traffic. BMAP-based queueing systems have been extensively analyzed by many authors with respect to the ATM traffic. Lucantoni (1993) provides a survey of the analysis of the BMAP/G/1 queue. As pointed out by Masuyama (2003), queues with batch Markovian arrival are so flexible that they can represent most of the queues studied in the past as special cases.

Many authors, including Heffes and Lucantoni (1986), Baiocchi et al. (1991), Yamada and Sumita (1991), and Li and Hwang (1993), have used MMPP(2) to model the superposed ATM traffic. Their analysis deals with MMPP/M/1 or MMPP/G/1 queueing systems. Fischer and Meier-Hellstern (1992) discuss other applications of MMPP.

Zhou and Gans (1999) have considered an M/MMPP/1 queue, which can be used to model a system that processes jobs from different sources. The time to process jobs from each source (or job type) is exponentially distributed, but each job type has a different mean service time. Moreover, after a job completion, the choice of the next job to be processed is governed by a Markov chain. Thus, while the aggregate arrival process is Poisson, the source from which a particular job comes is determined by an underlying Markov chain.

Muscariello et al. (2005) have used a hierarchical MMPP traffic model that very closely approximates the LRD characteristics of Internet traffic traces over relevant time scales. As stated earlier, LRD property of the Internet traffic means that values at any instant tend to be positively correlated with values at all future instants. This means that it has some sort of memory. However, long-term correlation properties, heavy tail distributions, and other characteristics are meaningful only over a limited range of time scale.

As discussed earlier, overflow traffic has been modeled using an IPP by Kuczura (1973) and Meier-Hellstern (1989). Min et al. (2001) analyzed adaptive wormhole-routed torus networks with IPP traffic input.

A survey of these traffic models is given by Bae and Suda (1991), Frost and Melamed (1994), Michiel and Laevens (1997), and Adas (1997).

12.8 Problems

12.1 Give the state transition rate diagram for the BMAP(2)/M/1 queue with internal rates α_{12} and α_{21}, external arrival rates λ_1 and λ_2, and service rate μ. Specify the infinitesimal generator, Q, if the batch size is equally likely to be 1, 2, or 3.

12.2 Consider the superposition of two identical IPPs with internal rates α and β and external arrival rate λ. Obtain the infinitesimal generator and arrival rate matrix for the superposed system.

12.3 Consider an MMBP(2)/Geo/1 queueing system, which is a single-server queueing system with a second-order Markov-modulated Bernoulli arrival process with external arrival parameters α and β and internal switching probabilities p and q, where $q = 1 - p$, and geometric service times with parameter γ. Give the state transition rate diagram.

12.4 Consider a queueing system in which the server is subject to breakdown and repair. When it is operational, the time until it fails is exponentially distributed with mean $1/\eta$. When it breaks down, the time until it is repaired and brought back to service is also exponentially distributed with mean $1/\gamma$. Customers arrive according to a Poisson process with rate λ. However, it has been found that the behavior of arriving customers depends on the state of the server. Specifically, when it is operational, all arriving customers stay in the system until they are served. But a customer that arrives when the server is down will balk (i.e., leave without receiving service) with probability p. Finally, the time to serve a customer when the system is operational is exponentially distributed with mean $1/\mu$. Give the state transition rate diagram of the process and determine the Q-matrix, identifying the A and D submatrices.

12.5 Consider an m-server queueing system that operates in the following manner. There are two types of customers: type 1 and type 2. Type 1 customers arrive according to a Poisson process with rate λ_1, and type 2 customers arrive according to a Poisson process with rate λ_2. All the m servers are identical, and the time each takes to serve a customer, regardless of its type, is exponentially distributed with mean $1/\mu$. As long as there is at least one idle server, all arriving customers are served without regard to their type. However, when all m servers are busy, type 2 customers are blocked; only type 1 customers may form a queue. When the number of customers in the system decreases to $k < m$ following an incidence of type 2 customer blocking, type 2 customers will once again be allowed to enter the system. Define the state of the system by (a, b), where a is the number of customers in the system and b is the phase of the system that takes the value 0 when both types of customers are allowed to enter the system and the value 1 when only type 1 customers are allowed. Give the state transition rate diagram of the process and determine the Q-matrix, including the D submatrices.

12.6 Consider a system whose environment changes according to a Markov chain. Specifically, Y_n is the state of the environment at the beginning of the nth period, where $Y = \{Y_n, n \geq 1\}$ is a Markov chain with a state transition probability matrix P. At the beginning of every period, a Bernoulli experiment is performed whose outcome depends on the state Y_n. Specifically, given that the process is in state Y_n, the probability of success is p_n and probability of failure is $q_n = 1 - p_n$. Thus, the outcome of the experiment, X_n, depends on the state of the environment. Assume that the conditional PMF of X_n is given by

$$p_{X_n}(x|Y) = \begin{cases} p_n & x = 1 \\ q_n & x = -1 \end{cases}$$

Define the random variable K_n as follows:

$$K_n = \begin{cases} 0 & n = 0 \\ X_1 + X_2 + \cdots + X_n & n \geq 1 \end{cases}$$

If we assume that a unit positive reward is associated with a success in the Bernoulli experiment, and a unit negative reward is associated with a failure, then K_n is the total reward at the end of the nth period. The bivariate process $\{(K_n, Y_n), n \geq 0\}$ is a Bernoulli-modulated Markov process. Consider the case where P is the matrix

$$P = \begin{bmatrix} p_{00} & p_{01} \\ p_{10} & p_{11} \end{bmatrix} = \begin{bmatrix} 1 - \alpha & \alpha \\ \beta & 1 - \beta \end{bmatrix}$$

Give the state transition rate diagram of the process, assuming that $p_n = p$.

13 Controlled Markov Processes

13.1 Introduction

Controlled Markov processes are a class of processes that deal with decision making under uncertainty. These processes, which include the Markov decision process (MDP), the semi-Markov decision process (SMDP), and the partially observable Markov decision process (POMDP), can be viewed as mathematical models that are concerned with optimal strategies of a decision maker who must make a sequence of decisions over time with uncertain outcomes. These three decision processes are the subject of this chapter.

13.2 Markov Decision Processes

In MDP, a decision maker or *agent* can influence the state of the system by taking a sequence of *actions* that causes the system to optimize a predefined performance criterion. To do this the agent observes the state of the system at specified points in time called *decision epochs* and gathers information necessary to choose actions that the agent expects will enable the desired performance criterion to be met. Each action that the agent takes incurs a cost or a reward, and the action affects the system state thereby affecting future actions. Thus, by applying a chosen action to the system the agent incurs an immediate cost and the system changes to a new state according to a transition probability distribution. In general, the immediate cost and transition probability distribution depend on the state and the chosen action.

If we denote the set of decision epochs by T, then the decision process can be classified as a discrete-time decision process or a continuous-time decision process, depending on whether T is discrete or continuous. In a discrete-time decision process, decisions are only made at the decision epochs. Similarly, in a continuous-time decision process, decision can be made continuously or at random points when certain predefined events occur. In discrete-time decision processes, the set of decision epochs, T, can be finite or infinite. When T is finite, we have that $T = \{1, 2, \ldots, N\}$ where $N < \infty$ and the elements of T are the decision epochs that are denoted by $t \in T$. When T is infinite, we have that $T = \{1, 2, \ldots\}$, which means that decisions will be made indefinitely. When N is finite, the decision process is called a *finite-horizon* (or *finite-stage*) decision process, otherwise, it is called an *infinite-horizon* (or *infinite-stage*) decision process.

The outcome of each decision is not fully predictable but can be anticipated to some extent before the next decision is made through the transition probability

Markov Processes for Stochastic Modeling. DOI: http://dx.doi.org/10.1016/B978-0-12-407795-9.00013-X

distribution. Also, as discussed earlier, the actions applied to the system have a long-term consequence because decisions made at the current decision epoch have an impact on decisions at the next decision epoch, and so on. Therefore, decisions cannot be viewed in isolation. Consequently, it is necessary to balance the desire for a low present cost against the undesirability of high future costs. Thus, good decision rules are needed to specify the actions that should be taken at any given decision epoch and state. A rule for making decisions at each decision epoch is called a *policy*. A policy used at decision epoch t could use the history of the system up to t (i.e., the system's sequence of observed states and sequence of actions). However, in practice, policies depend only on the observed state of the system at the decision epoch t. Thus, we can view a policy as a sequence of decision rules that prescribes the action to be taken at all decision epochs.

We denote a policy by $D = (d_1, d_2, \ldots, d_{N-1})$, where d_t is the action to be taken at the decision epoch $t \in T$. Policies can be classified as *stationary* or *nonstationary*. A stationary policy is the one in which the same action a_i is taken whenever the system is in a given state i. For example, consider a decision process where the states of the process are the outcomes of a flip of a coin. If the policy requires the agent to bet \$2 whenever the outcome is a head and \$1 whenever the outcome is a tail, then it is a stationary policy. A nonstationary policy is one in which different actions can be taken when the system is in a given state. The action taken might depend on the decision epoch. For example, for a finite-horizon process, we can take one action at the beginning of the horizon when the process is in state k and a different action toward the end of the horizon when the system is in state k again.

MDPs have been applied to a wide range of stochastic control problems, such as inspection—maintenance—replacement systems, inventory management, and economic planning. The topic is covered in several books including Bertsekas (1976, 1995a, 1995b), Borovkov (2003), Heyman and Sobel (1984), Howard (1960, 1971b), Kumar and Varaiya (1986), Puterman (1994), Ross (1970, 1983), and Tijms (1995). MDPs have been applied in modeling communication networks in Towsley et al. (2000), finance and dynamic options in Schal (2001), water reservoir in Lamond and Boukhtouta (2001), and medical treatment in Schaefer et al. (2004). We begin by presenting an overview of dynamic programming (DP).

13.2.1   Overview of DP

DP is a mathematical technique that is used for optimizing multistage decision problems. A multistage decision problem is a problem that can be separated into a number of stages (or steps), where each stage involves the optimization of exactly one variable. The computations at different stages are linked via a recursive algorithm that ensures that a feasible optimal solution to the entire problem is obtained when the last stage is reached. The optimization of each stage is based on a *decision* (which we defined earlier as an *action* taken), and a sequence of decisions is called a *policy*, as discussed earlier. Each stage has a number of states associated with it, where a state is any possible condition in which the system associated with the multistage problem can be in that stage. The number of states can be finite or infinite, and the effect of a decision at each stage is to transform the current state into a new state associated with the next stage.

A multistage decision problem usually has certain returns associated with each decision made; these returns can be costs or benefits. The objective of the solution to the problem is to determine the *optimal policy*, which is the policy that provides the best return. DP is based on the Bellman's *principle of optimality* in Bellman (1957), which states as follows:

Principle of optimality: An optimal policy has the property that, whatever the initial state and initial decision are, the remaining decisions must constitute an optimal policy with regard to the state resulting from the first decision.

This principle implies that given the current state, an optimal policy for the remaining stages is independent of the policy used in the previous stages. In other words, knowledge of the current state of the system embodies all the information about the past behavior that is necessary for determining the optimal policy from here on. This is essentially a Markovian property, as discussed in earlier chapters.

The implementation of this principle starts with finding the optimal policy for each state of the last stage. Then it moves backward stage by stage such that at each stage it determines the best policy for leaving each state of the stage using the results obtained in the previous stages.

13.2.2 Example of DP Problem

Consider a situation where we are given X units of a resource to be allocated to N activities. For example, we might want \$X to be used to fund N projects. Suppose we are also given a table that lists the return $r_i(x)$ to be realized from allocating x units of the resource to activity i, where $i = 1, 2, \ldots, N$ and $x = 0, 1, \ldots, X$. Then the problem we are confronted with becomes the following:

$$\text{Maximize } \sum_{i=1}^{N} r_i(x_i)$$

$$\text{Subject to } \sum_{i=1}^{N} x_i = X$$

$$0 \le x_i \le X \quad i = 1, 2, \ldots, N$$

To see this problem from the point of view of DP, we consider N stages labeled $1, 2, \ldots, N$, where each stage represents an activity and we first allocate x_1 units of the total resource to activity 1 at stage 1, then x_2 units of the remaining $X - x_1$ units of the resource to activity 2 at stage 2, and so on. To be able to optimize the remaining process, we must know the sum of units that have been allocated so far at each stage and the quantity left. We define the optimal value function $v_k(x)$ as the maximum return obtained from activities k through N, given that x units of the resource remain to be allocated. From the principle of optimality, we obtain the following recurrence relation

$$v_k(x) = \max[r_k(x_k) + v_{k+1}(x - x_k)] \quad x_k = 0, 1, \ldots, x$$

where x_k is the allocation to activity k and $x = 0, 1, \ldots, X$. The boundary condition is

$$v_N(x) = r_N(x)$$

and the solution to the problem is $v_1(x)$.

Example 13.1

For a numerical example, we consider $X = 6$ and $N = 3$. Table 13.1 gives the values of the returns $r_i(x)$.

Table 13.1 Data for the Numerical Example

x_1	$r_1(x_1)$	x_2	$r_2(x_2)$	x_3	$r_3(x_3)$
0	0	0	0	0	0
1	3	1	2	1	1
2	6	2	4	2	3
3	9	3	6	3	5
4	12	4	9	4	8
5	16	5	11	2	12
6	16	6	13	6	13

Solution

Because there are $N = 3$ activities, we define three stages with the following boundary conditions: $v_3(0) = 0, v_3(1) = 1, v_3(2) = 3, v_3(3) = 5, \ v_3(4) = 8, v_3(5) = 12,$ and $v_3(6) = 13$. These are essentially the values of $r_3(x_3)$. Let $m_k(x)$ denote the value of x_k that maximizes the right-hand side of the recurrence relation. Then using the recurrence relation, we obtain the following results:

$$v_2(0) = 0, \quad m_2(0) = 0$$

$$v_2(1) = \max[r_2(x_k) + v_3(1 - x_k)] = \max[r_2(0) + v_3(1), r_2(1) + v_3(0)]$$
$$= \max[0 + 1, 2 + 0] = 2, m_2(1) = 1$$

$$v_2(2) = \max[r_2(0) + v_3(2), r_2(1) + v_3(1), r_2(2) + v_3(0)]$$
$$= \max[0 + 3, 2 + 1, 4 + 0] = 4, m_2(2) = 2$$

$$v_2(3) = \max[r_2(0) + v_3(3), r_2(1) + v_3(2), r_2(2) + v_3(1), r_2(3) + v_3(0)]$$
$$= \max[0 + 5, 2 + 3, 4 + 1, 6 + 0] = 6, m_2(3) = 3$$

$$v_2(4) = \max[r_2(0) + v_3(4), r_2(1) + v_3(3), r_2(2) + v_3(2), r_2(3) + v_3(1), r_2(4) + v_3(0)]$$
$$= \max[0 + 8, 2 + 5, 4 + 7, 6 + 1, 9 + 0] = 9, m_2(4) = 4$$

$$v_2(5) = \max[r_2(0) + v_3(5), r_2(1) + v_3(4), r_2(2) + v_3(3), r_2(3) + v_3(2),$$
$$r_2(4) + v_3(1) + r_2(5) + v_3(0)]$$
$$= \max[0 + 12, 2 + 8, 4 + 5, 6 + 3, 9 + 1, 11 + 0] = 12, m_2(5) = 0$$

$$v_2(6) = \max[r_2(0) + v_3(6), r_2(1) + v_3(5), r_2(2) + v_3(4), r_2(3) + v_3(3),$$
$$r_2(4) + v_3(2) + r_2(5) + v_3(1), r_2(6) + v_3(0)]$$
$$= \max[0 + 13, 2 + 12, 4 + 8, 6 + 5, 9 + 3, 11 + 1, \ 13 + 0] = 14, m_2(6) = 1$$

Because we are starting with six units, we need only to compute $v_1(6)$, which is given by

$$v_1(6) = \max[r_1(0) + v_2(6), r_1(1) + v_2(5), r_1(2) + v_2(4), r_1(3) + v_2(3),$$
$$r_1(4) + v_2(2) + r_1(5) + v_2(1), r_1(6) + v_2(0)]$$
$$= \max[0 + 14, 2 + 12, 6 + 9, 9 + 6, 12 + 4, 16 + 2, 16 + 0] = 18, m_1(6) = 5$$

Thus, the optimal return from the 6-unit total allocation is 18. The number of allocations is as follows: because $m_1(6) = 5$, we allocate five units to activity 1 leaving one unit. Because $m_2(1) = 1$, we allocate one unit to activity 2, leaving a balance of 0; thus, we allocate no unit to activity 3. We can check to see that $r_1(5) + r_2(1) = 18$.

13.2.3   Markov Reward Processes

The Markov reward process (MRP) is an extension of the basic Markov process that associates each state of a Markov process with a reward. Specifically, let $\{X_n, n = 1, 2, \ldots, N\}$ be a discrete-time Markov chain with a finite state space $\{1, 2, \ldots, N\}$ and transition probability matrix P. Assume that when the process enters state i, it receives a reward r_{ij} when it makes a transition to state j, where r_{ij} can be positive or negative. Let the reward matrix R be defined as follows:

$$R = \begin{bmatrix} r_{11} & r_{12} & \cdots & r_{1N} \\ r_{21} & r_{22} & \cdots & r_{2N} \\ \vdots & \vdots & \ddots & \vdots \\ r_{N1} & r_{N2} & \cdots & r_{NN} \end{bmatrix}$$

That is, R is the matrix of the rewards. We define the process $\{X_n, R\}$ to be a discrete-time MRP.

Let $v_n(i)$ denote the expected total earnings in the next n transitions, given that the process is currently in state i. Assume that the process makes a transition to state j with probability p_{ij}. It receives an immediate reward of r_{ij}, where $j = 1, \ldots, N$. To compute $v_n(i)$, let the reward when there are n transitions to be made be represented by $\Re_n$, and let s_n denote the current state. Then we have

$$v_n(i) = E[\Re_n + \Re_{n-1} + \cdots + \Re_1 + \Re_0 | s_n = i]$$
$$= E[\{\Re_n | s_n = i\} + \{\Re_{n-1} + \cdots + \Re_1 + \Re_0 | s_n = i\}]$$
$$= E[\Re_n | s_n = i] + E[\Re_{n-1} + \cdots + \Re_1 + \Re_0 | s_n = i]$$
$$= \sum_{j=1}^{N} p_{ij} r_{ij} + \sum_{j=1}^{N} p_{ij} v_{n-1}(j)$$

The interpretation of the above equation is as follows. The first sum denotes the expected immediate reward that accrues from making a transition from state i to any state. When this transition takes place, the number of remaining transitions out of the n transitions is $n-1$. Thus, the second sum represents the expected total reward in these $n-1$ transitions given that the process is now in state j, $v_{n-1}(j)$, over all possible j that a transition from state i can be made.

If we define the parameter q_i by

$$q_i = \sum_{j=1}^{N} p_{ij} r_{ij}$$

then q_i is basically the expected reward in the next transition out of state i. Thus, we obtain

$$v_n(i) = q_i + \sum_{j=1}^{N} p_{ij} v_{n-1}(j)$$

If we define the column vector $v_n = [v_n(1), v_n(2), \ldots, v_n(N)]^{\mathrm{T}}$ and the column vector $q = [q_1, q_2, \ldots, q_N]^{\mathrm{T}}$, then we can rewrite that equation in following matrix form:

$$v_n = q + P v_{n-1}$$

which is equivalent to the following:

$$v_{n+1} = q + P v_n$$

Finally, if we denote the z-transform of v_n by $G_{v_n}(z)$, then taking the z-transform on both sides of the above equation, we obtain

$$z^{-1}[G_{v_n}(z) - v_0] = \frac{1}{1-z} q + P G_{v_n}(z) \Rightarrow [I - zP]G_{v_n}(z) = \frac{z}{1-z} q + v_0$$

$$G_{v_n}(z) = \frac{z}{1-z}[I - zP]^{-1} q + [I - zP]^{-1} v_0$$

where $v_0 = [v_0(1), v_0(2), \ldots, v_0(N)]^{\mathrm{T}}$. From the nature of the problem, we can determine v_0 and thus obtain the solution. Note that $v_0(i)$ is the terminal cost incurred when the process ends up at state i.

Recall that in Chapter 4, it was stated that the inverse transform can be expressed in the form

$$[I - Pz]^{-1} = \frac{1}{1-z} C + B(z)$$

where the constant term C has the characteristic that all the n rows are identical, and the elements of the rows are the limiting-state probabilities of the system whose transition probability matrix is P. Thus, if $b(n)$ is the sequence whose z-transform is $B(z)$, we have

$$G_{v_n}(z) = \frac{z}{1-z}[I-zP]^{-1}q + [I-zP]^{-1}v_0$$

$$= \frac{z}{(1-z)^2}Cq + \frac{z}{1-z}B(z)q + \frac{1}{1-z}Cv_0 + B(z)v_0$$

$$= \frac{z}{(1-z)^2}Cq + \left\{\frac{1}{1-z} - 1\right\}B(z)q + \frac{1}{1-z}Cv_0 + B(z)v_0$$

$$= \frac{z}{(1-z)^2}Cq + \frac{1}{1-z}Cv_0 + B(z)[v_0 - q] + \frac{1}{1-z}B(z)q$$

From this, we obtain the solution

$$v_n = nCq + Cv_0 + b(n)[v_0 - q] + q\sum_{k=0}^{n}b(k)$$

If we define $g = Cq$, we obtain the solution

$$v_n = ng + Cv_0 + b(n)[v_0 - q] + q\sum_{k=0}^{n}b(k)$$

13.2.4 MDP Basics

MDP is an extension of both MRP and DP in which an agent takes a set of actions that can be used to control the system at each state with a view to maximizing the expected reward. MDP is a discrete-time probabilistic system that can be represented by the tuple (S, A, R, P), where

- S is a finite set of N states, that is, $S = \{1, 2, \ldots, N\}$. In practice, the state of a system is a set of parameters that can be used to describe the system. For example, the state of a robot can be the coordinates of the robot.
- A is a finite set of K actions that can be taken at any state; that is, $A = \{a_1, a_2, \ldots, a_K\}$. In the case of a robot that can move in discrete steps, for example, an action can be a statement like "go east" or "go west."
- R is the reward matrix, which can vary with the action taken. Thus, for action $a \in A$, we denote the reward associated with a transition from state i to state j when action a is taken by $r_{ij}(a)$.
- P is the transition probability matrix, which can be different for each action. Thus, for action $a \in A$, we denote the probability that the process moves from state i to state j when action a is taken by $p_{ij}(a)$.

For such a system, we can see that

$$P[S_{n+1} = j | S_0, a_0, S_1, a_1, \ldots, S_n = i, a_n = a] = P[S_{n+1} = j | S_n = i, a_n = a] = p_{ij}(a)$$

Thus, the transition probabilities and reward functions are functions only of the last state and the subsequent action. Any homogeneous Markov chain $\{S_n\}$ whose transition probabilities are $p_{ij}(a)$ is called an MDP, where

$$\sum_j p_{ij}(a) = 1 \quad \text{for all } i \in S \text{ and all } a \in A$$

As stated earlier, the actions taken at each state are usually chosen according to a well-defined policy. Thus, a policy D is a mapping from S to A; that is, we can formally define a policy as a rule for taking actions at each state during a decision epoch. Because the objective in the decision process is to maximize the expected value of the sum of the returns (called expected total return) over a given time span, we define the *optimal policy* as the policy that maximizes the total expected return for each starting state i and number of transitions n. We are interested in stationary policies where, as we defined earlier, a stationary policy is a policy that assigns to each state i a fixed action $a = R_i$ that is always used whenever the process is in that state. Note that many texts use π to represent policy. However, in this book, we use D because in earlier chapters π has been used to represent limiting-state probabilities of discrete-time Markov chains.

To solve the problem, we consider the decision epochs to be stages in the decision-making process. Let $v_n(i, d)$ denote the expected total return in the next n stages, given that the process is in state i and policy d is used. $v_n(i, d)$ is sometimes called the *value function*. To derive the recursive equation for $v_n(i, d)$, we consider Figure 13.1.

From the figure we observe that for a finite-stage system, the recursive equation relating $v_n(i, d)$ and $v_{n-1}(i, d)$ is given by

$$v_n(i, d) = \sum_{j=1}^{N} p_{ij}(d) r_{ij}(d) + \sum_{j=1}^{N} p_{ij}(d) v_{n-1}(j, d) = q_i(d) + \sum_{j=1}^{N} p_{ij}(d) v_{n-1}(j, d)$$

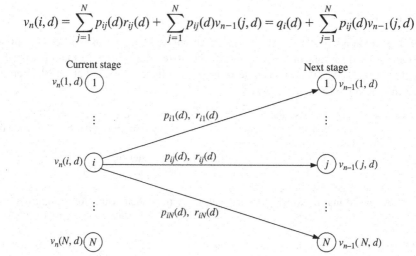

Figure 13.1 Stage-based expected return.

where

$$q_i(d) = \sum_{j=1}^{N} p_{ij}(d)r_{ij}(d)$$

is the expected return in the next transition out of state i using policy d. Thus, applying the Bellman's principle of optimality, the optimal return with respect to state i is given by

$$v_n(i) = \max_d \left\{ q_i(d) + \sum_{j=1}^{N} p_{ij}(d)v_{n-1}(j) \right\} \quad i = 1, 2, \ldots, N$$

13.2.5   MDPs with Discounting

In many economic systems, it is important to take the cost of money into consideration by introducing discounting. This is due to the fact that the value of \$1 now is not the same as its value in 3 years' time. This difference can be accounted for by introducing the so-called *discounted return*. Let the *discount factor* β denote the value at the beginning of a transition interval of a unit return received at the end of the transition, where $0 \le \beta \le 1$. As previously defined, let the random variable $\Re_n$ be the reward when there are n transitions to be made. Then for the finite-stage system, $v_n(i, d, \beta)$ is given by

$$v_n(i, d, \beta) = E[\Re_n + \beta\Re_{n-1} + \beta^2\Re_{n-2} + \cdots + \beta^{n-1}\Re_1 + \beta^n\Re_0 | s_n = i]$$

$$= E[\{\Re_n | s_n = i\} + \beta\{\Re_{n-1} + \beta\Re_{n-2} + \cdots + \beta^{n-2}\Re_1 + \beta^{n-1}\Re_0\} | s_n = i]$$

$$= E[\{\Re_n | s_n = i\}] + \beta E[\{\Re_{n-1} + \beta\Re_{n-2} + \cdots + \beta^{n-2}\Re_1 + \beta^{n-1}\Re_0\} | s_n = i]$$

$$= \sum_{j=1}^{N} p_{ij}(d)r_{ij}(d) + \beta \sum_{j=1}^{N} p_{ij}(d)v_{n-1}(j, d, \beta)$$

$$= q_i(d) + \beta \sum_{j=1}^{N} p_{ij}(d)v_{n-1}(j, d, \beta)$$

Thus, the only difference is the introduction of the discount factor to the future returns.

13.2.6   Solution Methods

The solution to any decision process is a sequence of actions that optimizes a given value function. There are three general methods of solving MDP problems. These are the *value iteration* method, which is used for finite-horizon problems; the *policy iteration* method, which is used for infinite-horizon problems; and the *linear programming* method, which is also used for infinite-horizon problems but will not be discussed here. The value iteration method is sometimes called the method of *successive approximations*.

The policy iteration method will find the stationary policy that optimizes the value function both when no discounting is used and when discounting is used. The value iteration method might not give the optimal policy using a finite number of iterations. However, compared to the policy iteration method, it has the advantage that it does not require the solution of a system of simultaneous equations, as the policy iteration and linear programming methods do. Thus, with the value iteration method, each iteration can be performed simply and quickly.

Value Iteration Method

The value–iteration method computes recursively a sequence of value functions approximating the optimal cost per unit time. It is essentially an extension of the technique used to solve the deterministic DP problem and thus utilizes a backward recursive relationship as follows:

$$v_n(i, d, \beta) = q_i(d) + \beta \sum_{j=1}^{N} p_{ij}(d) v_{n-1}(j, d, \beta)$$

$$v_n(i, \beta) = \max_d \left\{ q_i(d) + \sum_{j=1}^{N} p_{ij}(d) v_{n-1}(j, \beta) \right\} \quad i = 1, 2, \ldots, N$$

It starts by choosing a set of values of $v_0(i, d, \beta)$ as follows:

$$v_0(1, d, \beta) = v_0(2, d, \beta) = \cdots = v_0(N, d, \beta) = 0$$

Thus, we can obtain $v_1(i, \beta)$ as follows:

$$v_1(i, \beta) = \max_d \{ q_i(d) \} \quad i = 1, 2, \ldots, N$$

This gives the expected total return in stage 1, given that the process is in state i at that stage when the optimal policy is used. Using this set of $v_1(i, \beta)$, we obtain $v_2(i, \beta)$, the expected total return in stage 2, as follows:

$$v_2(i, \beta) = \max_d \left\{ q_i(d) + \sum_{j=1}^{N} p_{ij}(d) v_1(j, \beta) \right\} \quad i = 1, 2, \ldots, N$$

This process continues backward until we reach the first stage. The solution to the problem is $v_T(i)$, where T is the number of decision epochs, which corresponds to the number of stages. Thus, the process can be summarized as shown in Figure 13.2.

Figure 13.2 Recursive procedure for the expected return.

Example 13.2

Consider an equipment that is inspected daily at the end of each work day. It can be in one of three states at the inspection time, which are Good, Acceptable, and Bad. It has been found that the condition at the time of inspection on a given day depends probabilistically on the condition at the time of inspection on the previous day as follows. Given that it is Good on a given day, it will be Good the following day with probability 0.6, Acceptable with probability 0.3, and Bad with probability 0.1. Similarly, given that it is Acceptable on a given day, it will be Good the following day with probability 0, Acceptable with probability 0.6, and Bad with probability 0.4. Finally, given that it is Bad on a given day, it will be Good the following day with probability 0, Acceptable with probability 0, and Bad with probability 1.0. The possible maintenance actions are as follows:

a. Do nothing, and thus follow the transition probabilities defined above.
b. Overhaul, which is equally likely to bring it to the Good condition or the Acceptable condition; the cost of an overhaul is $500.
c. Replace the equipment, which automatically brings it to the Good condition; the cost of a new equipment is $2000.

Assume that when the equipment is operating in Good condition, the company makes $1000. When it is operating in Acceptable condition, the company makes $500, and when it is operating in Bad condition, the company loses $500. We consider the following policies:

a. Replace the equipment only when it is in Bad condition and nothing in other states.
b. Replace the equipment when it is in Bad condition and overhaul when it is in Acceptable condition.
c. Replace the equipment when it is in Acceptable condition and when it is in Bad condition.

Determine the optimal operating cost of the equipment when $T = 4$.

Solution

Let the states be defined as follows:

$$1 \equiv \text{Good}$$
$$2 \equiv \text{Acceptable}$$
$$3 \equiv \text{Bad}$$

The Markov chains for the different policies are shown in Figure 13.3.

We transform the problem into a maximization problem by assigning negative values to the costs incurred by the company. We take $500 as the baseline cost, that is, we assume

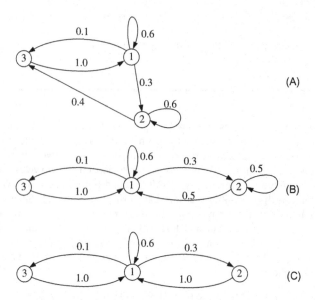

Figure 13.3 Markov chains for the policies 1−3 (A−C) of Example 13.2.

that \$500 is the unit cost. If we denote the state transition matrix, reward matrix, and immediate expected return matrix for policy d by $P(d)$, $R(d)$, and $Q(d)$, respectively, then the state transition, reward, and immediate expected return matrices for the different policies are as follows:

$$P(1) = \begin{bmatrix} 0.6 & 0.3 & 0.1 \\ 0.0 & 0.6 & 0.4 \\ 1.0 & 0.0 & 0.0 \end{bmatrix} \quad R(1) = \begin{bmatrix} 2 & 1 & -1 \\ 0 & 1 & -1 \\ -2 & 0 & 0 \end{bmatrix} \quad Q(1) = \begin{bmatrix} 1.4 \\ 0.2 \\ -2 \end{bmatrix}$$

$$P(2) = \begin{bmatrix} 0.6 & 0.3 & 0.1 \\ 0.5 & 0.5 & 0.0 \\ 1.0 & 0.0 & 0.0 \end{bmatrix} \quad R(2) = \begin{bmatrix} 2 & 1 & -1 \\ 1 & 0 & 0 \\ -2 & 0 & 0 \end{bmatrix} \quad Q(2) = \begin{bmatrix} 1.4 \\ 0.5 \\ -2 \end{bmatrix}$$

$$P(3) = \begin{bmatrix} 0.6 & 0.3 & 0.1 \\ 1.0 & 0.0 & 0.0 \\ 1.0 & 0.0 & 0.0 \end{bmatrix} \quad R(3) = \begin{bmatrix} 2 & 1 & -1 \\ 1 & 0 & 0 \\ -2 & 0 & 0 \end{bmatrix} \quad Q(3) = \begin{bmatrix} 1.4 \\ 1 \\ -2 \end{bmatrix}$$

Note that the elements of $R(1), R(2)$, and $R(3)$ take into account the cost of replacing and overhauling the equipment. For example, when the process is in state 2, under policy 2, the equipment is to be overhauled at a cost of one unit (or \$500), which will be subtracted from the gain made in the state into which the next transition takes place. Thus, the entries indicate the net rewards.

To continue the solution, we proceed in the following stages. We start with $v_0(i, d) = 0$, which gives $v_1(i, d) = q_i(d)$.

Stage 1

	$v_1(i, d) = q_i(d)$			Optimal Solution	
				$v_1(i)$	d^*
i	$d = 1$	$d = 2$	$d = 3$		
1	1.4	1.4	1.4	1.4	1,2,3
2	0.2	0.5	1	1	3
3	−2	−2	−2	−2	1,2,3

Stage 2

	$v_2(i, d) = q_i(d) + p_{i1}(d)v_1(1) + p_{i2}(d)v_1(2) + p_{i3}(d)v_1(3)$			Optimal Solution	
i	$d = 1$	$d = 2$	$d = 3$	$v_2(i)$	d^*
1	2.34	2.34	2.34	2.34	1,2,3
2	0	1.7	2.4	2.4	3
3	−0.6	−0.6	−0.6	−0.6	1,2,3

Stage 3

	$v_3(i, d) = q_i(d) + p_{i1}(d)v_2(1) + p_{i2}(d)v_2(2) + p_{i3}(d)v_2(3)$			Optimal Solution	
i	$d = 1$	$d = 2$	$d = 3$	$v_2(i)$	d^*
1	3.464	3.464	3.464	3.464	1,2,3
2	0.80	2.87	3.34	3.34	3
3	0.34	0.34	0.34	0.34	1,2,3

Stage 4

	$v_4(i, d) = q_i(d) + p_{i1}(d)v_3(1) + p_{i2}(d)v_3(2) + p_{i3}(d)v_3(3)$			Optimal Solution	
i	$d = 1$	$d = 2$	$d = 3$	$v_2(i)$	d^*
1	4.5144	4.5144	4.5144	4.5144	1,2,3
2	2.340	3.902	4.464	4.464	3
3	1.464	1.464	1.464	1.464	1,2,3

The optimal solution shows that the user should do nothing when the equipment is Good in any year, replace the equipment if it is Acceptable in any year, and replace the equipment if it is Bad in any year. The total expected return after 4 years is $v_4(1) = 4.5144$ units if the equipment is Good in the first year, $v_4(2) = 4.464$ units if it is Acceptable in the first year, and $v_4(3) = 1.464$ units if it is Bad in the first year.

Policy Iteration Method

As discussed earlier, the policy iteration method is used for infinite-horizon problems. The method requires the iteration of two steps: a *value determination* step followed by a *policy improvement* step. The value determination step is achieved by arbitrarily selecting an initial policy d and then solving the equation to obtain the long-run value function per unit time. Then using the policy improvement method, a better policy is selected and the value determination step is repeated. This step is continued until two successive iterations that lead to identical policies are reached when the procedure stops because an optimal policy has been obtained. After each step of the policy improvement scheme, an optimality test is carried out. If the test result is negative, an improvement is produced by another step, otherwise, the procedure stops and an optimal solution is obtained.

Recall that the value function at the nth decision epoch given that the process is in state i and policy d is used is given by

$$v_n(i, d) = q_i(d) + \sum_{j=1}^{N} p_{ij}(d) v_{n-1}(j, d)$$

Howard (1960) has shown that for an ergodic Markov process with a stationary policy, $v_n(i, d)$ has the asymptotic form

$$v_n(i, d) = ng(d) + v_i(d) \quad i = 1, 2, \ldots, N$$

where $g(d)$ and $v_i(d)$ depend on the policy used. $g(d)$ is called the *gain of the system* under policy d and is given by

$$g(d) = \sum_{i=1}^{N} \pi_i \sum_{j=1}^{N} p_{ij}(d) r_{ij}(d) = \sum_{i=1}^{N} \pi_i q_i(d)$$

where π_i are the limiting-state probabilities of the Markov chain. Thus, substituting for $v_n(i)$, we obtain

$$v_n(i, d) = ng(d) + v_i(d) = q_i(d) + \sum_{j=1}^{N} p_{ij}(d)\{(n-1)g(d) + v_j(d)\}$$

$$= q_i(d) + (n-1)g(d)\sum_{j=1}^{N} p_{ij}(d) + \sum_{j=1}^{N} p_{ij}(d) v_j(d)$$

$$= q_i(d) + (n-1)g(d) + \sum_{j=1}^{N} p_{ij}(d) v_j(d)$$

where the last equality follows from the fact that

$$\sum_{j=1}^{N} p_{ij}(d) = 1$$

Thus, we obtain

$$ng(d) + v_i(d) = q_i(d) + (n-1)g(d) + \sum_{j=1}^{N} p_{ij}(d)v_j(d)$$

From this, we obtain

$$g(d) + v_i(d) = q_i(d) + \sum_{j=1}^{N} p_{ij}(d)v_j(d) \quad i = 1, 2, \ldots, N$$

which is a set of N linear simultaneous equations. Thus, using the model in Howard (1960), we can summarize the solution algorithm as follows:

1. *Value-determination operation*: Use p_{ij} and q_i for a given policy to solve

$$g + v_i = q_i + \sum_{j=1}^{N} p_{ij}v_j \quad i = 1, 2, \ldots, N$$

for all relative values v_i and q_i by setting $v_N = 0$.
2. *Policy-improvement routine*: For each state i, find the alternative policy d^* that maximizes

$$q_i(d) + \sum_{j=1}^{N} p_{ij}(d)v_j$$

using the relative values v_i of the previous policy. Then d^* becomes the new policy in state i, $q_i(d^*)$ becomes q_i, and $p_{ij}(d^*)$ becomes p_{ij}.
3. *Stopping rule*: The optimal policy is reached (i.e., g is maximized) when the policies on two successive iterations are identical. Thus, if the current value of d is not the same as the previous value of d, go back to step 1, otherwise, stop.

The policy iteration method has the following properties:

- The problem reduces to solving sets of linear simultaneous equations and subsequent comparisons.
- Each succeeding policy has a higher gain than the previous one.
- The iteration cycle terminates on the policy that has the largest gain.

Example 13.3

Consider an MDP with the following parameters for two policies:

$$P(1) = \begin{bmatrix} 0.4 & 0.6 \\ 0.2 & 0.8 \end{bmatrix} \quad R(1) = \begin{bmatrix} 4 & 8 \\ -2 & 0 \end{bmatrix} \quad Q(1) = \begin{bmatrix} 6.4 \\ -0.4 \end{bmatrix}$$

$$P(2) = \begin{bmatrix} 0.8 & 0.2 \\ 0.5 & 0.5 \end{bmatrix} \quad R(2) = \begin{bmatrix} 4 & 6 \\ -1 & -1 \end{bmatrix} \quad Q(2) = \begin{bmatrix} 4.4 \\ -1 \end{bmatrix}$$

We are required to obtain the long-run optimal operating policy for the problem.

Solution

From the values of $q_i(d)$, it seems reasonable to start with $d_1 = 1$ for both states. Thus, the system of equations to be solved is as follows:

$$g + v_1 = q_1(1) + p_{11}(1)v_1 + p_{12}(1)v_2$$
$$g + v_2 = q_2(1) + p_{12}(1)v_1 + p_{22}(1)v_2$$

Because there are three unknown and two equations, we set $v_2 = 0$ and obtain the following system of equations:

$$g + v_1 = q_1(1) + p_{11}(1)v_1 = 6.4 + 0.4v_1$$
$$g = q_2(1) + p_{12}(1)v_1 = -0.4 + 0.2v_1$$

From these, we obtain the relative values $v_1 = 8.5, v_2 = 0, g = 1.3$. Applying these values to the policy improvement routine, we obtain

	$q_i(d) + p_{i1}(d)v_1 + p_{i2}(d)v_2 = q_i(d) + p_{i1}(d)v_1$		**Improved Policy**
i	$d = 1$	$d = 2$	d^*
1	$6.4 + (0.4)(8.5) = 9.8$	$4.4 + (0.8)(8.5) = 11.2$	2
2	$-0.4 + (0.2)(8.5) = 1.3$	$-1 + (0.5)(8.5) = 3.25$	2

This means that the improved policy used in the next round of the policy determination operation is policy 2 for both states 1 and 2. Thus, the system of equations to be solved is as follows:

$$g + v_1 = q_1(2) + p_{11}(2)v_1 + p_{12}(2)v_2$$
$$g + v_2 = q_2(2) + p_{12}(2)v_1 + p_{22}(2)v_2$$

As before, we set $v_2 = 0$ and obtain the following system of equations:

$$g + v_1 = q_1(2) + p_{11}(2)v_1 = 4.4 + 0.8v_1$$
$$g = q_2(2) + p_{12}(2)v_1 = -1 + 0.5v_1$$

From these equations, we obtain the relative values $v_1 = 7.71, v_2 = 0, g = 2.85$. Applying these values to the policy improvement routine, we obtain

	$q_i(d) + p_{i1}(d)v_1 + p_{i2}(d)v_2 = q_i(d) + p_{i1}(d)v_1$		**Improved Policy**
i	$d = 1$	$d = 2$	d^*
1	$6.4 + (0.4)(7.71) = 9.484$	$4.4 + (0.8)(7.71) = 10.568$	2
2	$-0.4 + (0.2)(7.71) = 1.142$	$-1 + (0.5)(7.71) = 2.855$	2

The improved policy is policy 2 for both states 1 and 2, which is the same as the current policy. Thus, it is the optimal policy, and the procedure is terminated. As stated earlier, the asymptotic solution under a fixed policy is

$$v_n(i) = ng + v_i \quad i = 1, 2, \dots, N$$

For the preceding example, $g = 2.85$ and $v_1 - v_2 = 7.71$.

Policy Iteration Method with Discounting

The solution procedure for the policy iteration method with discounting is similar to that without discounting. However, there are two areas of difference between the two. First, a discount factor, β, is introduced. Second, as shown by Howard (1960), in the value determination operation, we use the p_{ij} and q_i for the selected policy to solve the set of equations

$$v_i(\beta) = q_i + \beta \sum_{j=1}^{N} p_{ij}v_j(\beta) \quad i = 1, 2, \dots, N$$

Note that the gain g does not appear in the equation. Thus, we have N equations in N unknowns, which means that values of the $v_i(\beta)$ are not relative values but exact values.

The policy improvement routine is different from that of the system without discount by the introduction of the discount factor. That is, for each state i, we find the policy d^* that maximizes

$$q_i(d) + \beta \sum_{j=1}^{N} p_{ij}(d)v_j(\beta)$$

using the current values of $v_i(\beta)$ from the previous policy.

Example 13.4

Consider the problem of Example 13.3, and assume that $\beta = 0.9$. That is, we consider an MDP with the following parameters for two policies:

$$P(1) = \begin{bmatrix} 0.4 & 0.6 \\ 0.2 & 0.8 \end{bmatrix} \quad R(1) = \begin{bmatrix} 4 & 8 \\ -2 & 0 \end{bmatrix} \quad Q(1) = \begin{bmatrix} 6.4 \\ -0.4 \end{bmatrix}$$

$$P(2) = \begin{bmatrix} 0.8 & 0.2 \\ 0.5 & 0.5 \end{bmatrix} \quad R(2) = \begin{bmatrix} 4 & 6 \\ -1 & -1 \end{bmatrix} \quad Q(2) = \begin{bmatrix} 4.4 \\ -1 \end{bmatrix}$$

As in Example 13.3, we are required to obtain the long-run optimal operating policy for the problem.

Solution

As in Example 13.3, we start with $d_1 = 1$ for both states. Thus, the system of equations to be solved is as follows:

$$v_1 = q_1(1) + 0.9p_{11}(1)v_1 + 0.9p_{12}(1)v_2 = 6.4 + 0.9(0.4)v_1 + 0.9(0.6)v_2$$
$$= 6.4 + 0.36v_1 + 0.54v_2$$
$$v_2 = q_2(1) + 0.9p_{21}(1)v_1 + 0.9p_{22}(1)v_2 = -0.4 + 0.9(0.2)v_1 + 0.9(0.8)v_2$$
$$= -0.4 + 0.18v_1 + 0.72v_2$$

From these, we obtain the solution $v_1 = 19.22, v_2 = 10.93$. Applying these values to the policy improvement routine, we obtain

	$q_i(d) + \beta p_{i1}(d)v_1 + \beta p_{i2}(d)v_2$		Improved Policy
i	$d=1$	$d=2$	d^*
1	$6.4 + 0.9(0.4)(19.22) + 0.9(0.6)$ $(10.93) = 19.22$	$4.4 + 0.9(0.8)(19.22) + 0.9(0.2)$ $(10.93) = 20.21$	2
2	$-0.4 + 0.9(0.2)(19.22) + 0.9(0.8)$ $(10.93) = 10.93$	$-1 + 0.9(0.5)(19.22) + 0.9(0.5)$ $(10.93) = 12.57$	2

This means that the improved policy used in the next round of the policy determination operation is policy 2 for both states 1 and 2. Thus, the system of equations to be solved is as follows:

$$v_1 = q_1(2) + 0.9p_{11}(2)v_1 + 0.9p_{12}(2)v_2 = 4.4 + 0.9(0.8)v_1 + 0.9(0.2)v_2$$
$$= 4.4 + 0.72v_1 + 0.184v_2$$
$$v_2 = q_2(2) + 0.9p_{21}(2)v_1 + 0.9p_{22}(2)v_2 = -1 + 0.9(0.5)v_1 + 0.9(0.5)v_2$$
$$= -1 + 0.45v_1 + 0.45v_2$$

The solution to these equations is $v_1 = 30.69, v_2 = 23.29$. Applying these values to the policy improvement routine, we obtain

$$q_i(d) + \beta p_{i1}(d)v_1 + \beta p_{i2}(d)v_2$$

			Improved Policy
i	$d = 1$	$d = 2$	d^*
1	$6.4 + 0.9(0.4)(30.69) + 0.9(0.6)$ $(23.29) = 30.025$	$4.4 + 0.9(0.8)(30.69) + 0.9(0.2)$ $(23.29) = 30.689$	2
2	$-0.4 + 0.9(0.2)(30.69) + 0.9(0.8)$ $(23.29) = 21.893$	$-1 + 0.9(0.5)(30.69) + 0.9(0.5)$ $(23.29) = 23.291$	2

This shows that the improved policy is policy 2 for both states 1 and 2. Because the improved policy is the same as the current policy, the procedure is terminated as the optimal policy has been obtained. The present values of the states 1 and 2 under the optimal policy are 30.689 and 23.291, respectively.

13.3   Semi-MDPs

In Chapter 6, we defined a semi-Markov process (SMP) as a process that makes transitions from state to state like a Markov process but in which the amount of time spent in each state before a transition to the next state occurs is an arbitrary random variable that depends on the next state the process will enter. This means that at transition instants, an SMP behaves like a Markov process. Recall that when the process enters state i, it chooses its next state as state j with probability p_{ij}. In the case of a discrete-time SMP, after choosing j, the process spends a time H_{ij} called the holding time before making the transition, where H_{ij} is a positive, integer-valued random variable with the probability mass function (PMF) $p_{H_{ij}}(m) = P[H_{ij} = m]$, $m = 1, 2, \ldots$. Similarly, in the case of a continuous-time SMP, after choosing state j, the time H_{ij} that the process spends in state i until the next transition, which is the holding time for a transition from i to j, has the PDF $f_{H_{ij}}(t), t \geq 0$.

The SMDP is an extension of the SMP in which we control the state transitions through a set of actions and associate each action with a set of rewards. Thus, to describe an SMDP, we augment the description of an SMP by stating that whenever the system enters a state a set of possible actions will be taken, and associated with each action are the transition probability to the next state, the holding time, and the reward accruing from the transition to the next state.

The SMDP is to the SMP what MDP is to the Markov chain. The main difference between MDP and SMDP can be explained as follows. In MDP, it is assumed that decisions are made at specific epochs and an action taken at epoch t affects only the state where the action was taken and the reward at epoch t. In SMDP, the intervals between decisions are usually random. Alternatively, actions can take variable amounts of time to complete. In practice, decisions are made when a change of state occurs. For example, in a queueing system, decisions can be made when a customer arrives at the system or when a customer leaves the system. Thus, an SMDP model includes an additional parameter that defines the duration of an action or the interval between actions. For this reason, an SMDP is a discrete-time probabilistic system that can be represented by the 5-tuple (S, A, R, P, H) where

- S is a finite set of N states, as in the MDP; that is, $S = \{1, 2, \ldots, N\}$.
- A is a finite set of K actions that can be taken at any state; that is, $A = \{a_1, a_2, \ldots, a_K\}$.
- R is the reward matrix, which can vary with the action taken. Thus, for action $a \in A$ we denote the reward associated with a transition from state i to state j when action a is taken by $r_{ij}(a)$.
- P is the transition probability matrix, which can be different for each action. Thus, for action $a \in A$ we denote the probability that the system moves from state i to state j when action a is taken by $p_{ij}(a)$, which is independent of the history of the process up to the time the action was taken.
- H is the holding time distribution. $H_{ij}(a)$ is the holding time for a transition from state i to state j, which is the time the system spends in state i, given that upon leaving state i the next state it goes to is state j when action a is taken in state i. The PDF of $H_{ij}(a)$ is $f_{H_{ij}}(t, a), t \geq 0$, where $E[H_{ij}(a)] = h_{ij}(a) < \infty$.

Thus, while the only relevant feature in MDP is the sequential nature of the decision process, not the time that elapses between decision epochs, in SMDP the time between one decision epoch and the next is a random variable that can be real or integer-valued. For this reason, there are different types of SMDP that depend on whether the system state and the intervals between decision epochs are discrete or continuous. Specifically, there can be discrete-state SMDPs and continuous-state SMDPs, and discrete-decision-interval SMDPs and continuous-decision-interval SMDPs. From this classification, we observe that an MDP is an SMDP with discrete-decision intervals, where the interval is constant at one time unit. In a discrete-decision-interval system, the times between decisions are governed by a discrete random variable with a specified PMF. In a continuous-decision-interval system, the time between decisions is a random variable with a specified PDF. In the remainder of this chapter, we consider both discrete-state continuous-decision-interval SMDPs and discrete-state discrete-decision-interval SMDPs.

Sometimes a parameter T is listed as a component of the definition of the SMDP. T is defined to be the probability distribution of the times between actions. Thus, $T_i(a)$ denotes the time until the next decision epoch given that action a is taken in the current state i; its PDF is $f_{T_i}(x, a)$, and mean $E[T_i(a)]$, where $0 < E[T_i(a)] < \infty$. However, the holding time $H_{ij}(a)$ is related to $T_i(a)$ as follows:

$$T_i(a) = \sum_{j=1}^{N} p_{ij} H_{ij}(a)$$

Thus, there is no need for an additional parameter in the specification of the SMDP. Finally, we assume that stationary policies are used, which means that the same action a is taken whenever the system is in a given state i.

13.3.1 Semi-Markov Reward Model

In this section, we consider continuous-time semi-Markov reward models. Using the method proposed by Ross (1970), we assume the following reward structure.

When the system enters state i, it incurs an immediate reward B_{ij} and an extra reward accumulates at the rate of $b_{ij}(t)$ per unit time until the transition to state j occurs, where $B_{ij} < \infty$ and $b_{ij}(t) < \infty$. Howard (1971b) refers to B_{ij} as a bonus and $b_{ij}(t)$ as the "yield rate" of state i when the next state is j. Thus, if the system spends a time τ in state i before making a transition out of the state, the total expected accrued reward is

$$r_i(\tau) = \sum_{j=1}^{N} p_{ij} B_{ij} + \sum_{j=1}^{N} p_{ij} \int_{x=0}^{\tau} \int_{t=0}^{x} b_{ij}(t) f_{H_{ij}}(x) dt\, dx$$

$$= B_i + \sum_{j=1}^{N} p_{ij} \int_{x=0}^{\tau} \int_{t=0}^{x} b_{ij}(t) f_{H_{ij}}(x) dt\, dx$$

where

$$B_i = \sum_{j=1}^{N} p_{ij} B_{ij}$$

Let $v_i(t)$ denote the total expected reward that will accrue by time t given that the process enters state i at time 0. Then, to compute $v_i(t)$, we observe that either by time t the process is still in state i or it has moved out of state i. In the first case, the holding time in state i is greater than t, whereas in the second case it is less than or equal to t. If the holding time in state i is less than t, then we assume that the process made a transition from state i to some state j at time τ and spent the remaining time $t - \tau$ in state j. Thus, we have

$$v_i(t) = B_i + \sum_{j=1}^{N} p_{ij} \int_{\tau=t}^{\infty} \int_{x=0}^{\tau} b_{ij}(x) f_{H_{ij}}(\tau) dx\, d\tau$$

$$+ \sum_{j=1}^{N} p_{ij} \int_{\tau=0}^{t} \left\{ \int_{x=0}^{\tau} b_{ij}(x) dx + v_j(t-\tau) \right\} f_{H_{ij}}(\tau) dx\, d\tau$$

$$= r_i(t) + \sum_{j=1}^{N} p_{ij} \int_{\tau=0}^{t} v_j(t-\tau) f_{H_{ij}}(\tau) d\tau \quad i = 1, 2, \ldots, N; \quad t \geq 0$$

where

$$r_i(t) = B_i + \sum_{j=1}^{N} p_{ij} \int_{\tau=t}^{\infty} \int_{x=0}^{\tau} b_{ij}(x) f_{H_{ij}}(\tau) dx\, d\tau + \sum_{j=1}^{N} p_{ij} \int_{\tau=0}^{t} \int_{x=0}^{\tau} b_{ij}(x) f_{H_{ij}}(\tau) dx\, d\tau$$

Thus, $r_i(t)$ is the total expected reward that will accrue by time t given that the process enters state i at time 0. Note that the expected total reward at state i from the instant the process enters the state until it leaves the state is given by

$$r_i(\infty) = B_i + \sum_{j=1}^{N} p_{ij} \int_{\tau=0}^{\infty} \int_{x=0}^{\tau} b_{ij}(x) f_{H_{ij}}(\tau) dx\, d\tau$$

In some applications, such as in the insurance industry, there is a fixed additional reward that is paid at the end of the process' holding time in a state. We will refer to this reward as the *terminal reward*. If we denote the terminal reward in state i by $v_i(0)$, then we have

$$v_i(t) = B_i + \sum_{j=1}^{N} p_{ij} \int_{\tau=t}^{\infty} \left\{ \int_{x=0}^{\tau} b_{ij}(x)dx + v_i(0) \right\} f_{H_{ij}}(\tau)d\tau$$

$$+ \sum_{j=1}^{N} p_{ij} \int_{\tau=0}^{t} \left\{ \int_{x=0}^{\tau} b_{ij}(x)dx + v_j(t-\tau) \right\} f_{H_{ij}}(\tau)dx\,d\tau$$

$$= r_i(t) + \sum_{j=1}^{N} p_{ij} \int_{\tau=0}^{t} v_j(t-\tau) f_{H_{ij}}(\tau)d\tau + v_i(0) \sum_{j=1}^{N} p_{ij}\{1 - F_{H_{ij}}(t)\}$$

Note that the terminal reward in state i is only associated with the case where the system still occupies state i at time t. Thus, if we are interested in terminating the process after time t, then $v_i(0)$ will be the additional reward. Note also that the preceding model includes many types of reward that a real system might not include at the same time. For example, while $b_{ij}(t)$ might be permitted in many models, we are likely to have either B_{ij} or $v_i(0)$, but not both, in a model.

13.3.2 Discounted Reward

Let $u(t)$ denote the rate at which rewards will be accumulated t units from now, $t \geq 0$. The present value of this reward is defined by

$$\int_0^{\infty} u(t)\,e^{-\beta t}dt = U^*(\beta)$$

where $U^*(s)$ is the Laplace transform of $u(t)$ and β is the continuous-time discount rate. Thus, when discounting is used, the parameters $r_i(\tau)$ and $v_i(t)$ will be denoted by $r_i(\tau, \beta)$ and $v_i(t, \beta)$ and are given by

$$r_i(t, \beta) = B_i + \sum_{j=1}^{N} p_{ij} \int_{\tau=t}^{\infty} \int_{x=0}^{\tau} e^{-\beta x} b_{ij}(x) f_{H_{ij}}(\tau)dx\,d\tau$$

$$+ \sum_{j=1}^{N} p_{ij} \int_{\tau=0}^{t} \int_{x=0}^{\tau} e^{-\beta x} b_{ij}(x) f_{H_{ij}}(\tau)dx\,d\tau$$

$$r_i(\infty, \beta) = B_i + \sum_{j=1}^{N} p_{ij} \int_{\tau=0}^{\infty} \int_{x=0}^{\tau} e^{-\beta x} b_{ij}(x) f_{H_{ij}}(\tau)dx\,d\tau$$

$$v_i(t, \beta) = r_i(t, \beta) + \sum_{j=1}^{N} p_{ij} \int_{\tau=0}^{t} e^{-\beta \tau} v_j(t-\tau) f_{H_{ij}}(\tau)d\tau$$

where B_i is as previously defined. For the case, where a terminal reward is paid out at the end of time t, we have

$$v_i(t,\beta) = r_i(t,\beta) + \sum_{j=1}^{N} p_{ij} \int_{\tau=0}^{t} e^{-\beta\tau} v_j(t-\tau) f_{H_{ij}}(\tau) d\tau + v_i(0) \sum_{j=1}^{N} p_{ij} \int_{\tau=t}^{\infty} e^{-\beta\tau} f_{H_{ij}}(\tau) d\tau$$

13.3.3 Analysis of the Continuous-Decision-Interval SMDPs

We first consider an SMDP without discounting. Following the model developed in the previous section, we assume that if the action a is taken when the system enters state i, it incurs an immediate cost or reward $B_{ij}(a)$, and an extra reward (or cost) accumulates at the rate of $b_{ij}(t,a)$ per unit time until the transition to state j occurs, where $B_{ij}(a) < \infty$ and $b_{ij}(t,a) < \infty$. In the remainder of the chapter, we assume that there is no terminal reward, that is, $v_i(0) = 0$ for all i. We modify the results obtained earlier and obtain $r_i(t,a)$, the total expected reward that will accrue by time t given that the process enters state i at time 0 and action a is taken, as follows:

$$r_i(t,a) = B_i(a) + \sum_{j=1}^{N} p_{ij}(a) \int_{\tau=t}^{\infty} \int_{x=0}^{\tau} b_{ij}(x,a) f_{H_{ij}}(\tau,a) dx\, d\tau$$

$$+ \sum_{j=1}^{N} p_{ij}(a) \int_{\tau=0}^{t} \int_{x=0}^{\tau} b_{ij}(x,a) f_{H_{ij}}(\tau,a) dx\, d\tau$$

where

$$B_i(a) = \sum_{j=1}^{N} p_{ij}(a) B_{ij}(a) \quad i = 1,2,\ldots,N$$

Therefore, the expected total reward between two decision epochs, given that the system occupies state i at the first epoch and action a was chosen at state i, is given by

$$r_i(\infty,a) = r_i(a) = B_i(a) + \sum_{j=1}^{N} p_{ij}(a) \int_{\tau=0}^{\infty} \int_{x=0}^{\tau} b_{ij}(x,a) f_{H_{ij}}(\tau,a) dx\, d\tau$$

Similarly, let $v_i(t,a)$ denote the total expected reward that will accrue by time t given that the process enters state i at time 0 and action a is used. Modifying the results obtained earlier, we have

$$v_i(t,a) = r_i(t,a) + \sum_{j=1}^{N} p_{ij}(a) \int_{\tau=0}^{t} v_j(t-\tau,a) f_{H_{ij}}(\tau,a) d\tau$$

Thus, if we denote the long-run expected total return when the initial state is i and action a is taken by $v_i(a)$, we obtain

$$v_i(a) = r_i(a) + \sum_{j=1}^{N} p_{ij}(a) v_j(a) \quad i = 1, 2, \ldots, N$$

The result can be expressed in a matrix form as follows. Let

$$V(a) = [\, v_1(a) \quad v_2(a) \quad \ldots \quad v_N(a)\,]^{\mathrm{T}}$$
$$R(a) = [\, r_1(a) \quad r_2(a) \quad \ldots \quad r_N(a)\,]^{\mathrm{T}}$$
$$P(a) = \begin{bmatrix} p_{11}(a) & p_{12}(a) & \ldots & p_{1N}(a) \\ p_{21}(a) & p_{22}(a) & \ldots & p_{2N}(a) \\ \vdots & \vdots & \ddots & \vdots \\ p_{N1}(a) & p_{N2}(a) & \ldots & p_{NN}(a) \end{bmatrix}$$

Then we obtain

$$V(a) = R(a) + P(a)V(a) \Rightarrow V(a) = [I - P(a)]^{-1} R(a)$$

Finally, the optimal long-run expected total return

$$v_i^* = \max_a \left\{ r_i(a) + \sum_{j=1}^{N} p_{ij}(a) v_j^* \right\} \quad i = 1, 2, \ldots, N$$

can be obtained by exhaustively solving for all the $v_i(a)$ and choosing a policy with the maximum expected return. However, this is usually an inefficient method particularly when either the state space or action space or both are large. A more efficient solution method is the policy iteration method.

13.3.4 Solution by Policy Iteration

The continuous-decision-interval SMDP is not easily amenable to analysis via the value iteration method. Thus, we discuss only the policy iteration method. But we must first define the average gain for the SMDP. Let $h_i(a)$ denote the expected time until the next decision epoch given that action a is chosen in the present state i. That is,

$$h_i(a) = \sum_{j=1}^{N} p_{ij}(a) h_{ij}(a) \quad i = 1, 2, \ldots, N$$

Let $Z(t)$ denote the total reward that accrues up to time t, $t \geq 0$. We state the following theorem:

Theorem 13.1 Let the embedded Markov chain associated with a policy d be denoted by $\{X_n | n = 0, 1, \ldots, \}$, where X_n is the state of the system at the nth decision epoch. If $\{X_n\}$ has no two disjoint closed communicating classes, then for each initial state $X_0 = i$, the limit

$$\lim_{t \to \infty} \left\{ \frac{Z(t)}{t} \right\} = g(d)$$

exists and is independent of the initial state i. Moreover, if $\{\pi_j(d)\}$ denotes the set of limiting state probabilities of $\{X_n\}$, then

$$g(d) = \frac{\sum_j r_j(d)\pi_j(d)}{\sum_j h_j(d)\pi_j(d)} = \frac{E[r(d)]}{E[h(d)]}$$

The proof of this theorem can be found in Ross (1970) and Tijms (2003). $g(d)$ is the long-run average reward per unit time when policy d is used and can also be defined as the ratio of the expected reward per occupancy to the expected time between transitions in the steady state.

Recall from our discussion on MDP that $v_i(t, a)$ can be expressed in terms of $g(a)$ as follows:

$$v_i(t, a) = tg(a) + v_i(a)$$

When t is large, $v_i(t, a)$ can be expressed as follows:

$$v_i(t, a) = r_i(a) + \sum_{j=1}^{N} p_{ij}(a) \int_{\tau=0}^{\infty} v_j(t - \tau, a) f_{H_{ij}}(\tau, a) d\tau$$

Substituting for $v_i(t, a)$ and $v_j(t - \tau, a)$ in the limiting state gives

$$tg(a) + v_i(a) = r_i(a) + \sum_{j=1}^{N} p_{ij}(a) \int_{\tau=0}^{\infty} \{(t - \tau)g(a) + v_j(a)\} f_{H_{ij}}(\tau, a) d\tau$$

$$= r_i(a) + tg(a) - g(a) \sum_{j=1}^{N} p_{ij}(a) \int_{\tau=0}^{\infty} \tau f_{H_{ij}}(\tau, a) d\tau + \sum_{j=1}^{N} p_{ij}(a) v_j(a)$$

$$= r_i(a) + tg(a) - g(a) \sum_{j=1}^{N} p_{ij}(a) h_{ij}(a) + \sum_{j=1}^{N} p_{ij}(a) v_j(a)$$

This gives

$$v_i(a) = r_i(a) - g(a) \sum_{j=1}^{N} p_{ij}(a)h_{ij}(a) + \sum_{j=1}^{N} p_{ij}(a)v_j(a) \quad i = 1, \ldots, N$$

If we define the mean holding time in state i, $h_i(a)$, by

$$h_i(a) = \sum_{j=1}^{N} p_{ij}(a)h_{ij}(a),$$

we obtain the result

$$v_i(a) + g(a)h_i(a) = r_i(a) + \sum_{j=1}^{N} p_{ij}(a)v_j(a) \quad i = 1, \ldots, N$$

That is,

$$g(a) = \frac{1}{h_i(a)} \left\{ r_i(a) + \sum_{j=1}^{N} p_{ij}(a)v_j(a) - v_i(a) \right\} \quad i = 1, \ldots, N$$

Because there are $N + 1$ unknowns (N $v_i(a)$ and 1 $g(a)$), we set one of the unknowns to zero and thus obtain only relative values.

As in MDP, we can summarize the solution algorithm as follows:

1. *Value-determination operation*: Use p_{ij} and r_i for the present policy to solve

$$g(d)h_i(d) + v_i(a) = r_i(d) + \sum_{j=1}^{N} p_{ij}(d)v_j(d) \quad i = 1, 2, \ldots, N$$

for all relative values v_i and r_i by setting $v_N = 0$.
2. *Policy-improvement routine*: For each state i, find the alternative policy d^* that maximizes

$$r_i(d) + \frac{1}{h_i(d)} \left\{ \sum_{j=1}^{N} p_{ij}(d)v_j(d) - v_i(d) \right\}$$

using the relative values v_i of the previous policy. Then d^* becomes the new policy in state i, $r_i(d^*)$ becomes r_i, and $p_{ij}(d^*)$ becomes p_{ij}.
3. *Stopping rule*: The optimal policy is reached (i.e., g is maximized) when the policies on two successive iterations are identical. Thus, if the current value of d is not the same as the previous value of d, go back to step 1, otherwise, stop.

13.3.5 SMDP with Discounting

In many applications, the expected discounted reward that the process will generate in some time interval might be important to the decision maker. For these applications, we deal with the expected present values instead of the total expected reward. Let $r_i(t, a, \beta)$ be the equivalent of $r_i(t, a)$ when the discount factor is β. Similarly, let $v_i(t, a, \beta)$ be the equivalent of $v_i(t, a)$ when the discount factor is β. Then from earlier results, we have

$$r_i(t, a, \beta) = B_i(a) + \sum_{j=1}^{N} p_{ij}(a) \int_{\tau=t}^{\infty} \int_{x=0}^{\tau} e^{-\beta x} b_{ij}(x, a) f_{H_{ij}}(\tau, a) dx \, d\tau$$

$$+ \sum_{j=1}^{N} p_{ij}(a) \int_{\tau=0}^{t} \int_{x=0}^{\tau} e^{-\beta x} b_{ij}(x, a) f_{H_{ij}}(\tau, a) dx \, d\tau$$

$$v_i(t, a, \beta) = r_i(t, a, \beta) + \sum_{j=1}^{N} p_{ij}(a) \int_{\tau=0}^{t} e^{-\beta \tau} v_j(t - \tau, a, \beta) f_{H_{ij}}(\tau, a) d\tau$$

The long-run values are given by

$$r_i(a, \beta) = B_i(a) + \sum_{j=1}^{N} p_{ij}(a) \int_{\tau=0}^{\infty} \int_{x=0}^{\tau} e^{-\beta x} b_{ij}(x, a) f_{H_{ij}}(\tau, a) dx \, d\tau$$

$$v_i(a, \beta) = r_i(a, \beta) + \sum_{j=1}^{N} p_{ij}(a) \int_{\tau=0}^{\infty} e^{-\beta \tau} f_{H_{ij}}(\tau, a) v_j(a, \beta) d\tau$$

$$= r_i(a, \beta) + \sum_{j=1}^{N} p_{ij}(a) M_{H_{ij}}(a, \beta) v_j(a, \beta)$$

where $M_{H_{ij}}(a, s)$ is the s-transform of $f_{H_{ij}}(\tau, a)$. If we define

$$\theta_{ij}(a, \beta) = p_{ij}(a) M_{H_{ij}}(a, \beta)$$

we obtain the result

$$v_i(a, \beta) = r_i(a, \beta) + \sum_{j=1}^{N} \theta_{ij}(a, \beta) v_j(a, \beta)$$

This result can be expressed in a matrix form as follows. Let

$$V(a, \beta) = [\, v_1(a, \beta) \quad v_2(a, \beta) \quad \cdots \quad v_N(a, \beta) \,]^{\mathrm{T}}$$
$$R(a, \beta) = [\, r_1(a, \beta) \quad r_2(a, \beta) \quad \cdots \quad r_N(a, \beta) \,]^{\mathrm{T}}$$

$$\Theta(a, \beta) = \begin{bmatrix} \theta_{11}(a, \beta) & \theta_{12}(a, \beta) & \cdots & \theta_{1N}(a, \beta) \\ \theta_{21}(a, \beta) & \theta_{22}(a, \beta) & \cdots & \theta_{2N}(a, \beta) \\ \vdots & \vdots & \ddots & \vdots \\ \theta_{N1}(a, \beta) & \theta_{N2}(a, \beta) & \cdots & \theta_{NN}(a, \beta) \end{bmatrix}$$

Then we have

$$V(a, \beta) = R(a, \phi) + \Theta(a, \beta)V(a, \beta) \Rightarrow V(a, \beta) = [I - \Theta(a)]^{-1}R(a, \beta)$$

Thus, for a given discount factor and a given action, we can obtain the expected discounted reward from the preceding matrix form. The optimal expected total long-run return is given by

$$v_i^*(\beta) = \max_a \left\{ r_i(a, \beta) + \sum_{j=1}^{N} \theta_{ij}(a, \beta)v_j^*(\beta) \right\} \quad i = 1, 2, \ldots, N$$

As discussed earlier, we can use the matrix method to obtain the result by an exhaustive search of all the possible solutions, which is not an efficient method of solving the problem particularly when either the state space or the action space or both are large. However, a more efficient solution is the policy iteration method.

13.3.6 Solution by Policy Iteration When Discounting Is Used

When t is large, we have

$$v_i(t, a, \beta) = r_i(a, \beta) + \sum_{j=1}^{N} p_{ij}(a) \int_{\tau=0}^{\infty} v_j(t - \tau, a, \beta) \, e^{-\beta\tau} f_{H_{ij}}(\tau, a)d\tau$$

Substituting for $v_i(t, a, \beta)$ and $v_j(t - \tau, a, \beta)$ in the limiting state gives

$$tg(a) + v_i(a, \beta) = r_i(a, \beta) + \sum_{j=1}^{N} p_{ij}(a) \int_{\tau=0}^{\infty} \{(t - \tau)g(a) + v_j(a, \beta)\} \, e^{-\beta\tau} f_{H_{ij}}(\tau, a)d\tau$$

$$= r_i(a, \beta) + tg(a) - g(a)\sum_{j=1}^{N} p_{ij}(a) \int_{\tau=0}^{\infty} \tau \, e^{-\beta\tau} f_{H_{ij}}(\tau, a)d\tau$$

$$+ \sum_{j=1}^{N} p_{ij}(a)\left\{ \int_{\tau=0}^{\infty} e^{-\beta\tau} f_{H_{ij}}(\tau, a)d\tau \right\} v_j(a, \beta)$$

$$= r_i(a, \beta) + tg(a) + g(a)\sum_{j=1}^{N} p_{ij}(a) \frac{dM_{H_{ij}}(a, \beta)}{d\beta}$$

$$+ \sum_{j=1}^{N} p_{ij}(a)v_j(a, \beta)M_{H_{ij}}(a, \beta)$$

This gives

$$v_i(a, \beta) = r_i(a, \beta) + g(a)\sum_{j=1}^{N} p_{ij}(a) \frac{dM_{H_{ij}}(a, \beta)}{d\beta} + \sum_{j=1}^{N} p_{ij}(a)v_j(a, \beta)M_{H_{ij}}(a, \beta)$$

That is,

$$g(a) = \frac{1}{\sum_{j=1}^{N} p_{ij}(a)(dM_{H_{ij}}(a,\beta)/d\beta)} \left\{ v_i(a,\beta) - r_i(a,\beta) - \sum_{j=1}^{N} p_{ij}(a)v_j(a,\beta)M_{H_{ij}}(a,\beta) \right\}$$

Note that the calculation involves $g(d_k)$, which makes it different from the MDP with discounting where $g(a)$ does not appear in the calculations thereby enabling us to obtain the absolute values of $v_i(a,\beta)$ because we have N equations and N unknowns. In the SMDP with discounting, we can obtain only the relative values of the $v_i(a,\beta)$ because we have N equations and $N + 1$ unknowns. Thus, the solution method is similar to that of the undiscounted system, which is the reason why we have to set one of the $v_i(a,\beta)$ to 0.

13.3.7 Analysis of the Discrete-Decision-Interval SMDPs with Discounting

The development of the discrete-decision-interval SMDP parallels that of its continuous counterpart. The primary difference is the fact that the holding times have PMFs instead of PDFs, consequently, summations replace integrations. Also, the discounting process is defined slightly differently. The present value of a reward r_n received n time units from now is $\beta^n r_n$.

Assume that the system has just entered state i with n time periods remaining. Let $v_i(n, a, \beta)$ denote the total expected reward that will accrue in the next n time periods after the process enters state i, action a is taken, and the discount factor is β. Then, as in the discussion on SMP rewards, to compute $v_i(n, a, \beta)$, we observe that either by the next n time periods the process is still in state i or it has moved out of state i. In the first case the holding time in state i is greater than n, whereas in the second case it is less than or equal to n. If the holding time in state i is less than n, then we assume that the process made a transition from state i to some state j at time $m < n$ and spent the remaining time $n - m$ in state j. Thus, if we denote the contribution of the first case by $v_G(n, a, \beta)$ and the contribution of the second case by $v_L(n, a, \beta)$, we have

$$v_G(n, a, \beta) = \sum_{j=1}^{N} p_{ij}(a) \sum_{m=n+1}^{\infty} p_{H_{ij}}(m, a) \sum_{k=1}^{n} \beta^k b_{ij}(k, a)$$

$$v_L(n, a, \beta) = \sum_{j=1}^{N} p_{ij}(a) \sum_{m=1}^{n} p_{H_{ij}}(m, a) \left\{ \sum_{k=1}^{n} \beta^k b_{ij}(k, a) + \beta^m v_j(n - m, a, \beta) \right\}$$

$$v_i(n, a, \beta) = B_i(a) + v_G(n, a, \beta) + v_L(n, a, \beta) \quad i = 1, 2, \ldots, N; \quad n = 1, 2, \ldots$$

If we define $v_i(n, \beta)$ as the maximum expected present value of the future rewards that can accrue during an interval of length n given that the process has just entered state i and the discount factor is β, we obtain

$$v_i(n, \beta) = \max_a \{B_i(a) + v_G(n, a, \beta) + v_L(n, a, \beta)\} \quad i = 1, 2, \ldots, N$$

13.3.8 Continuous-Time Markov Decision Processes

A continuous-time MDP is a semi-Markov MDP in which the times between decision epochs are exponentially distributed. This means that the PDFs $f_{H_{ij}}(a, \tau)$ are exponential functions:

$$f_{H_{ij}}(a, \tau) = \lambda_{ij}^a e^{-\lambda_{ij}^a \tau} \quad \tau \geq 0$$

Thus,

$$E[H_{ij}(a)] = h_{ij}(a) = \frac{1}{\lambda_{ij}^a}$$

$$h_i(a) = \sum_{j=1}^{N} p_{ij}(a) h_{ij}(a) = \sum_{j=1}^{N} \left\{ \frac{p_{ij}(a)}{\lambda_{ij}^a} \right\}$$

$$M_{H_{ij}}(a, \beta) = \frac{\lambda_{ij}^a}{\beta + \lambda_{ij}^a}$$

$$\frac{dM_{H_{ij}}(a, \beta)}{d\beta} = -\frac{\lambda_{ij}^a}{(\beta + \lambda_{ij}^a)^2}$$

$$\Theta(a, \beta) = \begin{bmatrix} \dfrac{p_{11}(a)\lambda_{11}^a}{\beta + \lambda_{11}^a} & \dfrac{p_{12}(a)\lambda_{12}^a}{\beta + \lambda_{12}^a} & \cdots & \dfrac{p_{1N}(a)\lambda_{1N}^a}{\beta + \lambda_{1N}^a} \\[2ex] \dfrac{p_{21}(a)\lambda_{21}^a}{\beta + \lambda_{21}^a} & \dfrac{p_{22}(a)\lambda_{22}^a}{\beta + \lambda_{22}^a} & \cdots & \dfrac{p_{2N}(a)\lambda_{2N}^a}{\beta + \lambda_{2N}^a} \\[2ex] \vdots & \vdots & \vdots & \vdots \\[2ex] \dfrac{p_{N1}(a)\lambda_{N1}^a}{\beta + \lambda_{N1}^a} & \dfrac{p_{N2}(a)\lambda_{N2}^a}{\beta + \lambda_{N2}^a} & \cdots & \dfrac{p_{NN}(a)\lambda_{NN}^a}{\beta + \lambda_{NN}^a} \end{bmatrix}$$

13.3.9 Applications of SMDPs

SMDPs are used to model admission control in a G/M/1 queueing system. In this system, an agent regulates the system load by accepting or rejecting arriving customers. A rejected customer leaves the system without receiving service. An example is call arrival at a switchboard. Service times are assumed to be exponentially distributed, and each arriving customer brings a reward R. The system incurs a holding cost $c(n)$ when there are n customers in the system.

Another application is the service rate control in an M/G/1 queueing system. Here, an agent regulates the system load by varying the service rate; faster service rate is more expensive than a slower service rate. An agent changes the service rate upon completion of a service or upon the arrival of a customer to an empty system. There is a fixed cost K associated with switching the service rate and a holding cost $c(n)$ when there are n customers.

SMDPs are also used to model optimal control of queues. In this case, a service system has c identical servers, and the number of servers active at any time is controlled by an agent. Thus, it can be likened to a system of c service channels where a channel can be activated by turning it on and deactivated by turning it off. Examples include supermarket checkout counters, bank tellers, and production facilities. An active channel can service only one request at a time. A nonnegative cost $K(m,n)$ is incurred when the number of active channels is changed from m to n. There is an operation cost rate $r \geq 0$ per unit time associated with an active channel. Similarly, a holding cost $h \geq 0$ is incurred for each time unit a customer is in the system until its service is completed. The objective is to find a rule for controlling the number of active service channels to minimize the long-run average cost per unit time. Decision epochs are instants of service completion or epochs of a new arrival.

13.4   Partially Observable MDPs

In MDPs, the sequence of actions taken to make decisions assumes that the environment is completely observable and the effects of actions taken are deterministic. That is, in MDP, it is assumed that at the decision epoch the state i, transition probabilities $\{p_{ij}(a)\}$, and immediate returns $\{r_i(a)\}$ are all known. However, the real world is not always completely observable, which means that the effects of actions taken are often nondeterministic. Decision making in such environments can be modeled by a POMDP. That is, in POMDP, $\{p_{ij}(a)\}$ and $\{r_i(a)\}$ are all known at the decision epoch, but the state is not known precisely. Instead the agent has some observations from which to infer the probability of the system being in some state. From these observations, the agent takes an action that results in a reward. The reward received after an action is taken provides information on how good the action was.

As stated earlier, in a POMDP, the agent chooses and executes an action at the decision epoch based on information from the past observations and past actions and the current observation. Unfortunately, the amount of memory required to store past observations and past actions can be large thereby making it difficult to maintain past information after a long period of time. This difficulty is usually overcome by maintaining the agent's *belief state* instead of its past information.

A belief state is the probability distribution over the hidden states of the Markov process given the past history of observations and action. Thus, the belief state captures all the information contained in past information and current observation that is useful for selecting an action. Because the number of possible states of the environment is finite, maintaining the belief state is simpler than keeping track of all past information. Note that the fact that the belief state is defined in terms of probability distribution implies that the agent's knowledge is incomplete. Also, using the concept of belief state allows the POMDP to satisfy the Markov property because knowing the current belief state is all we need to predict the future. When the agent observes the current state of the environment, it updates its belief state.

POMDP had its origin in the field of operations research but has now been applied in many other disciplines. For example, it has been applied in machine maintenance and quality control, medical diagnosis, and the search for moving objects, which are areas where the observation of the state of the underlying Markov process is not always possible. Drake (1962) formulated the first explicit POMDP problem. According to this model, the decision maker gains some probabilistic information about the state after every decision he makes. For a given probability distribution on the initial state the decision maker can revise the distribution according to Bayes' rule. Dynkin (1965) and Aoki (1967) have also shown that using this Bayesian approach to estimate the state the process occupies enables the problem to be transformed to an MDP with complete information. In Hauskrecht and Fraser (2000), POMDP is applied to the planning of the treatment of ischemic heart disease. Cassandra et al. (1994), Cassandra (1998), Dean et al. (1995), Shatkay (1999), and Simmons and Koenig (1995) have used it to model robotic navigation within buildings. Jonas (2003) has also used it to model speech recognition. Different algorithms and applications of POMDP are discussed by Zhang (2001).

One important application of POMDP is the *reinforcement learning* problem. As defined by Sutton and Barto (1998), reinforcement learning is learning what to do, without the intervention of a supervisor, to maximize a numerical reward. Thus, the agent is not told what to do, as in *supervised learning*, but discovers which actions yield the most reward by trying them. We begin our discussion on POMDP by discussing the partially observable Markov process (POMP).

13.4.1 Partially Observable Markov Processes

Consider a discrete-state Markov process that can be in one of two states: S_1 and S_2. Given that it is currently in state S_1, it will enter state S_1 again with probability p_{11} and state S_2 next with probability $p_{12} = 1 - p_{11}$. Similarly, given that it is currently in state S_2, it will enter state S_2 again with probability p_{22}, and state S_1 next with probability $p_{21} = 1 - p_{22}$. Assume that the dynamics of the process is being observed through an imperfect medium that allows us to

State space

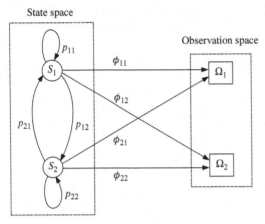

observe two states: Ω_1 and Ω_2. Let the conditional probability ϕ_{ij} be the probability that the process is actually in state S_i given that the observable state is $\Omega_j, i, j = 1, 2$, where $\phi_{i1} + \phi_{i2} = 1, i = 1, 2$. Figure 13.4 represents the state transition diagram of the Markov process with partial observability, which is called the POMP.

There are two processes involved in POMP: the *core* process and the *observation* process. The core process is the underlying Markov process whose states are the S_i and the transition probabilities are the p_{ij}. The observation process is the process whose states Ω_i are in the observation space. In the preceding example, one can interpret Ω_i by the statement "the core process seems to be in state S_i." Note that the preceding example assumes that there is a one-to-one correspondence between the core process and the observation process, even though we are not able to link Ω_i with certainty to S_i. In a more general case, the core process can have n states, whereas the observation process has m states, where $m \neq n$.

The goal in the analysis of the POMP is to estimate the state of the Markov process given an observation or a set of observations. We assume that the observation space has no memory, that is, the observations are not correlated and are thus made independently of one another. The estimation can be based on considering each observation independently, using the Bayes' rule. For the problem shown in Figure 13.4, assume that the steady-state probability that the underlying Markov process is in state S_i at a random time n is $P[S_1(n)] = P_1$ and $P[S_2(n)] = P_2 = 1 - P_1$. Let the function

$$\arg \max_y \{z\}$$

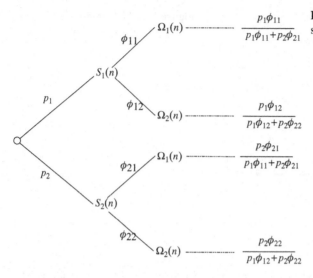

Figure 13.5 Sample space of state estimation process.

denote the argument y that corresponds to the maximum of the expression z. Also, let $\Omega_j(n)$ denote the event that the nth observation is Ω_j, and let $\hat{S}(\Omega_j(n))$ denote our estimate of the state as a result of $\Omega_j(n)$. Then the decision criterion becomes

$$\hat{S}(\Omega_j(n)) = \arg\max_{S_i}\{P[S_i(n)|\Omega_j(n)]\} = \arg\max_{S_i}\left\{\frac{P[S_i(n)]P[\Omega_j(n)|S_i(n)]}{P[\Omega_j(n)]}\right\}$$

$$= \arg\max_{S_i}\left\{\frac{P[S_i(n)]P[\Omega_j(n)|S_i(n)]}{P[S_i(n)]P[\Omega_j(n)|S_i(n)] + P[\bar{S}_i(n)]P[\Omega_j(n)|\bar{S}_i(n)]}\right\}$$

$$= \arg\max_{S_i}\left\{\frac{P_i\phi_{ij}}{P_i\phi_{ij} + (1 - P_i)\phi_{kj}}\right\}$$

where S_k is the state of the underlying Markov process with steady-state probability $P_k = 1 - P_i$. Applying this to the preceding example, we obtain the sample space as shown in Figure 13.5.

If we assume that $\phi_{11} > \phi_{12}, \phi_{22} > \phi_{21}$ and $P_1 = P_2$, then the decoding rule becomes that when the observed state is Ω_1 we consider the state of the core process to be S_1, and when the observed state is Ω_2 we assume that the state of the core process is S_2.

13.4.2 POMDP Basics

In POMDP, the decision maker or agent has to solve two problems simultaneously, namely, a control problem like that of a standard MDP, and an identification problem for the unobserved states. Each time the agent takes an action, the transition to a new state implicitly provides new information about the underlying state of the process. This new knowledge can enable the agent to make the next decision. Thus, starting with an initial probability distribution, the agent revises the distribution after every transition to take into consideration the new information provided by the observation resulting from the transition. This revised distribution is called the *posterior distribution*, which is used to identify the unobserved state and to control the system at the next decision epoch.

More formally, a POMDP is a probabilistic model that can be represented by the 6-tuple $(S, A, \Omega, P, \Phi, R)$, where

- S is a finite set of N states of the core process; that is, $S = \{s_1, s_2, \ldots, s_N\}$. The state at time t is denoted by S_t.
- A is a finite set of K actions that can be taken at any state; that is, $A = \{a_1, a_2, \ldots, a_K\}$. The action taken at time t is denoted by A_t.
- Ω is the a finite set of M observations that can be made; that is, $\Omega = \{o_1, o_2, \ldots, o_M\}$. The observation at time t is denoted by Ω_t.
- P is the transition probability matrix, which can be different for each action. As in the case of MDP, for action $a \in A$, we denote the probability that the system moves from state s_i to state s_j when action a is taken by $p_{ij}(a)$, which is independent of the history of the process up to the time the action was taken. That is,

$$p_{ij}(a) = P[S_{t+1} = j | S_t = i, A_t = a]$$

As stated earlier, it is assumed that the $p_{ij}(a)$ are known, but state s_i is not known at the decision epoch; it is inferred from the observation.

- Φ is the set of observation probabilities that describe the relationship between the observations, states of the core process and actions. We let $\phi_{ij}(a)$ denote the probability of observing the state $o_j \in \Omega$ after action a is taken and the core process enters state s_i. That is,

$$\phi_{ij}(a) = P[\Omega_t = o_j | S_t = s_i, A_{t-1} = a]$$

- R is the reward function, which can vary with the action taken. The reward at time t is denoted by R_t. The reward that the agent receives by taking action $a \in A$ in state s_i that results in a transition to state s_j is denoted by $r_{ij}(a)$. The total reward associated with action a in state is

$$r_i(a) = \sum_{s_j \in S} r_{ij}(a) p_{ij}(a)$$

Assume that the core process is in state S_t at time t. Because POMDP is based on a core process that is a Markov process, the current state S_t is sufficient to predict the future independently of the past states $\{S_0, S_1, \ldots, S_{t-1}\}$. As stated earlier, the state S_t is not directly observable but can only be inferred from the observations $\{\Omega_1, \Omega_2, \ldots, \Omega_t\}$. To help in making the determination of the state of the system, the agent keeps a complete trace of all observations and all actions it has taken and uses this information to choose actions to take next. The joint trace of actions and observations constitutes a *history* at time t, which is denoted by H_t, and defined by

$$H_t = \{A_0, \Omega_1, A_1, \Omega_2, \ldots, A_{t-1}, \Omega_t\}$$

Fortunately, this history does not need to be represented explicitly but can be summarized via a *belief distribution* $b_t(s)$, which is defined by

$$b_t(s) = P[S_t = s | \Omega_t, A_{t-1}, \Omega_{t-1}, \ldots, A_0, b_0]$$

Thus, $0 \le b_t(s) \le 1$ is the probability that the process is in state $S_t = s$ given the belief distribution b. That is, $b_t(s_j)$ is the agent's estimate that the core process is in state $S_t = s_j$. Therefore, based on the current belief state, the agent chooses an action a and receives the reward $r_{jk}(a)$ and the core process makes a transition to state s_k that leads to the observation o_m. This is illustrated in Figure 13.6, where the component labeled SE is the state estimator that takes as input the last belief state, the most recent action and the most recent observation and returns an updated belief state. The component labeled "d" represents the policy.

The initial state probability distribution, which defines the probability that the system is in state s at time $t = 0$, is given by

$$b_0(s) = P[S_0 = s]$$

Methods used to solve POMDPs are sometimes called reinforcement learning algorithms because the only feedback provided to the agent is a scalar reward signal at each step. One important feature of b_t is the fact that because it is a sufficient statistic for the history of the process, we can use it as a criterion for selecting actions. Also, it is computed recursively, using only the immediate past value, b_{t-1}, together with the most recent action A_{t-1} and the observation Ω_t. If we denote the

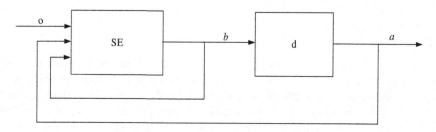

Figure 13.6 Structure of the state estimation process.

belief state for state $S_t = s_k$ at time t by $b_t(s_k)$, then based on A_{t-1} and Ω_t, the belief distribution is updated via the following Bayes' rule:

$$
\begin{aligned}
b_t(s_k) &= P[S_t = s_k | \Omega_t = o_m, A_{t-1} = a, \ldots, A_0] = P[s_k | o_m, a, b_{t-1}(s_j)] \\
&= \frac{P[s_k, o_m, a, b_{t-1}(s_j)]}{P[o_m, a, b_{t-1}(s_j)]} = \frac{P[o_m | s_k, a, b_{t-1}(s_j)]P[s_k, a, b_{t-1}(s_j)]}{P[o_m | a, b_{t-1}(s_j)]P[a, b_{t-1}(s_j)]} \\
&= \frac{P[o_m | s_k, a, b_{t-1}(s_j)]P[s_k | a, b_{t-1}(s_j)]P[a, b_{t-1}(s_j)]}{P[o_m | a, b_{t-1}(s_j)]P[a, b_{t-1}(s_j)]} \\
&= \frac{P[o_m | s_k, a, b_{t-1}(s_j)]P[s_k | a, b_{t-1}(s_j)]}{P[o_m | a, b_{t-1}(s_j)]} \\
&= \frac{P[o_m | s_k, a] \sum_{s \in S} P[s_k | a, b_{t-1}(s), s]P[s | a, b_{t-1}(s)]}{P[o_m | a, b_{t-1}(s_j)]} \\
&= \frac{P[o_m | s_k, a] \sum_{s_j \in S} P[s_k | a, s_j]b_{t-1}(s_j)}{P[o_m | a, b_{t-1}(s_j)]} \\
&= \frac{\phi_{km}(a) \sum_{s_j \in S} p_{jk}(a)b_{t-1}(s_j)}{P[o_m | a, b_{t-1}(s_j)]}
\end{aligned}
$$

The denominator is independent of s_k and can be regarded as a normalizing factor. The numerator contains the observation function, the transition probability, and the current belief state. Thus, we can write

$$
b_t(s_k) = \gamma \phi_{km}(a) \sum_{s_j \in S} p_{jk}(a)b_{t-1}(s_j)
$$

where γ is a normalizing constant. Because the belief b_t at time t is computed recursively using the belief b_{t-1} as well as the most recent observation Ω_t and the most recent action A_{t-1}, we can define the belief update procedure by the following operation:

$$
b_t(s) = \tau(b_{t-1}, A_{t-1}, \Omega_t)
$$

where $\tau(b_{t-1}, A_{t-1}, \Omega_t)$ is called the *belief updating function*. This shows that given a belief state, its successor belief state is determined by the action and observation.

13.4.3 Solving POMDPs

POMDP can be considered an MDP that is defined over the belief state space. Thus, when the belief state is updated via the action and observation, the solution methods used in MDP can then be used to obtain the optimal solution. In particular, the immediate reward associated with action a and belief state b is given by

$$r(b,a) = \sum_{s_i \in S} \sum_{s_j \in S} r_{ij}(a) p_{ij}(a) b(s_i) = \sum_{s_i \in S} r_i(a) b(s_i)$$

Let $v_t(b)$ denote the value function associated with belief b at time $t = 0, 1, \ldots$. If β is the discount factor, then using the method described by Sondik (1971), we apply multiple iterations of DP to compute more accurate values of the value function for each belief state. Thus, we have

$$v_0^*(b) = \max_{a \in A} r(b,a)$$

$$v_t^*(b) = \max_{a \in A} \left\{ r(b,a) + \beta \sum_{o \in \Omega} P[o|b,a] v_{t-1}^*(\tau(b,a,o)) \right\}$$

$$= \max_{a \in A} \left\{ r(b,a) + \beta \sum_{s_k \in S} P[s_k|s_i,a] \sum_{o_j \in \Omega} P[o_j|s_k,a] b(s_i) v_{t-1}^*(\tau(b,a,o)) \right\}$$

$$= \max_{a \in A} \left\{ r(b,a) + \beta \sum_{s_k \in S} p_{ik}(a) \sum_{o_j \in \Omega} \phi_{kj}(a) b(s_i) v_{t-1}^*(\tau(b,a,o)) \right\}$$

where $\tau(b,a,o)$ is the belief updating function. Unfortunately, the state space for this MDP is an $|S|$-dimensional continuous space that is more complex than traditional MDPs with discrete state space.

13.4.4 Computing the Optimal Policy

As defined earlier, a policy is a sequence of decisions, and an optimal policy is a policy that maximizes the expected discounted return. Recall that we denote a policy by d, and for a given belief state b, a policy is of the form $d(b) \rightarrow a \in A$. The optimal policy is given by

$$d^*(b_t) = \arg\max_{a \in A} \left\{ r(b,a) + \beta \sum_{o \in \Omega} P[o|b,a] v_{t-1}^*(\tau(b,a,o)) \right\}$$

$$= \arg\max_{a \in A} \left\{ r(b,a) + \beta \sum_{s_k \in S} p_{ik}(a) \sum_{o_j \in \Omega} \phi_{kj}(a) b(s_i) v_{t-1}^*(\tau(b,a,o)) \right\}$$

where β is the discount factor.

13.4.5   Approximate Solutions of POMDP

As pointed out by Papadimitrios and Tsitsiklis (1987) and Madani et al. (2003), the exact solution of POMDP is usually hard except for very simple cases. The reasons why POMDP problems are more difficult than fully observable MDP problems are as follows. First, in a fully observable MDP, an agent knows exactly the current state of its environment, which means that information from the past (i.e., its past observations and actions) is irrelevant to the current decision. This is precisely the Markov property. Because the agent does not fully observe the state of its environment in POMDP, past information becomes very relevant as it can help the agent to better estimate the current state of its environment. Unfortunately, the number of possible states of past information increases with time, and this presents computational difficulties.

Second, the effects of actions in MDP are fully observable at the next decision epoch. In POMDP, the effects of an action are not fully observable at the next decision epoch, which means that we cannot clearly tell the effects of the current action from those of the agent's future behavior. The ability to make this distinction requires looking into the future and considering the combination of each action with each of the agent's possible behaviors in a potentially large number of steps. This again becomes computationally involved because the number of ways that the agent can behave can be exponential in the number of future steps considered.

A variety of algorithms have been developed for solving POMDPs. Unfortunately, most of these techniques do not scale well to problems involving more than a few states due to their computational complexity. As a result of this problem, different approximate solutions have been proposed. These approximate solutions can be grouped into two classes, namely:

- Those solutions where approximation takes place in the process of solving the POMDP.
- Those solutions that use model approximation. Such solutions approximate POMDP itself by another problem that is easier to solve and use the solution of the latter to construct an approximate solution to the original POMPD. The approximation can be in different forms, including developing a more informative observation model, a more deterministic action model, a simpler state space, or a combination of two or all of the three alternatives.

More detailed information on these approximate solutions can be found in Monahan (1982), Lovejoy (1991), Zhang and Liu (1997), Yu (2006), and Wang (2007).

13.5   Problems

13.1 A recent college graduate is presented with N job offers, one after another. After looking at an offer, she must either accept it and thus terminate the process or reject it. A rejected offer is lost forever. The only information she has at any time is the relative rank of the current offer compared to the previous one. Assuming that the N jobs are

offered to her in a random order, which means that orderings of the offers are equally
likely, define the problem as a sequential decision problem where the objective is to
maximize the probability of selecting the best offer.

13.2 The price of a certain stock is fluctuating among $10, $20, and $30 from month to
month. Market analysis indicates that given that the stock is at $10 in the current
month, then in the following month it will be at $10 with probability 0.8 and at $20
with probability 0.2. Similarly, given that the stock is at $20 in the current month, then
in the following month it will be at $10 with probability 0.25, at $20 with probability
0.50, and at $30 with probability 0.25. Finally, given that the stock is at $30 in the current month, then in the following month it will be at $20 with probability 0.625 and at
$30 with probability 0.375. Given a discount factor of 0.9, use the policy improvement
method to determine when to sell and when to hold the stock to maximize the expected
long-run total discounted profit.

13.3 A farmer is considering the optimal course of action for his farm each year. The two
options are to fertilize the farm and not to fertilize the farm. Optimality is defined such
that the farmer will realize the highest expected revenue at the end of 4 years. The conditions (or states) of the farm are good (state 1), fair (state 2), and poor (state 3). If P_k
and R_k ($k = 1, 2$), which represent the transition probability matrix and the reward function matrix, respectively, are given by

$$P_1 = \begin{bmatrix} 0.2 & 0.5 & 0.3 \\ 0.0 & 0.5 & 0.5 \\ 0.0 & 0.0 & 1.0 \end{bmatrix} \quad R_1 = \begin{bmatrix} 7 & 6 & 3 \\ 0 & 5 & 1 \\ 0 & 0 & 1 \end{bmatrix}$$

$$P_2 = \begin{bmatrix} 0.2 & 0.6 & 0.1 \\ 0.1 & 0.6 & 0.3 \\ 0.1 & 0.4 & 0.5 \end{bmatrix} \quad R_2 = \begin{bmatrix} 6 & 5 & -1 \\ 7 & 4 & 0 \\ 6 & 3 & -2 \end{bmatrix}$$

determine the optimal expected revenue and the optimal decision in each of the 4
years.

13.4 Solve Problem 13.3, assuming a discount factor of 0.9.

13.5 Consider a salesman who has offices in two towns called town 1 and town 2. He can
be in one of these towns on any particular day but cannot split his time on any day
between the two. On any day that he works in town 1, the probability of making a sale
is 0.4; similarly, on any day that he is in town 2 the probability that he makes a sale is
0.25. The reward for making any sale is $100 and the cost of switching towns is $50.
The salesman is considering two operating alternatives:
a. Stay in each town until he makes a sale and then go to the next town.
b. Work in one town one day and then go to the next town the next day whether or
not a sale is made.

Define the four possible policies for the problem and find the salesman's long-run
expected profit per day.

14 Hidden Markov Models

14.1 Introduction

With the exception of partially observable Markov processes, all the Markov models we have considered up until now have visible states in the sense that the state sequence of the processes is known. Thus, we can refer to these models as *visible Markov models*. In this chapter, we consider a process in which the state sequence that the process passes through is not known but can only be guessed through a sequence of observations of the dynamics of the process. In the previous chapter, we considered a slightly different aspect of this model that we defined as the partially observable Markov decision process (POMDP). We devote this chapter to discussing another aspect of the model, which is called the *hidden Markov model* (HMM). POMDP differs from HMM by the fact that in POMDP we have control over the state transitions, whereas in HMM we do not have this control.

As in POMDP, an HMM assumes that the underlying process is a Markov chain whose internal states are hidden from the observer. It is usually assumed that the number of states of the system and the state-transition probabilities are known. Thus, there are two parameters associated with each state of the Markov chain:

- Symbol emission probabilities that describe the probabilities of the different possible outputs from the state.
- Transition probabilities that describe the probability of entering a new state from the current state.

The visible Markov models have limited power in modeling many applications. Their limitation arises from the fact that they assume perfect knowledge of the system's internal dynamics and/or that a decision maker can control the system evolution through some well-defined policy. Unfortunately, many applications do not conform to either of these two assumptions. For such applications, the HMM can be used. HMMs are used in a variety of applications, but the two most important application areas are speech recognition and biological sequence analysis like deoxyribonucleic acid (DNA) sequence modeling.

HMMs were first applied to speech recognition in the early 1970s. The use of HMMs in speech recognition is reported in Levinson et al. (1983) and Juang and Rabiner (1991). In many languages, when the same word is pronounced in different contexts, at different times, and by different people, the sound can be extremely

Markov Processes for Stochastic Modeling. DOI: http://dx.doi.org/10.1016/B978-0-12-407795-9.00014-1

variable. In speech recognition, HMMs are used to characterize the sound signal in a word in probabilistic terms. A speech signal can be represented as a long sequence of about 256 category labels, such as the phonemes, that are valid for a particular language. From this set, a speech recognition system has to determine what word was spoken. A well-trained speech recognition system assigns high probability to all sound sequences that are the likely utterances of the word it models and low probability to any other sequence.

Applications of HMMs in bioinformatics have been reported in Thompson (1983), Churchill (1989, 1992), Guttorp et al. (1990), Baldi et al. (1994), and Krogh et al. (1994). Another area of application of HMMs is financial time series modeling like the stock market. Ryden et al. (1998) used them to model temporal and distributional properties of daily data from speculative markets. Elliott and van der Hoek (1997) applied them to asset allocation problems.

Also, Albert (1991) and Le et al. (1992) have used HMMs to model time series of epileptic seizure counts. Similarly, Leroux and Puterman (1992) apply them to the pattern movement of a fetal lamb. HMMs have also been applied to hydroclimatology by Zucchini and Guttorp (1991).

HMMs have been used to model different communication environments. For example, they have been used to model fading communication channels by Turin and Sondhi (1993), Turin and van Nobelen (1998), Turin (2000), and Chao and Yao (1996). They are also used to model Internet traffic by Costamagna et al. (2003).

The theory and methodology for HMMs are described in many sources. Tutorials on HMMs are given by Rabiner (1989) and Ephraim and Merhav (2002). Books on HMMs include Rabiner and Juang (1993), Elliott et al. (1995), MacDonald and Zucchini (1997), Durbin et al. (1997), Koski (2001), and Cappe et al. (2005).

The relationship between the HMM and other Markov models is summarized in Table 14.1. The difference lies in whether or not the states are completely observable and whether or not the process can proceed without the intervention of an agent. Specifically, pure Markov chains have completely observable states, and the transitions are not under the control of an agent. Markov decision processes have completely observable states and the transitions are under the control of an agent. POMDPs have partially observable states, and the transitions are under the control of an agent. Finally, the HMM has partially observable states, and the states are not under the control of an agent.

Table 14.1 How HMM Is Related to Other Markov Models

Markov Models	Are the States Completely Observable?	Do We Have Control Over State Transitions?
Markov chains	Yes	No
Markov decision processes	Yes	Yes
POMDP	No	Yes
HMM	No	No

14.2 HMM Basics

An HMM is a doubly stochastic process in which an underlying stochastic process that is not observable (i.e., it is hidden) can only be observed through another stochastic process that produces a sequence of observations. Thus, if $S = \{S_n, n = 1, 2, \ldots\}$ is a Markov process and $\Omega = \{\Omega_k, k = 1, 2, \ldots\}$ is a function of S, then S is a hidden Markov process (or HMM) that is observed through Ω, and we can write $\Omega_k = f(S_k)$ for some function f. In this way, we can regard S as the *state process* that is hidden and Ω as the *observation process* that can be observed.

An HMM is usually defined as a 5-tuple $(S, \Omega, P, \Phi, \pi)$, where

- $S = \{s_1, s_2, \ldots, s_N\}$ is a finite set of N states;
- $\Omega = \{o_1, o_2, \ldots, o_M\}$ is a finite set of M possible symbols;
- $P = \{p_{ij}\}$ is the set of state-transition probabilities, where p_{ij} is the probability that the system goes from state s_i to state s_j;
- $\Phi = \{\phi_i(o_k)\}$ are the observation probabilities, where $\phi_i(o_k)$ is the probability that the symbol o_k is emitted when the system is in state s_i;
- $\pi = \{\pi_i\}$ are the initial state probabilities; that is, π_i is the probability that the system starts in state s_i.

Because the states and output sequence are understood, it is customary to denote the parameters of an HMM by

$$\lambda = (P, \Phi, \pi)$$

As an illustration of HMMs, consider Bill whose mood changes with the weather in the fall. Assume that the fall weather can be in one of three states: Sunny, Cloudy, and Rainy. Given that it is Sunny on a given day, then the next day it will be Sunny with probability 0.5, Cloudy with probability 0.3, and Rainy with probability 0.2. Similarly, given that it is Cloudy on a given day, then the next day it will be Sunny with probability 0.4, Cloudy with probability 0.4, and Rainy with probability 0.2. Finally, given that it is Rainy on a given day, then the next day it will be Sunny with probability 0.2, Cloudy with probability 0.4, and Rainy with probability 0.4. A study of Bill's mood swings shows that he is in a Good mood on a Sunny day, a So-so mood on a Cloudy day, and a Bad mood on a Rainy day. Thus, we can model the fall weather conditions and hence Bill's mood in the fall by a discrete-time Markov chain whose state-transition diagram is shown in Figure 14.1 along with the associated moods of Bill.

We can convert the process into an HMM as follows. Assume now that the weather still follows the probabilistic rules described earlier. However, Bill's mood can change with any weather condition. Specifically, when the weather is Sunny, he will be in a Good mood with probability 0.6, in a So-so mood with probability 0.3, and in a Bad mood with probability 0.1. Similarly, when the weather is Cloudy, he will be in a Good mood with probability 0.3, in a So-so mood with probability 0.5, and in a Bad mood with probability 0.2. Finally, when the weather is Rainy, he will be in a Good mood with probability 0.1, in a So-so mood with probability 0.3, and in a Bad mood with probability 0.6. The transition diagram for the new scheme is shown in Figure 14.2.

The problem with the new scheme is that when Bill is in, say, a So-so mood, we cannot know with certainty what the weather condition is. That is, we can no longer uniquely identify the state that a given mood was emitted from. Thus, if we observe Bill in the sequence of moods *Good−Good−Bad−Bad−So-so*, we cannot say exactly what weather state sequence produced the observed sequence of Bill's moods. For this reason, we say that the state sequence is "hidden" from us. However, we can calculate certain attributes of the model, such as the most likely state sequence that produced the observed sequence of Bill's moods. We will use this example to illustrate the analysis of HMMs in the remainder of this chapter. We can more formally represent an HMM as shown in Figure 14.3, where the S_i are the hidden states that we would like to estimate and the Ω_i are the observation

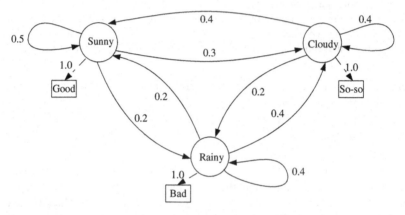

Figure 14.1 State-transition diagram for the fall weather.

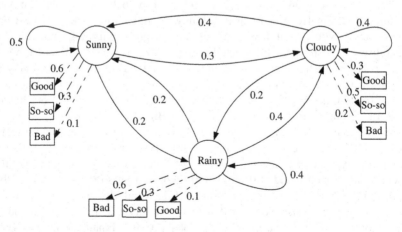

Figure 14.2 Example of HMM.

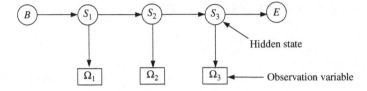

Figure 14.3 General structure of an HMM.

random variables from which the S_i are to be estimated. The letters B and E indicate the *Beginning* and *End* of the sequence of states.

14.3   HMM Assumptions

Let $Q = \{q_t\}_{t=0}^T$ be the hidden state sequence in the interval $0 \le t \le T$, where $q_t \in S$. There are three major assumptions that are made in the analysis of HMM problems, which are as follows:

1. *The Markov assumption*: This assumption states that the next state depends only on the current state, which means that the transition probabilities are defined by

$$P[q_{t+1} = j \mid q_t = i, q_{t-1} = l, \ldots, q_0 = n] = P[q_{t+1} = j \mid q_t = i] = p_{ij}$$

   In practice, the next state might depend on the past k states thereby giving rise to a kth-order HMM. However, such models are more difficult to analyze than the preceding first-order HMMs.

2. *The stationarity assumption*: This assumption states that the state-transition probabilities are independent of the actual time that the transitions take place. Thus, for any two times t_1 and t_2,

$$P[q_{t_1+1} = j \mid q_{t_1} = i] = P[q_{t_2+1} = j \mid q_{t_2} = i] = p_{ij}$$

3. *The observation independence assumption*: This assumption states that the current observation or output is statistically independent of previous observations. Thus, if we have the observation sequence $O = v_1, v_2, \ldots, v_T$, then

$$P[O \mid q_1, q_2, \ldots, q_T, \lambda] = \prod_{t=1}^{T} P[v_t \mid q_t, \lambda]$$

   With these assumptions, we may obtain the joint probability distribution $P[Q, O]$ by

$$P[Q, O] = \prod_{t=1}^{T} P[q_t \mid q_{t-1}] P[v_t \mid q_t]$$

   where it is understood that $P[q_1 \mid q_0] = P[q_1]$.

14.4   Three Fundamental Problems

There are three fundamental problems in HMM:

1. *The evaluation problem*: Given a model $\lambda = (P, \Phi, \pi)$ and an observation sequence $O = v_1, v_2, \ldots, v_T$ of length T, where $v_i \in \Omega$, how do we efficiently compute the probability that the model generated the observation sequence; that is, what is $P[O|\lambda]$?
2. *The decoding problem*: Given a model $\lambda = (P, \Phi, \pi)$, what is the most likely sequence of hidden states that could have generated a given observation sequence? Thus, we would like to find $Q^* = \arg \max_Q P[Q, O|\lambda]$, where Q is the hidden state sequence, as defined earlier.
3. *The learning problem*: Given a set of observation sequences, find the HMM that best explains the observation sequences; that is, find the values for λ that maximize $P[O|\lambda]$ or $\lambda^* = \arg \max_\lambda P[O|\lambda]$. Stated differently, the problem is to estimate the most likely HMM parameters for a given observation sequence.

14.5   Solution Methods

There are different methods of solving HMM problems, depending on which of the three fundamental problems we would like to solve. The evaluation problem is usually solved by the *forward algorithm* and the *backward algorithm*. The decoding problem is usually solved by the *Viterbi algorithm*. Finally, the learning problem is solved by the *Baum–Welch algorithm*. These algorithms are described in the remainder of this section.

14.5.1   The Evaluation Problem

Consider a model $\lambda = (P, \Phi, \pi)$ and a given observation sequence $O = v_1, v_2, \ldots, v_T$. We would like to compute $P[O|\lambda]$, the probability of the observation sequence given the model. $P[O|\lambda]$ is given by

$$P[O|\lambda] = \sum_Q P[O|Q, \lambda] P[Q|\lambda]$$

where $Q = q_1, q_2, \ldots, q_T$ is a fixed sequence, $P[O|Q, \lambda]$ is the probability of the observation sequence O for the specific state sequence Q, and $P[Q|\lambda]$ is the probability of the sequence Q for a given model. Because we assume that the observations are independent, the two probabilities are given by

$$P[O|Q, \lambda] = \prod_{t=1}^{T} P[o_t|q_t, \lambda] = \phi_{q_1}(o_1)\phi_{q_2}(o_2)\ldots\phi_{q_T}(o_T)$$

$$P[Q|\lambda] = \pi_{q_1} p_{q_1 q_2} p_{q_2 q_3} \cdots p_{q_{T-1} q_T}$$

Thus, we obtain

$$P[O|\lambda] = \sum_Q P[O|Q, \lambda] P[Q | \lambda]$$

$$= \sum_{q_1 \ldots q_T} \pi_{q_1} \phi_{q_1}(o_1) p_{q_1 q_2} \phi_{q_2}(o_2) p_{q_2 q_3} \cdots p_{q_{T-1} q_T} \phi_{q_T}(o_T)$$

We make the following observations on the preceding result. First, the number of possible paths of length T is N^T, which means that the number of equations required to obtain the solution is exponential in T. Also, using this direct method to obtain $P[O|\lambda]$ requires on the order of $2TN^T$ calculations. Thus, even for small values of N and T, the number of calculations is computationally large. For example, if we assume that $N = 3$ and $T = 100$, which can be associated with the problem with Bill's mood changes, the number of required calculations is on the order of $2 \times 100 \times 3^{100} \approx 10^{50}$. For this reason, we need a more efficient algorithm to solve the evaluation problem. One such algorithm is the forward algorithm, which is discussed next.

The Forward Algorithm

One important observation in the calculation of $P[O|\lambda]$ by the direct method is that it requires many redundant calculations that are not saved and reused. To reduce the computational complexity, we cache calculations. The caching is implemented as a trellis of states at each time step, as illustrated in Figure 14.4. A trellis can record the probability of all initial subpaths of the HMM that end in a certain state at a certain time. This allows the probability of longer subpaths to be worked out in terms of shorter subpaths.

A *forward probability variable* $\alpha_t(i)$ is defined as follows:

$$\alpha_t(i) = P[o_1, o_2, \ldots, o_{t-1}, o_t, q_t = s_i | \lambda] \quad t = 1, \ldots, T; \quad i = 1, \ldots, N$$

That is, $\alpha_t(i)$ is the probability of being in state s_i at time t after having observed the sequence $\{o_1, o_2, \ldots, o_t\}$. It is calculated by summing probabilities for all incoming arcs at a trellis node. This follows from the fact that

$$\alpha_t(i) = P[o_1, o_2, \ldots, o_{t-1}, o_t, q_t = s_i | \lambda]$$

$$= P[o_t | o_1, o_2, \ldots, o_{t-1}, q_t = s_i, \lambda] P[o_1, o_2, \ldots, o_{t-1}, q_t = s_i | \lambda]$$

$$= P[o_t | q_t = s_i, \lambda] P[o_1, o_2, \ldots, o_{t-1}, q_t = s_i | \lambda]$$

$$= P[o_t | q_t = s_i, \lambda] \sum_{s_j \in S} P[q_t = s_i | q_{t-1} = s_j, \lambda] P[o_1, o_2, \ldots, o_{t-1}, q_{t-1} = s_j | \lambda]$$

$$= \phi_i(o_t) \sum_{j=1}^{N} p_{ji} \alpha_{t-1}(j)$$

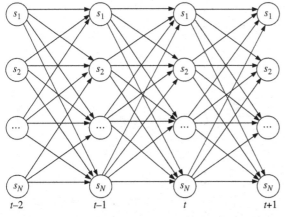

Figure 14.4 Trellis for forward algorithm.

s_1 s_1 s_1 s_1

s_2 s_2 s_2 s_2

$\cdots$ $\cdots$ $\cdots$ $\cdots$

s_N s_N s_N s_N

$t-2$ $t-1$ t $t+1$

where we have assumed that the observations are independent. Thus, if we work through the trellis filling in the values of the $\alpha_t(i)$, the sum of the final column of the trellis is the probability of the observation sequence. The forward algorithm works as follows:

1. Initialization:

$$\alpha_1(i) = \pi_i\phi_i(o_1) \quad 1 \le i \le N$$

2. Induction:

$$\alpha_{t+1}(j) = \left\{ \sum_{i=1}^{N} p_{ij}\alpha_t(i) \right\} \phi_j(o_{t+1}) \quad 1 \le t \le T-1, \ 1 \le j \le N$$

This step is the key to the algorithm and can be represented as shown in Figure 14.5.
3. Update time: Set $t = t + 1$. If $t < T$, go to step 2; otherwise, go to step 4.
4. Termination:

$$P[O|\lambda] = \sum_{i=1}^{N} \alpha_T(i) = \sum_{i=1}^{N} P[O, q_T = s_i|\lambda]$$

The forward algorithm requires $N(N+1)(T-1) + N$ multiplications and $N(N-1)(T-1)$ additions, giving a complexity on the order of N^2T rather than $2TN^T$. For example, in the Bill's mood changes example, we considered the case when $N = 3$ and $T = 100$, which requires on the order of 900 calculations with the forward algorithm compared to the order of 10^{50} calculations required for the direct method.

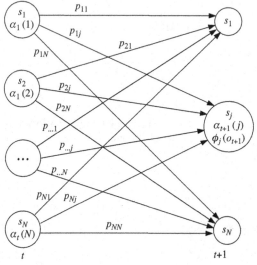

Figure 14.5 The induction step of the forward algorithm.

Example 14.1

Consider Bill's mood change problem illustrated in Figure 13.2. Assume that we observed Bill in the following sequence of moods: *Good, Good, So-so, Bad, Bad.* We are required to find the probability that the model generated such a sequence of moods using the forward algorithm.

Solution

Let S denotes Sunny state, C denotes Cloudy, and R denotes Rainy. Similarly, let G denotes Good mood, SS denotes So-so mood, and B denotes Bad mood. We assume that the process is equally likely to start in any state. That is, we assume that $\pi_S = \pi_C = \pi_R = 1/3$. Also, we have that $T = 5$. Then the initialization step becomes

$$\alpha_1(S) = \pi_S \phi_S(o_1) = \pi_S \phi_S(G) = \frac{1}{3}(0.6) = 0.2$$

$$\alpha_1(C) = \pi_C \phi_C(o_1) = \pi_C \phi_C(G) = \frac{1}{3}(0.3) = 0.1$$

$$\alpha_1(R) = \pi_R \phi_R(o_1) = \pi_R \phi_R(G) = \frac{1}{3}(0.1) = 0.033$$

The induction step for $t = 2$ is given by

$$\alpha_2(j) = \left\{ \sum_{i=1}^{N} p_{ij}\alpha_1(i) \right\} \phi_j(o_2) = \left\{ \sum_{i=1}^{N} p_{ij}\alpha_1(i) \right\} \phi_j(G)$$

$$\alpha_2(S) = \{p_{SS}\alpha_1(S) + p_{CS}\alpha_1(C) + p_{RS}\alpha_1(R)\}\phi_S(G)$$
$$= \{(0.5)(0.2) + (0.4)(0.1) + (0.2)(0.033)\}(0.6)$$
$$= 0.088$$

$$\alpha_2(C) = \{p_{SC}\alpha_1(S) + p_{CC}\alpha_1(C) + p_{RC}\alpha_1(R)\}\phi_C(G)$$
$$= \{(0.3)(0.2) + (0.4)(0.1) + (0.4)(0.033)\}(0.3)$$
$$= 0.034$$

$$\alpha_2(R) = \{p_{SR}\alpha_1(S) + p_{CR}\alpha_1(C) + p_{RR}\alpha_1(R)\}\phi_R(G)$$
$$= \{(0.2)(0.2) + (0.2)(0.1) + (0.4)(0.033)\}(0.1)$$
$$= 0.007$$

The induction step for $t = 3$ is given by

$$\alpha_3(j) = \left\{ \sum_{i=1}^{N} p_{ij}\alpha_2(i) \right\} \phi_j(o_3) = \left\{ \sum_{i=1}^{N} p_{ij}\alpha_2(i) \right\} \phi_j(SS)$$

$$\alpha_3(S) = \{p_{SS}\alpha_2(S) + p_{CS}\alpha_2(C) + p_{RS}\alpha_2(R)\}\phi_S(SS)$$
$$= \{(0.5)(0.088) + (0.4)(0.034) + (0.2)(0.007)\}(0.3)$$
$$= 0.018$$

$$\alpha_3(C) = \{p_{SC}\alpha_2(S) + p_{CC}\alpha_2(C) + p_{RC}\alpha_2(R)\}\phi_C(SS)$$
$$= \{(0.3)(0.088) + (0.4)(0.034) + (0.4)(0.007)\}(0.5)$$
$$= 0.021$$
$$\alpha_3(R) = \{p_{SR}\alpha_2(S) + p_{CR}\alpha_2(C) + p_{RR}\alpha_2(R)\}\phi_R(SS)$$
$$= \{(0.2)(0.088) + (0.2)(0.034) + (0.4)(0.007)\}(0.3)$$
$$= 0.008$$

The induction step for $t = 4$ is given by

$$\alpha_4(j) = \left\{\sum_{i=1}^{N}p_{ij}\alpha_3(i)\right\}\phi_j(o_4) = \left\{\sum_{i=1}^{N}p_{ij}\alpha_3(i)\right\}\phi_j(B)$$

$$\alpha_4(S) = \{p_{SS}\alpha_3(S) + p_{CS}\alpha_3(C) + p_{RS}\alpha_3(R)\}\phi_S(B)$$
$$= \{(0.5)(0.018) + (0.4)(0.021) + (0.2)(0.008)\}(0.1)$$
$$= 0.002$$
$$\alpha_4(C) = \{p_{SC}\alpha_3(S) + p_{CC}\alpha_3(C) + p_{RC}\alpha_3(R)\}\phi_C(B)$$
$$= \{(0.3)(0.018) + (0.4)(0.021) + (0.4)(0.008)\}(0.2)$$
$$= 0.003$$
$$\alpha_4(R) = \{p_{SR}\alpha_3(S) + p_{CR}\alpha_3(C) + p_{RR}\alpha_3(R)\}\phi_R(B)$$
$$= \{(0.2)(0.018) + (0.2)(0.021) + (0.4)(0.008)\}(0.6)$$
$$= 0.007$$

The final induction step for $t = 5$ is given by

$$\alpha_5(j) = \left\{\sum_{i=1}^{N}p_{ij}\alpha_4(i)\right\}\phi_j(o_5) = \left\{\sum_{i=1}^{N}p_{ij}\alpha_4(i)\right\}\phi_j(B)$$

$$\alpha_5(S) = \{p_{SS}\alpha_4(S) + p_{CS}\alpha_4(C) + p_{RS}\alpha_4(R)\}\phi_S(B)$$
$$= \{(0.5)(0.002) + (0.4)(0.003) + (0.2)(0.007)\}(0.1)$$
$$= 0.0004$$
$$\alpha_5(C) = \{p_{SC}\alpha_4(S) + p_{CC}\alpha_4(C) + p_{RC}\alpha_4(R)\}\phi_C(B)$$
$$= \{(0.3)(0.002) + (0.4)(0.003) + (0.4)(0.007)\}(0.2)$$
$$= 0.0009$$
$$\alpha_5(R) = \{p_{SR}\alpha_4(S) + p_{CR}\alpha_4(C) + p_{RR}\alpha_4(R)\}\phi_R(B)$$
$$= \{(0.2)(0.002) + (0.2)(0.003) + (0.4)(0.007)\}(0.6)$$
$$= 0.0023$$

Thus, at the termination of the algorithm, we obtain the solution as

$$P[O = G, G, SS, B, B|\lambda] = \sum_{i=1}^{N}\alpha_T(i) = \alpha_T(S) + \alpha_T(C) + \alpha_T(R) = 0.0036$$

The trellis for the problem is shown in Figure 14.6. From the figure, we can see one of the advantages of the trellis: it enables us to obtain such intermediate results as the probability that the model generated the sequence *Good, Good, So-so, Bad*, with the intermediate result $\alpha_4(S) + \alpha_4(C) + \alpha_4(R) = 0.012$; the probability that the model generated the sequence *Good, Good, So-so*, which is $\alpha_3(S) + \alpha_3(C) + \alpha_3(R) = 0.047$; and the probability that the model generated the sequence *Good, Good*, which is $\alpha_2(S) + \alpha_2(C) + \alpha_2(R) = 0.129$. All these results are valid for the assumption that the hidden process is equally likely to start from any of the three states.

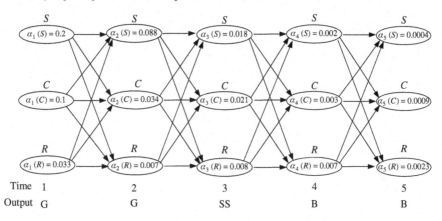

Figure 14.6: Time axis:

Time	1	2	3	4	5
Output	G	G	SS	B	B

Figure 14.6 Trellis for Example 14.1.

The Backward Algorithm

The backward algorithm is a dual method to solve the evaluation problem. It starts by defining a *backward probability variable* $\beta_t(i)$ as follows:

$$\beta_t(i) = P[o_{t+1}, o_{t+2}, \ldots, o_{T-1}, o_T | q_t = s_i, \lambda] \quad t = 1, \ldots, T; \ s_i \in S$$

That is, $\beta_t(i)$ is the conditional probability of the partial observation $o_{t+1}, o_{t+2}, \ldots, o_{T-1}, o_T$ given that the model is in state s_i at time t. Note that $\beta_t(i)$ is given by

$$\beta_t(i) = P[o_{t+1}, o_{t+2}, \ldots, o_{T-1}, o_T | q_t = s_i, \lambda]$$

$$= \sum_{s_j \in S} P[o_{t+1}, o_{t+2}, \ldots, o_{T-1}, o_T, q_{t+1} = s_j | q_t = s_i, \lambda]$$

$$= \sum_{s_j \in S} P[o_{t+1} | q_{t+1} = s_j] P[o_{t+2}, \ldots, o_{T-1}, o_T, q_{t+1} = s_j | q_t = s_i, \lambda]$$

$$= \sum_{s_j \in S} P[o_{t+1} | q_{t+1} = s_j] P[o_{t+2}, \ldots, o_{T-1}, o_T | q_{t+1} = s_j] P[q_{t+1} = s_j | q_t = s_i, \lambda]$$

$$= \sum_{j=1}^{N} \phi_j(o_{t+1}) \beta_{t+1}(j) p_{ij} \quad t = 1, \ldots, T; i = 1, \ldots, N$$

The backward algorithm works from right to left through the same trellis as follows:

1. Initialization:

$$\beta_T(i) = 1 \quad 1 \leq i \leq N$$

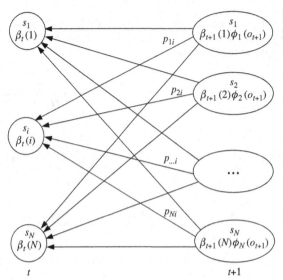

Figure 14.7 The induction step of the backward algorithm.

2. Induction:

$$\beta_t(i) = \sum_{j=1}^{N} p_{ij}\beta_{t+1}(j)\phi_j(o_{t+1}) \quad 1 \le t \le T-1, \quad 1 \le i \le N$$

The induction step is illustrated in Figure 14.7.

3. Update time: Set $t = t - 1$. If $t > 0$, go to step 2; otherwise, go to step 4.
4. Termination:

$$P[O|\lambda] = \sum_{i=1}^{N} \beta_1(i)\alpha_1(i) = \sum_{i=1}^{N} \beta_1(i)\pi_i\phi_1(o_1)$$

The so-called forward–backward algorithm is obtained from the observation that for any t, $1 \le t \le T$, it can be shown that

$$P[O|\lambda] = \sum_{i=1}^{N} \beta_t(i)\alpha_t(i)$$

Example 14.2

Consider Bill's mood change problem illustrated in Figure 14.2. Assume that we observed Bill in the following sequence of moods: *Good, Good, So-so, Bad, Bad*. We are required to find the probability that the model generated such a sequence of moods using the backward algorithm.

Solution

We use the same notation as in Example 14.1. Because $T = 5$, the initialization step is as follows:

$$\beta_5(S) = \beta_5(C) = \beta_5(R) = 1$$

The induction step for $t = 4$ is given by

$$\beta_4(i) = \sum_{j=1}^{N} p_{ij}\beta_5(j)\phi_j(o_5)$$

$$\begin{aligned}
\beta_4(S) &= p_{SS}\beta_5(S)\phi_S(o_5) + p_{SC}\beta_5(C)\phi_C(o_5) + p_{SR}\beta_5(R)\phi_R(o_5) \\
&= p_{SS}\beta_5(S)\phi_S(B) + p_{SC}\beta_5(C)\phi_C(B) + p_{SR}\beta_5(R)\phi_R(B) \\
&= (0.5)(1)(0.1) + (0.3)(1)(0.2) + (0.2)(1)(0.6) = 0.23
\end{aligned}$$

$$\begin{aligned}
\beta_4(C) &= p_{CS}\beta_5(S)\phi_S(o_5) + p_{CC}\beta_5(C)\phi_C(o_5) + p_{CR}\beta_5(R)\phi_R(o_5) \\
&= p_{CS}\beta_5(S)\phi_S(B) + p_{CC}\beta_5(C)\phi_C(B) + p_{CR}\beta_5(R)\phi_R(B) \\
&= (0.4)(1)(0.1) + (0.4)(1)(0.2) + (0.2)(1)(0.6) = 0.24
\end{aligned}$$

$$\begin{aligned}
\beta_4(R) &= p_{RS}\beta_5(S)\phi_S(o_5) + p_{RC}\beta_5(C)\phi_C(o_5) + p_{RR}\beta_5(R)\phi_R(o_5) \\
&= p_{RS}\beta_5(S)\phi_S(B) + p_{RC}\beta_5(C)\phi_C(B) + p_{RR}\beta_5(R)\phi_R(B) \\
&= (0.2)(1)(0.1) + (0.4)(1)(0.2) + (0.4)(1)(0.6) = 0.34
\end{aligned}$$

The induction step for $t = 3$ is given by

$$\beta_3(i) = \sum_{j=1}^{N} p_{ij}\beta_4(j)\phi_j(o_4)$$

$$\begin{aligned}
\beta_3(S) &= p_{SS}\beta_4(S)\phi_S(o_4) + p_{SC}\beta_4(C)\phi_C(o_4) + p_{SR}\beta_4(R)\phi_R(o_4) \\
&= p_{SS}\beta_4(S)\phi_S(B) + p_{SC}\beta_4(C)\phi_C(B) + p_{SR}\beta_4(R)\phi_R(B) \\
&= (0.5)(0.23)(0.1) + (0.3)(0.24)(0.2) + (0.2)(0.34)(0.6) = 0.0667
\end{aligned}$$

$$\begin{aligned}
\beta_3(C) &= p_{CS}\beta_4(S)\phi_S(o_4) + p_{CC}\beta_4(C)\phi_C(o_4) + p_{CR}\beta_4(R)\phi_R(o_4) \\
&= p_{CS}\beta_4(S)\phi_S(B) + p_{CC}\beta_4(C)\phi_C(B) + p_{CR}\beta_4(R)\phi_R(B) \\
&= (0.4)(0.23)(0.1) + (0.4)(0.24)(0.2) + (0.2)(0.34)(0.6) = 0.0692
\end{aligned}$$

$$\begin{aligned}
\beta_3(R) &= p_{RS}\beta_4(S)\phi_S(o_4) + p_{RC}\beta_4(C)\phi_C(o_4) + p_{RR}\beta_4(R)\phi_R(o_4) \\
&= p_{RS}\beta_4(S)\phi_S(B) + p_{RC}\beta_4(C)\phi_C(B) + p_{RR}\beta_4(R)\phi_R(B) \\
&= (0.2)(0.23)(0.1) + (0.4)(0.24)(0.2) + (0.4)(0.34)(0.6) = 0.1054
\end{aligned}$$

The induction step for $t = 2$ is given by

$$\beta_2(i) = \sum_{j=1}^{N} p_{ij}\beta_3(j)\phi_j(o_3)$$

$$\begin{aligned}
\beta_2(S) &= p_{SS}\beta_3(S)\phi_S(o_3) + p_{SC}\beta_3(C)\phi_C(o_3) + p_{SR}\beta_3(R)\phi_R(o_3) \\
&= p_{SS}\beta_3(S)\phi_S(SS) + p_{SC}\beta_3(C)\phi_C(SS) + p_{SR}\beta_3(R)\phi_R(SS) \\
&= (0.5)(0.0667)(0.3) + (0.3)(0.0692)(0.5) + (0.2)(0.1054)(0.3) = 0.0267
\end{aligned}$$

$$\beta_2(C) = p_{CS}\beta_3(S)\phi_S(o_3) + p_{CC}\beta_3(C)\phi_C(o_3) + p_{CR}\beta_3(R)\phi_R(o_3)$$
$$= p_{CS}\beta_3(S)\phi_S(SS) + p_{CC}\beta_3(C)\phi_C(SS) + p_{CR}\beta_3(R)\phi_R(SS)$$
$$= (0.4)(0.0667)(0.3) + (0.4)(0.0692)(0.5) + (0.2)(0.1054)(0.3) = 0.0282$$
$$\beta_2(R) = p_{RS}\beta_3(S)\phi_S(o_3) + p_{RC}\beta_3(C)\phi_C(o_3) + p_{RR}\beta_3(R)\phi_R(o_3)$$
$$= p_{RS}\beta_3(S)\phi_S(SS) + p_{RC}\beta_3(C)\phi_C(SS) + p_{RR}\beta_3(R)\phi_R(SS)$$
$$= (0.2)(0.0667)(0.3) + (0.4)(0.0692)(0.5) + (0.4)(0.1054)(0.3) = 0.0305$$

The induction step for $t = 1$ is given by

$$\beta_1(i) = \sum_{j=1}^{N} p_{ij}\beta_2(j)\phi_j(o_2)$$

$$\beta_1(S) = p_{SS}\beta_2(S)\phi_S(o_2) + p_{SC}\beta_2(C)\phi_C(o_2) + p_{SR}\beta_2(R)\phi_R(o_2)$$
$$= p_{SS}\beta_2(S)\phi_S(G) + p_{SC}\beta_2(C)\phi_C(G) + p_{SR}\beta_2(R)\phi_R(G)$$
$$= (0.5)(0.0267)(0.6) + (0.3)(0.0282)(0.3) + (0.2)(0.0305)(0.1) = 0.0112$$
$$\beta_1(C) = p_{CS}\beta_2(S)\phi_S(o_2) + p_{CC}\beta_2(C)\phi_C(o_2) + p_{CR}\beta_2(R)\phi_R(o_2)$$
$$= p_{CS}\beta_2(S)\phi_S(G) + p_{CC}\beta_2(C)\phi_C(G) + p_{CR}\beta_2(R)\phi_R(G)$$
$$= (0.4)(0.0267)(0.6) + (0.4)(0.0282)(0.3) + (0.2)(0.0305)(0.1) = 0.0104$$
$$\beta_1(R) = p_{RS}\beta_2(S)\phi_S(o_2) + p_{RC}\beta_2(C)\phi_C(o_2) + p_{RR}\beta_2(R)\phi_R(o_2)$$
$$= p_{RS}\beta_2(S)\phi_S(G) + p_{RC}\beta_2(C)\phi_C(G) + p_{RR}\beta_2(R)\phi_R(G)$$
$$= (0.2)(0.0267)(0.6) + (0.4)(0.0282)(0.3) + (0.4)(0.0305)(0.1) = 0.0078$$

Thus, at the termination of the algorithm, we obtain the solution as

$$P[O = G, G, SS, B, B | \lambda] = \sum_{i=1}^{N} \beta_1(i)\alpha_1(i) = \sum_{i=1}^{N} \beta_1(i)\pi_i\phi_1(o_1)$$
$$= \beta_1(S)\pi_S\phi_S(G) + \beta_1(C)\pi_C\phi_C(G) + \beta_1(R)\pi_R\phi_R(G)$$
$$= \frac{1}{3}\{(0.0112)(0.6) + (0.0104)(0.3) + (0.0078)(0.1)\}$$
$$= 0.00354$$

This result is consistent with the result obtained using the forward algorithm.

14.5.2   *The Decoding Problem and the Viterbi Algorithm*

The second HMM problem is the decoding problem, which seeks to find the best (or optimal) state sequence associated with a given observation sequence O of a given model λ. The first step is to define what we mean by an optimal state sequence because there are several possible optimality criteria. One possible definition is the state sequence that has the highest probability of producing the given observation sequence. Thus, we find the state sequence Q that maximizes $P[Q|O, \lambda]$. Unfortunately, for an observation sequence of T symbols and a system with N states, there are N^T possible sequences for Q. For our example of Bill's mood swings with $N = 3$ and $T = 100$, there are 3^{100} possible sequences.

Consider the case where we find the most likely states individually rather than as a whole sequence. For each time t, $1 \leq t \leq T$, we define the variable $\gamma_t(i)$ as follows:

$$\gamma_t(i) = P[q_t = s_i | O, \lambda] = \frac{P[q_t = s_i, O | \lambda]}{P[O|\lambda]} = \frac{P[q_t = s_i, o_1, o_2, \ldots, o_T | \lambda]}{P[O|\lambda]}$$

$$= \frac{P[q_t = s_i, o_1, o_2, \ldots, o_t, o_{t+1}, \ldots, o_T | \lambda]}{P[O|\lambda]}$$

$$= \frac{P[o_1, o_2, \ldots, o_t, o_{t+1}, \ldots, o_T | q_t = s_i, \lambda] P[q_t = s_i | \lambda]}{P[O|\lambda]}$$

$$= \frac{P[o_1, o_2, \ldots, o_t | o_{t+1}, \ldots, o_T, q_t = s_i, \lambda] P[o_{t+1}, \ldots, o_T | q_t = s_i, \lambda] P[q_t = s_i | \lambda]}{P[O|\lambda]}$$

$$= \frac{P[o_1, o_2, \ldots, o_t | q_t = s_i, \lambda] P[q_t = s_i | \lambda] P[o_{t+1}, \ldots, o_T | q_t = s_i, \lambda]}{P[O|\lambda]}$$

$$= \frac{P[o_1, o_2, \ldots, o_t, q_t = s_i | \lambda] P[o_{t+1}, \ldots, o_T | q_t = s_i, \lambda]}{P[O|\lambda]}$$

$$= \frac{\alpha_t(i)\beta_t(i)}{\sum_{i=1}^{N} \beta_t(i)\alpha_t(i)}$$

where the equality of the fifth line is due to the observation independence assumption and the last equality follows from our earlier definitions of $\alpha_t(i)$ and $\beta_t(i)$ and the statement that

$$P[O|\lambda] = \sum_{i=1}^{N} \beta_t(i)\alpha_t(i)$$

Note that

$$\sum_{i=1}^{N} \gamma_t(i) = 1$$

which makes $\gamma_t(i)$ a true conditional probability. Thus, the individual most likely state at time t is

$$q_t^* = \arg \max_{1 \leq i \leq N} \{\gamma_t(i)\} \quad 1 \leq t \leq T$$

Thus, the method generates the most likely state sequence $Q^* = \{q_1^*, q_2^*, \ldots, q_T^*\}$ for the given observation sequence $O = \{o_1, o_2, \ldots, o_T\}$. Unfortunately, the scheme might generate an unlikely state sequence because it does not take state-transition probabilities into consideration. For example, if we have a sequence that includes two neighboring states s_i and s_j in the sequence whose transition probability $p_{ij} = 0$, then the result is an invalid state sequence. An efficient method that avoids such unlikely sequences is the Viterbi algorithm, which is based on dynamic programming.

The Viterbi Algorithm

The Viterbi algorithm was originally designed for decoding convolutional codes and is now applied in many other areas. In HMMs, it is used to find the most likely state sequence $Q^* = \{q_1^*, q_2^*, \ldots, q_T^*\}$ for a given observation sequence $O = \{o_1, o_2, \ldots, o_T\}$. As defined earlier, let the function

$$\arg\max_y \{z\}$$

denote the argument y that corresponds to the maximum of the expression z. The Viterbi algorithm simultaneously maximizes both the joint probability $P[q, O]$ and the conditional probability $P[q|O]$ due to the fact that

$$\arg\max_Q \{P[Q|O, \lambda]\} = \arg\max_Q \left\{ \frac{P[Q, O|\lambda]}{P[O|\lambda]} \right\} = \arg\max_Q \{P[Q, O|\lambda]\}$$

The algorithm defines the variable $\delta_t(i)$ as follows:

$$\delta_t(i) = \max_{q_1, q_2, \ldots, q_{t-1}} P[q_1, q_2, \ldots, q_{t-1}, q_t = s_i, o_1, o_2, \ldots, o_{t-1}, o_t | \lambda]$$

That is, $\delta_t(i)$ is the largest probability along a single path that accounts for the first t observations and ends in state s_i. Thus, it is the probability of the most likely state path for the partial observation sequence. Another variable $\psi_t(j)$ stores the node of the incoming arc that leads to this most probable path. That is,

$$\psi_t(j) = \arg\max_{1 \leq j \leq N} \{\delta_{t-1}(i)p_{ij}\}$$

The details of the algorithm are as follows:

1. Initialization:

$$\delta_1(i) = \pi_i \phi_i(o_1)$$
$$\psi_1(i) = 0 \qquad\qquad 1 \leq i \leq N$$

2. Recursion:

$$\delta_t(j) = \max_{1 \leq i \leq N} \{\delta_{t-1}(i)p_{ij}\}\phi_j(o_t)$$
$$\psi_t(j) = \arg\max_{1 \leq i \leq N} \{\delta_{t-1}(i)p_{ij}\} \quad 1 \leq j \leq N, \; 2 \leq t \leq T$$

Note that this step is similar to the induction step of the forward algorithm. The main difference between the two is that the forward algorithm uses summation over previous states, whereas the Viterbi algorithm uses minimization.

3. Update time: Set $t = t + 1$. If $t < T$, go to step 2; otherwise, go to step 4.
4. Termination:

$$P^* = \max_{1 \leq i \leq N} \{\delta_T(i)\}$$

$$q_T^* = \arg \max_{1 \leq i \leq N} \{\delta_T(i)\}$$

5. Path (or state sequence) backtracking:

$$q_t^* = \psi_{t+1}(q_{t+1}^*) \quad t = T - 1, T - 2, \ldots, 1$$

The backtracking step allows the best state sequence to be found from the back pointers stored in the recursion step.

Example 14.3

Consider Bill's mood change problem illustrated in Figure 14.2. Assume that we observed Bill in the following sequence of moods: *Good, Good, So-so, Bad, Bad*. We are required to find the most likely state sequence that generated such a sequence of moods using the Viterbi algorithm.

Solution

We use the same notation and assumptions of initial distribution as in Example 14.1. The initialization step is as follows:

$$\delta_1(S) = \pi_S \phi_S(o_1) = \pi_S \phi_S(G) = \frac{1}{3}(0.6) = 0.2$$

$$\delta_1(C) = \pi_C \phi_C(o_1) = \pi_C \phi_C(G) = \frac{1}{3}(0.3) = 0.1$$

$$\delta_1(R) = \pi_R \phi_R(o_1) = \pi_R \phi_R(G) = \frac{1}{3}(0.1) = 0.033$$

$$\psi_1(S) = \psi_1(C) = \psi_1(R) = 0$$

The recursion step for $t = 2$ is given by

$$\delta_2(S) = \max\{\delta_1(S)p_{SS}, \delta_1(C)p_{CS}, \delta_1(R)p_{RS}\}\phi_S(o_2)$$
$$= \max\{\delta_1(S)p_{SS}, \delta_1(C)p_{CS}, \delta_1(R)p_{RS}\}\phi_S(G)$$
$$= \max\{(0.2)(0.5), (0.1)(0.4), (0.033)(0.2)\}(0.6)$$
$$= \max\{0.1, 0.04, 0.066\}(0.6) = 0.06$$
$$\psi_2(S) = S$$

$$\delta_2(C) = \max\{\delta_1(S)p_{SC}, \delta_1(C)p_{CC}, \delta_1(R)p_{RC}\}\phi_C(o_2)$$
$$= \max\{\delta_1(S)p_{SC}, \delta_1(C)p_{CC}, \delta_1(R)p_{RC}\}\phi_C(G)$$
$$= \max\{(0.2)(0.3), (0.1)(0.4), (0.033)(0.4)\}(0.3)$$
$$= \max\{0.06, 0.04, 0.0132\}(0.3) = 0.018$$
$$\psi_2(C) = S$$

$$\delta_2(R) = \max\{\delta_1(S)p_{SR}, \delta_1(C)p_{CR}, \delta_1(R)p_{RR}\}\phi_R(o_2)$$
$$= \max\{\delta_1(S)p_{SR}, \delta_1(C)p_{CR}, \delta_1(R)p_{RR}\}\phi_S(G)$$
$$= \max\{(0.2)(0.2), (0.1)(0.2), (0.033)(0.4)\}(0.1)$$
$$= \max\{0.04, 0.02, 0.0132\}(0.1) = 0.004$$
$$\psi_2(R) = S$$

The recursion step for $t = 3$ is given by

$$\delta_3(S) = \max\{\delta_2(S)p_{SS}, \delta_2(C)p_{CS}, \delta_2(R)p_{RS}\}\phi_S(o_3)$$
$$= \max\{\delta_2(S)p_{SS}, \delta_2(C)p_{CS}, \delta_2(R)p_{RS}\}\phi_S(SS)$$
$$= \max\{(0.06)(0.5), (0.036)(0.4), (0.004)(0.2)\}(0.3)$$
$$= \max\{0.03, 0.0144, 0.0008\}(0.3) = 0.009$$
$$\psi_3(S) = S$$

$$\delta_3(C) = \max\{\delta_2(S)p_{SC}, \delta_2(C)p_{CC}, \delta_2(R)p_{RC}\}\phi_C(o_3)$$
$$= \max\{\delta_2(S)p_{SC}, \delta_2(C)p_{CC}, \delta_2(R)p_{RC}\}\phi_C(SS)$$
$$= \max\{(0.06)(0.3), (0.036)(0.4), (0.004)(0.4)\}(0.5)$$
$$= \max\{0.018, 0.0144, 0.0016\}(0.5) = 0.009$$
$$\psi_3(C) = S$$

$$\delta_3(R) = \max\{\delta_2(S)p_{SR}, \delta_2(C)p_{CR}, \delta_2(R)p_{RR}\}\phi_R(o_3)$$
$$= \max\{\delta_2(S)p_{SR}, \delta_2(C)p_{CR}, \delta_2(R)p_{RR}\}\phi_S(SS)$$
$$= \max\{(0.06)(0.2), (0.036)(0.2), (0.004)(0.4)\}(0.3)$$
$$= \max\{0.012, 0.0072, 0.0016\}(0.3) = 0.0036$$
$$\psi_3(R) = S$$

The recursion step for $t = 4$ is given by

$$\delta_4(S) = \max\{\delta_3(S)p_{SS}, \delta_3(C)p_{CS}, \delta_3(R)p_{RS}\}\phi_S(o_4)$$
$$= \max\{\delta_3(S)p_{SS}, \delta_3(C)p_{CS}, \delta_3(R)p_{RS}\}\phi_S(B)$$
$$= \max\{(0.009)(0.5), (0.009)(0.4), (0.0036)(0.2)\}(0.1)$$
$$= \max\{0.0045, 0.0036, 0.00072\}(0.1) = 0.00045$$
$$\psi_4(S) = S$$

$$\delta_4(C) = \max\{\delta_3(S)p_{SC}, \delta_3(C)p_{CC}, \delta_3(R)p_{RC}\}\phi_C(o_4)$$
$$= \max\{\delta_3(S)p_{SC}, \delta_3(C)p_{CC}, \delta_3(R)p_{RC}\}\phi_C(B)$$
$$= \max\{(0.009)(0.3), (0.009)(0.4), (0.0036)(0.4)\}(0.2)$$
$$= \max\{0.0027, 0.0036, 0.00144\}(0.2) = 0.00072$$
$$\psi_4(C) = C$$

$$\delta_4(R) = \max\{\delta_3(S)p_{SR}, \delta_3(C)p_{CR}, \delta_3(R)p_{RR}\}\phi_R(o_4)$$
$$= \max\{\delta_3(S)p_{SR}, \delta_3(C)p_{CR}, \delta_3(R)p_{RR}\}\phi_R(B)$$
$$= \max\{(0.009)(0.2), (0.009)(0.2), (0.0036)(0.4)\}(0.6)$$
$$= \max\{0.0018, 0.0018, 0.00144\}(0.6) = 0.00108$$
$$\psi_4(R) = S, C$$

The recursion step for $t = 5$ is given by

$$\delta_5(S) = \max\{\delta_4(S)p_{SS}, \delta_4(C)p_{CS}, \delta_4(R)p_{RS}\}\phi_S(o_5)$$
$$= \max\{\delta_4(S)p_{SS}, \delta_4(C)p_{CS}, \delta_4(R)p_{RS}\}\phi_S(B)$$
$$= \max\{(0.00045)(0.5), (0.00072)(0.4), (0.00108)(0.2)\}(0.1)$$
$$= \max\{0.000225, 0.000288, 0.000216\}(0.1) = 0.0000288$$
$$\psi_5(S) = C$$

$$\delta_5(C) = \max\{\delta_4(S)p_{SC}, \delta_4(C)p_{CC}, \delta_4(R)p_{RC}\}\phi_C(o_5)$$
$$= \max\{\delta_4(S)p_{SC}, \delta_4(C)p_{CC}, \delta_4(R)p_{RC}\}\phi_C(B)$$
$$= \max\{(0.00045)(0.3), (0.00072)(0.4), (0.00108)(0.4)\}(0.2)$$
$$= \max\{0.000135, 0.000288, 0.000432\}(0.2) = 0.0000864$$
$$\psi_5(C) = R$$

$$\delta_5(R) = \max\{\delta_4(S)p_{SR}, \delta_4(C)p_{CR}, \delta_4(R)p_{RR}\}\phi_R(o_5)$$
$$= \max\{\delta_4(S)p_{SR}, \delta_4(C)p_{CR}, \delta_4(R)p_{RR}\}\phi_R(B)$$
$$= \max\{(0.00045)(0.2), (0.00072)(0.2), (0.00108)(0.4)\}(0.6)$$
$$= \max\{0.00009, 0.000144, 0.000432\}(0.6) = 0.0002592$$
$$\psi_5(R) = R$$

The termination step is given by

$$P^* = \max\{\delta_5(S), \delta_5(C), \delta_5(R)\}$$
$$= \max\{0.0000288, 0.0000864, 0.0002592\} = 0.0002592$$
$$q_T^* = \arg\max\{\delta_5(S), \delta_5(C), \delta_5(R)\} = R$$

The path backtracking step is as follows:

$$q_t^* = \psi_{t+1}(q_{t+1}^*)$$
$$q_4^* = \psi_5(q_5^*) = \psi_5(R) = R$$
$$q_3^* = \psi_4(q_4^*) = \psi_4(R) = S, C$$

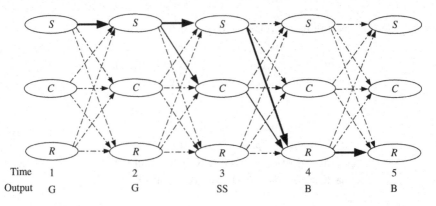

Figure 14.8 Trellis for the Viterbi algorithm.

$$q_2^* = \psi_3(q_3^*) = \psi_3(S) = \psi_3(C) = S$$
$$q_1^* = \psi_2(q_2^*) = \psi_2(S) = S$$

Thus, one of the two most likely state sequences is $Q^* = \{S, S, S, R, R\}$. This path is illustrated in Figure 14.8. The other most likely state sequence is $Q^* = \{S, S, C, R, R\}$, which is differentiated from the previous sequence by the thin solid lines between times 2 and 3 and times 3 and 4 in Figure 14.8.

14.5.3 The Learning Problem and the Baum–Welch Algorithm

The learning problem deals with how we can adjust the HMM parameters so that the given set of observations, which is usually referred to as the *training set*, is represented by the model in the best way for the intended application. Because we are looking for the "best" way to represent the observation, we are solving an optimization problem, and we must define the criterion for optimization. The most commonly used optimization criterion is the maximum likelihood criterion, which seeks to find the parameters of the HMM that maximize the probability of a given observation sequence. That is, we find the following solution:

$$\lambda^* = \arg \max_{\lambda}\{P[O|\lambda]\}$$

Unfortunately, this problem is so complex that there is no known method to analytically obtain λ that maximizes $P[O|\lambda]$, but we can choose the model parameters in such a way that $P[O|\lambda]$ is locally maximized. This method is an iterative solution called the *Baum–Welch algorithm*, which is sometimes called the *forward–backward algorithm* and is a special case of the *expectation maximization* (EM) method.

The Baum–Welch Algorithm

The algorithm starts by setting the parameters P, Φ, and π to some initial values that can be chosen from some prior knowledge or from some uniform distribution.

Then using the current model, all possible paths for each training set are considered to get new estimates $\hat{P}, \hat{\Phi}$, and $\hat{\pi}$. The procedure is repeated until there are insignificant changes in the parameters of the current model.

As a forward–backward algorithm, the Baum–Welch algorithm uses the same forward probability variable $\alpha_t(i)$ and backward probability variable $\beta_t(i)$ used in the evaluation problem that were defined earlier as follows:

$$\alpha_t(i) = P[o_1, o_2, \ldots, o_t, q_t = s_i | \lambda]$$
$$\beta_t(i) = P[o_{t+1}, o_{t+2}, \ldots, o_T | q_t = s_i, \lambda]$$

where $t = 1, \ldots, T$; $i = 1, \ldots, N$. Recall that $\alpha_t(i)$ is the probability of being in state s_i at time t after having observed the sequence $\{o_1, o_2, \ldots, o_t\}$ and $\beta_t(i)$ is the conditional probability of the partial observation $\{o_{t+1}, o_{t+2}, \ldots, o_T\}$ given that the model is in state s_i at time t. Also, recall that these variables are computed inductively as follows:

$$\alpha_t(i) = \pi_i \phi_i(o_1) \qquad\qquad 1 \le i \le N$$
$$\alpha_{t+1}(j) = \left\{\sum_{i=1}^{N} p_{ij} \alpha_t(i)\right\} \phi_j(o_{t+1}) \quad 1 \le t \le T-1, \ 1 \le j \le N$$
$$\beta_T(i) = 1 \qquad\qquad\qquad 1 \le i \le N$$
$$\beta_t(i) = \sum_{j=1}^{N} p_{ij} \beta_{t+1}(j) \phi_j(o_{t+1}) \quad 1 \le t \le T-1, \ 1 \le j \le N$$

As in the Viterbi algorithm, we define the probability variable $\gamma_t(i)$ as follows:

$$\gamma_t(i) = \frac{\alpha_t(i)\beta_t(i)}{P[O|\lambda]} = \frac{\alpha_t(i)\beta_t(i)}{\sum_{i=1}^{N} \beta_t(i)\alpha_t(i)}$$

This is the probability of being in state s_i at time t given the entire observation sequence and the model. Summing $\gamma_t(i)$ over t gives the expected number of transitions made from state s_i. Finally, we define the variable $\xi_t(i,j)$ as the probability of being in state s_i at time t and in state s_j at time $t+1$ given the observation sequence and the model; that is,

$$\xi_t(i,j) = P\left[q_t = s_i, q_{t+1} = s_j | O, \lambda\right] = \frac{P[q_t = s_i, q_{t+1} = s_j, O | \lambda]}{P[O|\lambda]}$$
$$= \frac{\alpha_t(i) p_{ij} \phi_j(o_{t+1}) \beta_{t+1}(j)}{\sum_{i=1}^{N} \beta_t(i)\alpha_t(i)} = \frac{\alpha_t(i) p_{ij} \phi_j(o_{t+1}) \beta_{t+1}(j)}{\sum_{i=1}^{N} \sum_{j=1}^{N} \alpha_t(i) p_{ij} \phi_j(o_{t+1}) \beta_{t+1}(i)}$$

Note that $\gamma_t(i)$ and $\xi_t(i,j)$ are related as follows:

$$\gamma_t(i) = \sum_{j=1}^{N} \xi_t(i,j)$$

Summing $\xi_t(i,j)$ over t gives a value that can be interpreted as the expected number of transitions from state s_i to s_j. Now, we can estimate p_{ij} as the expected number of transitions from state s_i to s_j normalized by the expected number of transitions from state s_i; that is,

$$\overline{p}_{ij} = \frac{\sum_{t=1}^{T-1} \xi_t(i,j)}{\sum_{t=1}^{T-1} \gamma_t(i)}$$

Similarly, we can estimate the probability that the output symbol $o_t = k$ is emitted at time t when the system is in state s_j as the ratio of the expected number of times the system is in state s_j and observing the symbol k to the expected number of times it is in state s_j; that is,

$$\overline{\phi}_j(k) = \frac{\sum_{t=1, o_t=k}^{T} \gamma_t(j)}{\sum_{t=1}^{T} \gamma_t(j)}$$

The details of the algorithm are as follows:

1. Obtain the estimate of the initial state distribution for state s_i as the expected frequency with which state is visited at time $t = 1$; that is

$$\overline{\pi}_i = \gamma_1(i)$$

2. Obtain the estimates $\overline{p}_{ij}$ and $\overline{\phi}_j(k)$ as defined earlier.
3. Let the current model be $\lambda = (P, \Phi, \pi)$ that is used to compute the values of $\overline{p}_{ij}$ and $\overline{\phi}_j(k)$. Let the reestimated model be $\overline{\lambda} = (\overline{P}, \overline{\Phi}, \overline{\pi})$. Using the updated model $\overline{\lambda} = (\overline{P}, \overline{\Phi}, \overline{\pi})$, we perform a new iteration until convergence.
4. If $P[O|\overline{\lambda}] - P[O|\lambda] < \varepsilon$, stop, where ε is a predefined threshold value.

The EM theory states that after each iteration, one of two things can happen:

a. $\overline{\lambda}$ is more likely than λ in the sense that $P[O|\overline{\lambda}] > P[O|\lambda]$ or
b. we have reached a stationary point of the likelihood function at which $\overline{\lambda} = \lambda$.

It must be emphasized that the algorithm is not guaranteed to converge at the global maximum, which is the main problem with the algorithm. This is because many local maxima of the target function might exist. One way to deal with this problem is to run the algorithm several times, each time with different initial values for λ. This problem notwithstanding, the algorithm has been found to yield good results in practice.

14.6 Types of HMMs

HMMs can be classified according to the nature of the distribution of the output probabilities $\phi_i(o_k)$. If the observations o_k are discrete quantities, as we have

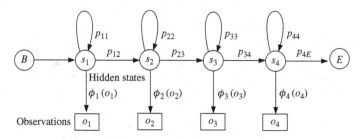

Figure 14.9 Example of left-to-right HMM.

assumed up until now, then $\phi_i(o_k)$ are probability mass functions (PMFs), and the HMM is called a *discrete* HMM. If the observations are continuous random variables, then the HMM is called a *continuous* HMM. In this case, $\phi_i(o_k)$ are probability distribution functions (PDFs) and we have a continuous observation space.

Another popular model is the *left-to-right HMM*. A left-to-right HMM has a left-to-right transition to the next state as well as a self-transition. The self-transition is used to model contiguous features in the same state. It is popularly used to model speech as a time sequence of distinct events that start at an initial state, which is usually labeled *Begin*, and end at a final state, which is usually labeled *End*. The model is also used in *profile HMMs* (PHMMs) that will be discussed later. An example of a left-to-right HMM is illustrated in Figure 14.9, where the states labeled B and E denote Begin and End, respectively, of a sequence.

14.7 HMMs with Silent States

Silent states are special states that do not emit any symbols. They are usually introduced to enhance the clarity of the HMM. In particular, they are used to reduce the number of transitions in a model. For example, if every state is connected to many other states, silent states can be used to skip any state that emits symbols, as shown in Figure 14.10 where m denotes an emitting state (or state that emits symbols) and s denotes a silent state. An emitting state is also called a *match* state. The silent states enable any or all match states to be skipped.

14.8 Extensions of HMMs

Different extensions of the HMM have been proposed by adding flexibility to the model, either through introducing additional sets of new features, developing dependencies among existing feature sets or creating additional relationships between existing features. In this section, we provide a brief description of five of

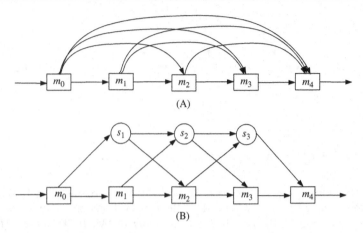

Figure 14.10 The use of silent states: (A) without and (B) with silent states.

these extensions, which are the hierarchical HMM (HHMM), factorial HMM (FHMM), coupled HMM (CHMM), hidden semi-Markov model (HSMM), and PHMM.

14.8.1   Hierarchical Hidden Markov Model

HHMM was proposed by Fine et al. (1998) to extend the standard HMM in a hierarchical manner to a hierarchy of hidden states. Alternatively, it can be considered a structured multilevel model that makes each hidden state in the standard HMM an HHMM as well. This means that each state can emit sequences rather than single symbols. There are two types of states: the "normal" HMM states $S = \{s_1, s_2, \ldots, s_N\}$, which are called *production states*, and *internal* states $I = \{i_1, i_2, \ldots, i_M\}$ that can connect to other states but cannot produce observations. Only the production states can produce observations. There are *end states* at every level from where control is returned to the immediate upper level internal state from where the transition to the sub-HMM originated. That is, entering an end state causes a sub-HMM to terminate, and a transition to an end state could be triggered by some environmental condition.

An example of the HHMM is illustrated in Figure 14.11, where i_{kl} is an internal state l, $l = 1, 2, \ldots$, at level k, $k = 0, 1, \ldots$; q_{kl} is a production state l at level k; and e_{kl} is an end state k at level l. The output states are $o_k, 1, 2, \ldots$.

HHMM is useful in modeling domains with hierarchical structures. For example, it has been used by Fine et al. (1998) to model handwriting, by Ivanov and Bobick (2000) to model visual action recognition, and by Bui et al. (2001) to model spatial navigation.

One of the limitations of HHMM is its computational complexity, which is known to be $O(T^3 N b^2)$, where T is the length of the observation sequence, N is the

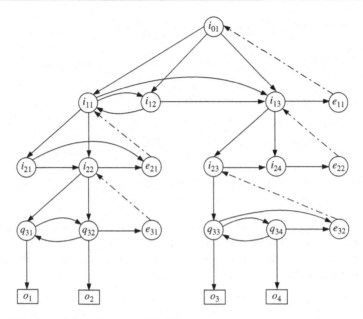

Figure 14.11 Example of a four-level HHMM.

total number of hidden states, and b is the maximum number of substates of each state. Also, the state hierarchy in the original HHMM is restricted to the tree structure. However, Bui et al. (2004) presents a general HHMM in which the state hierarchy can be a lattice that permits arbitrary sharing of substructures.

14.8.2 Factorial Hidden Markov Model

FHMM was proposed by Ghahramani and Jordan (1997). In a regular HMM, information about the past is conveyed through a single discrete variable, which is the hidden state. FHMM permits the state to be factored into multiple state variables and is therefore represented in a distributed manner. Thus, FHMM can be used to represent a combination of multiple signals produced independently where the characteristics of each signal are described by a distinct Markov chain. For example, Kadirkamanathan and Varga (1991) used one chain to represent speech and another chain to represent some dynamic noise source. Similarly, Logan and Moreno (1998) used two chains to represent two underlying concurrent subprocesses governing the realization of an observation vector in speech processing. Jacobs et al. (2002) developed a generalized backfitting algorithm that computes customized error signals for each hidden Markov chain of an FHMM and then trains each chain one at a time using conventional techniques. Figure 14.12 represents an FHMM with two underlying Markov chains governing two subprocesses.

While FHMM enhances the representation power of hidden states by using multiple hidden state chains for one HMM, it also makes the model training difficult

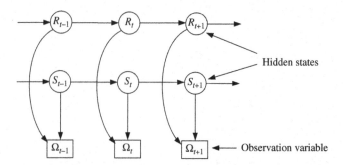

Figure 14.12 FHMM with two underlying Markov chains.

and sometimes impossible when the number of hidden state chains is large. Thus, the combinatorial nature of the model renders the exact algorithm intractable. Consequently, only approximate inference can be obtained using Gibbs sampling or variational methods.

14.8.3   Coupled Hidden Markov Model

CHMM was introduced by Brand (1996) and Brand et al. (1997) to solve one of the limitations of regular HMM, which is its strong restrictive assumption about the system generating the signal. HMM essentially assumes that there is a single process with a small number of states and an extremely limited state memory. The single process model is often inappropriate for vision, speech, and other applications that are composed of multiple interacting processes. CHMM provides an efficient way to resolve many of these problems by coupling HMMs to model interacting processes. It is particularly useful for modeling multimedia applications that integrate multiple streams of data. In this case, one HMM can be used to model one data stream and the model becomes a collection of HMMs.

The simplest type of CHMM consists of two HMM chains with separate observation alphabets, say A and B. Each state has two parents, one from each chain, and the state variable at time t depends on the states of both chains at time $t - 1$. In this way, we are able to capture the temporal relationship between the two chains. An example of a two-chain CHMM is illustrated in Figure 14.13.

CHMMs have been applied in several areas. Brand et al. (1997) demonstrated their superiority to regular HMMs in a vision task classifying two-handed actions. Rezek et al. (2002) derived the maximum *a posteriori* equations for the EM algorithm for CHMM and applied the model to a variety of biomedical signal analysis problems. Kwon and Murphy (2000) used CHMM to model freeway traffic. Xie and Liu (2006) used CHMM for speech animation approach. CHMM permitted the authors to model the asynchrony, different discriminative abilities, and temporal coupling between the audio speech and the visual speech, which are important factors for animations to look natural.

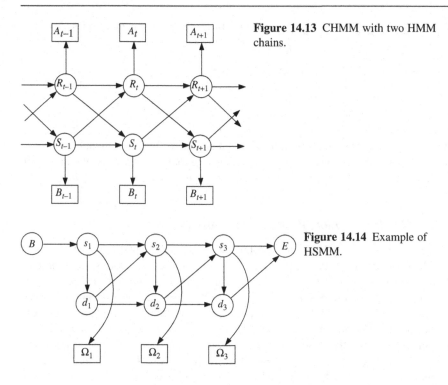

Figure 14.13 CHMM with two HMM chains.

Figure 14.14 Example of HSMM.

14.8.4   Hidden Semi-Markov Models

Just as the semi-Markov process attempts to generalize the Markov process by permitting a generally distributed holding time at each state instead of the exponential or geometric holding time, the HSMM is an HMM in which the number of symbols emitted when the process is at a given state before it moves to a new state is a random variable with some mean and variance. Thus, each state can emit a sequence of observations. In Ferguson (1980) and Levinson (1986), HSMM is called "HMM with variable duration," whereas in Mitchell et al. (1995) it is called "HMM with explicit duration." The model was first investigated by Ferguson (1980).

As stated earlier, a hidden state can emit a string of symbols rather than a single symbol. A hidden state does not have a self-transition because a self-transition defines a geometric distribution over the holding time at the state. A good graphical representation of the model, which is given in Murphy (2002), is shown in Figure 14.14. The states s_k are the regular states that emit symbols, and the states d_k are used to capture the remaining duration of the process in state s_k. When the process enters state s_k, the value of the duration in d_k is chosen according to the probability distribution associated with s_k. When the time in d_k counts down to zero, the state is free to change. Details of the model are given by Murphy (2002). Note that the Ω_k are strings of symbols, i.e., $\Omega = \{o_1, o_2, \ldots, o_m\}$.

Because a string of symbols might be emitted from a hidden state, one of the problems that needs to be solved in addition to the standard HMM problems is to

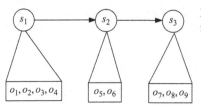

Figure 14.15 Example of output sequence of model in Figure 14.14.

calculate the duration distribution of a given state. For the model shown in Figure 14.14, an example of the time-series output sequence is illustrated in Figure 14.15, where $\Omega_1 = \{o_1, o_2, o_3, o_4\}, \Omega_2 = \{o_5, o_6\}$, and $\Omega_3 = \{o_7, o_8, o_9\}$.

14.8.5 PHMMs for Biological Sequence Analysis

DNA, ribonucleic acid (RNA), and proteins are the fundamental building blocks of life. The three are large molecules. DNA is composed of four bases: *adenine* (A), *cytosine* (C), *guanine* (G), and *thymine* (T). Similarly, RNA has four bases: *adenine* (A), *cytosine* (C), *guanine* (G), and *uracil* (U). Thus, one major difference between DNA and RNA is that RNA has uracil instead of thymine. Proteins are more diverse in structure and function than the other kinds of molecules and are built from an alphabet of 20 smaller molecules known as *amino acids* whose single letter representations are A, V, L, I, F, P, M, D, E, K, R, S, T, C, N, Q, H, Y, W, and G. The molecules are usually connected in a linear sequence such that a DNA molecule, RNA molecule, or protein molecule is represented as a sequence of letters. Such sequences are called *biological sequences*.

This simple sequence representation of the molecules enables them to be compared in a simple way. Thus, it is possible to match or align two sequences letter by letter to see how they pair up. One of the reasons for making such a comparison is to find the evolutionary relation between species on a molecular level. The use of computers has enabled efficient *sequence alignment* methods that are now commonly used in bioinformatics and molecular biology.

Early research in molecular biology and bioinformatics was motivated by protein sequence analysis. However, due to the human genome project and other high-throughput projects, there is a dramatic increase in many types of biological data available. This has extended the scope of bioinformatics research to include topics such as protein classification, RNA analysis, structural and functional predictions, and gene prediction.

We can make an analogy between speech recognition and protein sequence analysis. Both attempt to determine what a sequence represents based on a set of symbols from some alphabet. The alphabet in speech recognition can be a set of valid phonemes for a particular language, while in protein sequence analysis the alphabet is the set of 20 amino acids from which protein molecules are constructed. As in speech recognition, a good stochastic model for a set of proteins is one that assigns high probability to sequences in that particular set and low probability to any other sequence.

HMMs have become one of the most statistically powerful methods used to model sequence alignment. A special type of left-to-right HMM called PHMM is commonly used to model multiple alignments. The architecture of PHMM was introduced by Krogh et al. (1994). PHMM is well suited to the popular "profile" methods for searching databases using multiple sequence alignments instead of single query sequences. It has three types of states: *match states* that are represented by squares labeled *m*, *insert states* that are represented by diamonds labeled *i*, and *delete states* that are represented by circles labeled *d*.

Match states generate amino acids according to a probability distribution for the 20 amino acids, and different probability distributions apply to different match states. They thus correspond to positions in a protein or columns in multiple alignments. The amino acids emitted in these states are the same as those in the common ancestor, and if not, then they are the result of substitutions. We assume that there are M match states and match state m_k generates amino acid x with probability $P[x|m_k], k = 1, \ldots, M$.

Delete states are silent states that do not generate amino acids and are used for sequences from the family in which the amino acid from such a column has been deleted. They are "dummy" states that are used to skip the match states. For each match state m_k, there is a corresponding delete state d_k used to skip m_k. The match–delete pair is sometimes called a *fat state* that is visited exactly once on every path from *Begin* to *End*.

Insert states also generate amino acids according to some probability distribution and represent sequences with one or more inserted amino acids between columns of multiple sequences. That is, insert states are used to represent possible amino acids that are not found in most of the sequences in the family being modeled and are thus the result of insertion. There are $M + 1$ insert states that generate amino acid x with probability $P[x|i_k], k = 1, \ldots, M + 1$, and they permit self-transitions. Figure 14.16 illustrates the architecture of the PHMM.

Note that there are at most three transitions into each state and three transitions out of each state. Thus, when the Viterbi algorithm is used to analyze the model, the computational complexity is $O(nt)$, where n is the number of states and t is the observation sequence length. For a traditional HMM, the computational complexity is $O(n^2 t)$.

PHMM Training

HMM training is the estimation of the emission and transition probabilities. For PHMM, these parameters are obtained from multiple alignment sequences in a protein, DNA, or RNA sequence family. If there is any sequence whose components are known, then it can be used for the training. In general, the emission probabilities are the maximum likelihood estimates of the letters in each column. Similarly, the transition probabilities are obtained by counting the number of times each transition would be taken.

Multiple alignments mean taking a group of three or more sequences and identifying the amino acids that are homologous (or structurally and functionally

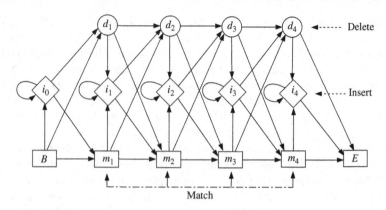

Figure 14.16 Architecture of PHMM.

similar). Proteins and nucleic acid sequences and their interrelationships can be demonstrated by multiple alignments of the sequences. The information from the multiple alignments is usually condensed into a single sequence called a *consensus sequence* that shows which symbols are conserved (i.e., are always the same) and which symbols are variable. Multiple alignments are performed by arranging the sequences in a matrix such that each row of the matrix consists of one sequence padded by gaps, and the individual columns represent homologous characters. The columns of the matrix highlight similarity (or *residue conservation*) between positions of each sequence. An optimal multiple alignment is one that has the highest degree of similarity.

Consider the following sequences:

    ACAATC
    TCAACTATC
    ACACAGC
    AGAATG
    ACCGATC

Because the sequences are of different lengths, the first step is to introduce gaps to make them of the same length as follows:

    A C A – – – A T C
    T C A A C T A T C
    A C A C – – A G C
    A G A – – – A T G
    A C C G – – A T C

Thus, we can create a PHMM for this multiple sequence alignments as follows. Three columns were introduced to equalize the number of columns, and these constitute the insert state. There are six other columns that constitute the match states. The first column consists of two distinct letters with the following frequencies of occurrence: A (4) and T (1), which means that the emission probabilities in the first state are $4/5 = 0.8$ for A and $1/5 = 0.2$ for T. These probabilities are used to populate match state m_1. Similarly, in column 2, the emission probability is $4/5 = 0.8$ for C and $1/5 = 0.2$ for G. These probabilities are used to populate match state m_2 and so on for the other match states. The transition probability from m_1 to m_2 is 1 and from m_2 to m_3 is also 1. Two of the insertion rows contain only gaps, which means that the probability of a direct transition from m_3 to m_4 is $2/5 = 0.4$, and the probability of a transition from m_3 to the insert state i_3 is $3/5 = 0.6$. Also, the transition probability from m_4 to m_5 is 1 and from m_5 to m_6 is also 1. Five letters can be emitted in the insert state, and their emission probabilities are as follows: $1/5 = 0.2$ each for A, G, and T, and $2/5 = 0.4$ for C. Finally, the insert state requires two self-transitions: A to C and C to T. Thus, since two of the five letters that are emitted in this state are due to self-transition action, the probability of a self-transition is $2/5 = 0.4$. With all the necessary parameters defined, we can construct the PHMM as shown in Figure 14.17.

Note that sometimes a match state is defined as a column in which the number of gaps is no more than half the number of elements in the column. Thus, in this case, column 4 of the preceding example would be a match state, and we would have an insert state i_4 and a delete state d_4 associated with it. The delete state would permit a transition from column 3, which is m_3, to column 7, which would be m_5. While PHMM enhances the modeling capability of the standard HMM, it utilizes the solution methodologies of the standard HMM.

Scoring a Sequence with PHMM

Any sequence can be represented by a path through the model. The probability of any sequence, given the model, is computed by multiplying the emission and

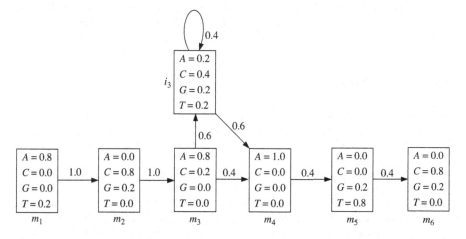

Figure 14.17 Example of PHMM construction.

transition probabilities along the path. Given a PHMM, the probability of a sequence is the product of the emission and transition probabilities along the path of the sequence. For example, the probability of *AGCATG*, given the PHMM in Figure 14.17, is

$$0.8 \times 1.0 \times 0.2 \times 1.0 \times 0.8 \times 0.4 \times 1.0 \times 1.0 \times 0.8 \times 1.0 \times 0.2 = 0.008192$$

The probability of a sequence is used to calculate a *score* for the sequence. Because multiplication of fractions is computationally expensive and prone to floating point errors such as underflow, the calculation is simplified by taking the logarithm of the score, thereby replacing multiplication by addition. The resulting number is the *log score* of a sequence. Applying this method to the previous calculation, we obtain the log score as follows:

$$3 \log_e(0.8) + 4 \log_e(1) + 2 \log_e(0.2) + \log_e(0.4) = -4.8046$$

Because a score measures the probability that a sequence belongs to a given family, a high score implies that the sequence of interest is probably a member of the class, whereas a low score implies that it is probably not a member.

14.9   Other Extensions of HMM

We have discussed five extensions of the basic HMM: HHMM, FHMM, CHMM, HSMM, and PHMM. However, many other extensions of HMM have also been proposed but are not discussed in this book. These include the *buried Markov model*, which was introduced by Bilmes (2003), and the *partially hidden Markov model*, which was introduced by Forchhammer and Rissanen (1996). The *partly hidden Markov model* discussed by Kobayashi et al. (1999) and Ogawa and Kobayashi (2005) is an extension of the partially hidden Markov model.

14.10   Problems

14.1 Consider an HMM with two states 1 and 2 and emits two symbols: A and B. The state-transition diagram is shown in Figure 14.18.
   a. Use the Viterbi algorithm to obtain the most likely state sequence that produced the observation sequence {ABBAB}.
   b. Estimate the probability that the sequence {BAABA} was emitted by the preceding system.

14.2 Consider the HMM shown in Figure 14.19 that has three hidden states 1, 2, and 3, and emits two output symbols: *U* and *V*. When it is in state 1, it is equally likely to emit either symbol. When it is in state 2, the probability that it emits the symbol *U* is 0.1, and the probability that it emits the symbol *V* is 0.9. Finally, when it is in state 3, the

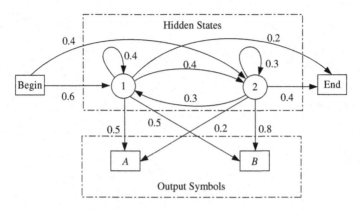

Figure 14.18 Figure for Problem 14.1.

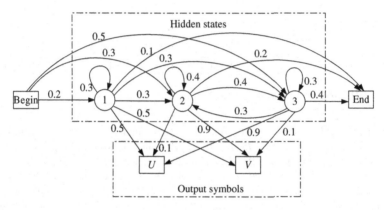

Figure 14.19 Figure for Problem 14.2.

probability of emitting the symbol U is 0.9 and the probability of emitting the symbol V is 0.1.

a. If the output symbol is $\{UUV\}$, estimate the most likely transition path through the system.

b. Convert the HMM into an HMM with silent states.

14.3 Construct the PHMM for the following variable length sequences DOR, DM, DAP, VGBLM. (Hint: Use the following alignment to identify the match, insert, and delete states.)

$$D\ O\ -\ -\ R$$
$$D\ -\ -\ -\ M$$
$$D\ A\ -\ -\ P$$
$$V\ G\ B\ L\ M$$

14.4 Consider three coins labeled 1, 2, and 3. When coin 1 is tossed, the probability that it comes up heads is 0.75 and the probability that it comes up tails is 0.25. Similarly,

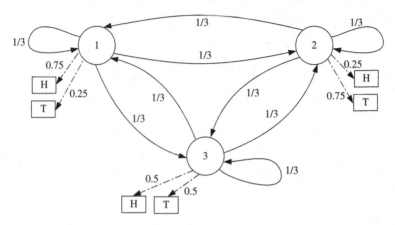

Figure 14.20 Figure for Problem 14.4.

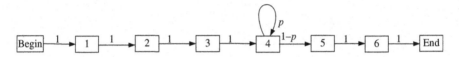

Figure 14.21 Figure for Problem 14.5.

when coin 2 is tossed, the probability that it comes up heads is 0.25 and the probability that it comes up tails is 0.75. Finally, when coin 3 is tossed, the probability that it comes up heads is 0.5 and the probability that it comes up tails is 0.5. Assume that in an experiment that involves a sequence of tosses of these coins, the experimenter is equally likely to choose any coin during the next toss. Thus, if we denote "heads" by H and "tails" by T, the experiment can be modeled by the Markov chain as shown in Figure 14.20. Assume that the experimenter is equally likely to start a sequence of tosses with any coin.

a. What is the probability that the observation sequence {HTTHT} was emitted by the model?

b. Use the Viterbi algorithm to obtain the most likely state sequence that produced the observation sequence {HTTHT}.

14.5 Consider a system that can be modeled by an array of six states labeled $1, 2, \ldots, 6$. Apart from state 4, which makes a transition to itself with probability p, every other state is visited only once in each experiment that starts in the Begin state and ends when the End state is reached. The model is illustrated in Figure 14.21.

Let L be a random variable that denotes the length of time that the process spends in state 4 when the process reaches that state in an experiment.

a. What is the PMF of L?

b. What is the expected value of L?

14.6 Consider a system that can be modeled by an array of six states labeled $1, 2, \ldots, 6$. Every state makes a transition to itself with probability p and makes a transition to the next higher state with probability $1 - p$. An experiment starts in the Begin state and ends when the End state is reached. The model is illustrated in Figure 14.22.

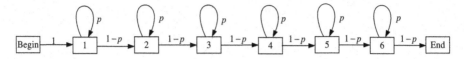

Figure 14.22 Figure for Problem 14.6.

Let L be a random variable that denotes the length of time that the process takes to pass from state 1 to the End state in an experiment.
a. What is the PMF of L?
b. What is the expected value of L?

15 Markov Point Processes

15.1 Introduction

Point processes are stochastic processes that are used to model events that occur at random intervals relative to the time axis or the space axis. Thus, there are two types of point processes: *temporal point processes* and *spatial point processes*. The representation of physical events as point processes is based on two major assumptions. The first is that the physical events must be pointlike in the sense of occupying a small area in the relevant domain. The second is that the events must be discrete entities so that there will be no ambiguity when they occur. For this reason, a point process can be considered as a set of discrete events that occur at well-defined but random points in time or space.

A temporal point pattern is basically a list of times of events. Many real phenomena produce data that can be represented as a temporal point pattern. Usually complex mechanisms are behind these seemingly random times, e.g., earthquakes cause new earthquakes in the form of aftershocks. An essential tool for dealing with these mechanisms, e.g., in predicting future events, is a stochastic process modeling the point patterns: a *temporal point process*. The term point is used since we may think of an event as being an instant and thus we can represent it as a point on the time line. For the same reason, the words point and event can be used interchangeably. The lengths of the time intervals between subsequent events are known as *interevent times*.

While temporal point processes deal with events that are observed over time as a time series, spatial point processes describe the locations of objects in a d-dimensional space R^d, where $d = 2$ or $d = 3$ in many applications of interest. Spatial point processes are used in modeling in a variety of scientific disciplines including agriculture, astronomy, bacteriology, biology, climatology, ecology, epidemiology, forestry, geography, geology, and seismology.

Let X be a point process defined in a given bounded set S, and let $x = \{x_1, x_2, \ldots, x_n\}$ be a configuration of X. Let $N_X(B)$ denote the random variable that represents the number of points of X in the finite set (or region) $B \subseteq S$; that is,

$$N_X(B) = \#\{x_i \in B\}$$

where the symbol # is used to denote the number of points in the set following it. For temporal point processes, $B \subseteq S \subseteq R$, where R is the real line; and for spatial processes, $B \subseteq S \subseteq R^d$, where typically $d = 2$ or $d = 3$.

A point process is called a *simple point process* if, with probability one, all points are distinct. It is called an *orderly point process* if for any t,

Markov Processes for Stochastic Modeling. DOI: http://dx.doi.org/10.1016/B978-0-12-407795-9.00015-3

$$\lim_{\Delta t \to 0} \left\{ \frac{P[N_X(t, t + \Delta t) > 1]}{\Delta t} \right\} = 0$$

Thus, an orderly point process is one that does not allow multiple simultaneous event occurrences. Another way to mathematically define an orderly point process is as follows:

$$P[N_X(t, t + \Delta t) > 1] = o(\Delta t) \quad \forall\, t \in R$$

A *self-exciting point process* is one in which $\text{Cov}[N_X(A), N_X(B)] > 0$ for any two adjacent sets A and B. A self-exciting point process is also called a *clustered* or *underdispersed* process. A *self-correcting point process* is one in which $\text{Cov}[N_X(A), N_X(B)] < 0$. It is also called an *inhibitory* or *overdispersed* process. Thus, in a self-exciting point process, the occurrence of a point (or event) enables other events to occur, whereas events in a self-correcting point process tend to inhibit the occurrence of other events. Self-exciting point process models are often used in epidemiology and seismology to model events that are clustered together in time and space.

15.2   Temporal Point Processes

A temporal point process is a stochastic process where the time points of occurrence of events consist of the times $\{T_k\}$ of isolated events scattered in time, where $0 \le T_1 \le T_2 \le \cdots$. There are two parts to the process: a *counting process* that deals with the number of events in a fixed time interval and an *interval process* that deals with the time intervals between subsequent events. Thus, the process X can be represented in a number of ways including

- Counting measure $N_X(0, t)$, which denotes the number of events over the interval $(0, t)$. Then we can write

$$N_X(0, t) = \sum_{k=1}^{\infty} I(T_k \le t) = \#\{0 < T_k \le t\}$$

where $I(a)$ is an indicator function that is 1 if the statement a is true and 0 otherwise.
- Interevent intervals $\{W_k\}$, which are given by $W_k = T_k - T_{k-1} \ge 0$. This means that the random variables W_k represent distances between points.

Thus, we can alternatively define a temporal point process on R_+, where R_+ is the set of positive real numbers, as a sequence $\{T_k, k \ge 0\}$ of nonnegative random variables with the following properties:

a. $T_0 = 0$
b. $0 < T_1 < T_2 < \cdots$
c. $\lim_{k \to \infty} T_k = \infty$

Thus, the distances between points, W_k, are given by $W_1 = T_1$ and

$$W_k = T_k - T_{k-1} \quad k = 2, 3, \ldots$$

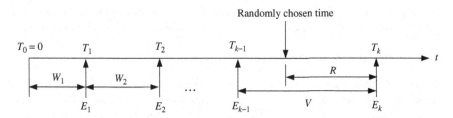

Figure 15.1 Relationship between W and R.

A temporal point process $\{X\}$ is defined to be a *time-stationary point process* if $N_X(B)$ has the same distribution as $N_X(B + t)$ for some $t \in R$, where R is the set of real numbers, and the sequence $\{W_k\}$ corresponding to $N_X(B)$ has the same distribution as $\{W_{k+l}\}$ for some $l \in Z$, where Z is the set of integers. Thus, for a stationary temporal point process X, the time origin can be any arbitrary point, and the expected value of the counting measure $N_X(B)$ is directly proportional to $|B|$ and is given by

$$E[N_X(B)] = \Lambda |B|$$

where $|B|$ is the Lebesgue measure of B and Λ is a constant of proportionality called the *intensity measure* or the *rate* of the process. (Lebesgue measure coincides with the standard measure of length, area, or volume in one, two, and three dimensions, respectively.)

In the one-dimensional case, assume that the intervals, W_k, between events in a stationary temporal point process are identically distributed, and let $f_W(w)$ and $F_W(w)$ denote the probability distribution function (PDF) and cumulative distribution function (CDF), respectively, of W. Assume also that $E[W] = \eta$. Let the random variable R denotes the forward-recurrence time, which is the time from now until the next event occurs, and let $f_R(r)$ and $F_R(r)$ denote the PDF and CDF, respectively, of R. These random variables are shown in Figure 15.1, which is similar to Figure 6.3.

Suppose we select an interval between events of length V at random and choose the point indicated in Figure 15.1 at random also. As discussed in Chapter 6,

$$f_V(v) = \frac{v f_W(v)}{E[W]} = \frac{v f_W(v)}{\eta}$$

$$f_{RV}(r, v) = \frac{f_W(v)}{\eta} \qquad 0 \le r \le v < \infty$$

$$f_R(r) = \frac{1 - F_W(r)}{\eta} \qquad r \ge 0$$

15.3 Specific Temporal Point Processes

There are many examples of temporal point processes, which include the Poisson point process, the renewal point process, and the Cox point process, all of which we briefly discuss in the following sections. The Poisson process is the most important point process and is widely used as a building block for more complex models.

15.3.1 Poisson Point Processes

The general characteristics of the Poisson process are discussed in Chapter 2. In this section, we summarize those characteristics that are pertinent to the theory of temporal point processes. A Poisson process X with rate $\Lambda(B)$ is a stochastic process whose probability mass function (PMF) for the number of events in the interval $(t_1, t_2]$ is given by

$$p_{N_X(t_1,t_2)}(n) = P[N_X(t_1, t_2) = n] = \frac{[\Lambda(\tau)]^n}{n!} e^{-\Lambda(\tau)} \quad n = 0, 1, \ldots$$

where $\tau = t_2 - t_1$. For a stationary Poisson process, $\Lambda(\tau) = \lambda\tau$ and $E[N_X(t_1, t_2)] = \lambda\tau$. Also, the PDF of the times between events (or the waiting times) for a stationary Poisson process with rate λ is given by

$$f_W(w) = \lambda e^{-\lambda w} \quad w \geq 0$$

That is, the times between events are independent and exponentially distributed with a mean of $1/\lambda$.

A nonhomogeneous Poisson point process is a Poisson process with a variable rate $\lambda(t)$. It is used to model Poisson arrival processes where arrival occurrence epochs depend on time, such as time of the day or time of the year.

15.3.2 Cox Point Processes

A Cox point process is sometimes called the *doubly stochastic Poisson process*, because it is an extension of the Poisson process. It can be obtained by first randomly generating the intensity measure $\Lambda = \lambda$, then generating a Poisson process with the intensity measure λ. Thus, a point process X is a Cox point process if, conditional on the intensity measure $\Lambda = \lambda$, X is a Poisson process with rate λ. This means that for a single realization of X, it is not possible to distinguish a Cox point process from its corresponding Poisson point process.

Since the intensity measure $\Lambda(t)$ is a random variable with a predefined probability distribution, we can also define a Cox point process as a Poisson point process with a variable intensity measure that is itself a stochastic process, which we refer to as the intensity process. The Cox point process is stationary if and only if the intensity process is stationary.

15.4 Spatial Point Processes

Spatial point processes are used to model events that occur in space (or on a plane). Thus, a spatial point process is a finite subset S of a d-dimensional space R^d or the entire R^d; that is, $S \subseteq R^d$. One important class of spatial point processes is the *stationary* or *homogeneous* point process. A stationary (or homogeneous) spatial point process is a spatial point process whose distribution is invariant under translation; that is, for an integer k and regions $B_i, i = 1, \ldots, k$, the joint distribution of $N_X(B_1), \ldots, N_X(B_k)$ is equal to the joint distribution of $N_X(B_1 + y), \ldots, N_X(B_k + y)$ for an arbitrary y. Another class of spatial point processes is the *isotropic* point process, which is a spatial point process whose distribution is invariant under rotation through an arbitrary angle; that is, there is no directional effect.

Spatial point processes are used in many applications. They are used to model multihop radio networks in Cheng and Robertazzi (1990). They have also been used to model defensive strategies in Kornak et al. (2006). In Ayala et al. (2006), they are used in clustering of spatial point patterns where the interest is in finding groups of images corresponding with groups of spatial point patterns.

Spatial point processes are commonly characterized by their moment measures. For a spatial point process X, the intensity measure is given by

$$\Lambda(B) = E[N_X(B)]$$

The first- and second-order intensities are used to determine the mean and dependency structure of the data. The intensity measure is related to the *first-order intensity function* $\lambda(x)$ as follows:

$$\Lambda(B) = \int_B \lambda(x) dx$$

For a spatial point process that is stationary and isotropic, the intensity function is a constant, λ. In the case where $S \subseteq R^2$, the first-order intensity function is defined as the number of events per unit area. That is,

$$\lambda(x) = \lim_{|\Delta x| \to 0} \left\{ \frac{E[N_X(\Delta x)]}{|\Delta x|} \right\}$$

The *second-order intensity function* of a spatial point process is defined by

$$\lambda_2(x_1, x_2) = \lim_{\substack{|\Delta x_1| \to 0 \\ |\Delta x_2| \to 0}} \left\{ \frac{E[N_X(\Delta x_1) N_X(\Delta x_2)]}{|\Delta x_1| \, |\Delta x_2|} \right\}$$

The quantity $\lambda_2(x_1, x_2) dx_1 \, dx_2$ is the approximate probability that there is at least one point of X in each of the regions dx_1 and dx_2. In a homogeneous case, we have that

$$\lambda_2(x_1, x_2) = \lambda_2(x_1 - x_2)$$

The second-order intensity function is also called the *second-order product density*. For a stationary and isotropic spatial process, we have that

$$\lambda_2(x_1, x_2) = \lambda_2(u)$$

where $u = |x_1 - x_2|$. The covariance density of a spatial point process is given by

$$\gamma(x_1, x_2) = \lambda_2(x_1, x_2) - \lambda(x_1)\lambda(x_2)$$

The *pair correlation function* is the normalized second-order intensity function and is defined by

$$g(x_1, x_2) = \frac{\lambda_2(x_1, x_2)}{\lambda(x_1)\lambda(x_2)} \Rightarrow \gamma(x_1, x_2) = \lambda(x_1)\lambda(x_2)\{g(x_1, x_2) - 1\}$$

For a spatial Poisson point process, which is a completely random process, the pair correlation function $g(x_1, x_2) = 1$. If $g(x_1, x_2) > 1$, it means that for a relatively small interpoint distance $u = |x_1 - x_2|$, the interdistance is more frequent than in a random point pattern, which implies that the points in X tend to cluster relative to a Poisson process with the same intensity function as the process X. Similarly, when $g(x_1, x_2) < 1$, the points tend to repel relative to a Poisson process with the same intensity function.

Another common second-order characteristic of a stationary isotropic spatial point process is the K-function (or *reduced second-moment measure*), which is defined as follows:

$$K(r) = \frac{1}{\lambda} \int_{\theta=0}^{2\pi} \int_{x=0}^{r} \lambda_2(x, \theta) x \, dx \, d\theta$$

where λ is the rate of the process. $K(r)$ measures the expected relative rate of events within a distance r of an arbitrary event. If a point process is clustered, then each event is likely to be surrounded by more events from the same cluster, which means that $K(r)$ will be relatively large for small r. Similarly, if a process is randomly distributed in space, then each event is likely to be surrounded by an empty space, which means that $K(r)$ will be relatively small for small r. For a stationary Poisson process, $K(r) = \pi r^2$. When a process has $K(r) > \pi r^2$, it means that the points tend to cluster, and when $K(r) < \pi r^2$, it means that the points tend to repel each other.

Several examples of applications of spatial point processes in urban public systems are discussed by Larson and Odoni (1981).

15.5 Specific Spatial Point Processes

As in the case of temporal point processes, examples of spatial point processes include the Poisson process, the renewal process, the Cox point process, and the Gibbs process.

15.5.1 Spatial Poisson Point Processes

A spatial point process X defined in a given bounded region S is a Poisson process with rate λ if the number $N_X(B)$ of points of X in the region $B \subseteq S$ has a Poisson distribution with mean $\lambda|B|$. Thus, the PMF of $N_X(B)$ is given by

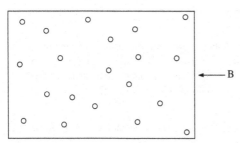

Figure 15.2 Realization of two-dimensional Poisson process.

$$p_{N_X(B)}(n) = P[N_X(B) = n] = \frac{[\lambda|B|]^n}{n!} e^{-\lambda|B|} \quad \lambda|B| \geq 0, \quad n = 0, 1, \ldots$$

The homogeneous spatial Poisson point process exhibits what is known as *complete spatial randomness* (CSR), which means that the events are independent and uniformly distributed over the region B.

For a nonhomogeneous Poisson process, the rate λ will not be a constant but some function of position, $\lambda(B)$. Also, in the case where $S \subseteq R^2$, the Lebesgue measure $|B|$ will be the area of the region B. In this case, if we denote the probability that a point occurs in the region $\{(x, x + dx), (y, y + dy)\}$ by $\lambda(x, y)dx\, dy$, then we have that

$$\lambda|B| = \int_B \int \lambda(x, y)dx\, dy$$

As stated earlier, because of the CSR property of the spatial Poisson point process, given that $N_X(B) = n$, the locations of the n points in B are independent and uniformly distributed random variables. Thus, one realization of the process is shown in Figure 15.2.

For a stationary and isotropic Poisson point process, the first-order intensity function is given by

$$\lambda(x) = \frac{E[N_X(B)]}{|B|} = \frac{\lambda|B|}{|B|} = \lambda$$

which is a constant for all B. Similarly, the second-order intensity function depends only on the distance between locations x_1 and x_2, i.e.,

$$\lambda_2(x_1, x_2) = \lambda_2(||x_1 - x_2||)$$

Thus, the covariance density of the process is given by

$$\gamma(x_1, x_2) = \lambda_2(||x_1 - x_2||) - \lambda^2 = \gamma(||x_1 - x_2||)$$

Example 15.1

Consider a city in which police cars are distributed according to a Poisson process with a rate of η cars per square mile. Assume that an incident requiring police presence occurs somewhere in the city. What is the PDF of the distance L between the location of the incident and the nearest police car, assuming an Euclidean travel distance?

Solution

This is an example of a class of problems called the *nearest-neighbor* problems. Let the point of the incident be (x, y). For an Euclidean travel distance, we construct a circle of radius r centered at (x, y) so that the Lebesgue measure is the area of the circle, i.e., $|B| = \pi r^2$. Let $M(r)$ denote the number of police cars within a circle of radius r. Then the PMF of $M(r)$ is

$$p_{M(r)}(m) = P[M(r) = m] = \frac{(\eta \pi r^2)^m \exp(-\eta \pi r^2)}{m!} \quad m = 0, 1, \ldots$$

The CDF of L is given by

$$F_L(l) = P[L \leq l] = 1 - P[L > l] = 1 - P[M(l) = 0]$$
$$= 1 - \exp(-\eta \pi l^2) \quad l \geq 0$$

Thus, the PDF of L is

$$f_L(l) = \frac{d}{dl} F_L(l) = 2l\eta\pi \exp(-\eta \pi l^2) \quad l \geq 0$$

15.5.2 Spatial Cox Point Processes

As discussed in an earlier section, the Cox process X is a doubly stochastic Poisson process. Thus, conditional on $\Lambda = \lambda$, the process becomes a spatial Poisson process with mean $\lambda |B|$. One property of Cox processes is that their variances always exceed those of the stationary Poisson processes with the same intensity. This can be demonstrated in the following manner using the method used by Kingman (1993), as follows:

$$E[\{N_X(B)\}^2] = E[E[\{N_X(B)\}^2 | \lambda]]$$
$$= E[\text{Var}(\text{Poi}(\lambda(B))) + \{E[\text{Poi}(\lambda(B))]\}^2]$$
$$= E[\lambda(B) + \{\lambda(B)\}^2] = E[\lambda(B)] + E[\{\lambda(B)\}^2]$$
$$= E[\lambda(B)] + \text{Var}(\lambda(B)) + \{E[\lambda(B)]\}^2$$

where $\text{Var}(\text{Poi}(\lambda(B)))$ is the variance of a Poisson process with mean $\lambda(B)$. From this we obtain

$$E[\{N_X(B)\}^2] - \{E[\lambda(B)]\}^2 = \text{Var}(N_X(B)) = E[\lambda(B)] + \text{Var}(\lambda(B))$$

Because $E[\lambda(B)] \geq 0$, we have that $\text{Var}(N_X(B)) \geq \text{Var}(\lambda(B))$ and also $\text{Var}(N_X(B)) \geq E[\lambda(B)]$. Thus, the count measure $N_X(B)$ has a greater variance than a Poisson random variable with the same mean. For this reason, all Cox processes are said to be "overdispersed" relative to the Poisson processes.

15.5.3 Spatial Gibbs Processes

The Gibbs process originated in statistical physics and is described in terms of forces acting on and between particles. It enables the total potential energy associated with a given configuration of particles to be decomposed into terms representing the external force field on individual particles and terms representing interactions between particles taken in pairs, triplets, etc.

Gibbs processes are not universal models that apply to all situations. Instead, their distributions are defined according to the application of interest. More importantly, they do not perform well in applications with strong regularity; they are good for applications with some degree of regularity. Also, they are more applicable when the system to be modeled contains only a finite number of points in a bounded region B.

One advantage of the Gibbs point process over the Poisson point process is that the Poisson point process is not able to account for interactions between points, whereas the Gibbs point process can. In fact, the Gibbs point process can be regarded as a pairwise interacting process.

The PDF of the Gibbs process X is given by

$$f_X(x) = f_{X_1 X_2 \ldots X_n}(x_1, x_1, \ldots, x_n) = \frac{1}{Z} \exp\{-\beta U(x)\}$$

$$= \frac{1}{Z} \exp\{-\beta U(x_1, x_2, \ldots, x_n)\}$$

where $U(x)$ is called the *energy function*; $\beta = 1/kT$, where T is the *temperature* (in absolute degrees) and k is the *Boltzmann's constant*; and Z is a normalizing constant called the *partition function* and is given by

$$Z = \int_B \exp\{-\beta U(x_1, x_2, \ldots, x_n)\} dx_1 dx_2 \ldots dx_n$$

$U(x)$ is usually defined in such a manner as to match the application. However, it is generally in the form of a series as follows:

$$U(x) = \sum_{i=1}^{n} V_1(x_i) + \sum_{i_1 > i_2} V_2(x_{i_1} - x_{i_2}) + \sum_{i_1 > i_2 > i_3} V_3(x_{i_1} - x_{i_2}, x_{i_1} - x_{i_3})$$

$$+ \cdots + \sum_{i_1 > i_2 > \cdots > i_k} V_3(x_{i_1} - x_{i_2}, x_{i_1} - x_{i_3}, \ldots, x_{i_1} - x_{i_k})$$

where the $V_1(\cdot)$ are called the *potential functions* and $V_i(\cdot), i > 1$, is ith-order interaction potential function. In many applications, it is assumed that only the first-order and second-order interactions are significant, i.e., $k = 2$. In this case, we obtain

$$U(x_1,\ldots,x_n) = \sum_{i=1}^{n} V_1(x_i) + \sum_{i_1 > i_2} V_2(x_{i_1} - x_{i_2})$$

$$= \sum_{i=1}^{n} V_1(x_i) + \sum_{i=1}^{n-1} \sum_{j=i+1}^{n} V_2(x_i - x_j)$$

Discussions on the different restrictions on the values of the potential functions can be found in Cox and Isham (2000) and Daley and Vere-Jones (2003).

15.6 Spatial–Temporal Point Processes

A spatial–temporal (or spatio-temporal or space–time) point process is a random collection of points whose coordinates represent the time and location of an event. If the space component describes the locations of objects in a d-dimensional space R^d, then $S \subseteq R \times R^d$. For the case of $d = 2$, the points in a spatial–temporal process are represented in the form (t, x, y), where t denotes the time of occurrence of the event, x denotes the location on the x-axis, and y denotes the location on the y-axis. Figure 15.3 illustrates a spatial–temporal point process for a one-dimensional space.

Spatial–temporal processes are used in many disciplines including the study of earthquakes, epidemiology, and occurrence of fire and lightning strikes. One important issue in the analysis of spatial–temporal processes is the interaction between the space and time components. They may be noninteracting or interact in any number of ways.

A simple illustration of the spatial–temporal process X is the following. Consider a region $B \subseteq S \subseteq R^d$, where events occur according to a Poisson process with rate $\lambda|B|$ events per unit time. Let $N_X(B,t)$ denote the number of events in the region B over a time interval t. Then the PMF of $N_X(B,t)$ is given by

$$
\begin{aligned}
p_{N_X(B,t)}(n,t) &= P[N_X(B,t) = n] \\
&= \frac{\{\lambda t |B|\}^n}{n!} \exp\{-\lambda t |B|\} \qquad \lambda|B| \geq 0; \ t \geq 0; \ n = 0, 1, \ldots
\end{aligned}
$$

An application of the spatial–temporal point process to model Poisson processes with birth, death, and movement is discussed by Cox and Isham (2000). The

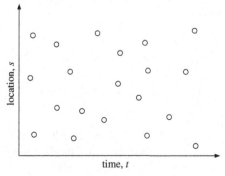

Figure 15.3 Illustration of spatial–temporal point process.

process is also used to model earthquakes in Rathbun (1996), Ogata (1999), and Choi and Hall (1999). It is used in Rathbun and Cressie (1994) to analyze longleaf pines in southern Georgia.

The intensity measures are functions of both time and space. Thus, the first-order intensity function is given by

$$\lambda(x, t) = \lim_{\substack{|\Delta x| \to 0 \\ |\Delta t| \to 0}} \left\{ \frac{E[N_X(\Delta x, \Delta t)]}{|\Delta x| \, |\Delta t|} \right\}$$

where Δx is an infinitesimal disk containing the location x and Δt is an infinitesimal interval containing the time t. The *marginal first-order spatial intensity function* is given by

$$\lambda(x, -) = \int_T \lambda(x, t) \mathrm{d}t$$

where integration is over all time T. Similarly, the *marginal first-order temporal intensity function* is given by

$$\lambda(-, t) = \int_A \lambda(x, t) \mathrm{d}x$$

where integration is over the region A. These marginal intensity functions permit us to view one component while ignoring the other. The *conditional first-order spatial intensity function* is given by

$$\lambda(x|t = t_0) = \lim_{|\Delta x| \to 0} \left\{ \frac{E[N_X(\Delta x, t_0)]}{|\Delta x|} \right\}$$

Similarly, the *conditional first-order temporal intensity function* is given by

$$\lambda(t|x = x_0) = \lim_{|\Delta t| \to 0} \left\{ \frac{E[N_X(x_0, \Delta t)]}{|\Delta t|} \right\}$$

A spatial–temporal point process is defined to be *first-order stationary in space* if

$$\lambda(t|x) = \lambda(t)$$

Similarly, a spatial–temporal point process is defined to be *first-order stationary in time* if

$$\lambda(x|t) = \lambda(x)$$

Thus, the conditional first-order temporal intensity function of a stationary temporal–spatial point process is independent of location, and the conditional first-order spatial intensity function is independent of time.

We can also obtain the second-order intensity function by

$$\lambda_2(x_1, x_2, t_1, t_2) = \lim_{\substack{|\Delta x_1| \to 0, \, |\Delta x_2| \to 0 \\ |\Delta t_1| \to 0, \, |\Delta t_2| \to 0}} \left\{ \frac{E[N_X(\Delta x_1, \Delta t_1) N_X(\Delta x_2, \Delta t_2)]}{|\Delta x_1| \, |\Delta x_2| \, |\Delta t_1| \, |\Delta t_2|} \right\}$$

The *marginal second-order spatial intensity function* and the *marginal second-order temporal intensity function* are given, respectively, by

$$\lambda_2(x_1, x_2, -, -) = \int_T \int_T \lambda_2(x_1, x_2, t_1, t_2) dt_1 \, dt_2$$

$$\lambda_2(-, -, t_1, t_2) = \int_{A_1} \int_{A_2} \lambda_2(x_1, x_2, t_1, t_2) dx_1 \, dx_2$$

Details on how to compute other statistics of the spatial–temporal point process can be found in Dorai-Raj (2001).

15.7   Operations on Point Processes

Sometimes new point processes can be constructed from old ones to fit the environment of interest. There are several methods used to generate new processes from old ones, but we consider only three of them, which are:

- thinning,
- superposition, and
- clustering.

15.7.1   Thinning

Thinning is an operation on a point process that essentially reduces the average density but leaves the correlation function intact. The operation is similar to the filtering operation used to generate the filtered Poisson process discussed in Chapter 2 and works as follows. Given a point process X with the intensity $\Lambda(B)$, obtain a new process in the following manner. For each point x_k in the configuration $x = \{x_1, \ldots, x_n\}$, independently of other points, either retain the point with probability q_k or delete it with probability $1 - q_k$, where $0 \leq q_k \leq 1$. The new process is a point process with intensity $q\Lambda(B)$, where $q = \{q_1, \ldots, q_n\}$.

Thinning can be used to generate a *hard-core point process*, which is a point process in which the points are not allowed to lie closer than a predefined minimum distance. One type of hard-core process, called the *Matern hard-core process*, is obtained by applying thinning to a stationary Poisson process. In this case, the points in the Poisson process are randomly marked, and points that are within a distance less than $R/2$ from a marked point are deleted, where R is the predefined minimum distance between points.

15.7.2 Superposition

The superposition of independent point processes $X_1, \ldots, X_K$ is the union of these processes. That is, if the process X is generated from the superposition of these processes, then

$$X = \bigcup_{k=1}^{K} X_k$$

We refer to X as the *pooled process*. Let $N_{X_k}(B)$ denote the counting measure of the process X_k, and let $\Lambda_k(B)$ denote its rate. Then the rate of the pooled process is given by

$$N_X(B) = \sum_{k=1}^{K} N_{X_k}(B)$$

$$\Lambda(B) = \sum_{k=1}^{K} \Lambda_k(B)$$

The probability mass function of $N_X(B)$ is the K-fold convolution of the probability mass functions of the $N_{X_k}(B)$. That is,

$$p_{N_X}(x) = p_{N_{X_1}}(x) * p_{N_{X_2}}(x) * \cdots * p_{N_{X_K}}(x)$$

where the symbol $*$ denotes convolution operation.

15.7.3 Clustering

In a clustering operation, every point in a given point process, called the *parent point process*, is used to generate a cluster of points, called *child points*. Each cluster is generated independently of other clusters, however, the same construction rules apply to all clusters. Thus, within a cluster, the child points are placed independently according to the density function of the cluster points.

Each cluster can be regarded as being within a disk of radius $R > 0$ with the parent point as the center of the disk. Thus, there are three parameters that characterize a clustering operation: the intensity $\Lambda(B)$ of the parent point process, which defines the locations of the centers of the clusters; the disk radius R; and the cluster intensity $\Lambda_1(B_R)$, where B_R is the region within a cluster.

For most of the commonly used clusters, the parent point process is a homogeneous (stationary) Poisson point process. These clusters differ primarily in the way the child points are generated and placed within the cluster. Examples of cluster processes whose parent point processes are homogeneous Poisson point processes include the following:

- *Matern cluster process*, where the number of points per cluster follows a Poisson process, and these child points are uniformly placed within a disk of radius R centered about the cluster, where R is the same for all clusters.
- *Thomas cluster process*, where the number of points per cluster is a Poisson process, and the child points in each cluster are distributed independently according to a symmetric normal distribution around the cluster origin.
- *Neyman−Scott cluster process*, where the number of points in a cluster is an independent and identically distributed random variable. The points are also placed uniformly and independently within a disk of radius R around each cluster's center.
- *Hawkes* (or *self-exciting*) *cluster process*, where a parent point produces a cluster of child points, and each child point further produces its own cluster of child points, and so on.

15.8 Marked Point Processes

As stated earlier, a point process is a stochastic system that places points in the plane. Often there is more information that can be associated with an event. This information is known as a *mark*. The marks may be of separate interest or may simply be included to make a more realistic model of the event times. For example, it is of practical relevance to know the position and magnitude of an earthquake, not just its time. Thus, if each point of a point process has a mark (generally, a real number or a set of real numbers) associated with it, the process is called a marked point process. Let X be a point process on $S \subseteq R^d$. Given some space L, if a random mark $m_k \in M$ is attached to a point $x_k \in X$, where M is a set defined on L, then the process

$$Y = \{(x_k, m_k) | x_k \in X\}$$

is called a marked point process with points in S and mark space L, which can be a finite set or $L \subseteq R^p, p \geq 1$. Thus, a marked point process can be defined as a random collection of points, where each point has associated with it a further random variable called a mark. The process X is called the *ground process* and S is the *ground space*.

Marked point processes are useful for describing many physical systems. In general, they are commonly used for representing a finite number of events located in space and time. For example, consider a queueing system in which the nth customer arrives at time x_n and brings with it an amount of service m_n. The process $\{(x_n, m_n), n \geq 1\}$ is a marked point process that can be used in the performance analysis of the system. Another example is a collection of the arrival times and locations of hurricanes along with the dollar amount of damage attributed to each hurricane. In these cases, we have a marked spatial–temporal point processes.

Marked point processes have been used by Vere-Jones (1995), Ogata (1998), and Holden et al. (2003) to model earthquakes. They have also been used by Smith (1993) to model raindrop-size distributions. In Descombes and Zerubia (2002), they are used in image analysis and in Prigent (2001) to model option pricing. They have also been used in ecological and forestry studies by Gavrikov and Stoyan (1995), and Stoyan and Penttinen (2000) present a summary of the applications of marked point processes in forestry. They are used by McBride (2002) to model the source proximity effect in the indoor environment. In Stoica et al. (2000) and van Lieshout and Stoica (2003), a marked point process model for line segments called the Candy model is presented as a prior distribution for the image analysis problem of extracting linear networks, such as roads and rivers, from images.

Because M is a process on a bounded set L, we can interpret the marked point process Y as an ordinary point process in $R^d \times L$. If X is defined in the finite region $B \subseteq S$ and $N_Y(B \times L)$ denotes the random variable that represents the number of points of X with marks in M, then the intensity measure of N_Y is given by

$$\Lambda(B \times L) = E[N_Y(B \times L)]$$

There are two types of marks. In one case, the marks are independent and identically distributed random variables that are independent of the point process. In another case, the marks depend on the point process. We first consider the case where the marks are independent of the ground process. Let the random variable $M(B)$ denote the number of marks in the region B. Then given that $N_X(B) = n$, $M(B)$ is the sum of n independent and identically distributed random variables, and as shown by Ibe (2005) we have that

$$E[M(B)] = E[N_X(B)]E[M] = \Lambda|B|E[M]$$
$$\mathrm{Var}\{M(B)\} = \Lambda|B|\sigma_M^2 + (E[M])^2\mathrm{Var}\{N_X(B)\}$$
$$\mathrm{Cov}\{N_X(B), M(B)\} = E[M]\mathrm{Var}\{N_X(B)\}$$

The case when the marks depend on the point process can only be analyzed on a case-by-case basis, because there is no general solution.

15.9 Introduction to Markov Random Fields

A random field is essentially a stochastic process defined on a set of spatial nodes (or sites). Specifically, let $S = \{1, \ldots, N\}$ be a finite set and let $\{X(s), s \in S\}$ be a collection of random variables on the sample space Ω. Let $X(s_1) = x_1, \ldots, X(s_m) = x_m,$

where $x_i \in \Omega$. Then the joint event $x = \{x_1, \ldots, x_m\}$ is called a *configuration* of $X(s)$, which corresponds to a realization of the random field.

Random fields can also be classified according to their spatial variability. The term *homogeneous random field* is used to denote a random field in which the statistical values of the point properties are constant and the statistical value of the cross-point properties, namely autocorrelation and autocovariance, depends only on the separation between the points. For nonhomogeneous random fields, the statistical properties depend on the space origin. Finally, random fields can be classified by their memory property. In this case, we have *Markov random fields* (MRFs) and *non-MRFs*. If for $s_1 < s_2 < \cdots < s_m$, we have that the joint PDF

$$f_{X_1 X_2 \ldots X_m}(s_1, s_2, \ldots, s_m) = f_{X_1}(s_1) f_{X_2|X_1}(s_2|s_1) \cdots f_{X_m|X_{m-1}}(s_m|s_{m-1})$$

the random field is defined to be an MRF. In the case of the non-MRF, we have that for $s_1 < s_2 < \cdots < s_m$, the random variables $X_1(s_1), X_2(s_2), \ldots, X_m(s_m)$ are independent, i.e.,

$$f_{X_1 X_2 \ldots X_m}(s_1, s_2, \ldots, s_m) = \prod_{i=1}^{m} f_{X_i}(s_i)$$

A random field is defined to be strictly stationary if for any finite set of sites $s_1, \ldots, s_m$ and any $v \in S$, the joint distribution of $\{X_1(s_1), \ldots, X_m(s_m)\}$ and that of $\{X_1(s_1 + v), \ldots, X_m(s_m + v)\}$ are the same. A stationary random field in which the covariance function depends only on the absolute distance between the points is said to be *isotropic*, otherwise it is said to be *anisotropic*. That is, $\{X(s), s \in S\}$ is defined to be isotropic if

$$C_{XX}(s, u) = E\left[\left\{X(s) - \overline{X(s)}\right\}\left\{X(u) - \overline{X(u)}\right\}\right] = C_{XX}(\tau)$$

where $\tau = ||s - u||$. Sometimes the sites are points on a lattice and are, therefore, spatially regular. For such cases, we consider an $m \times n$ lattice where $S = \{(i,j)| 1 \le i \le m, 1 \le j \le n\}$. Also, there might be an interrelationship between sites, which is captured by the concept of a *neighborhood system* that is discussed in the next section.

15.9.1 MRF Basics

MRFs were originally used in statistical mechanics to model systems of particles interacting in a two-dimensional or three-dimensional lattice. More recently, they have been widely used in statistics and image analysis. They were introduced into image segmentation by Geman and Geman (1984). Since then the MRF theory has become the basic framework for statistical image analysis where images are modeled as data organized in lattices and represented as pixels or voxels. Thus, pixels and voxels play the role of particles in the physical system. MRFs are also called *Markov networks* or *undirected graphical models*.

The idea behind the concept of MRFs is that when a process is at a particular location, it is more likely to be influenced by events at other points that are relatively nearer the location than events at points that are farther away. Thus, the process attempts to define the concept of a neighborhood of a point within which it is affected by the points of the process and outside of which the impact of points is considered to be negligible and hence ignored. For example, in the case of image analysis, regions in real images are often homogeneous in the sense that neighboring pixels usually have similar properties, such as intensity, color, and texture. These contextual constraints are captured by the Markov property of MRF.

More formally, in an MRF, the sites in S are related to one another through a neighborhood system denoted by $\aleph = \{\aleph(i), i \in S\}$, where $\aleph(i)$ is the set of sites that are neighbors of i, $i \notin \aleph(i)$. The neighborhood relation is symmetrical, which means that $i \in \aleph(j) \Leftrightarrow j \in \aleph(i)$. Thus, for a finite set of sites $S = \{1, \ldots, N\}$, an MRF is a family of random variables $X_i, i \in S$, with probability functions that satisfy the following conditions relative to the neighborhood system $\aleph$:

a. $P[X = x] > 0$
b. $P[X_i = x_i | X_j = x_j, j \neq i] = P[X_i = x_i | X_j = x_j, j \in \aleph(i)]$

The first condition is called the *positivity property*, which ensures that all configurations (or possible realizations of X) have a chance of occurring. The second is usually called the *Markovianity property* that establishes the local characteristics of X; that is, only neighboring sites have direct interactions on each other. This property is sometimes expressed as follows:

$$P[X_i | X_{S-\{i\}}] = P[X_i | X_{\aleph(i)}]$$

It is this ability to describe local properties that makes MRF useful for image processing because, as we stated earlier, regions in real images are often homogeneous in the sense that neighboring pixels tend to have similar properties, such as intensity, color, and texture. MRF is also widely used in speech recognition, neural networks, and coding.

However, the specification of MRF via local conditional probabilities has some disadvantages. First, there is no direct method for deducing the joint probability distribution $P[X_1, \ldots, X_N]$ from the conditional probabilities $P[X_i | X_j, j \in \aleph(i)]$. This is an important issue because it is the joint probability mass function and not the conditional PMFs that contains the complete system representation. Also, the equilibrium conditions of a random process are usually specified in terms of the joint probability function rather than the conditional probabilities. Later in this section, we will see how this problem is resolved through the Hammersley–Clifford theorem, which is proved by Besag (1974). Note that the probability function is the probability mass function when the X_i are discrete random variables and the probability density function when they are continuous random variables. In the remainder of the discussion, we use the generic term *probability function* except where there is an explicit need to specify a PMF or PDF.

Another way to define an MRF is through the concept of *conditional independence*. Two random variables X and Y are conditionally independent given the random variable Z, written $X \perp Y | Z$, if and only if

$$P[X, Y | Z] = P[X | Z] P[Y | Z]$$

A random field $\{X(s), s \in S\}$ is defined to be an MRF if the random variable $X(s)$ is conditionally independent of all other sites in S, given its values in $\aleph(s)$; that is,

$$X(s) \perp X(S - \{s \cup \aleph(s)\}) | X(\aleph(s))$$

As long as $X(s)$ satisfies the positivity condition, the joint probability distribution can be obtained from the conditional distributions as follows. Consider two configurations $x = \{x_1, x_2, \ldots, x_n\} \in S^n$ and $y = \{y_1, x_2, \ldots, y_n\} \in S^n$. Then we have that

$$P[x_1, x_2, \ldots, x_{n-1}, x_n] = P[x_n | x_1, x_2, \ldots, x_{n-1}] P[x_1, x_2, \ldots, x_{n-1}]$$
$$P[x_1, x_2, \ldots, x_{n-1}, y_n] = P[y_n | x_1, x_2, \ldots, x_{n-1}] P[x_1, x_2, \ldots, x_{n-1}]$$

From this we obtain

$$P[x_1, x_2, \ldots, x_{n-1}] = \frac{P[x_1, x_2, \ldots, x_{n-1}, x_n]}{P[x_n | x_1, x_2, \ldots, x_{n-1}]} = \frac{P[x_1, x_2, \ldots, x_{n-1}, y_n]}{P[y_n | x_1, x_2, \ldots, x_{n-1}]}$$

Thus,

$$P[x_1, x_2, \ldots, x_{n-1}, x_n] = \frac{P[x_n | x_1, x_2, \ldots, x_{n-1}]}{P[y_n | x_1, x_2, \ldots, x_{n-1}]} P[x_1, x_2, \ldots, x_{n-1}, y_n]$$

Similarly,

$$P[x_1, \ldots, x_{n-1}, y_n] = P[x_{n-1} | x_1, x_2, \ldots, x_{n-2}, y_n] P[x_1, x_2, \ldots, x_{n-2}, y_n]$$
$$P[x_1, \ldots, x_{n-2}, y_{n-1}, y_n] = P[y_{n-1} | x_1, x_2, \ldots, x_{n-2}, y_n] P[x_1, x_2, \ldots, x_{n-2}, y_n]$$

From this we obtain

$$P[x_1, x_2, \ldots, x_{n-2}, y_n] = \frac{P[x_1, \ldots, x_{n-1}, y_n]}{P[x_{n-1} | x_1, x_2, \ldots, x_{n-2}, y_n]} = \frac{P[x_1, \ldots, x_{n-2}, y_{n-1}, y_n]}{P[y_{n-1} | x_1, x_2, \ldots, x_{n-2}, y_n]}$$

which gives

$$P[x_1, x_2, \ldots, x_{n-1}, y_n] = \frac{P[x_{n-1} | x_1, x_2, \ldots, x_{n-2}, y_n]}{P[y_{n-1} | x_1, x_2, \ldots, x_{n-2}, y_n]} P[x_1, \ldots, x_{n-2}, y_{n-1}, y_n]$$

Combining the two results, we obtain

$$P[x_1, x_2, \ldots, x_{n-1}, x_n]$$
$$= \frac{P[x_n | x_1, x_2, \ldots, x_{n-1}]}{P[y_n | x_1, x_2, \ldots, x_{n-1}]} \frac{P[x_{n-1} | x_1, x_2, \ldots, x_{n-2}, y_n]}{P[y_{n-1} | x_1, x_2, \ldots, x_{n-2}, y_n]} P[x_1, \ldots, x_{n-2}, y_{n-1}, y_n]$$

Proceeding inductively, we obtain

$$\frac{P[x_1, x_2, \ldots, x_{n-1}, x_n]}{P[y_1, y_2, \ldots, y_{n-1}, y_n]} = \frac{P[x]}{P[y]} = \prod_{i=1}^{n} \frac{P[x_i | x_1, x_2, \ldots, x_{i-1}, y_{i+1}, \ldots, y_n]}{P[y_i | x_1, x_2, \ldots, x_{i-1}, y_{i+1}, \ldots, y_n]}$$

Thus, the ratio $P[x]/P[y]$ is determined by the conditional probabilities. Assume now that $X(s)$ is an MRF relative to the neighborhood $\aleph$, and let $\aleph(i)^+$ denote the set of neighbors X_j of site X_i such that $i < j$. Similarly, let $\aleph(i)^-$ denote the set of neighbors X_j of site X_i such that $i > j$. Then we obtain

$$\frac{P[x_1, x_2, \ldots, x_{n-1}, x_n]}{P[y_1, y_2, \ldots, y_{n-1}, y_n]} = \prod_{i=1}^{n} \frac{P[x_i | x_j \quad \text{for all } X_j \in \aleph(i)^-, \ y_j \quad \text{for all } Y_j \in \aleph(i)^+]}{P[y_i | x_j \quad \text{for all } X_j \in \aleph(i)^-, \ y_j \quad \text{for all } Y_j \in \aleph(i)^+]}$$

Thus, we obtain the ratio $P[x]/P[y]$ in terms of the conditional probabilities over the neighborhood system. Alternatively, we can write

$$P[x] = P[y] \prod_{i=1}^{n} \frac{P[x_i | x_j \quad \text{for all } X_j \in \aleph(i)^-, \ y_j \quad \text{for all } Y_j \in \aleph(i)^+]}{P[y_i | x_j \quad \text{for all } X_j \in \aleph(i)^-, \ y_j \quad \text{for all } Y_j \in \aleph(i)^+]}$$

15.9.2 Graphical Representation

An important characteristic of image data is the special nature of the statistical dependence of the gray level at a lattice site on those of its neighbors. For this reason, it is important to understand the neighborhood structure in MRFs. This neighborhood structure can be represented by a graph with sites as the *nodes*, and two sites are connected by an *edge* if and only if they are neighbors. Let $\aleph(i)$ and $\aleph$ be as defined earlier. Then we consider two types of neighborhood systems on the regular rectangular lattice: *first-order neighborhood* and *second-order neighborhood*. Relative to node (i, j), the diagonal nodes are not considered as its first-order neighbors. Thus, the first-order neighborhood system of node (i, j) is given by

$$\aleph_1(i, j) = \{(i - 1, j), (i, j - 1), (i + 1, j), (i, j + 1)\}$$

Therefore, if node (i, j) is not at the boundary of the lattice, it has four neighbors, as illustrated in Figure 15.4.

An example of a first-order neighborhood for a linear graph is illustrated in Figure 15.5 and is characterized by the relationship:

$$P[X_i | X_j, j \notin i] = P[X_i | X_{i-1}, X_{i+1}] \Rightarrow X_{\aleph(i)} = \{X_{i-1}, X_{i+1}\}$$

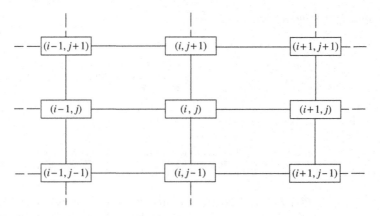

Figure 15.4 Example of a first-order neighborhood system.

Figure 15.5 Example of a first-order neighborhood system on a linear graph.

In the case of a second-order neighborhood, the diagonal nodes are considered to be neighbors, which means that the second-order neighborhood of (i,j) is given by

$$\aleph_2(i,j) = \{(i-1,j),(i-1,j-1),(i-1,j+1),(i,j-1),(i,j+1),(i+1,j-1),$$
$$(i+1,j),(i+1,j+1)\}$$

Thus, if node (i,j) is not at the boundary of the lattice, it has eight neighbors, as illustrated in Figure 15.6.

In general, the cth-order neighborhood system for node (i,j) is given by

$$\aleph_c(i,j) = \{(k,l)|0 < (k-i)^2 + (l-j)^2 \leq c\}$$

A graph, $G = (V,E)$, is a pair of sets V (or $V(G)$) and E (or $E(G)$) called vertices (or nodes) and edges (or arcs), respectively, where the edges join different pairs of nodes. The vertices are represented by points and the edges are represented by lines joining the nodes. A graph is a mathematical concept that is used to represent the notion of relationships such that an edge exists between two nodes if there is a direct interaction between them. If an edge exists between nodes i and j, we define them to be neighbors and write $i \in \Gamma(j)$, where $\Gamma(j)$ denotes the set of neighbors of node j. We consider undirected graphs in which $i \in \Gamma(j) \Leftrightarrow j \in \Gamma(i)$. A subgraph of G is a graph H such that $V(H) \subseteq V(G)$ and $E(H) \subseteq E(G)$, and the endpoints of an edge $e \in E(H)$ are the same endpoints in G. A complete graph is a graph in which all the

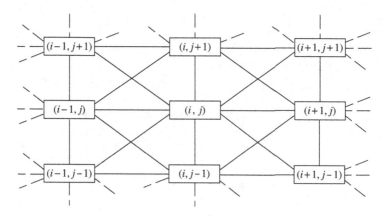

Figure 15.6 Examples of second-order neighborhood system.

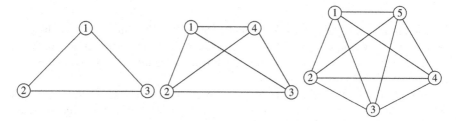

Figure 15.7 Examples of complete graphs.

nodes are neighbors of one another. Figure 15.7 illustrates examples of complete graphs.

A *clique* of a graph G is a single node or a complete subgraph of G. That is, a clique is a subgraph of G in which every site is a neighbor of all other sites. Figure 15.8 shows examples of cliques.

The local conditional probability $P[X_i|X_j, j \in \Gamma(i)]$ is represented through cliques, because every node in a clique is a neighbor to all other nodes in the clique. Thus, the MRF model consists of a set of cliques.

15.9.3 Gibbs Random Fields and the Hammersley–Clifford Theorem

As stated earlier, the specification of MRFs via local conditional probabilities has the disadvantage that it does not provide a direct method for deducing the joint probability distribution $P[X_1, \ldots, X_N]$ from the conditional probabilities $P[X_i|X_j, j \in \aleph(i)]$. Fortunately, this problem is resolved via the *Hammersley–Clifford theorem* that states as follows:

Theorem 15.1 The random field X is an MRF if and only if X is a Gibbs random field.

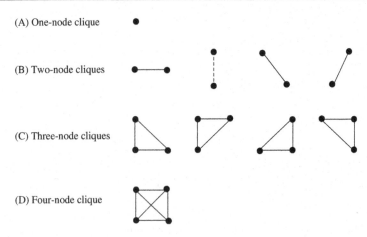

Figure 15.8 Examples of cliques.

The proof of this theorem is given by Besag (1974). The implication of the theorem is that any conditional MRF distribution has a joint distribution; that is, a Gibbs distribution. The theorem thus establishes a connection between the local and global specifications of an MRF.

The Gibbs distribution originated in statistical mechanics where the large-scale properties of a lattice system are to be deduced from local models. The approach was pioneered by Ising (1925), who modeled the behavior of ferromagnetic material by considering only the interactions of spins of neighboring atoms. Two neighboring atoms of opposite spin were considered to have a positive *potential*, and the state of the lattice was characterized by its *energy*, which is computed as the sum of the potentials. In this system, configurations of low energy are defined to be more stable than those of high energy.

A random variable X is defined to have the Gibbs distribution if its distribution function is of the following form:

$$P[X = x] = \frac{1}{Z}\exp\{-\beta U(x)\}$$

$U(x)$ is called the *energy function*, which is such that the higher the energy of the configuration, the smaller the probability, and β is a nonnegative parameter called the *spatial smoothness parameter* that characterizes the label-scale variability in an image. The value $\beta = 0$ corresponds to a uniform distribution on the configuration space; small values of β indicate small and broken structures, whereas large values imply large and bloblike structures that make it more likely to observe largely clustered configurations corresponding to large $U(x)$. The denominator, Z, is a normalizing constant called the *partition function* and is given by

$$Z = \sum_{x} \exp\{-\beta U(x)\}$$

Note that $\beta = 1/kT$ in the definition of the Gibbs distribution used in statistical mechanics, where T is the *temperature* (in absolute degrees) and k is the *Boltzmann's constant*. $U(x)$ is usually defined in terms of the cliques and given by

$$U(x) = \sum_{c \in C} V_c(x) + \sum_{i \in \aleph_1} V_{\aleph_1}(x_i) + \sum_{(i,j) \in \aleph_2} V_{\aleph_2}(x_i, x_j) + \cdots$$

where the sum is over all cliques, and $V_c(x)$ is called the *potential function* that is associated with clique $c \in C$, and C is the family of cliques. The function $V_{\aleph_k}$ is called the *potential of order k*.

A random field $X(s)$ is defined to be a Gibbs random field on S with respect to $\aleph$ if and only if its joint probability distribution is the Gibbs distribution; that is, a Gibbs random field is a family of random variables that have the Gibbs distribution.

As stated earlier, an MRF is characterized by its local property, according to the Markovianity property, whereas the Gibbs random field is characterized by its global property, which is the Gibbs distribution. The equivalence between the two fields is established by the Hammersley–Clifford theorem, which fundamentally states that a random field X is an MRF if and only if X is a Gibbs random field; that is, X is an MRF if and only if it has a Gibbs distribution with potentials defined on the cliques of the neighborhood system $\aleph$. The importance of the theorem lies in the fact that it provides a simple way to specify the joint probability distribution, which is by defining the clique potential functions. That is, an MRF can be specified via the clique potentials. Thus, we have that the energy function is a sum of *clique potentials* $V_c(x)$ over all possible cliques $c \in C$:

$$U(x) = \sum_{c \in C} V_c(x)$$

The conditional probability $P[X_i | X_j, j \in \aleph(i)]$ is given by

$$P\left[X_i | X_j, j \in \aleph(i)\right] = \frac{1}{Z_i} \exp\left\{ -\beta \sum_{c:i \in c} V_c(x) \right\}$$

where

$$Z_i = \sum_{w \in \Omega} \exp\left\{ -\beta \sum_{c:i \in c} V_c(x | X_i = w) \right\}$$
$$x = \{x_1, \ldots, x_i, \ldots, x_N\}$$
$$x | \{X_i = w\} = \{x_1, \ldots, w, \ldots, x_N\}$$

One important feature of the Gibbs random field is that its definition is rather general in the sense that the only restriction on the potential field $V_c(x)$ is its dependence on the values associated with the clique c. There is no consistency condition among different values of the potential functions with the result that their choice represents a basic modeling decision that needs to be made when using MRFs.

MRF models are popularly used in several areas of image analysis because of their flexibility. Image analysis seeks to find an adequate representation of the intensity distribution of a given image. An image is typically represented by two-dimensional scalar data whose gray level variations are defined over a rectangular or square lattice. One important characteristic of image data is the statistical dependence of the gray level at a lattice point on those of its neighbors. By placing specific restrictions on the membership of the neighbor set, different representations are obtained.

15.10 Markov Point Processes

Markov point processes were introduced by Ripley and Kelly (1977) as an extension of MRFs with the capability to express inhibition between neighboring points on a finite graph. Thus, they are popular models for point patterns with interaction between the points. A good discussion of Markov point processes is given by Baddeley and Moller (1989), van Lieshout (2000), and by Moller and Waagepetersen (2004). The interactions between points are usually local with respect to a neighborhood system.

Markov point processes provide a rich source for point processes featuring interaction between points. Since the introduction of Markov point processes, attention has focused on a special class called the *pairwise interaction point process*. In the pairwise interaction process, the interaction between points is local with respect to a neighborhood system. Thus, for a point process X with a configuration $x = \{x_1, \ldots, x_n\}$, where $x_i \in X, i = 1, \ldots, n$, the configuration interacts only via pairs of points from this configuration. Pairwise interaction models are particularly useful for imposing inhibition and clustering. We denote the neighborhood relation by $\lozenge$ and define two points $x_i, x_j \in S$ to be neighbors, written $x_i \sim x_j$, if for a number $r > 0, x_j \lozenge x_i < r$. Then, both the *positivity* and *Markovianity* properties hold, i.e.,

$$P[X = x] > 0$$
$$P[X_i = x_i | X_j = x_j, j \neq i] = P[X_i = x_i | X_j = x_j, j \in \aleph(i)]$$

where $X = \{X_1, \ldots, X_n\}$ and $\aleph(i)$ is the set of neighbors of point x_i. Another important general property of Markov point processes is the *hereditary* property, which states that for any finite configuration x,

$$f_X(x) > 0 \Rightarrow f_X(y) > 0 \quad \forall y \subset x$$

For a pairwise interaction process, the PDF of the configuration X is of the form

$$f_X(x) = \alpha \prod_{i=1}^{n} \lambda(x_i) \prod_{x_i \sim x_j} h(x_i \lozenge x_j)$$

where h is the interaction function, n is the cardinality of x, $\lambda(x)$ is the intensity function, and α is a normalizing constant. A pairwise interaction function is defined to be homogeneous if $\lambda(x) = \lambda$, a constant, and $h(x_i \lozenge x_j)$ is invariant under motions, which permits us to define the neighborhood relation by the Euclidean distance, i.e., $x_i \lozenge x_j = \|x_i - x_j\|$. Thus, in this case, the set of neighbors of point x_i is given by

$$\aleph(i) = \{x_j | x_i \lozenge x_j < r, j \neq i\} = \{x_j | \|x_i - x_j\| < r, j \neq i\}$$

We denote the neighborhood system by $\aleph = \{\aleph(i)\}$. The neighborhood relation is symmetrical in the sense that $x_i \in \aleph(j) \Leftrightarrow x_j \in \aleph(i)$, i.e., $x_i \sim x_j \Leftrightarrow x_j \sim x_i$. Also, the relation is reflexive in the sense that $x_i \sim x_i$. In the remainder of the discussion, we assume that the pairwise interaction process, and hence the Markov point process, is homogeneous; that is,

$$f_X(x) = \alpha \lambda^n \prod_{x_i \sim x_j} h(\|x_i - x_j\|)$$

The PDF $f_X(x)$ can be expressed in the form of the Gibbs distribution using a version of the Hammersley–Clifford theorem as follows:

$$f_X(x) = \frac{1}{Z} \exp\{-\beta U(x)\}$$

where, as defined earlier, $U(x)$ is the *energy function* of the configuration x, Z is the partition function (or the normalizing constant), $\beta = 1/kT$, T is the absolute temperature, and k is the Boltzmann's constant. The simplest form of the energy function for pairwise interaction processes is given by

$$U(x) = V_0 + \sum_{i=1}^{n} V_1(x_i) + \sum_{1 \leq i < j \leq n} V_2(x_i, x_j)$$

where the function V_k is called the *potential of order k*. Thus, the energy is computed by taking one point at a time or two points at a time. For the process to be homogeneous, we need $V_1(u)$ to be a constant, such as $V_1(u) = a_1$; similarly, we need $V_2(x_i, x_j) = V_2(\|x_i - x_j\|)$. If we let $V_0 = a_0$, we obtain

$$f_X(x) = \frac{1}{Z} \exp\left\{-\beta U(x)\right\} = \frac{1}{Z} e^{-\beta a_0} e^{-n\beta a_1} \exp\left\{-\beta \sum_{1 \leq i < j \leq n} V_2(\|x_i - x_j\|)\right\}$$

$$\equiv ab^n \exp\left\{-\beta \sum_{1 \leq i < j \leq n} V_2(\|x_i - x_j\|)\right\}$$

Because earlier we have obtained $f_X(x)$ as

$$f_X(x) = \alpha\lambda^n \prod_{x_i \sim x_j} h(||x_i - x_j||) = \alpha\lambda^n \exp\left\{\sum_{x_i \sim x_j} \log[h(||x_i - x_j||)]\right\}$$

which is similar to the result obtained via the Gibbs distribution, it means that every homogeneous Markov point process is a Gibbs point process. We consider the case where the second-order (or pair) potential is of the form

$$V_2(u) = \begin{cases} \infty & u = 0 \\ q & 0 < u \le \mu \\ 0 & u > \mu \end{cases}$$

where $q > 0$ and $\mu > 0$. We obtain

$$f_X(x) = ab^n \exp\{-\beta\mu n_\mu(x)\} = ab^n \gamma^{n_\mu(x)}$$

where $n_\mu(x)$ is the number of point pairs with an Euclidean distance of less than μ and $\gamma = \exp\{-\beta\mu\}$. This is the PDF of a point process called the *Strauss point process*.

15.11 Markov Marked Point Processes

As discussed earlier, if X is a point process on $S \subseteq R^d$, and a random mark $m_k \in M$ is attached to each point $x_k \in X$, where M is a set defined on some space L, then the process

$$Y = \{(x_k, m_k)|x_k \in X\}$$

is called a marked point process with points in S and mark space L. In the remainder of this section, we assume that the marks are independent and identically distributed random variables that are independent of the point process.

A marked point process Y is defined to be a Markov marked point process if the following conditions hold:

$$P[Y = y] > 0$$
$$P[y] > 0 \Rightarrow P[z] > 0 \quad \forall z \subset y$$
$$P[Y_i = y_i|Y_j = y_j, j \ne i] = P[Y_i = y_i|Y_j = y_j, j \in \aleph(i)]$$

These are the positivity, hereditary, and Markovianity properties, respectively. The discussion on Markov point processes can be extended to the Markov marked point process fairly easily. In particular, using the pairwise interaction point process model, we have that if $y = \{(y_1, m_1), \ldots, (y_n, m_n)\}$ is a configuration of Y, then the PDF of Y is given by

$$f_Y(y) = \alpha \prod_{i=1}^{n} \lambda(y_i, m_i) \prod_{(y_i, m_j) \sim (y_j, m_j)} h(||(y_i, m_i) - (y_j, m_j)||)$$

where the neighborhood relation has to be appropriately defined. It must also be stated that the process is also a Markov point process with respect to the neighborhood relation

$$(y_i, m_i) \sim (y_j, m_j) \quad \text{iff} \quad ||y_i - y_j|| < r$$

for some $r > 0$.

15.12 Applications of Markov Point Processes

Point processes, marked point processes, and Markov point processes are used to model many applications that include earthquakes, raindrop-size distributions, image analysis, option pricing, and ecological and forestry studies. The fundamental principles of these processes have been discussed in this chapter. More comprehensive discussion on the principles and applications of point processes and their derivatives are discussed by Bremaud (1981), Ripley (1988), Reiss (1993), Stoyan et al. (1995), Cox and Isham (2000), Daley and Vere-Jones (2003), and Jacobsen (2006).

The analysis of point processes, marked point processes, and Markov point processes is usually complex and is generally done via Markov chain Monte Carlo simulation. Also, a major aspect of the analysis deals with parameter estimation, which is not covered in this chapter and can be found in most of the references cited. Even in the case of parameter estimation, the preferred method is the Markov chain Monte Carlo maximum likelihood estimate method for missing data models proposed by Gelfand and Carlin (1993).

15.13 Problems

15.1 Two classes of customers arrive at Paul's barber shop: class 1 and class 2. Class 1 customers arrive according to a Poisson process with rate λ_1 customers per hour, and class 2 customers arrive according to a Poisson process with rate λ_2 customers per hour.
  a. What is the probability that no customer arrives over a 2-h period?
  b. What is the mean time between customer arrivals at Paul's shop?
  c. Given that a customer has just arrived at the shop, what is the probability that the next customer to arrive at the shop is a class 2 customer?

15.2 A hard-core point process is produced from a Poisson process of rate λ by deleting any point within distance v_0 of another point, regardless of whether that point has itself already been deleted. Prove that the rate of the hard-core process is $\lambda \exp(-\lambda \pi v_0^2)$.

15.3 Let $\{X_1, X_2, \ldots\}$ be a sequence of independent and identically distributed random variables with PDF $f_X(x)$, and let N be an integer-valued random variable with PMF $p_N(n)$, where N and the X_i are independent. Consider a process in which events occur at times $X_1, X_1 + X_2, \ldots, X_1 + X_2 + \cdots + X_N$.

a. Calculate the mean, variance, and s-transform of the PDF of the time of the last event.

b. What is the expected number of events in the interval $(0,t)$?

15.4 A restaurant has two entrances A and B. Customers arrive at the restaurant through entrance A according to a Poisson process with a rate of five customers per hour, and customers arrive through entrance B according to a Poisson process with a rate of seven customers per hour.

a. What is the probability that no new customer arrives over a 2-h period?

b. What is the mean time between arrivals of new customers at the restaurant?

c. What is the probability that a given customer arrived through entrance B?

15.5 Passengers arrive at a train station according to a Poisson process with a rate of 25 customers per hour. It has been found that 60% of the passengers are females. What is the probability that no male customer arrives at the station over a 15-min period?

15.6 One important parameter in stationary point processes is the *autointensity function*. The autointensity function, $h_{NN}(t, v)$, of the process $N(t)$ is the conditional probability that a point occurs at time $t + v$ given that a point occurs at time t. Specifically,

$$h_{NN}(t, v) = \frac{P[dN(t + v) = 1 | dN(t) = 1]}{dv}$$

where

$$dN(t) = \begin{cases} 1 & \text{if a point in } (0, 0 + dt) \\ 0 & \text{otherwise} \end{cases}$$

Show that for a Poisson process the autointensity function is a constant. (Note: This demonstrates the nonpredictability of the Poisson process.)

References

Adas, A., 1997. Traffic models in broadband networks. *IEEE Communications Magazine*. July, pp. 82–89.

Albert, P.S., 1991. A two-state Markov mixture model for a time series of epileptic seizure counts. *Biometrics*. 47, pp. 1371–1381.

Andrey, P., Tarroux, P., 1998. Unsupervised segmentation of Markov random field modeled textured images using selectionist relaxation. *IEEE Transactions on Pattern Analysis and Machine Intelligence*. 20, 252–262.

Aoki, M., 1967. *Optimization of Stochastic Systems*. Academic Press, New York.

Arakawa, K., Krotkov, E., 1994. Modeling of natural terrain based on fractal geometry. *Systems and Computers in Japan*. 25, 99–113.

Argyrakis, P., Kehr, K.W., 1992. Mean number of distinct sites visited by correlated walks ii: disordered lattices. *Journal of Chemical Physics*. 97, 2718–2723.

Ayala, G., Epifanio, I., Simó, A., Zapater, V., 2006. Clustering of spatial point patterns. *Computational Statistics & Data Analysis*. 50, 1016–1032.

Applebaum, D., 2004a. Levy processes—from probability to finance and quantum groups. *Notices of the AMS*. 51, 1336–1347.

Applebaum, D., 2004b. *Levy Processes and Stochastic Calculus*. Cambridge University Press, Cambridge, England.

Bachelier, L., 1900. *Theorie de la Speculation*. Gauthier-Villars, Paris.

Baddeley, A., Moller, J., 1989. Nearest-neighbor Markov point processes and random sets. *International Statistical Review*. 57, 89–121.

Bae, J.J., Suda, T., 1991. Survey of traffic control schemes and protocols in ATM networks. *Proceedings of the IEEE*. 79, 170–189.

Baiocchi, A., Mellazzi, N.B., Listani, M., 1991. Loss performance analysis of an ATM multiplexer loaded with high-speed ON–OFF sources. *IEEE Journal on Selected Areas in Communications*. 9, 388–393.

Baldi, P., Chauvin, Y., Hunkapiller, T., McClure, M.A., 1994. Hidden Markov models for biological primary sequence information. *Proceedings of the National Academy of Science*. 91, 1059–1063.

Ball, C.A., Torous, W.N., 1983. A simplified Jump process for common stock returns. *Journal of Financial and Quantitative Analysis*. 18, 53–65.

Bandyopadhyay, S., Coyle, E.J., Falck, T., 2006. Stochastic properties of mobility models in mobile ad hoc networks. *Proceedings of the 40th Annual Conference Information Sciences Systems*, pp. 1205–1211.

Barndorff-Nielsen, O.E., Mikosch, T., Resnick, S.I., 2001. *Levy Processes: Theory and Applications*. Birkhauser, Boston, MA.

Bellman, R., 1957. *Dynamic Programming*. Princeton University Press, Princeton, NJ (Reprinted by Dover Publications, 2003.).

Benth, F.E., 2003. On Arbitrage-free pricing of weather derivatives based on fractional Brownian motion. *Applied Mathematical Finance*. 10, 303–324.

482

References

Benth, F.E., 2004. *Option Theory with Stochastic Analysis: An Introduction to Mathematical Finance*. Springer, New York, NY.

Bertsekas, D.P., 1976. *Dynamic Programming and Stochastic Control*. Academic Press, New York, NY.

Bertsekas, D.P., 1995a. *Dynamic Programming and Optimal Control Volume One*. Athena Scientific, Belmont, MA.

Bertsekas, D.P., 1995b. *Dynamic Programming and Optimal Control Volume Two*. Athena Scientific, Belmont, MA.

Besag, J., 1974. Spatial interaction and the statistical analysis of lattice systems. *Journal of the Royal Statistical Society, Series B*. 36, 192–236.

Besag, J., 1986. On the statistical analysis of dirty pictures. *Journal of the Royal Statistical Society, Series B*. 48, 259–302.

Bianchini, M., Gori, M., Scarselli, F., 2005. Inside PageRank. *ACM Transactions on Internet Technology*. 5, 92–128.

Bilmes, J.A., 2003. Buried Markov models: a graphical-modeling approach to automatic speech recognition. *Computer Speech and Language*. 17, 213–231.

Black, F., Scholes, M., 1973. The pricing options and corporate liabilities. *Journal of Political Economy*. 81, 637–654.

Bohm, W., 2000. The correlated random walk with boundaries: a combinatorial solution. *Journal of Applied Probability*. 37, 470–479.

Bollobas, B., 1998. *Modern Graph Theory*. Springer, New York, NY.

Borovkov, K., 2003. *Elements of Stochastic Modeling*. World Scientific, Singapore.

Brand, M., 1996. Coupled Hidden Markov models for modeling interacting processes. Technical Report Number 405, MIT Media Laboratory for Perceptual Computing, Learning and Common Sense.

Brand, M., Oliver, N., Pentland, A., 1997. Coupled hidden Markov models for complex action recognition. *Proceedings of the 1997 Conference on Computer Vision and Pattern Recognition* (CVPR'97), pp. 994–999.

Brekke, K.A., Oksendal, B., 1991. The high contact principle as a sufficiency condition for optimal stopping. In: Lund, D., Oksendal, B. (Eds.), *Stochastic Models and Option Values*. North-Holland, New York, NY, pp. 187–208.

Bremaud, P., 1981. *Point Processes and Queues: Martingale Dynamics*. Springer, New York, NY.

Bremaud, P., 1999. *Markov Chains: Gibbs Fields, Monte Carlo Simulation, and Queues*. Springer, New York, NY.

Brody, D.C., Syroka, J., Zervos, M., 2002. Dynamical pricing of weather derivatives. *Quantitative Finance*. 2, 189–198.

Brooks, S.P., 1998. Markov Chain Monte Carlo method and its application. *The Statistician*. 47 (Part 1), 69–100.

Bui, H.H., Venkatesh, S., West, G., 2000. On the recognition of abstract Markov policies. *Proceedings of the National Conference on Artificial Intelligence*. AAAI-2000, Austin, Texas, pp. 524–530.

Bui, H.H., Venkatesh, S., West, G., 2001. Tracking and surveillance in wide-area spatial environments using the abstract hidden Markov model. *International Journal of Pattern Recognition and Artificial Intelligence*. 15, 177–195.

Bui, H.H., Phung, D.Q., Venkatesh, S., 2004. Hierarchical hidden Markov models with general state hierarchy. *Proceedings of the National Conference on Artificial Intelligence*. AAAI-2004, San Jose, California, pp. 324–329.

Byers, J.A., 2001. Correlated random walk equations of animal dispersal resolved by simulation. *Ecology*. 82, 1680–1690.

Capasso, V., Bakstein, D., 2005. *An Introduction to Continuous-Time Stochastic Processes: Theory, Models and Applications to Finance, Biology and Medicine.* Birkhauser, Boston , MA.

Cappe, O., Moulines, E., Ryden, T., 2005. *Inference in Hidden Markov Models.* Springer, New York, NY.

Casella, G., George, E.I., 1992. Explaining the Gibbs sampler. *The American Statistician.* 46 (3), 167–174.

Cassandra, A.R., 1998. Exact and approximate algorithms for partially observable Markov decision processes. Ph.D. Thesis, Department of Computer Science, Brown University, Providence, RI.

Cassandra, A.R., Kaelbling, L.P., Littman, M.L., 1994. Acting optimally in partially observable stochastic domains. *Proceedings of the 12th National Conference on Artificial Intelligence*, Seattle, Washington, DC, pp. 1023–1028.

Cerny, V., 1985. Thermodynamical approach to the traveling salesman problem: An efficient simulation algorithm. *Journal of Optimization Theory and Applications.* 45, 41–51.

Chao, C.C., Yao, L.H., 1996. Hidden Markov models for burst error statistics of Viterbi decoding. *IEEE Transactions on Communications.* 44, 1620–1622.

Charniak, E., 1991. *Bayesian networks without tears.* AI Magazine, Winter, pp. 50–63.

Chellappa, R., Chatterjee, S., 1985. Classification of textures using Gaussian Markov random fields. *IEEE Transactions on Acoustics, Speech, and Signal Processing.* 33, 959–963.

Cheng, Y.-C., Robertazzi, T.G., 1990. A New spatial point process for multihop radio network modeling. *Proceedings of the IEEE International Conference on Communications (ICC 90).* 3, 1241–1245.

Chib, S., Greenberg, E., 1995. Understanding the Metropolis–Hastings algorithm. *The American Statistician.* 49 (4), 327–335.

Choi, E., Hall, P., 1999. Nonparametric approach to analysis of space-time data on earthquake occurrences. *Journal of Computational and Graphical Statistics.* 8, 733–748.

Christakos, G., 1992. *Random Field Models in Earth Sciences.* Dover Publications, Mineola, NY.

Churchill, G.A., 1989. Stochastic models for heterogeneous DNA sequences. *Bulletin of Mathematical Biology.* 51, 79–94.

Churchill, G.A., 1992. Hidden Markov Chains and the analysis of genome structure. *Computers and Chemistry.* 16, 107–115.

Cipra, B.A., 1987. An introduction to the Ising model. *The American Mathematical Monthly.* 94, 937–959.

Costamagna, E., Favalli, L., Tarantola, F., 2003. Modeling and analysis of aggregate and single stream internet traffic. *Proceedings of the IEEE Global Telecommunications Conference (GLOBECOM2003).* 22, 3830–3834, December 2003.

Cox, D.R., Isham, V., 2000. *Point Processes.* Chapman & Hall/CRC, Boca Raton, FL.

Cox, D.R., Miller, H.D., 2001. *The Theory of Stochastic Processes.* CRC Press, Boca Raton, FL.

Crovella, M.E., Bestavros, A., 1997. Self-similarity in world wide web traffic: evidence and possible causes. *IEEE/ACM Transactions on Networking.* 5, 835–846.

Daley, D.J., Vere-Jones, D., 2003. *An Introduction to the Theory of Point Processes: Volume I: Elementary Process and Methods.* second ed. Springer, New York, NY.

Daniels, H.E., 1969. The minimum of a stationary Markov process superimposed on a U-shaped trend. *Journal of Applied Probability.* 6, 399–408.

Dasgupta, A., 1998. Fractional Brownian motion: its properties and applications to stochastic integration. Ph.D. Thesis, University of North Carolina.

Dean, T., Kaelbling, L.P., Kirman, J., Nicholson, A., 1995. Planning under time constraints in stochastic domains. *Artificial Intelligence*. 76, 35−74.

Descombes, X., Kruggel, F., von Cramon, D.Y., 1998. Spatio-temporal fMRI analysis using Markov random fields. *IEEE Transactions on Medical Imaging*. 17, 1028−1039.

Descombes, X., Zerubia, I., 2002. Marked point processes in image analysis. *IEEE Signal Processing Magazine*. September, 77−84.

Di Nardo, E., Nobile, A.G., Pirozzi, E., Ricciardi, L.M., 2003. Towards the modeling of neuronal firing by Gaussian processes. *Scientiae Mathematicae Japonicae*. 58, 255−264.

Diligenti, M., Gori, M., Maggini, M., 2004. A unified probabilistic framework for web page scoring systems. *IEEE Transactions on Knowledge and Data Engineering*. 16, 4−16.

Dogandzic, A., Zhang, B., 2006. Distributed Estimation and detection for sensor networks using hidden Markov random field models. *IEEE Transactions on Signal Processing*. 54, 3200−3215.

Dorai-Raj, S.S., 2001. First- and second-order properties of spatiotemporal point processes in the space-time and frequency domains. Ph.D. Thesis, Department of Statistics, Virginia Polytechnic Institute and State University, Blacksburg, VA.

Doyle, P.G., Snell, J.L., 1984. *Random Walks and Electric Networks*. The Mathematical Association of America, Washington, DC.

Drake, A.W., 1962. Observation of a Markov process through a noisy channel. Sc.D. Thesis, Massachusetts Institute of Technology, Cambridge, MA.

Durbin, R., Mitchison, G., Krogh, A.S., Eddy, S.R., 1997. *Biological Sequence Analysis: Probabilistic Models of Proteins and Nucleic Acids*. Cambridge University Press, Cambridge, England.

Dynkin, E.B., 1965. Controlled random sequences. *Theory of Probability*. 10, 1−14.

Einstein, A., 1905. On the movement of small particles suspended in a stationary liquid demanded by the molecular-Kinetic theory of heat, Reprinted In: Furth, R. (Ed.), *Albert Einstein: Investigations on the Theory of the Brownian Movement*, 1956. Dover Publications, New York, NY, pp. 1−18.

Elliott, R.J., Aggoun, L., Moore, J.B., 1995. *Hidden Markov Models: Estimation and Control*. Springer, New York, NY.

Elliot, R.J., van der Hoek, J., 1997. An application of hidden Markov models to asset allocation problems. *Finance and Stochastics*. 1, 229−238.

Elliott, R.J., van der Hoek, J., 2003. A general fractional white noise theory and applications to finance. *Mathematical Finance*. 13, 301−330.

Ephraim, Y., Merhav, N., 2002. Hidden Markov processes. *IEEE Transactions on Information Theory*. 48, 1518−1569.

Feinberg, S.W., 1970. A note on the diffusion approximation for single neuron firing problem. *Kybernetic*. 7, 227−229.

Ferguson, J.D., 1980. Variable duration models for speech. *Proceedings of the Symposium on the Application of Hidden Markov Models to Text and Speech*, pp. 143−179.

Fine, S., Singer, Y., Tishby, N., 1998. The hierarchical hidden Markov model: analysis and application. *Machine Learning*. 32, 41−62.

Fischer, W., Meier-Hellstern, K., 1992. The Markov-modulated poisson process (MMPP) cookbook. *Performance Evaluation*. 18, 149−171.

Fjortoft, R., Delignon, Y., Pieczynski, W., Sigelle, M., Tupin, F., 2003. Unsupervised classification of radar images using hidden Markov chains and hidden Markov random fields. *IEEE Transactions on Geoscience and Remote Sensing*. 41, 675−686.

Forbes, F., Peyrard, N., 2003. Hidden Markov random field model selection criteria based on mean field-like approximations. *IEEE Transactions on Pattern Analysis and Machine Intelligence.* 25, 1089−1101.

Forchhammer, S., Rissanen, J., 1996. Partially hidden Markov models. *IEEE Transactions on Information Theory.* 42, 1253−1256.

Francois, O., Ancelet, S., Guillot, G., 2006. Bayesian clustering using hidden Markov random fields in spatial population genetics. *Genetics.* 174, 805−816.

Frost, V.S., Melamed, B., 1994. Traffic modeling for telecommunications networks. *IEEE Communications Magazine.* March, 70−81.

Gallager, R.G., 1996. *Discrete Stochastic Processes.* Kluwer Academic Publishers, Boston, MA.

Gavrikov, V., Stoyan, D., 1995. The use of marked point processes in ecological and environmental forest studies. *Environmental and Ecological Statistics.* 2, 331−344.

Gelfand, A.E., Smith, A.F.M., 1990. Sampling-based approaches to calculating marginal densities. *Journal of the American Statistical Association.* 85, 398−409.

Gelfand, A.E., Carlin, B.P., 1993. Maximum likelihood estimation for constrained or missing data models. *Canadian Journal of Statistics.* 21, 303−311.

Gelfand, A.E., 2000. Gibbs sampling. *Journal of the American Statistical Association.* 95, 1300−1304.

Gelman, A., Rubin, D., 1992. Inference from iterative simulation using multiple sequences. *Statistical Science.* 7, 457−472.

Geman, S., Geman, D., 1984. Stochastic relaxation, Gibbs distribution, and the Bayesian restoration of images. *IEEE Transactions on Pattern Analysis and Machine Intelligence.* 6, 721−741.

Ghahramani, Z., Jordan, M.I., 1997. Factorial hidden Markov models. *Machine Learning.* 29, 245−275.

Gillespie, D.T., 1996. The mathematics of Brownian motion and Johnson noise. *American Journal of Physics.* 64, 225−240.

Gillis, J., 1955. Correlated random walk. *Proceedings of the Cambridge Philosophical Society.* 51, 639−651.

Goldstein, S., 1951. On diffusion by discontinuous movements, and on the telegraph equation. *Quarterly Journal of Mechanics.* 4, 129−156.

Greco, G., Greco, S., Zumprano, E., 2001. A probabilistic approach for distillation and ranking of web pages. *World Wide Web.* 4, 189−207.

Green, P.J., 1995. Reversible jump MCMC computation and Bayesian model determination. *Biometrika.* 82, 711−732.

Griffeath, D., 1976. Introduction to random fields. In: Kemeny, J.G., Snell, J.L., Knapp, A.W. (Eds.), *Denumerable Markov Chains.* Springer-Verlag, New York, NY.

Grimmett, G., Welsh, D., 1986. *Probability: An Introduction.* Oxford University Press, Oxford, England.

Grimmett, G., Stirzaker, D., 2001. *Probability and Random Processes.* third ed. Oxford University Press, Oxford, England.

Grinstead, C.M., Snell, J.L., 1997. *Introduction to Probability.* American Mathematical Society, Providence, RI.

Guttorp, P., Newton, M.A., Abkowitz, J.L., 1990. A stochastic model for haematopoiesis in cats. *IMA Journal of Mathematical Medicine and Biology.* 7, 125−143.

Hanneken, J.W., Franceschetti, D.R., 1998. Exact distribution function for discrete time correlated random walks in one dimension. *Journal of Chemical Physics.* 109, 6533−6539.

Hastings, W.K., 1970. Monte Carlo sampling methods using Markov chains and their applications. *Biometrika.* 57, 97−109.

Hauskrecht, M., Fraser, H., 2000. Planning the treatment of ischemic heart disease with partially observable Markov decision processes. *Artificial Intelligence in Medicine.* 18, 221−244.

Hayes, B., 1998. How to avoid yourself. *American Scientist.* 86, 314−319.

Hazel, G.G., 2000. Multivariate gaussian MRF for multispectral scene segmentation and anomaly detection. *IEEE Transactions on Geoscience and Remote Sensing.* 38, 1199−1211.

Heffes, H., Lucantoni, D.M., 1986. A Markov modulated characterization of packetized voice and data traffic and related statistical multiplexer performance. *IEEE Journal on Selected Areas in Communications.* SAC-4, 856−868.

Held, K., Kops, E.R., Krause, B.J., Wells III, W.M., Kikinis, R., Muller-Gartner, H.-W., 1997. Markov random field segmentation of brain MR images. *IEEE Transactions on Medical Imaging.* 16, 878−886.

Helmstetter, A., Sornette, D., 2002. Diffusion of earthquake aftershocks, Omori's law, and generalized continuous-time random walk models. *Physics Review E.* 66, 061104.

Henzinger, M., 2001. Hyperlink analysis for the web. *IEEE Internet Journal.* 1, 45−50.

Heyman, D.P., Sobel, M.J., 1982. *Stochastic Models in Operations Research, Volume 1: Stochastic Processes and Operating Characteristics.* McGraw-Hill, New York, NY.

Heyman, D.P., Sobel, M.J., 1984. *Stochastic Models in Operations Research, Volume II: Stochastic Optimization.* Mc-Graw Hill, New York, NY.

Holden, L., Sannan, S., Bungum, H., 2003. A stochastic marked process model for earthquakes. *Natural Hazards and Earth System Sciences.* 3, 95−101.

Howard, R.A., 1960. *Dynamic Programming and Markov Processes.* MIT Press, Cambridge, MA.

Howard, R.A., 1971a. *Dynamic Probabilistic Systems Volume I: Markov Models.* John Wiley, New York, NY.

Howard, R.A., 1971b. Dynamic Probabilistic Systems Volume II: Semi-Markov and Decision Processes. John Wiley, New York, NY.

Ibe, O.C., 1997. *Essentials of ATM Networks and Services.* Addison-Wesley, Reading, MA.

Ibe, O.C., 2005. *Fundamentals of Applied Probability and Random Processes.* Academic Press, New York, NY.

Ibe, O.C., 2011. *Fundamentals of Stochastic Networks.* John Wiley, Hoboken, NJ.

Iosifescu, M., 1980. *Finite Markov Processes and Their Applications.* John Wiley, Chichester, England.

Ising, E., 1925. Beitrag zur theorie des ferromagnetismus. *Zeit. Phys.* 31, 253−258.

Ivanov, Y., Bobick, A., 2000. Recognition of visual activities and interactions by stochastic parsing. *IEEE Transactions on Pattern Recognition and Machine Intelligence.* 22, 852−872.

Jackson, Q., Landgrebe, D.A., 2002. Adaptive Bayesian contextual classification based on Markov random fields. *IEEE Transactions on Geoscience and Remote Sensing.* 40, 2454−2463.

Jacobs, R.A., Jiang, W., Tanner, M.A., 2002. Factorial hidden Markov models and the generalized backfitting algorithm. *Neural Computation.* 14, 2415−2437.

Jacobsen, M., 2006. *Point Process Theory and Applications: Marked Point and Piecewise Deterministic Processes.* Birkhauser, Boston, MA.

Jensen, F.V., 2001. *Bayesian Networks and Decision Graphs.* Springer, New York, NY.

Johannesma, P.I.M., 1968. Diffusion models for the stochastic activity of neurons. In: Caianello, E.R. (Ed.), *Neural Networks.* Springer, Berlin, pp. 116−144.

Jonas, M., 2003. Modeling speech using partially observable Markov processes. Ph.D. Thesis, Department of Computer Science, Tufts University, Medford, MA.

Jonsen, I.D., Flemming, J.M., Myers, R.A., 2005. Roburst state-space modeling of animal movement data. *Ecology*. 86, 2874–2880.

Juang, B.H., Rabiner, L.R., 1991. Hidden Markov models for speech recognition. *Technometrics*. 33, 251–272.

Kadirkamanathan, M., Varga, A.P., 1991. Simultaneous model re-estimation from contaminated data by compressed hidden Markov modeling. *Proceedings of the IEEE International Conference on Acoustics, Speech and Signal Processing*, pp. 897–900.

Kahn, J.D., Linial, N., Nisan, N., Saks, M.E., 1989. On the cover time of random walks on graphs. *Journal of Theoretical Probability*. 2, 121–128.

Kareiva, P.M., Shigesada, N., 1983. Analyzing insect movement as a correlated random walk. *Oecologia*. 56, 234–238.

Kehr, K.W., Argyrakis, P., 1986. Mean number of distinct sites visited by correlated walks I: perfect lattices. *Journal of Chemical Physics*. 84, 5816–5823.

Kemeny, J.G., Snell, J.L., 1976. *Finite Markov Chains*. Springer-Verlag, New York, NY.

Kindermann, R., Snell, J.L., 1980. *Markov Random Fields and Their Applications*. American Mathematical Society, Providence, RI.

Kingman, J.F.C., 1993. *Poisson Processes*. Oxford University Press, Oxford, England.

Kirkpatrick, S., Gelatt, C.D., Vecchi, M.P., 1983. Optimization by simulated annealing. *Science*. 220, 67–680.

Klebaner, F.C., 2005. *Introduction to Stochastic Calculus with Applications*. second ed. Imperial College Press, London.

Kleinrock, L., 1975. *Queueing Systems Volume 1: Theory*. John Wiley, New York, NY.

Klemm, A., Lindemann, C., Lohmann, M., 2003. Modeling IP traffic using the batch Markovian arrival process. *Performance Evaluation*. 54, 149–173.

Kliewer, J., Goertzand, N., Mertins, A., 2006. Iterative source-channel decoding with Markov random field source models. *IEEE Transactions on Signal Processing*. 54, 3688–3701.

Kobayashi, T., Masumitsu, K., Furuyama, J., 1999. Partly hidden Markov model and its application to speech recognition. *Proceedings of the IEEE International Conference on Signal and Speech Processing (ICASSP99)*. vol. 1, pp. 121–124.

Kornak, J., Irwin, M., Cressie, N., 2006. Spatial point process models of defensive strategies: detecting changes. *Statistical Inference for Stochastic Processes*. 9, 31–46.

Koski, T., 2001. *Hidden Markov Models for Bioinformatics*. Kluwer Academic Publishers, Dordrecht, Germany.

Kou, S.G., 2002. A jump-diffusion model for option pricing. *Management science*. 48, 1086–1101.

Kou, S.G., 2008. Jump-diffusion models for asset pricing in financial engineering. In: Birge, J.R., Linetsky, V. (Eds.), *Handbooks in OR & MS*, vol. 15. Elsevier (Chapter 2).

Krogh, A., Brown, M., Mian, I.S., Sjolander, K., Haussler, D., 1994. Hidden Markov models in computational biology: applications to protein modeling. *Journal of Molecular Biology*. 235, 1501–1531.

Kuczura, A., 1973. The interrupted Poisson process as an overflow process. *Bell System Technical Journal*. 52, 437–448.

Kulkarni, V.G., 2010. *Modeling and Analysis of Stochastic Systems*. second ed. CRC Press, Boca Raton, FL.

Kumar, P.R., Varaiya, P., 1986. *Stochastic Systems: Estimation, Identification, amd Adaptive Control*. Prentice-Hall, Englewood Cliffs, NJ.

Kunsch, H., Geman, S., Kehagias, A., 1995. Hidden Markov random fields. *The Annals of Applied Probability.* 5, 577−602.

Kwon, J., Murphy, K., 2000. Modeling freeway traffic with coupled HMMs. Technical Report. University of California, Berkeley, CA.

Lal, R., Bhat, U.N., 1989. Some explicit results for correlated random walks. *Journal of Applied Probability.* 27, 756−766.

Lamond, B.F., Boukhtouta, A., 2001. Water reservoir applications of Markov decision processes. In: Feinberg, E.A., Shwartz, A. (Eds.), *Handbook of Markov Decision Processes: Methods and Applications.* Kluwer Academic Publishers, Boston, MA.

Larson, R.C., Odoni, A.R., 1981. *Urban Operations Research.* Prentice-Hall, Englewood Cliffs, NJ.

Latouche, G., Ramaswami, V., 1999. *Introduction to Matrix Analytic Methods in Stochastic Modeling.* Siam Press, Philadelphia, PA.

Le, N.D., Leroux, B.G., Puterman, M.L., 1992. Reader reaction: exact likelihood evaluation of a Markov mixture model for time series of seizure counts. *Biometrics.* 48, 317−323.

Leland, W.E., Taqqu, M.S., Willinger, W., Wilson, V., 1994. On the self-similar nature of Ethernet traffic (Extended Version). *IEEE/ACM Transactions on Networking.* 2, 1−15.

Leroux, B.G., Puterman, M.L., 1992. Maximum-penalized-likelihood estimation for independent and Markov-dependent mixture models. *Biometrics.* 48, 545−558.

Levinson, S.E., 1986. Continuously variable duration hidden Markov models for automatic speech recognition. *Computer Speech and Language.* 1, 29−45.

Levinson, S.E., Rabiner, L.R., Sondi, M.M., 1983. An introduction to the application of the theory of probabilistic functions of a Markov process to automatic speech recognition. *Bell System Technical Journal.* 62, 1035−1074.

Lewis, P.A., Goodman, A.S., Miller, J.M., 1969. A pseudo-random number generator for the system/360. *IBM System Journal.* 8, 136−146.

Li, S.-Q., Hwang, C.-L., 1993. Queue response to input correlation functions: continuous spectral analysis. *IEEE/ACM Transactions on Networking.* 1, 678−692.

Limnios, N., Oprisan, G., 2001. *Semi-Markov Processes and Reliability.* Birkhauser, Boston, MA.

Liu, N., Ulukus, S., 2006. Optimal distortion-power tradeoffs in sensor networks: Gauss−Markov random processes. *Proceedings of the IEEE International Conference on Communications.* June 11−15, Istanbul, Turkey.

Little, J.D.C., 1961. A proof for the queueing formula $L = \lambda W$. *Operations Research.* 9, 383−387.

Logan, B., Moreno, P.J., 1998. Factorial hidden Markov models for acoustic modeling. *Proceedings of the IEEE International Conference on Acoustics, Speech and Signal Processing,* pp. 813−816.

Lovejoy, W.S., 1991. A survey of algorithmic methods for partially observable Markov decision processes. *Annals of Operations Research.* 28, 47−65.

Lucantoni, D., 1993. The BMAP/G/1 queue: A tutorial. In: Donatiello, L., Nelson, R. (Eds.), *Models and Techniques for Performance Evaluation of Computer and Communication Systems.* pp. 330−358.

Lucantoni, D.M., 1991. New results on the single server queue with a batch arrival process. *Stochastic Models.* 7, 1−46.

Lucantoni, D.M., Meier-Hellstern, K.S., Neuts, M.F., 1990. A Single-server queue with server vacations and a class of non-renewal arrival processes. *Advances in Applied Probability.* 22, 676−705.

MacDonald, I.L., Zucchini, W., 1997. *Hidden Markov and Other Models for Discrete-valued Time Series.* Chapman and Hall, London.

Madani, O., Hanks, S., Condon, A., 2003. On the undecidability of probabilistic planning and related stochastic optimization problems. *Artificial Intelligence*. 147, 5−34.

Mandelbroth, B.B., van Ness, J.W., 1968. Fractional Brownian motion, fractional noises and applications. *SIAM Review*. 10, 422−437.

Manjunath, B.S., Chellappa, R., 1991. Unsupervised texture segmentation using Markov random field models. *IEEE Transactions on Pattern Analysis and Machine Intelligence*. 13, 478−482.

Masoliver, J., Montero, M., 2003. Continuous-time random-walk model for financial distributions. *Physical Review E*. 67, 021112.

Masoliver, J., Montero, M., Perello, J., Weiss, G.H., 2006. The continuous time random walk formalism in financial markets. *Journal of Economic Behavior & Organization*. 61, 577−598.

Masuyama, H., 2003. Studies on algorithmic analysis of queues with batch Markovian arrival streams. Ph.D. Thesis, Department of Applied Mathematics and Physics, Kyoto University, Kyoto, Japan.

McBride, S.J., 2002. A marked point process model for the source proximity effect in the indoor environment. *Journal of the American Statistical Association*. 97, 683−691.

McNeil, D.R., Schach, S., 1973. Central limit analogues for Markov population processes. *Journal of the Royal Statistical Society*. 35, 1−23.

Medard, M., 2000. The effect upon channel capacity in wireless communications of perfect and imperfect knowledge of the channel. *IEEE Transactions on Information Theory*. 46, 933−946.

Meier-Hellstern, K., 1989. The analysis of a queue arising in overflow models. *IEEE Transactions on Communications*. 37, 367−372.

Melgani, F., Serpico, S.B., 2003. A Markov random field approach to spatio-temporal contextual image classification. *IEEE Transactions on Geoscience and Remote Sensing*. 41, 2478−2487.

Merton, R.C., 1973. Theory of rational option pricing. *Bell Journal of Economics and Management Science*. 4, 141−183.

Metropolis, N., Rosenbluth, M.N., Teller, A.H., Teller, E., 1953. Equations of state calculations by fast computing machines. *The Journal of Chemical Physics*. 21, 1087−1092.

Michiel, H., Laevens, K., 1997. Teletraffic engineering in a broad-band era. *Proceedings of the IEEE*. 85, 2007−2033.

Mikosch, T., Resnick, S., Rootzen, H., Stegeman, A., 2002. Is network traffic approximated by stable Levy motion or fractional Brownian motion? *Annals of Applied Probability*. 12 (1), 23−68.

Min, G., Ferguson, J., Ould-Khaoua, M., 2001. Analysis of adaptive wormhole-routed torus networks with IPP input traffic. *Proceedings of the 2001 ACM Symposium on Applied Computing*, pp. 494−498.

Mitchell, C., Harper, M., Jamieson, L., 1995. On the complexity of explicit duration HMMs. *IEEE Transactions on Speech and Audio Processing*. 3, 213−217.

Mohan, C., 1955. The Gambler's ruin problem with correlation. *Biometrika*. 42, 486−493.

Moller, J., Waagepetersen, R.P., 2004. *Statistical Inference and Simulation for Spatial Point Processes*. Chapman & Hall/CRC, Boca Raton, FL.

Monahan, G.E., 1982. A survey of partially observable Markov decision processes: theory, models, and algorithms. *Management Science*. 28, 1−16.

Montroll, E.W., Weiss, G.H., 1965. Random walks on lattice II. *Journal of Mathematical Physics*. 6, 167−181.

Montroll, E.W., Shlensinger, M.F., 1984. In: Lebowitz, J.L., Montroll, E.W. (Eds.), *Nonequilibrium Phenomena II: From Stochastics to Hydrodynamics*. North-Holland, Amsterdam, pp. 1–121.

Morgenthal, G.W., 1961. The theory and application of simulations in operations research. In: Ackoff, R.L. (Ed.), *Progress in Operations Research*, John Wiley, New York.

Murphy, K., 2002. Dynamic Bayesian networks: representations, inference and learning. Ph.D. Thesis, Department of Computer Science, University of California, Berkeley, CA.

Muscariello, L., Mellia, M., Meo, M., Marsan, M.A., Cigno, R.L., 2005. Markov models of Internet traffic and a new hierarchical MMPP model. *Computer Communications*. 28, 1835–1851.

Neuts, M.F., 1981. *Matrix-Geometric Solutions in Stochastic Models: An Algorithmic Approach*. The Johns Hopkins University Press, Baltimore, MD.

Neuts, M.F., 1989. *Structured Stochastic Matrices of M/G/1 Type and Their Applications*. Marcel Dekker, New York, NY.

Neuts, M.F., 1992. Models based on the Markovian arrival process. *IEICE Transactions on Communications*. E75-B, 1255–1265.

Neuts, M.F., 1995. *Algorithmic Probability: A Collection of Problems*. Chapman & Hall, London.

Newell, G.F., 1971. *Applications of Queueing Theory*. Chapman & Hall, London.

Norris, J.R., 1997. *Markov Chains*. Cambridge University Press, Cambridge, England.

Ogata, Y., 1998. Space-time point process models for earthquake occurrence. *Annals of the Institute of Statistical Mathematics*. 50, 379–402.

Ogata, Y., 1999. Seismicity analysis through point-process modeling: a review. *Pure and Applied Geophysics*. 155, 471–507.

Ogawa, T., Kobayashi, T., 2005. An extension of the state-observation dependency in partly hidden Markov models and its application to continuous speech recognition. *Systems and Computers in Japan*. 36, 31–39.

Oksendale, B., 2005. *Stochastic Differential Equations*. sixth ed. Springer, New York, NY.

Onsager, L., 1944. A 2D model with an order–disorder transition. *Physical Review*. 65, 117–149.

Onural, L., 1991. Generating connected textured fractal patterns using Markov random fields. *IEEE Transactions on Pattern Analysis and Machine Intelligence*. 13, 819–825.

Osaki, S., 1985. *Stochastic System Reliability Modeling*. World Scientific, Singapore.

Ozekici, S., 1997. Markov modulated Bernoulli process. *Mathematical Methods of Operations Research*. 45, 311–324.

Ozekici, S., Soyer, R., 2003. Bayesian analysis of Markov modulated Bernoulli processes. *Mathematical Methods of Operations Research*. 57, 125–140.

Page, L., Brin, S., Motwani, R., Winograd, T., 1998. The PageRank citation ranking: bringing order to the web. Technical Report, Computer Science Department, Stanford University, Stanford, CA.

Panjwani, D.K., Healey, G., 1995. Markov random field models for unsupervised segmentation of textured color images. *IEEE Transactions on Pattern Analysis and Machine Intelligence*. 17, 939–954.

Papadimitrios, C., Tsitsiklis, J.N., 1987. The complexity of Markov decision processes. *Mathematics of Operations Research*. 12, 441–450.

Pearl, J., 1988. *Probabilistic Reasoning in Intelligent Systems: Networks of Plausible Inference*. Morgan Kaufman, San Francisco, CA.

Poggi, G., Scarpa, G., Zerubia, J.B., 2005. Supervised segmentation of remote sensing images based on a tree-structured MRF model. *IEEE Transactions on Geoscience and Remote Sensing*. 43, 1901–1911.

Prais, S.J., 1955. Measuring social mobility. *Journal of the Royal Statistical Society, Series A.* 118, 56–66.

Preston, C.J., 1977. Spatial birth-and-death processes. *Bulletin of the International Statistical Institute.* 46, 371–391.

Prigent, J.-L., 2001. Option pricing with a general marked point process. *Mathematics of Operations Research.* 26, 50–66.

Puterman, M.L., 1994. *Markov Decision Processes: Discrete Stochastic Dynamic Programming.* John Wiley, New York, NY.

Rabiner, L.R., 1989. A tutorial on hidden Markov processes and selected applications in speech recognition. *Proceedings of the IEEE.* 77, 257–286.

Rabiner, L.R., Juang, B.-H., 1993. *Fundamentals of Speech Recognition.* Prentice-Hall, Englewood Cliffs, NJ.

Rasmussen, C.E., Williams, C.K.I., 2006. *Gaussian Processes for Machine Learning.* MIT Press, Cambridge, MA.

Rathbun, S.L., 1996. Asymptotic properties of the maximum likelihood estimator for spatio-temporal point processes. *Journal of Statistical Planning and Inference.* 51, 55–74.

Rathbun, S.L., Cressie, N.A.C., 1994. A space-time survival point process for a longleaf pine forest in Southern Georgia. *Journal of the American Statistical Association.* 89, 1164–1174.

Renshaw, E., Henderson, R., 1981. The correlated random walk. *Journal of Applied Probability.* 18, 403–414.

Rezek, I., Gibbs, M., Roberts, S.J., 2002. Maximum a posteriori estimation of coupled hidden Markov models. *The Journal of VLSI Signal Processing.* 32, 55–66.

Ripley, B.D., 1977. Modeling spatial patterns. *Journal of the Royal Statistical Society, Series B.* 39, 172–212.

Ripley, B.D., 1988. *Statistical Inference for Spatial Processes.* Cambridge University Press, Cambridge, England.

Ripley, B.D., Kelly, F.P., 1977. Markov point processes. *Journal of the London Mathematical Society, Series 2.* 15, 188–192.

Robert, C.P., Casella, G., 1999. *Monte Carlo Statistical Methods.* Springer-Verlag, New York, NY.

Rogers, L.C.G., 1997. Arbitrage with fractional Brownian motion. *Mathematical Finance.* 7, 95–105.

Rogers, L.C.G., Williams, D., 2000. *Diffusions, Markov Processes and Martingales Volume 1: Foundations.* Cambridge University Press, Cambridge, England.

Romanow, A.L., 1984. A Brownian motion model for decision making. *Journal of Mathematical Sociology.* 10, 1–28.

Ross, S.M., 1970. *Applied Probability Models with Optimization Applications.* Holden-Day, San Francisco, CA.

Ross, S.M., 1983. *Introduction to Stochastic Dynamic Programming.* Academic Press, New York, NY.

Rue, H., Held, L., 2005. *Gaussian Markov Random Fields: Theory and Applications.* Chapman & Hall/CRC, Boca Raton, FL.

Ryden, T., Terasvirta, T., Asbrink, S., 1998. Stylized facts of daily returns and the hidden Markov model. *Journal of Applied Econometrics.* 13, 217–244.

Sarkar, A., Biswas, M.K., Kartikeyan, B., Kumar, V., Majumbar, K.L., Pal, D.K., 2002. A MRF model-based segmentation approach to classification for multispectral imagery. *IEEE Transactions on Geoscience and Remote Sensing.* 40, 1102–1113.

Scalas, E., 2006a. Five years of continuous-time random walks in econophysics. In: Namatame, A., Kaizouji, T., Aruga, Y. (Eds.), *The Complex Networks of Economic Interactions: Essays in Agent-Based Economics and Econophysics*. Springer, Tokyo, pp. 1−16.

Scalas, E., 2006b. The application of continuous-time random walks in finance and economics. *Physica A*. 362, 225−239.

Scher, H., Montroll, E.W., 1975. Anomalous transition-time dispersion in amorphous solids. *Physical Review B*. 12, 2455−2477.

Servi, L.D., 2002. Algorithmic solutions to two-dimensional birth-death processes with application to capacity planning. *Telecommunication Systems*. 21, 205−212.

Seth, A., 1963. The correlated unrestricted random walk. *Journal of the Royal Statistical Society, Series B*. 25, 394−400.

Schaefer, A.J., Bailey, M.D., Shechter, S.M., Roberts, M.S., 2004. Modeling medical treatment using Markov decision processes. In: Brandeau, M.L., Sainfort, F., Pierskalla, W.P. (Eds.), *Operations Research and Health Care: A Handbook of Methods and Applications*. Springer, New York, NY.

Schal, M., 2001. Markov decision processes in finance and dynamic options. In: Feinberg, E.A., Shwartz, A. (Eds.), *Handbook of Markov Decision Processes: Methods and Applications*. Kluwer Academic Publishers, Boston, MA.

Shatkay, H., 1999. Learning models for robot navigation. Ph.D. Thesis, Department of Computer Science, Brown University, Providence, RI.

Simmons, R., Koenig, S., 1995. Probabilistic robot navigation in partially observable environments. *Proceedings of the International Joint Conference on Artificial Intelligence*, pp. 1080−1087.

Smith, J.A., 1993. Marked point process models of raindrop-size distributions. *Journal of Applied Meterology*. 32, 284−296.

Smits, P.C., Dellepiane, S.G., 1997. Synthetic aperture radar image segmentation by a detail preserving Markov random field approach. *IEEE Transactions on Geoscience and Remote Sensing*. 35, 844−857.

Solberg, A.H.S., Taxt, T., Jain, A.K., 1996. A Markov random field model for classification of multisource satellite imagery. *IEEE Transactions on Geoscience and Remote Sensing*. 34, 100−113.

Sondik, E.J., 1971. The optimal control of partially observable Markov processes. Ph.D. Thesis, Stanford University.

Sottinen, T., 2001. Fractional Brownian motion, random walks and binary market models. *Finance and Stochastics*. 5, 343−355.

Steele, M.J., 2001. *Stochastic Calculus Financial Applications*, Springer, New York, NY.

Stirzaker, D., 2005. *Stochastic Processes & Models*. Oxford University Press, Oxford, England.

Stoica, R.S., Descombes, X., Zerubia, J., 2000. A Gibbs point process for road extraction in remotely sensed images. Research Report 3923, INRIA, Sophia Antipolis, France.

Stoyan, D., Penttinen, A., 2000. Recent applications of point process methods in forestry statistics. *Statistical Science*. 15, 61−78.

Stoyan, D., Kendall, W.S., Mecke, J., 1995. *Stochastic Geometry and its Applications*. second ed. John Wiley, Chichester, England.

Sutton, R.S., Barto, A.G., 1998. *Reinforcement Learning: An Introduction*. The MIT Press, Cambridge, MA.

Szummer, M., Jaakkola, T., 2001. Partially labeled classification with Markov random walks. In: Dietterich, T.G., Becker, S., Ghahramani, Z. (Eds.), *Advances in Neural Information Processing Systems*, vol. 14. Vancouver, British Columbia, Canada, pp. 945−952.

Tang, X.Z., Tracy, E.R., 1998. Data compression and information retrieval via symbolization. *Chaos.* 8, 688–696.

Thompson, E.A., 1983. Optimal sampling for pedigree analysis: parameter estimation and genotypic uncertainty. *Theoretical Population Biology.* 24, 39–58.

Tijms, H.C., 1986. *Stochastic Modeling and Analysis: A Computational Approach.* John Wiley, Chichester, England.

Tijms, H.C., 1995. *Stochastic Models: An Algorithmic Approach.* John Wiley, Chichester, England.

Tijms, H.C., 2003. *A First Course in Stochastic Models.* John Wiley, Chichester, England.

Towsley, D., Hwang, R.H., Kurose, J.F., 2000. MDP routing for multi-rate loss networks. *Computer Networks and ISDN Systems.* 34, 241–261.

Tran, T., Wehrens, R., Hoekman, D.H., Buydens, L.M.C., 2005. Initialization of Markov random field clustering of large remote sensing images. *IEEE Transactions on Geoscience and Remote Sensing.* 43, 1912–1919.

Tso, B.C.K., Mather, P.M., 1999. Classification of multisource remote sensing imagery using a genetic algorithm and Markov random fields. *IEEE Transactions on Geoscience and Remote Sensing.* 37, 1255–1260.

Turin, W., 2000. MAP decoding in channels with memory. *IEEE Transactions on Communications.* 48, 757–763.

Turin, W., Sondhi, M.M., 1993. Modeling error sources in digital channels. *IEEE Journal of Selected Areas in Communications.* 11, 340–347.

Turin, W., van Nobelen, R., 1998. Hidden Markov modeling of fading channels. *IEEE Journal of Selected Areas in Communications.* 16, 1809–1817.

van Lieshout, M.N.M., 2000. *Markov Point Processes and Their Applications.* Imperial College Press, London.

van Lieshout, M.N.M., Stoica, R.S., 2003. The Candy model: properties and inference. *Statistica Neerlandica.* 57, 177–206.

Vanmarcke, E., 1988. *Random Fields: Analysis and Synthesis.* MIT Press, Cambridge, MA.

Vere-Jones, D., 1995. Forecasting earthquakes and earthquake risk. *International Journal of Forecasting.* 11, 503–538.

Viswanathan, G.M., Buldyrev, S.V., Havlin, S., da Luz, M.G.E., Raposo, E.P., Stanley, H.E., 1999. Optimizing the success of random searches. *Nature.* 401, 911–914.

Viswanathan, G.M., da Luz, M.G.E., Raposo, E.P., Stanley, H.E., 2011. *The Physics of Foraging: An Introduction to Random Searches and Biological Encounters.* Cambridge University Press, Cambridge, England.

Viterbi, A.M., 1986. Approximate analysis of time-synchronous packet networks. *IEEE Journal on Selected Areas in Communications.* SAC-4, 879–890.

Wang, C., 2007. First order Markov decision processes. Ph.D. Thesis, Department of Computer Science, Tufts University, Medford, MA.

Wein, L.M., 1990. Brownian networks with discretionary routing. *Operations Research.* 39, 322–340.

Weiss, G.H., 1994. *Aspects and Applications of the Random Walk.* North-Holland, Amsterdam.

Weiss, G.H., Porra, J.M., Masoliver, J., 1998. Statistics of the depth probed by CW measurements of photons in a turbid medium. *Physical Review E.* 58, 6431–6439.

Wheeler, M.D., Ikeuchi, K., 1995. Sensor modeling, probabilistic hypothesis generation, and robust localization for object recognition. *IEEE Transactions on Pattern Analysis and Machine Intelligence.* 17, 252–265.

Wiener, N., 1923. Differential space. *Journal of Mathematics and Physics.* 2, 131–174.

Wilf, H.S., 1990. *Generating Functionology.* Academic Press, Boston, MA, p. 50.

Wolff, R.W., 1982. Poisson arrivals see time averages. *Operations Research.* 30, 223−231.

Xie, H., Pierce, L.E., Ulaby, F.T., 2002. SAR speckle reduction using Wavelet denoising and Markov Rrandom field modeling. *IEEE Transactions on Geoscience and Remote Sensing.* 40, 2196−2212.

Xie, L., Liu Z.-Q., 2006. Speech animation using coupled hidden Markov models. *Proceedings of the 18th International Conference on Pattern Recognition (ICPR'06),* pp. 1128−1131.

Yamada, H., Sumita, S., 1991. A traffic measurement method and its application for cell loss probability estimation in ATM networks. *IEEE Journal on Selected Areas in Communications.* 9, 315−324.

Yu, H., 2006. Approximate solution methods for partially observable Markov and semi-Markov decision processes. Ph.D. Thesis, Department of Electrical Engineering and Computer Science, Massachusetts Institute of Technology, Cambridge, MA.

Zhang, N.L., Liu, W., 1997. A model approximation scheme for planning in partially observable stochastic domains. *Journal of Artificial Intelligence Research.* 7, 199−230.

Zhang, W., 2001. Algorithms for partially observable Markov decision processes. Ph.D. Thesis, Department of Computer Science, The Hong Kong University of Science and Technology, Hong Kong.

Zhang, W., 2006. The role of correlation in communication over fading channels. PhD Thesis, Department of Electrical Engineering, University of Notre Dame, Notre Dame, IN.

Zhang, Y., Brady, M., Smith, S., 2001. Segmentation of brain MR images through a hidden Markov random field model and the expectation-maximization algorithm. *IEEE Transactions on Medical Imaging.* 20, 45−57.

Zhou, Y.-P., Gans, N., 1999. A single-server queue with Markov modulated service times. Technical Report 99-40-B, The Wharton School, University of Pennsylvania, Philadelphia, PA.

Zucchini, W., Guttorp, P., 1991. A hidden Markov model for space-time precipitation. *Water Resources Research.* 27, 1917−1923.

Printed in the United States
By Bookmasters